Accounting and Causal Effects

For further volumes:
www.springer.com/series/6192

Douglas A. Schroeder

Accounting and Causal Effects

Econometric Challenges

Springer

Douglas A. Schroeder
The Ohio State University
Columbus, OH 43210
USA
schroeder_9@fisher.osu.edu

ISSN 1572-0284
ISBN 978-1-4419-7224-8   e-ISBN 978-1-4419-7225-5
DOI 10.1007/978-1-4419-7225-5
Springer New York Dordrecht Heidelberg London

Library of Congress Control Number: 2010932012

© Springer Science+Business Media, LLC 2010
All rights reserved. This work may not be translated or copied in whole or in part without the written permission of the publisher (Springer Science+Business Media, LLC, 233 Spring Street, New York, NY 10013, USA), except for brief excerpts in connection with reviews or scholarly analysis. Use in connection with any form of information storage and retrieval, electronic adaptation, computer software, or by similar or dissimilar methodology now known or hereafter developed is forbidden.
The use in this publication of trade names, trademarks, service marks, and similar terms, even if they are not identified as such, is not to be taken as an expression of opinion as to whether or not they are subject to proprietary rights.

Printed on acid-free paper

Springer is part of Springer Science+Business Media (www.springer.com)

to Bonnie

Preface

In this book, we synthesize a rich and vast literature on econometric challenges associated with accounting choices and their causal effects. Identification and estimation of endogenous causal effects is particularly challenging as observable data are rarely directly linked to the causal effect of interest. A common strategy is to employ logically consistent probability assessment via Bayes' theorem to connect observable data to the causal effect of interest. For example, the implications of earnings management as equilibrium reporting behavior is a centerpiece of our explorations. Rather than offering recipes or algorithms, the book surveys our experiences with accounting and econometrics. That is, we focus on why rather than how.

The book can be utilized in a variety of venues. On the surface it is geared toward graduate studies and surely this is where its roots lie. If we're serious about our studies, that is, if we tackle interesting and challenging problems, then there is a natural progression. Our research addresses problems that are not well understood then incorporates them throughout our curricula as our understanding improves and to improve our understanding (in other words, learning and curriculum development are endogenous). For accounting to be a vibrant academic discipline, we believe it is essential these issues be confronted in the undergraduate classroom as well as graduate studies. We hope we've made some progress with examples which will encourage these developments. For us, the Tuebingen-style treatment effect examples, initiated by and shared with us by Joel Demski, introduced (to the reader) in chapter 8 and pursued further in chapters 9 and 10 are a natural starting point.

The layout of the book is as follows. The first two chapters introduce the philosophic style of the book — we iterate between theory development and numerical examples. Chapters three through seven survey standard econometric background along with some scattered examples. An appendix surveys standard asymptotic theory. Causal effects, our primary focus, are explored mostly in the latter chapters — chapters 8 through 13. The synthesis draws heavily and unabashedly from labor econometrics or microeconometrics, as it has come to be known. We claim no originality regarding the econometric theory synthesized in these pages and attempt to give credit to the appropriate source. Rather, our modest contribution primarily derives from connecting econometric theory to causal effects in various accounting contexts.

I am indebted to numerous individuals. Thought-provoking discussions with colloquium speakers and colleagues including Anil Arya, Anne Beatty, Steve Coslett, Jon Glover, Chris Hogan, Pierre Liang, Haijin Lin, John Lyon, Brian Mittendorf, Anup-menon Nandialath, Pervin Shroff, Eric Spires, Dave Williams, and Rick Young helped to formulate and refine ideas conveyed in these pages. In a very real sense, two events, along with a perceived void in the literature, prompted my attempts to put these ideas to paper. First, Mark Bagnoli and Susan Watts invited me to discuss these issues in a two day workshop at Purdue University during Fall 2007. I am grateful to them for providing this important opportunity, their hospitality and intellectual curiosity, and their continuing encouragement of this project. Second, the opportunity arose for me to participate in Joel Demski and John Fellingham's seminar at the University of Florida where many of these issues were discussed. I am deeply indebted to Joel and John for their steadfast support and encouragement of this endeavor as well as their intellectual guidance. I borrow liberally from their work for not only the examples discussed within these pages but in all facets of scholarly endeavors. I hope that these pages are some small repayment toward this debt but recognize that my intellectual debt to Joel and John continues to dwarf the national debt. Finally, and most importantly, this project would not have been undertaken without the love, encouragement, and support of Bonnie.

Doug Schroeder
Columbus, Ohio

Contents

Preface	vii
Contents	ix
List of Tables	xvii
List of Figures	xxv
1 Introduction	**1**
1.1 Problematic illustration	2
1.2 Jaynes' desiderata for scientific reasoning	4
1.2.1 Probability as logic illustration[1]	4
1.3 Overview	7
1.4 Additional reading	8
2 Accounting choice	**9**
2.1 Equilibrium earnings management	10
2.1.1 Implications for econometric analysis	11
2.2 Asset revaluation regulation	12
2.2.1 Numerical example	13
2.2.2 Implications for econometric analysis	14
2.3 Regulated report precision	14

---

[1] This example was developed from conversations with Anil Arya and Brian Mittendorf.

		2.3.1	Public precision choice	15
	2.4	2.3.2	Private precision choice	15
		2.3.3	Regulated precision choice and transaction design	16
		2.3.4	Implications for econometric analysis	16
	2.4		Inferring transactions from financial statements	17
		2.4.1	Implications for econometric analysis	17
	2.5		Additional reading	18

3 Linear models   19
- 3.1 Standard linear model (*OLS*) 19
- 3.2 Generalized least squares (*GLS*) 21
- 3.3 Tests of restrictions and *FWL* (Frisch-Waugh-Lovell) 22
- 3.4 Fixed and random effects 26
- 3.5 Random coefficients 31
  - 3.5.1 Nonstochastic regressors 31
  - 3.5.2 Correlated random coefficients 32
- 3.6 Ubiquity of the Gaussian distribution 33
  - 3.6.1 Convolution of Gaussians 35
- 3.7 Interval estimation 36
- 3.8 Asymptotic tests of restrictions: Wald, *LM*, *LR* statistics 38
  - 3.8.1 Nonlinear restrictions 41
- 3.9 Misspecification and *IV* estimation 41
- 3.10 Proxy variables 43
  - 3.10.1 Accounting and other information sources 45
- 3.11 Equilibrium earnings management 48
- 3.12 Additional reading 54
- 3.13 Appendix 55

4 Loss functions and estimation   59
- 4.1 Loss functions 59
  - 4.1.1 Quadratic loss 60
  - 4.1.2 Linear loss 61
  - 4.1.3 All or nothing loss 61
- 4.2 Nonlinear Regression 62
  - 4.2.1 Newton's method 62
  - 4.2.2 Gauss-Newton regression 63
- 4.3 Maximum likelihood estimation (*MLE*) 65
  - 4.3.1 Parameter estimation 65
  - 4.3.2 Estimated asymptotic covariance for *MLE* of $\hat{\theta}$ 66
- 4.4 James-Stein shrinkage estimators 70
- 4.5 Summary 75
- 4.6 Additional reading 76

5 Discrete choice models   77
- 5.1 Latent utility index models 77

5.2	Linear probability models	78
5.3	Logit (logistic regression) models	78
5.3.1	Binary logit	79
5.3.2	Multinomial logit	80
5.3.3	Conditional logit	80
5.3.4	*GEV* (generalized extreme value) models	81
5.3.5	Nested logit models	81
5.3.6	Generalizations	84
5.4	Probit models	86
5.4.1	Conditionally-heteroskedastic probit	86
5.4.2	Artificial regression specification test	87
5.5	Robust choice models	92
5.5.1	Mixed logit models	92
5.5.2	Semiparametric single index discrete choice models	92
5.5.3	Nonparametric discrete choice models	93
5.6	Tobit (censored regression) models	94
5.7	Bayesian data augmentation	94
5.8	Additional reading	95

6 Nonparametric regression — 97

6.1	Nonparametric (kernel) regression	97
6.2	Semiparametric regression models	99
6.2.1	Partial linear regression	99
6.2.2	Single-index regression	99
6.2.3	Partial index regression models	101
6.3	Specification testing against a general nonparametric benchmark	101
6.4	Locally linear regression	103
6.5	Generalized cross-validation (*GCV*)	104
6.6	Additional reading	105

7 Repeated-sampling inference — 107

7.1	Monte Carlo simulation	108
7.2	Bootstrap	108
7.2.1	Bootstrap regression	108
7.2.2	Bootstrap panel data regression	109
7.2.3	Bootstrap summary	111
7.3	Bayesian simulation	111
7.3.1	Conjugate families	111
7.3.2	*McMC* simulations	117
7.4	Additional reading	122

8 Overview of endogeneity — 123

8.1	Overview	124
8.1.1	Simultaneous equations	124
8.1.2	Endogenous regressors	126

		8.1.3	Fixed effects	127
		8.1.4	Differences-in-differences	129
		8.1.5	Bivariate probit	130
		8.1.6	Simultaneous probit	131
		8.1.7	Strategic choice model	135
		8.1.8	Sample selection	142
		8.1.9	Duration models	143
		8.1.10	Latent *IV*	146
	8.2	Selectivity and treatment effects		147
	8.3	Why bother with endogeneity?		148
		8.3.1	Sample selection example	148
		8.3.2	Tuebingen-style treatment effect examples	149
	8.4	Discussion and concluding remarks		155
	8.5	Additional reading		155

9 Treatment effects: ignorability   157

9.1	A prototypical selection setting		157
9.2	Exogenous dummy variable regression		158
9.3	Tuebingen-style examples		159
9.4	Nonparametric identification		164
9.5	Propensity score approaches		169
	9.5.1	*ATE* and propensity score	169
	9.5.2	*ATT*, *ATUT*, and propensity score	170
	9.5.3	Linearity and propensity score	172
9.6	Propensity score matching		172
9.7	Asset revaluation regulation example		175
	9.7.1	Numerical example	176
	9.7.2	Full certification	177
	9.7.3	Selective certification	183
	9.7.4	Outcomes measured by value x only	190
	9.7.5	Selective certification with missing "factual" data	193
	9.7.6	Sharp regression discontinuity design	196
	9.7.7	Fuzzy regression discontinuity design	198
	9.7.8	Selective certification setting	199
	9.7.9	Common support	201
	9.7.10	Summary	202
9.8	Control function approaches		203
	9.8.1	Linear control functions	203
	9.8.2	Control functions with expected individual-specific gain	203
	9.8.3	Linear control functions with expected individual-specific gain	204
9.9	Summary		204
9.10	Additional reading		204

10 Treatment effects: *IV*   207

10.1	Setup	207
10.2	Treatment effects	208
10.3	Generalized Roy model	210
10.4	Homogeneous response	211
	10.4.1 Endogenous dummy variable *IV* model	211
	10.4.2 Propensity score *IV*	212
10.5	Heterogeneous response and treatment effects	212
	10.5.1 Propensity score *IV* and heterogeneous response	213
	10.5.2 Ordinate control function *IV* and heterogeneous response	213
	10.5.3 Inverse Mills control function *IV* and heterogeneous response	214
	10.5.4 Heterogeneity and estimating *ATT* by *IV*	217
	10.5.5 *LATE* and linear *IV*	217
10.6	Continuous treatment	236
10.7	Regulated report precision	239
	10.7.1 Binary report precision choice	239
	10.7.2 Continuous report precision but observed binary	253
	10.7.3 *Observable* continuous report precision choice	266
10.8	Summary	273
10.9	Additional reading	273

11 Marginal treatment effects 275

11.1	Policy evaluation and policy invariance conditions	275
11.2	Setup	277
11.3	Generalized Roy model	277
11.4	Identification	278
11.5	*MTE* connections to other treatment effects	280
	11.5.1 Policy-relevant treatment effects vs. policy effects	282
	11.5.2 Linear *IV* weights	283
	11.5.3 *OLS* weights	284
11.6	Comparison of identification strategies	286
11.7	*LIV* estimation	286
11.8	Discrete outcomes	288
	11.8.1 Multilevel discrete and continuous endogenous treatment	289
11.9	Distributions of treatment effects	291
11.10	Dynamic timing of treatment	292
11.11	General equilibrium effects	293
11.12	Regulated report precision example	293
	11.12.1 Apparent nonnormality and *MTE*	293
11.13	Additional reading	300

12 Bayesian treatment effects 301

12.1	Setup	302
12.2	Bounds and learning	302
12.3	Gibbs sampler	303

xiv   CONTENTS

        12.3.1   Full conditional posterior distributions 303
  12.4   Predictive distributions 305
        12.4.1   Rao-Blackwellization 306
  12.5   Hierarchical multivariate Student t variation 306
  12.6   Mixture of normals variation 306
  12.7   A prototypical Bayesian selection example 307
        12.7.1   Simulation 308
        12.7.2   Bayesian data augmentation and *MTE* 309
  12.8   Regulated report precision example 311
        12.8.1   Binary choice 313
        12.8.2   Continuous report precision but observed binary selection 316
        12.8.3   Apparent nonnormality of unobservable choice 319
        12.8.4   Policy-relevant report precision treatment effect 326
        12.8.5   Summary 328
  12.9   Probability as logic and the selection problem 330
  12.10 Additional reading 331

13 Informed priors   **333**
  13.1   Maximum entropy 334
  13.2   Complete ignorance 336
  13.3   A little background knowledge 337
  13.4   Generalization of maximum entropy principle 337
  13.5   Discrete choice model as maximum entropy prior 340
  13.6   Continuous priors 342
        13.6.1   Maximum entropy 343
        13.6.2   Transformation groups 344
        13.6.3   Uniform prior 346
        13.6.4   Gaussian prior 347
        13.6.5   Multivariate Gaussian prior 348
        13.6.6   Exponential prior 349
        13.6.7   Truncated exponential prior 349
        13.6.8   Truncated Gaussian prior 350
  13.7   Variance bound and maximum entropy 351
  13.8   An illustration: Jaynes' widget problem 355
        13.8.1   Stage 1 solution 356
        13.8.2   Stage 2 solution 359
        13.8.3   Stage 3 solution 362
        13.8.4   Stage 4 solution 370
  13.9   Football game puzzle 370
  13.10 Financial statement example 371
        13.10.1 Under-identification and Bayes 371
        13.10.2 Numerical example 373
  13.11 Smooth accruals 376
        13.11.1 *DGP* 377
        13.11.2 Valuation results 377

		13.11.3	Performance evaluation	380

        13.11.4 Summary 382
   13.12 Earnings management 382
        13.12.1 Stochastic manipulation 382
        13.12.2 Selective earnings management 393
   13.13 Jaynes' A_p distribution 398
        13.13.1 Football game puzzle revisited 400
   13.14 Concluding remarks 401
   13.15 Additional reading 401
   13.16 Appendix 401

A Asymptotic theory   413
   A.1   Convergence in probability (laws of large numbers) 413
        A.1.1   Almost sure convergence 414
        A.1.2   Applications of convergence 415
   A.2   Convergence in distribution (central limit theorems) 417
   A.3   Rates of convergence 422
   A.4   Additional reading 423

Bibliography   425

Index   444

List of Tables

3.1	Multiple information sources case 1 setup	45
3.2	Multiple information sources case 1 valuation implications	46
3.3	Multiple information sources case 2 setup	47
3.4	Multiple information sources case 2 valuation implications	47
3.5	Multiple information sources case 3 setup	47
3.6	Multiple information sources case 3 valuation implications	48
3.7	Results for price on reported accruals regression	51
3.8	Results for price on reported accruals saturated regression	51
3.9	Results for price on reported accruals and propensity score regression	52
3.10	Results for price on reported accruals and estimated propensity score regression	53
3.11	Results for price on reported accruals and logit-estimated propensity score regression	54
5.1	Variations of multinomial logits	82
5.2	Nested logit with moderate correlation	84
5.3	Conditional logit with moderate correlation	85
5.4	Nested logit with low correlation	85
5.5	Conditional logit with low correlation	85
5.6	Nested logit with high correlation	85
5.7	Conditional logit with high correlation	85
5.8	Homoskedastic probit results with heteroskedastic DGP	90
5.9	BRMR specification test 1 with heteroskedastic DGP	90

5.10	BRMR specification test 2 with heteroskedastic DGP	91
5.11	BRMR specification test 3 with heteroskedastic DGP	91
5.12	Heteroskedastic probit results with heteroskedastic DGP	92
5.13	Homoskedastic probit results with homoskedastic DGP	92
5.14	BRMR specification test 1 with homoskedastic DGP	93
5.15	BRMR specification test 2 with homoskedastic DGP	93
5.16	BRMR specification test 3 with homoskedastic DGP	94
5.17	Heteroskedastic probit results with homoskedastic DGP	94
7.1	Conjugate families for univariate discrete distributions	113
7.2	Conjugate families for univariate continuous distributions	114
7.3	Conjugate families for multivariate discrete distributions	115
7.4	Conjugate families for multivariate continuous distributions	116
8.1	Strategic choice analysis for player B	138
8.2	Strategic choice analysis for player A	139
8.3	Parameter differences in strategic choice analysis for player B	140
8.4	Parameter differences in strategic choice analysis for player A	140
8.5	Production data: Simpson's paradox	149
8.6	Tuebingen example case 1: ignorable treatment	151
8.7	Tuebingen example case 1 results: ignorable treatment	152
8.8	Tuebingen example case 2: heterogeneous response	152
8.9	Tuebingen example case 2 results: heterogeneous response	152
8.10	Tuebingen example case 3: more heterogeneity	153
8.11	Tuebingen example case 3 results: more heterogeneity	153
8.12	Tuebingen example case 4: Simpson's paradox	154
8.13	Tuebingen example case 4 results: Simpson's paradox	154
9.1	Tuebingen example case 1: extreme homogeneity	159
9.2	Tuebingen example case 1 results: extreme homogeneity	160
9.3	Tuebingen example case 2: homogeneity	160
9.4	Tuebingen example case 2 results: homogeneity	161
9.5	Tuebingen example case 3: heterogeneity	162
9.6	Tuebingen example case 3 results: heterogeneity	163
9.7	Tuebingen example case 4: Simpson's paradox	163
9.8	Tuebingen example case 4 results: Simpson's paradox	164
9.9	Exogenous dummy variable regression example	165
9.10	Exogenous dummy variable regression results	166
9.11	Nonparametric treatment effect regression	167
9.12	Nonparametrically identified treatment effect: exogenous dummy variable regression results	168
9.13	Nonparametric treatment effect regression results	168
9.14	Investment choice and payoffs for no certification and selective certification	176
9.15	Investment choice and payoffs for full certification	177

9.16	OLS results for full certification setting	179
9.17	Average treatment effect sample statistics for full certification setting	179
9.18	Adjusted outcomes OLS results for full certification setting . . .	181
9.19	Propensity score treatment effect estimates for full certification setting	182
9.20	Propensity score matching average treatment effect estimates for full certification setting	183
9.21	OLS parameter estimates for selective certification setting	188
9.22	Average treatment effect sample statistics for selective certification setting	189
9.23	Reduced OLS parameter estimates for selective certification setting	189
9.24	Propensity score average treatment effect estimates for selective certification setting	190
9.25	Propensity score matching average treatment effect estimates for selective certification setting	190
9.26	OLS parameter estimates for Y=x in selective certification setting	191
9.27	Average treatment effect sample statistics for $Y = x$ in selective certification setting	192
9.28	Propensity score average treatment effect for $Y = x$ in selective certification setting	192
9.29	Propensity score matching average treatment effect for $Y = x$ in selective certification setting	192
9.30	OLS parameter estimates ignoring missing data for selective certification setting	193
9.31	Treatment effect OLS model estimates based on augmentation of missing data for selective certification setting	195
9.32	Sharp RD OLS parameter estimates for full certification setting .	196
9.33	Average treatment effect sample statistics for full certification setting	197
9.34	Sharp RD OLS parameter estimates for selective certification setting	197
9.35	Sharp RD OLS parameter estimates with missing data for selective certification setting	198
9.36	Fuzzy RD OLS parameter estimates for full certification setting .	199
9.37	Fuzzy RD 2SLS-IV parameter estimates for full certification setting	199
9.38	Fuzzy RD OLS parameter estimates for selective certification setting	200
9.39	Fuzzy RD 2SLS-IV parameter estimates for selective certification setting	200
9.40	Fuzzy RD OLS parameter estimates with missing data for selective certification setting	200

9.41	Fuzzy RD 2SLS-IV parameter estimates with missing data for selective certification setting	201
9.42	Fuzzy RD OLS parameter estimates for full certification setting	202
9.43	Average treatment effect sample statistics for full certification setting	202
10.1	Tuebingen IV example treatment likelihoods for case 1: ignorable treatment	223
10.2	Tuebingen IV example outcome likelihoods for case 1: ignorable treatment	223
10.3	Tuebingen IV example results for case 1: ignorable treatment	224
10.4	Tuebingen IV example treatment likelihoods for case 1b: uniformity fails	224
10.5	Tuebingen IV example treatment likelihoods for case 2: heterogeneous response	225
10.6	Tuebingen IV example outcome likelihoods for case 2: heterogeneous response	226
10.7	Tuebingen IV example results for case 2: heterogeneous response	226
10.8	Tuebingen IV example treatment likelihoods for case 2b: LATE = ATT	227
10.9	Tuebingen IV example outcome likelihoods for case 2b: LATE = ATT	227
10.10	Tuebingen IV example results for case 2b: LATE = ATT	228
10.11	Tuebingen IV example treatment likelihoods for case 3: more heterogeneity	228
10.12	Tuebingen IV example outcome likelihoods for case 3: more heterogeneity	229
10.13	Tuebingen IV example results for case 3: more heterogeneity	229
10.14	Tuebingen IV example treatment likelihoods for case 3b: LATE = ATUT	230
10.15	Tuebingen IV example outcome likelihoods for case 3b: LATE = ATUT	230
10.16	Tuebingen IV example results for case 3b: LATE = ATUT	231
10.17	Tuebingen IV example treatment likelihoods for case 4: Simpson's paradox	231
10.18	Tuebingen IV example outcome likelihoods for case 4: Simpson's paradox	232
10.19	Tuebingen IV example results for case 4: Simpson's paradox	232
10.20	Tuebingen IV example treatment likelihoods for case 4b: exclusion restriction violated	233
10.21	Tuebingen IV example outcome likelihoods for case 4b: exclusion restriction violated	233
10.22	Tuebingen IV example results for case 4b: exclusion restriction violated	234

10.23	Tuebingen IV example outcome likelihoods for case 5: lack of common support	234
10.24	Tuebingen IV example treatment likelihoods for case 5: lack of common support	235
10.25	Tuebingen IV example results for case 5: lack of common support	235
10.26	Tuebingen IV example outcome likelihoods for case 5b: minimal common support	236
10.27	Tuebingen IV example outcome likelihoods for case 5b: minimal common support	236
10.28	Tuebingen IV example results for case 5b: minimal common support	237
10.29	Report precision OLS parameter estimates for binary base case	242
10.30	Report precision average treatment effect sample statistics for binary base case	242
10.31	Report precision saturated OLS parameter estimates for binary base case	243
10.32	Report precision adjusted outcome OLS parameter estimates for binary base case	245
10.33	Report precision adjusted outcome OLS parameter estimates for binary heterogeneous case	247
10.34	Report precision average treatment effect sample statistics for binary heterogeneous case	247
10.35	Report precision poor 2SLS-IV estimates for binary heterogeneous case	248
10.36	Report precision weak 2SLS-IV estimates for binary heterogeneous case	249
10.37	Report precision stronger 2SLS-IV estimates for binary heterogeneous case	250
10.38	Report precision propensity score estimates for binary heterogeneous case	251
10.39	Report precision propensity score matching estimates for binary heterogeneous case	251
10.40	Report precision ordinate control IV estimates for binary heterogeneous case	252
10.41	Report precision inverse Mills IV estimates for binary heterogeneous case	253
10.42	Continuous report precision but observed binary OLS parameter estimates	255
10.43	Continuous report precision but observed binary average treatment effect sample statistics	255
10.44	Continuous report precision but observed binary propensity score parameter estimates	256
10.45	Continuous report precision but observed binary propensity score matching parameter estimates	256

10.46	Continuous report precision but observed binary ordinate control IV parameter estimates	257
10.47	Continuous report precision but observed binary inverse Mills IV parameter estimates	258
10.48	Continuous report precision but observed binary sample correlations	259
10.49	Continuous report precision but observed binary stronger propensity score parameter estimates	260
10.50	Continuous report precision but observed binary stronger propensity score matching parameter estimates	260
10.51	Continuous report precision but observed binary stronger ordinate control IV parameter estimates	261
10.52	Continuous report precision but observed binary stronger inverse Mills IV parameter estimates	262
10.53	Continuous report precision but observed binary OLS parameter estimates for Simpson's paradox DGP	264
10.54	Continuous report precision but observed binary average treatment effect sample statistics for Simpson's paradox DGP	264
10.55	Continuous report precision but observed binary ordinate control IV parameter estimates for Simpson's paradox DGP	265
10.56	Continuous report precision but observed binary inverse Mills IV parameter estimates for Simpson's paradox DGP	266
10.57	Continuous treatment OLS parameter estimates and average treatment effect estimates and sample statistics with only between individual variation	268
10.58	Continuous treatment 2SLS-IV parameter and average treatment effect estimates with only between individual variation	269
10.59	Continuous treatment OLS parameter and average treatment effect estimates for modest within individual report precision variation setting	270
10.60	Continuous treatment ATE and ATT sample statistics and correlation between treatment and treatment effect for modest within individual report precision variation setting	270
10.61	Continuous treatment 2SLS-IV parameter and average treatment effect estimates for modest within individual report precision variation setting	271
10.62	Continuous treatment OLS parameter and average treatment effect estimates for the more between and within individual report precision variation setting	272
10.63	Continuous treatment ATE and ATT sample statistics and correlation between treatment and treatment effect for the more between and within individual report precision variation setting	272
10.64	Continuous treatment 2SLS-IV parameter and average treatment effect estimates for the more between and within individual report precision variation setting	272

11.1	Comparison of identification conditions for common econometric strategies (adapted from Heckman and Navarro-Lozano's [2004] table 3)	285
11.2	Continuous report precision but observed binary OLS parameter estimates for apparently nonnormal DGP	295
11.3	Continuous report precision but observed binary average treatment effect sample statistics for apparently nonnormal DGP	295
11.4	Continuous report precision but observed binary ordinate control IV parameter estimates for apparently nonnormal DGP	295
11.5	Continuous report precision but observed binary inverse Mills IV parameter estimates for apparently nonnormal DGP	296
11.6	Continuous report precision but observed binary LIV parameter estimates for apparently nonnormal DGP	297
11.7	Continuous report precision but observed binary sample correlations for apparently nonnormal DGP	298
11.8	Continuous report precision but observed binary stronger ordinate control IV parameter estimates for apparently nonnormal DGP	299
11.9	Continuous report precision but observed binary average treatment effect sample statistics for apparently nonnormal DGP	299
11.10	Continuous report precision but observed binary stronger inverse Mills IV parameter estimates for apparently nonnormal DGP	299
11.11	Continuous report precision but observed binary stronger LIV parameter estimates for apparently nonnormal DGP	300
12.1	McMC parameter estimates for prototypical selection	310
12.2	McMC estimates of average treatment effects for prototypical selection	310
12.3	McMC average treatment effect sample statistics for prototypical selection	311
12.4	McMC MTE-weighted average treatment effects for prototypical selection	311
12.5	Binary report precision McMC parameter estimates for heterogeneous outcome	315
12.6	Binary report precision McMC average treatment effect estimates for heterogeneous outcome	315
12.7	Binary report precision McMC average treatment effect sample statistics for heterogeneous outcome	315
12.8	Binary report precision McMC MTE-weighted average treatment effect estimates for heterogeneous outcome	317
12.9	Continuous report precision but observed binary selection McMC parameter estimates	318
12.10	Continuous report precision but observed binary selection McMC average treatment effect estimates	318

12.11	Continuous report precision but observed binary selection McMC average treatment effect sample statistics	319
12.12	Continuous report precision but observed binary selection McMC MTE-weighted average treatment effect estimates	319
12.13	Continuous report precision but observed binary selection McMC parameter estimates for nonnormal DGP	322
12.14	Continuous report precision but observed binary selection McMC average treatment effect estimates for nonnormal DGP	322
12.15	Continuous report precision but observed binary selection McMC average treatment effect sample statistics for nonnormal DGP	322
12.16	Continuous report precision but observed binary selection McMC MTE-weighted average treatment effect estimates for nonnormal DGP	324
12.17	Continuous report precision but observed binary selection stronger McMC parameter estimates	324
12.18	Continuous report precision but observed binary selection stronger McMC average treatment effect estimates	324
12.19	Continuous report precision but observed binary selection stronger McMC average treatment effect sample statistics	325
12.20	Continuous report precision but observed binary selection stronger McMC MTE-weighted average treatment effect estimates	326
12.21	Policy-relevant average treatment effects with original precision cost parameters	327
12.22	Policy-relevant average treatment effects with revised precision cost parameters	328
13.1	Jaynes' widget problem: summary of background knowledge by stage	356
13.2	Jaynes' widget problem: stage 3 state of knowledge	364
13.3	Jaynes' widget problem: stage 3 state of knowledge along with standard deviation	366

List of Figures

3.1	Price versus reported accruals	50
8.1	Fixed effects regression curves	128
8.2	Strategic choice game tree	136
11.1	MTE and weight functions for other treatment effects	281
12.1	$MTE(u_D)$ versus $u_D = p_\nu$ for prototypical selection	312
12.2	$MTE(u_D)$ versus $u_D = p_\nu$ for binary report precision	316
12.3	$MTE(u_D)$ versus $u_D = p_\nu$ for continuous report precision but binary selection	320
12.4	$MTE(u_D)$ versus $u_D = p_\nu$ for nonnormal DGP	323
12.5	$MTE(u_D)$ versus $u_D = p_\nu$ with stronger instruments	325
12.6	$MTE(u_D)$ versus $u_D = p_\nu$ for policy-relevant treatment effect	329
13.1	"Exact" distributions for daily widget demand	369
13.2	Directed graph of financial statements	374
13.3	Spanning tree	375
13.4	Stochastic manipulation σ_d known	386
13.5	Incidence of stochastic manipulation and posterior probability	386
13.6	Stochastic manipulation σ_d unknown	393
13.7	Selective manipulation σ_d known	395
13.8	Incidence of selective manipulation and posterior probability . .	395
13.9	Selective manipulation σ_d unknown	398

1
Introduction

We believe progress in the study of accounting (perhaps any scientific endeavor) is characterized by attention to theory, data, and model specification. Understanding the role of accounting in the world typically revolves around questions of causal effects. That is, holding other things equal what is the impact on outcome (welfare) of some accounting choice. The ceteris paribus conditions are often awkward because of simultaneity or endogeneity. In these pages we attempt to survey some strategies for addressing these challenging questions and share our experiences. These shared experiences typically take the form of identifying the theory through stylized accounting examples, and exploring the implications of varieties of available data (to the analyst or social scientist). Theory development is crucial for careful identification of the focal problem. Progress can be seriously compromised when the problem is not carefully defined. Once the problem is carefully defined, identifying the appropriate data is more straightforward but, of course, data collection often remains elusive.[1] Recognizing information available to the economic agents as well as limitations of data available to the analyst is of paramount importance. While our econometric tool kit continues to grow richer, frequently there is no substitute for finding data better suited to the problem at hand. The combination of theory (problem identification) and data leads to model specification. Model specification and testing frequently lead us to revisit theory development and data collection. This three-legged, iterative strategy for "creating order from chaos" proceeds without end.

---

[1] We define empiricists. as individuals who have special talents in identification and collection of task-appropriate data. A skill we regard as frequently undervalued and, alas, one which we do not possess (or at least, have not developed).

1.1 Problematic illustration

The following composite illustration discusses some of our concerns when we fail to faithfully apply these principles.[2] It is common for analysts (social scientists) to deal with two (or more) alternative first order considerations (theories or framings) of the setting at hand. One theory is seemingly more readily manageable as it proceeds with a more partial equilibrium view and accordingly suppresses considerations that may otherwise enter as first order influences. The second view is more sweeping, more of a general equilibrium perspective of the setting at hand.

Different perspectives may call for different data (regressands and/or regressors in a conditional analysis). Yet, frequently in the extant literature some of the data employed reflect one perspective, some a second perspective, some both perspectives, and perhaps, some data reflect an alternate, unspoken theory. Is this cause for concern?

Consider asset valuation in public information versus private information settings. A *CAPM* (public information) equilibrium (Sharpe [1964], Lintner [1965], and Mossin [1966]; also see Lambert, Leuz, and Verrecchia [2007]) calls for the aggregation of risky assets into an efficient market portfolio and the market portfolio is a fundamental right-hand side variable. However, in a world where private information is a first order consideration, there exists no such simple aggregation of assets to form an efficient (market) portfolio (Admati [1985]). Hence, while diversification remains a concern for the agents in the economy it is less clear what role any market index plays in the analysis.[3]

Empirical model building (specification and diagnostic checking) seems vastly different in the two worlds. In the simpler *CAPM* world it is perhaps sensible to consider the market index as exogenous. However, its measurement is of critical importance (Roll [1977]).[4] Measures of the market index are almost surely inadequate and produce an errors-in-variables (in other words, correlated omitted

---

[2]The example is a composite critique of the current literature. Some will take offense at these criticisms even though no individual studies are referenced. The intent is not to place blame or dwell on the negative but rather to move forward (hopefully, by inventing new mistakes rather than repeating the old ones). Our work (research) is forever incomplete.

[3]Another simple example involving care in data selection comes from cost of capital analysis where, say, the focus is on cost of debt capital. Many economic analyses involve the influence of various (often endogenous) factors on the marginal cost of debt capital. Nevertheless, the analysts employ a historical weighted average of a firm's debt cost (some variant of reported interest scaled by reported outstanding debt). What does this tell us about influences on the firm's cost of raising debt capital?

[4]Arbitrage pricing (Ross [1976]) is a promising complete information alternative that potentially avoids this trap. However, identification of risk factors remains elusive.

variable) problem.[5] When experimental variables are added,[6] care is required as they may pick up measurement error in the market index rather than the effect being studied. In addition, it may be unwise to treat the factors of interest exogenously. Whether endogeneity arises as a first order consideration or not in this seemingly simpler setting has become much more challenging than perhaps was initially suspected.

In our alternate private information world, inclusion of a market index may serve as a (weak) proxy for some other fundamental factor or factors. Further, these missing fundamental variables may be inherently endogenous. Of course, the diagnostic checks we choose to employ depend on our perception of the setting (theory or framing) and appropriate data.

Our point is econometric analysis associated with either perspective calls for a careful matching of theory, data, and model specification. Diagnostic checking follows first order considerations outlined by our theoretical perspective and data choices. We hope evidence from such analyses provides a foundation for discriminating between theories or perspectives as a better first order approximation. In any case, we cannot investigate every possible source of misspecification but rather we must focus our attention on problematic issues to which our perspective (theory) guides us. While challenging, the iterative, three-legged model building strategy is a cornerstone of scientific inquiry.

In writing these pages (including the above discussion), we found ourselves to be significantly influenced by Jaynes' [2003] discussion of probability theory as the logic of science. Next, we briefly outline some of the principles he describes.

---

[5]Is the lack of a significant relation between individual stocks, or even portfolios of stocks, with the market index a result of greater information asymmetry (has there been a marked jump in the exploitation of privately informed-opportunism? – Enron, Worldcom, etc.), or the exclusion of more assets in the index (think of the financial engineering explosion) over the past twenty years?

[6]The quality of accounting information and how it affects some response variable (say, firm value) is often the subject of inquiry. Data is an enormous challenge here. We know from Blackwell [1953] (see also Blackwell and Girshick [1954], Marschak and Miyasawa [1968], and Demski [1973]), information systems, in general, are not comparable as fineness is the only generally consistent ranking metric and it is incomplete. This means that we have to pay attention to the context and are only able to make contextual comparisons of information systems. As accounting is not a myopic supplier of information, informational complementarities abound. What is meant by accounting quality surely cannot be effectively captured by vague proxies for relevance, reliability, precision, etc. that ignore other information and belie Blackwell comparability. Further, suppose we are able to surmount these issues, what is learned in say the valuation context may be of no consequence in a stewardship context (surely a concern in accounting). Demski [1994,2008] and Christensen and Demski [2003] provide numerous examples illustrating this point. Are we forgetting the idea of statistical sufficiency? A statistic is not designed to be sufficient for the data in the address of all questions but for a specific question (often a particular moment). Moving these discussions forward demands more creativity in identifying and measuring the data.

1.2 Jaynes' desiderata for scientific reasoning

Jaynes' discussion of probability as logic (the logic of science) suggests the following desiderata regarding the assessment of plausible propositions:
1. Degrees of plausibility are represented by real numbers;
2. Reasoning conveys a qualitative correspondence with common sense;
3. Reasoning is logically consistent.

Jaynes' [2003, p. 86] goes on to argue the fundamental principle of probabilistic inference is

> To form a judgment about the likely truth or falsity of any proposition A, the correct procedure is to calculate the probability that A is true
>
> $$\Pr(A \mid E_1, E_2, \ldots)$$
>
> conditional on all the evidence at hand.

Again, care in problem or proposition definition is fundamental to scientific inquiry.

In our survey of econometric challenges associated with analysis of accounting choice, we attempt to follow these guiding principles. However, the preponderance of extant econometric work on endogeneity is classical, our synthesis reflects this, and, as Jaynes points out, classical methods sometimes fail to consider all evidence. Therefore, where classical approaches may be problematic, we revisit the issue with a "more complete" Bayesian analysis. The final chapter synthesizes (albeit incompletely) Jaynes' thesis on probability as logic and especially informed, maximum entropy priors. Meanwhile, we offer a simple but provocative example of probability as logic.

1.2.1 Probability as logic illustration[7]

Suppose we only know a variable, call it X_1, has support from $(-1, 1)$ and a second variable, X_2, has support from $(-2, 2)$. Then, we receive an aggregate report — their sum, $Y = X_1 + X_2$, equals $\frac{1}{2}$. What do we know about X_1 and X_2? Jayne's maximum entropy principle (*MEP*) suggests we assign probabilities based on what we know but only what we know. Consider X_1 alone. Since we only know support, consistent probability assignment leads to the uniform density

$$f(X_1 : \{-1 < X_1 < 1\}) = \frac{1}{2}$$

Similarly, for X_2 we have

$$f(X_2 : -2 < X_2 < 2) = \frac{1}{4}$$

---

[7]This example was developed from conversations with Anil Arya and Brian Mittendorf.

Now, considered jointly we have[8]

$$f(X_1, X_2 : \{-1 < X_1 < 1, -2 < X_2 < 2\}) = \frac{1}{8}$$

What is learned from the aggregate report $y = \frac{1}{2}$? Bayesian updating based on the evidence suggests

$$f\left(X_1 \mid y = \frac{1}{2}\right) = \frac{f\left(X_1, y = \frac{1}{2}\right)}{f\left(y = \frac{1}{2}\right)}$$

and

$$f\left(X_2 \mid y = \frac{1}{2}\right) = \frac{f\left(X_2, y = \frac{1}{2}\right)}{f\left(y = \frac{1}{2}\right)}$$

Hence, updating follows from probability assignment of $f(X_1, Y)$, $f(X_2, Y)$, and $f(Y)$. Since we have $f(X_1, X_2)$ and $Y = X_1 + X_2$ plus knowledge of any two of (Y, X_1, X_2) supplies the third, we know

$$f\left(X_1, Y : \begin{array}{l} \{-3 < Y < -1, -1 < X_1 < Y+2\} \\ \{-1 < Y < 1, -1 < X_1 < 1\} \\ \{1 < Y < 3, Y-2 < X_1 < 1\} \end{array}\right) = \frac{1}{8}$$

and

$$f\left(X_2, Y : \begin{array}{l} \{-3 < Y < -1, -2 < X_2 < Y+1\} \\ \{-1 < Y < 1, -1 < X_2 < 1\} \\ \{1 < Y < 3, Y-1 < X_2 < 2\} \end{array}\right) = \frac{1}{8}$$

Further,

$$\begin{aligned} f(Y) &= \int f(X_1, Y) \, dX_1 \\ &= \int f(X_2, Y) \, dX_2 \end{aligned}$$

Hence, integrating out X_1 or X_2 yields

$$\int_{-1}^{Y+2} f(X_1, Y) \, dX_1 = \int_{-1}^{Y+1} f(X_2, Y) \, dX_2 \quad \text{for } -3 < Y < -1$$

$$\int_{-1}^{1} f(X_1, Y) \, dX_1 = \int_{-1}^{1} f(X_2, Y) \, dX_2 \quad \text{for } -1 < Y < 1$$

and

$$\int_{Y-2}^{1} f(X_1, Y) \, dX_1 = \int_{Y-1}^{1} f(X_2, Y) \, dX_2 \quad \text{for } 1 < Y < 3$$

---

[8] *MEP* treats X_1 and X_2 as independent random variables as we have no knowledge regarding their relationship.

Collectively, we have[9]

$$f(Y : \{-3 < Y < -1\}) = \frac{3+Y}{8}$$
$$f(Y : \{-1 < Y < 1\}) = \frac{1}{4}$$
$$f(Y : \{1 < Y < 3\}) = \frac{3-Y}{8}$$

Now, conditional probability assignment given $y = \frac{1}{2}$ is

$$f\left(X_1 : \{-1 < X_1 < 1\} \mid y = \frac{1}{2}\right) = \frac{\frac{1}{8}}{\frac{1}{4}} = \frac{1}{2}$$

and

$$f\left(X_2 : \{Y - 1 < X_2 < Y + 1\} \mid y = \frac{1}{2}\right) = \frac{\frac{1}{8}}{\frac{1}{4}}$$

or

$$f\left(X_2 : \left\{-\frac{1}{2} < X_2 < \frac{3}{2}\right\} \mid y\right) = \frac{1}{2}$$

Hence, the aggregate report tells us nothing about X_1 (our unconditional beliefs are unaltered) but a good deal about X_2 (support is cut in half). For instance, updated beliefs conditional on the aggregate report imply $E\left[X_1 \mid y = \frac{1}{2}\right] = 0$ and $E\left[X_2 \mid y = \frac{1}{2}\right] = \frac{1}{2}$. This is logically consistent as $E\left[X_1 + X_2 \mid y = \frac{1}{2}\right] = E\left[Y \mid y = \frac{1}{2}\right]$ must be equal to $\frac{1}{2}$.

On the other hand, if the aggregate report is $y = 2$, then revised beliefs are

$$f(X_1 : \{Y - 2 < X_1 < 1\} \mid y = 2) = \frac{\frac{1}{8}}{\frac{3-Y}{8}} = \frac{1}{3-2}$$

or

$$f(X_1 : \{0 < X_1 < 1\} \mid y = 2) = 1$$

---

[9]Likewise, the marginal densities for X_1 and X_2 are identified by integrating out the other variable from their joint density. That is

$$\int_{-2}^{2} f(X_1, X_2) \, dX_2$$
$$= f(X_1 : \{-1 < X_1 < 1\}) = \frac{1}{2}$$

and

$$\int_{-1}^{1} f(X_1, X_2) \, dX_1$$
$$= f(X_2 : \{-2 < X_2 < 2\}) = \frac{1}{4}$$

This consistency check brings us back to our starting point.

and
$$f(X_2 : \{Y - 1 < X_2 < Y + 1\} \mid y = 2) = \frac{1}{3-2}$$
or
$$f(X_2 : \{1 < X_2 < 2\} \mid y = 2) = 1$$

The aggregate report is informative for both variables, X_1 and X_2. For example, updated beliefs imply
$$E[X_1 \mid y = 2] = \frac{1}{2}$$
and
$$E[X_2 \mid y = 2] = \frac{3}{2}$$
and
$$E[X_1 + X_2 \mid y = 2] = 2$$

Following a brief overview of chapter organization, we explore probability as logic in other accounting settings.

1.3 Overview

The second chapter introduces several recurring accounting examples and their underlying theory including any equilibrium strategies. We make repeated reference to these examples throughout later chapters as well as develop other sparser examples. Chapter three reviews linear models including double residual regression (*FWL*) and linear instrumental variable estimation. Prominent examples survey some econometric issues which arise in the study of earnings management as equilibrium reporting behavior and econometric challenges associated with documenting information content in the presence of multiple sources of information.

Chapter four continues where we left off with linear models by surveying loss functions and estimation. The discussion includes maximum likelihood estimation, nonlinear regression, and James-Stein shrinkage estimators. Chapter five utilizes estimation results surveyed in chapter four to discuss discrete choice models — our point of entry for limited dependent variable models. Discrete choice models and other limited dependent variable models play a key role in many identification and estimation strategies associated with causal effects.

Distributional and structural conditions can sometimes be relaxed via nonparametric and semiparametric approaches. A brief survey is presented in chapter six. Nonparametric regression is referenced in the treatment effect discussions in chapters 8 through 12. In addition, nonparametric regression can be utilized to evaluate information content in the presence of multiple sources of information as introduced in chapter three. Chapter seven surveys repeated-sampling inference methods with special attention to bootstrapping and Bayesian simulation. Analytic demands of Bayesian inference are substantially reduced via Markov chain

Monte Carlo (*McMC*) methods which are briefly discussed in chapter seven and applied to the treatment effect problem in chapter 12.

Causal effects are emphasized in the latter chapters — chapters 8 through 13. A survey of econometric challenges associated with endogeneity is included in chapter eight. This is not intended to be comprehensive but a wide range of issues are reviewed to emphasize the breadth of extant work on endogeneity including simultaneous probit, strategic choice models, duration models, and selection analysis. Again, the Tuebingen-style treatment effect examples are introduced at the end of chapter eight.

Chapter nine surveys identification of treatment effects via ignorable treatment conditions, or selection on observables, including the popular and intuitively appealing propensity score matching. Tuebingen-style examples are extended to incorporate potential regressors and ask whether, conditional on these regressors, average treatment effects are identified. In addition, treatment effects associated with the asset revaluation regulation example introduced in chapter two are extensively analyzed.

Chapter ten reviews some instrumental variable (*IV*) approaches. *IV* approaches are a natural response when available data do not satisfy ignorable treatment conditions. Again, Tuebingen-style examples incorporating instruments are explored. Further, treatment effects associated with the report precision regulation setting introduced in chapter two are analyzed.

Chapter 11 surveys marginal treatment effects and their connection to other (average) treatment effects. The chapter also briefly mentions newer developments such as dynamics and distributions of treatment effects as well as general equilibrium considerations though in-depth exploration of these issues are beyond the scope of this book. Bayesian (*McMC*) analysis of treatment effects are surveyed in chapter 12. Analyses of marginal and average treatment effects in prototypical selection setting are illustrated and the regulated report precision setting is revisited.

Chapter 13 brings the discussion full circle. Informed priors are fundamental to probability as logic. Jayne's [2003] widget problem is a clever illustration of the principles of consistent reasoning in an uncertain setting. Earnings management as equilibrium reporting behavior is revisited with informed priors explicitly recognized. We only scratch the surface of potential issues to be addressed but hope that others are inspired to continue the quest for a richer and deeper understanding of causal effects associated with accounting choices.

1.4 Additional reading

Jaynes [2003] describes a deep and lucid account of probability theory as the logic of science. Probabilities are assigned based on the maximum entropy principle (*MEP*).

2
Accounting choice

Accounting is an information system design problem. An objective in the study of accounting is to understand its nature and its utility for helping organizations manage uncertainty and private information. As one of many information sources, accounting has many peculiar properties: it's relatively late, it's relatively coarse, it's typically aggregated, it selectively recognizes and reports information (or, equivalently, selectively omits information), however accounting is also highly structured and well disciplined against random errors, and frequently audited. Like other information sources accounting competes for resources. The initial features cited above may suggest that accounting is at a competitive disadvantage. However, the latter features (integrity) are often argued to provide accounting its comparative strength and its integrity is reinforced by the initial features (see Demski [1994, 2008] and Christensen and Demski [2003]).

Demski [2004] stresses endogenous expectations, that is, emphasis on microfoundations or choices (economic and social psychology) and equilibrium to tie the picture together. His remarks sweep out a remarkably broad path of accounting institutions and their implications beginning with a fair game *iid* dividend machine coupled with some report mechanism and equilibrium pricing. This is then extended to include earnings management, analysts' forecasts, regulation assessment studies, value-relevance studies, audit judgement studies, compensation studies, cost measurement studies, and governance studies. We continue this theme by focusing on a modest subset of accounting choices.

In this chapter we begin discussion of four prototypical accounting choice settings. We return to these examples repeatedly in subsequent chapters to illustrate and explore their implications for econometric analysis and especially endogenous causal effects. The first accounting choice setting evaluates equilibrium earnings management. The second accounting choice setting involves the ex ante impact of accounting asset revaluation regulation on an owner's investment decision and welfare. The third accounting choice setting involves the impact of the choice (discretionary or regulated) of costly accounting report precision on an owner's welfare for assets in place (this example speaks to the vast literature on "accounting quality").[1] A fourth accounting choice setting explores recovery of recognized transactions from reported financial statements.

2.1 Equilibrium earnings management

Suppose the objective is to track the relation between a firm's value P_t and its accruals z_t.[2] To keep things simple, firm value equals the present value of expected future dividends, the market interest rate is zero, current period cash flows are fully paid out in dividends, and dividends $\tilde{d}$ are Normal *iid* with mean zero and variance σ^2. Firm managers have private information $\tilde{y}_t^p$ about next period's dividend $\tilde{y}_t^p = \tilde{d}_{t+1} + \tilde{\varepsilon}_t$ where $\tilde{\varepsilon}$ are Normal *iid* with mean zero and variance σ^2.[3] If the private information is revealed, ex dividend firm value at time t is

$$P_t \equiv E\left[\tilde{d}_{t+1} \mid \tilde{y}_t^p = y_t^p\right]$$
$$= \frac{1}{2}y_t^p$$

Suppose management reveals its private information through income I_t (cash flows plus change in accruals) where fair value accruals

$$z_t = E\left[\tilde{d}_{t+1} \mid \tilde{y}_t^p = y_t^p\right] = \frac{1}{2}y_t^p$$

are reported. Then, income is

$$I_t = d_t + (z_t - z_{t-1})$$
$$= d_t + \frac{1}{2}\left(y_t^p - y_{t-1}^p\right)$$

and

$$P_t \equiv E\left[\tilde{d}_{t+1} \mid \tilde{d}_t = d_t, I_t = d_t + \frac{1}{2}\left(y_t^p - y_{t-1}^p\right)\right]$$
$$= E\left[\tilde{d}_{t+1} \mid \tilde{z}_t = \frac{1}{2}y_t^p\right]$$
$$= z_t$$

---

[1] An additional setting could combine precision choice and investment (such as in Dye and Sridar [2004, 2007]). Another could perhaps add accounting asset valuation back into the mix. But we leave these settings for later study.
[2] This example draws from Demski [2004].
[3] For simplicity, there is no other information.

There is a linear relation between price and fair value accruals.

Suppose the firm is owned and managed by an entrepreneur who, for intergenerational reasons, liquidates his holdings at the end of the period. The entrepreneur is able to misrepresent the fair value estimate by reporting, $z_t = \frac{1}{2}y_t^p + \theta$, where $\theta \geq 0$. Auditors are unable to detect any accrual overstatements below a threshold equal to $\frac{1}{2}\Delta$. Traders anticipate the entrepreneur reports $z_t = \frac{1}{2}y_t^p + \frac{1}{2}\Delta$ and the market price is

$$P_t = z_t - E[\theta] = z_t - \frac{1}{2}\Delta$$

Given this anticipated behavior, the entrepreneur's equilibrium behavior is to report as conjectured. Again, there is a linear relationship between firm value and reported fair value accruals.

Now, consider the case where the entrepreneur can misreport but with probability α; the probability of misreporting is common knowledge. Investors process the entrepreneur's report with misreporting in mind. The probability of misreporting given an accrual report of z_t is

$$\Pr(D \mid \tilde{z}_t = z_t) = \frac{\alpha \phi\left(\frac{z_t - 0.5\Delta}{\sqrt{0.5}\sigma}\right)}{\alpha \phi\left(\frac{z_t - 0.5\Delta}{\sqrt{0.5}\sigma}\right) + (1-\alpha) \phi\left(\frac{z_t}{\sqrt{0.5}\sigma}\right)}$$

where $\phi(\cdot)$ is the standard normal density function and $D = 1$ if there is misreporting ($\theta = \frac{1}{2}\Delta$) and $D = 0$ otherwise. In turn, the equilibrium price for the firm following the report is

$$P_t = E\left[\tilde{d}_{t+1} \mid \tilde{z}_t = z_t\right] = \frac{\alpha(z_t - 0.5\Delta) \phi\left(\frac{z_t - 0.5\Delta}{\sqrt{0.5}\sigma}\right) + (1-\alpha) z_t \phi\left(\frac{z_t}{\sqrt{0.5}\sigma}\right)}{\alpha \phi\left(\frac{z_t - 0.5\Delta}{\sqrt{0.5}\sigma}\right) + (1-\alpha) \phi\left(\frac{z_t}{\sqrt{0.5}\sigma}\right)}$$

Again, the entrepreneur's equilibrium reporting strategy is to misreport the maximum whenever possible and the accruals balance is $\alpha\left(\frac{1}{2}\Delta\right)$, on average. Price is no longer a linear function of reported fair value.

The example could be extended to address a more dynamic, multiperiod setting. A setting in which managers report discretion is limited by audited "cookie jar" accounting reserves. We leave this to future work.

2.1.1 Implications for econometric analysis

Econometric analysis must carefully attend to the connections between theory and data. For instance, in this setting the equilibrium behavior is based on investors' perceptions of earnings management which may differ from potentially observed (by the analyst) levels of earnings management. This creates a central role in our econometric analysis for the propensity score (discussed later along with discrete choice models). The evidence or data helps us distinguish between

12   2. Accounting choice

various earnings management propositions. In the stochastic or selective manipulation settings, manipulation likelihood is the focus.

Econometric analysis of equilibrium earnings management is pursed in chapters 3 and 13. Chapter 3 focuses on the relation between firm value and reported accruals. The discussion in chapter 13 first explores accruals smoothing in both valuation and evaluation contexts then focuses on separation of signal from noise in stochastically and selectively manipulated accruals. Informed priors and Bayesian analysis are central to these discussions in chapter 13.

2.2   Asset revaluation regulation

Our second example explores the ex ante impact of accounting asset revaluation policies on owners' investment decisions (and welfare) in an economy of, on average, price protected buyers.[4] Prior to investment, an owner evaluates both investment prospects and the market for resale in the event the owner becomes liquidity stressed. The payoff from investment I is distributed uniformly and centered at $\hat{x} = \frac{\beta}{\alpha} I^\alpha$ where $\alpha, \beta > 0$ and $\alpha < 1$. Hence, support for investment payoff is $x = \hat{x} \pm f = [\underline{x}, \overline{x}]$. A potential problem with the resale market is the owner will have private information — knowledge of the asset value. However, since there is some positive probability the owner becomes distressed, π, the market will not collapse (as in Dye [1985]). The equilibrium price is based on distressed sellers being forced to pool potentially healthy assets with non-distressed sellers' impaired assets. Regulators may choose to prop-up the price to aid distressed sellers by requiring certification of assets at cost k with values below some cutoff x_c.[5,6] The owner's ex ante expected payoff from investment I and certification cutoff x_c is

$$E[V \mid I, x_c] = \pi \frac{1}{2f} \left[\frac{1}{2} \left(x_c^2 - \underline{x}^2 \right) - k \left(x_c - \underline{x} \right) + P \left(\overline{x} - x_c \right) \right]$$
$$+ (1-\pi) \frac{1}{2f} \left[\frac{1}{2} \left(x_c^2 - \underline{x}^2 \right) + P \left(P - x_c \right) + \frac{1}{2} \left(\overline{x}^2 - P^2 \right) \right]$$
$$- I$$

The equilibrium uncertified asset price is

$$P = \frac{x_c + \sqrt{\pi} \overline{x}}{1 + \sqrt{\pi}}$$

---

[4]This example draws heavily from Demski, Lin, and Sappington [2008].

[5]This cost is incremental to normal audit cost. As such, even if audit fee data is available, k may be difficult for the analyst to observe.

[6]Owners never find it ex ante beneficial to voluntarily certify asset revaluation because of the certification cost. We restrict attention to targeted certification but certification could be proportional rather than targeted (see Demski, et al [2008] for details). For simplicity, we explore only targeted certification.

2.2 Asset revaluation regulation

This follows from the equilibrium condition

$$P = \frac{1}{4fq}\left[\pi\left(\bar{x}^2 - x_c^2\right) + (1-\pi)\left(P^2 - x_c^2\right)\right]$$

where

$$q = \frac{1}{2f}\left[\pi\left(\bar{x} - x_c\right) + (1-\pi)\left(P - x_c\right)\right]$$

is the probability that an uncertified asset is marketed. Further, the regulator may differentially weight the welfare $W(I, x_c)$ of distressed sellers compared with non-distressed sellers. Specifically, the regulator may value distressed seller's net gains dollar-for-dollar but value non-distressed seller's gains at a fraction, w, on the dollar.

$$W(I, x_c) = \pi \frac{1}{2f}\left[\frac{1}{2}\left(x_c^2 - \underline{x}^2\right) - k\left(x_c - \underline{x}\right) + P\left(\bar{x} - x_c\right)\right]$$
$$+ w(1-\pi)\frac{1}{2f}\left[\frac{1}{2}\left(x_c^2 - \underline{x}^2\right) + P(P - x_c) + \frac{1}{2}\left(\bar{x}^2 - P^2\right)\right]$$
$$- I\left[\pi + (1-\pi)w\right]$$

2.2.1 Numerical example

As indicated above, owners will choose to never certify assets if it's left to their discretion. Consider the following parameters

$$\left\{\alpha = \frac{1}{2}, \beta = 10, \pi = 0.7, k = 20, f = 150\right\}$$

Then never certify $(x_c = \underline{x})$ results in investment $I = 100$, owner's expected payoff $E[V \mid I, x_c] = 100$, and equilibrium uncertified asset price $P \approx 186.66$.

However, regulators may favor distressed sellers and require selective certification. Continuing with the same parameters, if regulators give zero consideration ($w = 0$) to the expected payoffs of non-distressed sellers, then the welfare maximizing certification cutoff $x_c = \bar{x} - \frac{(1+\sqrt{\pi})k}{(1-\sqrt{\pi})(1-w)} \approx 134.4$.[7] This induces investment $I = \left[\frac{\beta(2f+\pi k)}{2f}\right]^{\frac{1}{1-\alpha}} \approx 109.6$, owner's expected payoff approximately equal to 96.3, and equilibrium uncertified asset price $P \approx 236.9$ (an uncertified price more favorable to distressed sellers).

---

[7] This is optimal for k small; that is, $k < Z(w)$ where

$$Z(w) = \frac{2f\left(1 - \sqrt{\pi}\right)(1-w)}{\left(1+\sqrt{\pi}\right)\pi}\left[\pi - \frac{\pi\left(1+\sqrt{\pi}\right)c}{f\left(1-\sqrt{\pi}\right)(1-w)}\right]$$

and

$$c = [\pi + w(1-\pi)]\frac{1-\alpha}{\alpha} - \left(\frac{2f}{\alpha(2f+\pi k)} - 1\right)\left(1 + \frac{\pi k}{2f}\right)^{\frac{1}{1-\alpha}}\beta^{\frac{1}{1-\alpha}}$$

2.2.2 Implications for econometric analysis

For econometric analysis of this setting, we refer to the investment choice as the treatment level and any revaluation regulation (certification requirement) as policy intervention. Outcomes Y are reflected in exchange values[8] (perhaps less initial investment and certification cost if these data are accessible) and accordingly (as is typical) reflect only a portion of the owner's expected utility.

$$Y = P(I, x_c) = \frac{x_c + \sqrt{\pi \bar{x}}}{1 + \sqrt{\pi}}$$

Some net benefits may be hidden from the analysts' view; these may include initial investment and certification cost, and gains from owner retention (not selling the assets) where exchange prices represent lower bounds on the owner's outcome. Further, outcomes (prices) reflect realized draws whereas the owner's expected utility is based on expectations. The causal effect of treatment choice on outcomes is frequently the subject under study and almost surely is endogenous.

This selection problem is pursued in later chapters (chapters 8 through 12). Here, the data help us distinguish between various selection-based propositions. For instance, is investment selection inherently endogenous, is price response to investment selection homogeneous, or is price response to investment selection inherently heterogeneous? Econometric analysis of asset revaluation regulation is explored in chapter 9.

2.3 Regulated report precision

Our third example explores the impact of costly report precision on owner's welfare in an economy of price protected buyers.[9] Suppose a risk averse asset owner sees strict gains to trade from selling her asset to risk neutral buyers. However, the price the buyer is willing to pay is tempered by his perceived ability to manage the asset.[10] This perception is influenced by the reliability of the owner's report on the asset $s = V + \varepsilon_2$ where $\varepsilon_2 \sim N\left(0, \sigma_2^2\right)$. The gross value of the asset is denoted $V = \mu + \varepsilon_1$ where $\varepsilon_1 \sim N\left(0, \sigma_1^2\right)$ and ε_1 and ε_2 are independent. Hence, the price is

$$\begin{aligned} P &= E[V \mid s] - \beta Var[V \mid s] \\ &= \mu + \frac{\sigma_1^2}{\sigma_1^2 + \sigma_2^2}(s - \mu) - \beta \frac{\sigma_1^2 \sigma_2^2}{\sigma_1^2 + \sigma_2^2} \end{aligned}$$

---

[8]This may include a combination of securities along the debt-equity continuum.
[9]This example draws heavily from Chistensen and Demski [2007].
[10]An alternative interpretation is that everyone is risk averse but gains to trade arise due to differential risk tolerances and/or diversification benefits.

2.3 Regulated report precision

The owner chooses σ_2^2 (inverse precision) at a cost equal to $\alpha\left(b - \sigma_2^2\right)^2$ (where $\sigma_2^2 \in [a, b]$) and has mean-variance preferences[11]

$$E\left[U \mid \sigma_2^2\right] = E\left[P \mid \sigma_2^2\right] - \gamma Var\left[P \mid \sigma_2^2\right]$$

where

$$E\left[P \mid \sigma_2^2\right] = \mu - \beta \frac{\sigma_1^2 \sigma_2^2}{\sigma_1^2 + \sigma_2^2}$$

and

$$Var\left[P \mid \sigma_2^2\right] = \frac{\sigma_1^4}{\sigma_1^2 + \sigma_2^2}$$

Hence, the owner's expected utility from issuing the accounting report and selling the asset is

$$\mu - \beta \frac{\sigma_1^2 \sigma_2^2}{\sigma_1^2 + \sigma_2^2} - \gamma \frac{\sigma_1^4}{\sigma_1^2 + \sigma_2^2} - \alpha\left(b - \sigma_2^2\right)^2$$

2.3.1 Public precision choice

Public knowledge of report precision is the benchmark (symmetric information) case. Precision or inverse precision σ_2^2 is chosen to maximize the owner's expected utility. For instance, the following parameters

$$\{\mu = 1,000, \sigma_1^2 = 100, \beta = 7, \gamma = 2.5, \alpha = 0.02, b = 150\}$$

result in optimal inverse-precision $\sigma_2^{2*} \approx 128.4$. and expected utility approximately equal to 487.7. Holding everything else constant, $\alpha = 0.04$ produces $\sigma_2^{2*} \approx 140.3$. and expected utility approximately equal to 483.5. Not surprisingly, higher cost reduces report precision and lowers owner satisfaction.

2.3.2 Private precision choice

Private choice of report precision introduces asymmetric information. The owner chooses the Nash equilibrium precision level; that is, when buyers' conjectures $\bar{\sigma}_2^2$ match the owner's choice of inverse-precision σ_2^2. Now, the owner's expected utility is

$$\mu - \beta \frac{\sigma_1^2 \bar{\sigma}_2^2}{\sigma_1^2 + \bar{\sigma}_2^2} - \gamma \frac{\sigma_1^4 \left(\sigma_1^2 + \sigma_2^2\right)}{\left(\sigma_1^2 + \bar{\sigma}_2^2\right)^2} - \alpha\left(b - \sigma_2^2\right)^2$$

For the same parameters as above

$$\{\mu = 1,000, \sigma_1^2 = 100, \beta = 7, \gamma = 2.5, \alpha = 0.02, b = 150\}$$

---

[11] Think of a *LEN* model. If the owner has negative exponential utility (*CARA*; contstant absolute risk aversion), the outcome is linear in a normally distributed random variable(s), then we can write the certainty equivalent as $E[P(s)] - \frac{\rho}{2}Var[P(s)]$ as suggested.

the optimal inverse-precision choice is $\sigma_2^{2**} \approx 139.1$. and expected utility is approximately equal to 485.8. Again, holding everything else constant, $\alpha = 0.04$ produces $\sigma_2^{2**} \approx 144.8$. and expected utility is approximately equal to 483.3. Asymmetric information reduces report precision and lowers the owner's satisfaction.

2.3.3 Regulated precision choice and transaction design

Asymmetric information produces a demand or opportunity for regulation. Assuming the regulator can identify the report precision preferred by the owner σ_2^{2*}, full compliance with regulated inverse-precision $\hat{b}$ restores the benchmark solution. However, the owner may still exploit her private information even if it is costly to design transactions which appear to meet the regulatory standard when in fact they do not.

Suppose the cost of transaction design takes a similar form to the cost of report precision $\alpha_d \left(\hat{b} - \sigma_2^2 \right)^2$; that is, the owner bears a cost of deviating from the regulatory standard. The owner's expected utility is the same as the private information case with transaction design cost added.

$$\mu - \beta \frac{\sigma_1^2 \bar{\sigma}_2^2}{\sigma_1^2 + \bar{\sigma}_2^2} - \gamma \frac{\sigma_1^4 \left(\sigma_1^2 + \sigma_2^2 \right)}{\left(\sigma_1^2 + \bar{\sigma}_2^2 \right)^2} - \alpha \left(b - \sigma_2^2 \right)^2 - \alpha_d \left(\hat{b} - \sigma_2^2 \right)^2$$

For the same parameters as above

$$\{ \mu = 1,000, \sigma_1^2 = 100, \beta = 7, \gamma = 2.5, \alpha = 0.02, \alpha_d = 0.02, b = 150 \}$$

the Nash equilibrium inverse-precision choice, for regulated inverse-precision $\hat{b} = 128.4$, is $\sigma_2^{2***} \approx 133.5$. and owner's expected utility is approximately equal to 486.8. Again, holding everything else constant, $\alpha_d = 0.04$ produces $\sigma_2^{2***} \approx 131.7$. and owner's expected utility is approximately equal to 487.1. While regulation increases report precision and improves the owner's welfare relative to private precision choice, it also invites transaction design (commonly referred to as earnings management) which produces deviations from regulatory targets.

2.3.4 Implications for econometric analysis

For econometric analysis of this setting, we refer to the report precision choice as the treatment level and any regulation as policy intervention. Outcomes Y are reflected in exchange values[12] and accordingly (as is typical) reflect only a portion of the owner's expected utility.

$$Y = P\left(\bar{\sigma}_2^2\right) = \mu + \frac{\sigma_1^2}{\sigma_1^2 + \bar{\sigma}_2^2} (s - \mu) - \beta \frac{\sigma_1^2 \bar{\sigma}_2^2}{\sigma_1^2 + \bar{\sigma}_2^2}$$

---

[12]This may include a combination of securities along the debt-equity continuum.

In particular, cost is hidden from the analysts' view; cost includes the explicit cost of report precision, cost of any transaction design, and the owner's risk premia. Further, outcomes (prices) reflect realized draws from the accounting system s whereas the owner's expected utility is based on expectations and her knowledge of the distribution for (s, V). The causal effect of treatment choice on outcomes is frequently the subject under study and almost surely is endogenous. This selection problem is pursued in later chapters (chapters 8 through 12). Again, the data help us distinguish between various selection-based propositions. For instance, is report precision selection inherently endogenous, is price response to report precision selection homogeneous, or is price response to report precision selection inherently heterogeneous? Econometric analysis of regulated report precision is explored in chapters 10 and 12. Chapter 10 employs classical identification and estimation strategies while chapter 12 employs Bayesian analysis.

2.4 Inferring transactions from financial statements

Our fourth example asks to what extent can recognized transactions be recovered from financial statements.[13] Similar to the above examples but with perhaps wider scope, potential transactions involve strategic interaction of various economic agents as well as the reporting firm's and auditor's restriction of accounting recognition choices.

We denote accounting recognition choices by the matrix A, where journal entries make up the columns and the rows effectively summarize entries that change account balances (as with ledgers or T accounts). The changes in account balances are denoted by the vector x and the transactions of interest are denoted by the vector y. Then, the linear system describing the problem is

$$Ay = x$$

2.4.1 Implications for econometric analysis

Solving for y is problematic as A is not invertible — A is typically not a square matrix and in any case doesn't have linearly independent rows due to the balancing property of accounting. Further, y typically has more elements than x. Classical methods are stymied. Here we expressly lean on a Bayesian approach including a discussion of the merits of informed, maximum entropy priors. Financial statement data help us address propositions regarding potential equilibrium play. That is, the evidence may strongly support, weakly support, or refute anticipated equilibrium responses and/or their encoding in the financial statements. Evidence supporting either of the latter two may resurrect propositions that are initially considered unlikely. Econometric analysis of financial statements is explored in chapter 13.

---

[13] This example draws primarily from Arya et al [2000].

2.5 Additional reading

Extensive reviews and illuminating discussions are found in Demski [1994, 2008] and Christensen and Demski [2003]. Demski's American Accounting Association Presidential Address [2004] is particularly insightful.

3
Linear models

Though modeling endogeneity may involve a variety of nonlinear or generalized linear, nonparametric or semiparametric models, and maximum likelihood or Bayesian estimation, much of the intuition is grounded in the basic linear model. This chapter provides a condensed overview of linear models and establishes connections with later discussions.

3.1 Standard linear model (*OLS*)

Consider the data generating process (*DGP*):

$$Y = X\beta + \varepsilon$$

where $\varepsilon \sim (0, \sigma^2 I)$, X is $n \times p$ (with rank p), and $E\left[X^T \varepsilon\right] = 0$, or more generally $E\left[\varepsilon \mid X\right] = 0$.

The *Gauss-Markov theorem* states that $b = \left(X^T X\right)^{-1} X^T Y$ is the minimum variance estimator of β amongst linear unbiased estimators. Gauss' insight follows from a simple idea. Construct b (or equivalently, the residuals or estimated errors, e) such that the residuals are orthogonal to every column of X (recall the objective is to extract all information in X useful for explaining Y — whatever is left over from Y should be unrelated to X).

$$X^T e = 0$$

where $e = Y - Xb$. Rewriting the orthogonality condition yields

$$X^T (Y - Xb) = 0$$

3. Linear models

or the normal equations
$$X^T X b = X^T Y$$
Provided X is full column rank, this yields the usual *OLS* estimator
$$b = \left(X^T X\right)^{-1} X^T Y$$
It is straightforward to show that b is unbiased (conditional on the data X).

$$\begin{aligned}
E[b \mid X] &= E\left[\left(X^T X\right)^{-1} X^T Y \mid X\right] \\
&= E\left[\left(X^T X\right)^{-1} X^T \left(X\beta + \varepsilon\right) \mid X\right] \\
&= \beta + \left(X^T X\right)^{-1} X^T E[\varepsilon \mid X] = \beta + 0 = \beta
\end{aligned}$$

Iterated expectations yields $E[b] = E_X[E[b \mid X]] = E_X[\beta] = \beta$. Hence, unbiasedness applies unconditionally as well.

$$\begin{aligned}
Var[b \mid X] &= Var\left[\left(X^T X\right)^{-1} X^T Y \mid X\right] \\
&= Var\left[\left(X^T X\right)^{-1} X^T \left(X\beta + \varepsilon\right) \mid X\right] \\
&= E\left\{\beta + \left(X^T X\right)^{-1} X^T \varepsilon - \beta\right\}\left\{\left(X^T X\right)^{-1} X^T \varepsilon\right\}^T \mid X \\
&= \left(X^T X\right)^{-1} X^T E\left[\varepsilon \varepsilon^T\right] X \left(X^T X\right)^{-1} \\
&= \sigma\left(^2 X^T X\right)^{-1} X^T I X \left(X^T X\right)^{-1} \\
&= \sigma^2 \left(X^T X\right)^{-1}
\end{aligned}$$

Now, consider the stochastic regressors case,
$$Var[b] = Var_X[E[b \mid X]] + E_X[Var[b \mid X]]$$
The first term is zero since $E[b \mid X] = \beta$ for all X. Hence,
$$Var[b] = E_X[Var[b \mid X]] = \sigma^2 E\left[\left(X^T X\right)^{-1}\right]$$
the unconditional variance of b can only be described in terms of the average behavior of X.

To show that *OLS* yields the minimum variance linear unbiased estimator consider another linear unbiased estimator $b_0 = LY$ (L replaces $\left(X^T X\right)^{-1} X^T$). Since $E[LY] = E[LX\beta + L\varepsilon] = \beta$, $LX = I$.
Let $D = L - \left(X^T X\right)^{-1} X^T$ so that $DY = b_0 - b$.

$$\begin{aligned}
Var[b_0 \mid X] &= \sigma^2 \left[D + \left(X^T X\right)^{-1} X^T\right]\left[D + \left(X^T X\right)^{-1} X^T\right]^T \\
&= \sigma^2 \left\{ \begin{array}{l} DDT + \left(X^T X\right)^{-1} X^T D^T + DX \left(X^T X\right)^{-1} \\ + \left(X^T X\right)^{-1} X^T X \left(X^T X\right)^{-1} \end{array} \right\}
\end{aligned}$$

Since
$$LX = I = DX + (X^T X)^{-1} X^T X, DX = 0$$
and
$$Var[b_0 \mid X] = \sigma^2 \left(DD^T + (X^T X)^{-1} \right)$$

As DD^T is positive semidefinite, $Var[b]$ (and $Var[b \mid X]$) is at least as small as any other $Var[b_0]$ ($Var[b_0 \mid X]$). Hence, the Gauss-Markov theorem applies to both nonstochastic and stochastic regressors.

Theorem 3.1 *Rao-Blackwell theorem. If $\varepsilon \sim N(0, \sigma^2 I)$ for the above DGP, b has minimum variance of all unbiased estimators.*

Finite sample inferences typically derive from normally distributed errors and t (individual parameters) and F (joint parameters) statistics. Some asymptotic results related to the Rao-Blackwell theorem are as follows. For the Rao-Blackwell DGP, OLS is consistent and asymptotic normally (*CAN*) distributed. Since *MLE* yields b for the above *DGP* with normally distributed errors, *OLS* is asymptotically efficient amongst all *CAN* estimators. Asymptotic inferences allow relaxation of the error distribution and rely on variations of the laws of large numbers and central limit theorems.

3.2 Generalized least squares (*GLS*)

Suppose the *DGP* is $Y = X\beta + \varepsilon$ where $\varepsilon \sim (0, \Sigma)$ and $E[X^T \varepsilon] = 0$, or more generally, $E[\varepsilon \mid X] = 0$, X is $n \times p$ (with rank p). The *BLU* estimator is

$$b_{GLS} = (X^T \Sigma^{-1} X)^{-1} X^T \Sigma^{-1} Y$$

$$E[b_{GLS}] = \beta$$

$$Var[b_{GLS} \mid X] = (X^T \Sigma^{-1} X)^{-1}$$

and
$$Var[b_{GLS}] = E\left[(X^T \Sigma^{-1} X)^{-1}\right] = \sigma^2 E\left[(X^T \Omega^{-1} X)^{-1}\right]$$

where scale is extracted to construct $\Omega^{-1} = \frac{1}{\sigma^2} \Sigma^{-1}$.

A straightforward estimation approach involves Cholesky decomposition of Σ.

$$\Sigma = \Gamma \Gamma^T = L D^{1/2} D^{1/2} L^T$$

where D is a matrix with pivots on the diagonal.

$$\Gamma^{-1} Y = \Gamma^{-1} (X\beta + \varepsilon)$$

and
$$\Gamma^{-1} \varepsilon \sim (0, I)$$

since $\Gamma^{-1}0 = 0$ and $\Gamma^{-1}\Sigma\left(\Gamma^T\right)^{-1} = \Gamma^{-1}\Gamma\Gamma^T\left(\Gamma^T\right)^{-1} = I$. Now, *OLS* applied to the regression of $\Gamma^{-1}Y$ (in place of Y) onto $\Gamma^{-1}X$ (in place of X) yields

$$\begin{aligned} b_{GLS} &= \left(\left(\Gamma^{-1}X\right)^T \Gamma^{-1}X\right)^{-1}\left(\Gamma^{-1}X\right)^T \Gamma^{-1}Y \\ &= \left(X^T \left(\Gamma^{-1}\right)^T \Gamma^{-1}X\right)^{-1} X^T \left(\Gamma^{-1}\right)^T \Gamma^{-1}Y \\ b_{GLS} &= \left(X^T \Sigma^{-1} X\right)^{-1} X^T \Sigma^{-1} Y \text{ (Aitken estimator)} \end{aligned}$$

Hence, *OLS* regression of suitably transformed variables is equivalent to *GLS* regression, the minimum variance linear unbiased estimator for the above *DGP*.

OLS is unbiased for the above *DGP* (but inefficient),

$$E[b] = \beta + E_X\left[\left(X^T X\right)^{-1} X^T E[\varepsilon \mid X]\right] = \beta$$

However, $Var[b \mid X]$ is not the standard one described above. Rather,

$$Var[b \mid X] = \left(X^T X\right)^{-1} X^T \Sigma^{-1} X \left(X^T X\right)^{-1}$$

which is typically estimated by Eicker-Huber-White asymptotic heteroskedasticity consistent estimator

$$\begin{aligned} Est.Asy.Var[b] &= n\left(X^T X\right)^{-1} S_0 \left(X^T X\right)^{-1} \\ &= n^{-1}\left(n^{-1}X^T X\right)^{-1}\left(n^{-1}\sum_{i=1}^{n} e_i^2 x_i x_i^T\right)\left(n^{-1}X^T X\right)^{-1} \end{aligned}$$

where x_i is the ith row from X and $S_0 = 1/n \sum_{i=1}^{n} e_i^2 x_i x_i^T$, or the Newey-West autocorrelation consistent covariance estimator where S_0 is replaced by $S_0 + n^{-1}\sum_{l=1}^{L}\sum_{t=l+1}^{n} w_l e_i e_{t-l}\left(x_l x_{t-l}^T + x_{t-l} x_l^T\right)$, $w_l = 1 - \frac{l}{L+1}$, and the maximum lag L is set in advance.

3.3 Tests of restrictions and *FWL* (Frisch-Waugh-Lovell)

Causal effects are often the focus of accounting and economic analysis. That is, the question often boils down to what is the response to a change in one variable holding the others constant. *FWL* (partitioned regression or double residual regression) and tests of restrictions can help highlight causal effects in the context of linear models (and perhaps more broadly).

Consider the *DGP* for *OLS* where the matrix of regressors is partitioned $X = \begin{bmatrix} X_1 & X_2 \end{bmatrix}$ and X_1 represents the variables of prime interest and X_2 perhaps represents control variables.[1]

$$Y = X\beta + \varepsilon = X_1\beta_1 + X_2\beta_2 + \varepsilon$$

Of course, β can be estimated via *OLS* as b and b_1 (the estimate for β_1) can be extracted from b. However, it is instructive to remember that each β_k represents the response (of Y) to changes in X_k conditional on all other regressors X_{-k}. The *FWL* theorem indicates that b_1 can also be estimated in two steps. First, regress X_1 and Y onto X_2. Retain their residuals, e_1 and e_Y. Second, regress e_Y onto e_1 to estimate $b_1 = \left(e_1^T e_1\right)^{-1} e_1^T e_Y$ (a no intercept regression) and $Var\left[b_1 \mid X\right] = \sigma^2 \left(X^T X\right)_{11}^{-1}$, where $\left(X^T X\right)_{11}^{-1}$ refers to the upper left block of $\left(X^T X\right)^{-1}$.

FWL produces the following three results:

1.
$$\begin{aligned} b_1 &= \left(X_1^T \left(I - P_2\right) X_1\right)^{-1} X_1^T \left(I - P_2\right) Y \\ &= \left(X_1^T M_2 X_1\right)^{-1} X_1^T M_2 Y \end{aligned}$$

is the same as b_1 from the upper right partition of

$$b = \left(X^T X\right)^{-1} X^T Y$$

where $P_2 = X_2 \left(X_2^T X_2\right)^{-1} X_2^T$.

2.
$$\begin{aligned} Var\left[b_1\right] &= \sigma^2 \left(X_1^T \left(I - P_2\right) X_1\right)^{-1} \\ &= \sigma^2 \left(X_1^T M_2 X_1\right)^{-1} \end{aligned}$$

is the same as from the upper left partition of

$$Var\left[b\right] = \sigma^2 \left(X^T X\right)^{-1}$$

3. The regression or predicted values are

$$\begin{aligned} \widehat{Y} &= P_X Y = X \left(X^T X\right)^{-1} X^T Y = Xb \\ &= X_1 b_1 + X_2 b_2 = P_2 Y + \left(I - P_2\right) X_1 b_1 \\ &= P_2 Y + M_2 X_1 b_1 \end{aligned}$$

First, we demonstrate result 1. Since $e_1 = \left(I - P_2\right) X_1 = M_2 X_1$ and $e_Y = \left(I - P_2\right) Y = M_2 Y$,

$$\begin{aligned} b_1 &= \left(X_1^T \left(I - P_2\right) X_1\right)^{-1} X_1^T \left(I - P_2\right) Y \\ &= \left(X_1^T M_2 X_1\right)^{-1} X_1^T M_2 Y \end{aligned}$$

---

[1] When a linear specification of the control variables is questionable, we might employ partial linear or partial index regressions. For details see the discussion of these semi-parametric regression models in chapter 6. Also, a model specification test against a general nonparametric regression model is discussed in chapter 6.

3. Linear models

To see that this is the same as from standard (one-step) multiple regression derive the normal equations from

$$X_1^T X_1 b_1 = X_1^T Y - X_1^T X_2 b_2$$

and

$$P_2 X_2 b_2 = X_2 b_2 = P_2 (Y - X_1 b_1)$$

Substitute to yield

$$X_1^T X_1 b_1 = X_1^T Y - X_1^T P_2 (Y - X_1 b_1)$$

Combine like terms in the normal equations.

$$X_1^T (I - P_2) X_1 b_1 = X_1^T (I - P_2) Y$$
$$= X_1^T M_2 Y$$

Rewriting yields

$$b_1 = \left(X_1^T M_2 X_1\right)^{-1} X_1^T M_2 Y$$

This demonstrates 1.[2]

---

[2] A more constructive demonstration of *FWL* result 1 is described below. From Gauss,

$$b = \left(\begin{bmatrix} X_2^T \\ X_1^T \end{bmatrix} \begin{bmatrix} X_2 & X_1 \end{bmatrix} \right)^{-1} \begin{bmatrix} X_2^T \\ X_1^T \end{bmatrix} Y$$

(for convenience X is reordered as $\begin{bmatrix} X_2 & X_1 \end{bmatrix}$).

$$\left(\begin{bmatrix} X_2^T \\ X_1^T \end{bmatrix} \begin{bmatrix} X_2 & X_1 \end{bmatrix} \right)^{-1} = \begin{bmatrix} X_2^T X_2 & X_2^T X_1 \\ X_1^T X_2 & X_1^T X_1 \end{bmatrix}^{-1}$$

(by LDL^T block "rank-one" factorization)

$$= \left(\begin{bmatrix} I & 0 \\ X_1^T X_2 \left(X_2^T X_2\right)^{-1} & I \end{bmatrix} \begin{bmatrix} X_2^T X_2 & 0 \\ 0 & X_1^T (I - P_2) X_1 \end{bmatrix} \begin{bmatrix} I & \left(X_2^T X_2\right)^{-1} X_2^T X_1 \\ 0 & I \end{bmatrix} \right)^{-1}$$

$$= \begin{bmatrix} I & -\left(X_2^T X_2\right)^{-1} X_2^T X_1 \\ 0 & I \end{bmatrix} \begin{bmatrix} \left(X_2^T X_2\right)^{-1} & 0 \\ 0 & \left(X_1^T M_2 X_1\right)^{-1} \end{bmatrix}$$

$$\times \begin{bmatrix} I & 0 \\ -X_1^T X_2 \left(X_2^T X_2\right)^{-1} & I \end{bmatrix}$$

Multiply the first two terms and apply the latter inverse to $\begin{bmatrix} X_2 & X_1 \end{bmatrix}^T Y$

$$\begin{bmatrix} b_2 \\ b_1 \end{bmatrix} = \begin{bmatrix} \left(X_2^T X_2\right)^{-1} & -\left(X_2^T X_2\right)^{-1} X_2^T X_1 \left(X_1^T M_2 X_1\right)^{-1} \\ 0 & \left(X_1^T M_2 X_1\right)^{-1} \end{bmatrix} \begin{bmatrix} X_2^T Y \\ X_1^T (I - P_2) Y \end{bmatrix}$$

$b_1 = (x^T M_2 x)^{-1} x^T M_2 Y$. This demonstrates *FWL* result 1.

3.3 Tests of restrictions and *FWL* (Frisch-Waugh-Lovell)

FWL result 2 is as follows.

$$Var\,[b] = \sigma^2 \left(X^T X\right)^{-1} = \sigma^2 \begin{bmatrix} X_1^T X_1 & X_1^T X_2 \\ X_2^T X_1 & X_2^T X_2 \end{bmatrix}^{-1}$$

$$= \sigma^2 \begin{bmatrix} A_{11} & -\left(X_1^T X_1\right)^{-1} X_1^T X_2 A_{22} \\ -\left(X_2^T X_2\right)^{-1} X_2^T X_1 A_{11} & A_{22} \end{bmatrix}$$

where

$$A_{11} = \left(X_1^T X_1 - X_1^T X_2 \left(X_2^T X_2\right)^{-1} X_2^T X_1\right)^{-1}$$

and

$$A_{22} = \left(X_2^T X_2 - X_2^T X_1 \left(X_1^T X_1\right)^{-1} X_1^T X_2\right)^{-1}$$

Rewriting A_{11}, the upper left partition, and combining with σ^2 produces

$$\sigma^2 \left(X_1^T (I - P_2) X_1\right)^{-1}$$

This demonstrates *FWL* result 2.

To demonstrate *FWL* result 3

$$X_1 b_1 + X_2 b_2 = P_2 Y + (I - P_2) X_1 b_1$$

refer to the estimated model

$$Y = X_1 b_1 + X_2 b_2 + e$$

where the residuals e, by construction, are orthogonal to X. Multiply both sides by P_2 and simplify

$$\begin{aligned} P_2 Y &= P_2 X_1 b_1 + P_2 X_2 b_2 + P_2 e \\ &= P_2 X_1 b_1 + X_2 b_2 \end{aligned}$$

Rearranging yields

$$X_2 b_2 = P_2 \left(Y - X_1 b_1\right)$$

Now, add $X_1 b_1$ to both sides

$$X_1 b_1 + X_2 b_2 = X_1 b_1 + P_2 \left(Y - X_1 b_1\right)$$

Simplification yields

$$\begin{aligned} X_1 b_1 + X_2 b_2 &= P_2 Y + (I - P_2) X_1 b_1 \\ &= P_2 Y + M_2 X_1 b_1 \end{aligned}$$

This demonstrates *FWL* result 3.

3.4 Fixed and random effects

Often our data come in a combination of cross-sectional and time-series data, or panel data which can substantially increase sample size. A panel data regression then is

$$Y_{tj} = X_{tj}\beta + u_{tj}$$

where t refers to time and j refers to individuals (or firms). With panel data, one approach is multivariate regression (multiple dependent variables and hence multiple regressions as in, for example, seemingly unrelated regressions). Another common approach, and the focus here, is an error-components model. The idea is to model u_{tj} as consisting of three individual shocks, each independent of the others

$$u_{tj} = e_t + \nu_j + \varepsilon_{tj}$$

For simplicity, suppose the time error component e_t is independent across time $t = 1, \ldots, T$, the individual error component ν_j is independent across units $j = 1, \ldots, n$, and the error component ε_{tj} is independent across all time t and individuals j.

There are two standard regression strategies for addressing error components: (1) a fixed effects regression, and (2) a random effects regression. Fixed effects regressions model time effects e_t and/or individual effects ν_j conditionally. On the other hand, the random effects regressions are modeled unconditionally. That is, random effects regressions model time effects e_t and individual effects ν_j as part of the regression error. The trade-offs between the two involve the usual regression considerations. Since fixed effects regressions condition on e_t and ν_j, fixed effects strategies do not rely on independence between the regressors and the error components e_t and ν_j. On the other hand, when appropriate (when independence between the regressors and the error components e_t and ν_j is satisfied), the random effects model more efficiently utilizes the data. A Hausman test (Hausman [1978]) can be employed to test the consistency of the random effects model by reference to the fixed effects model.

For purposes of illustration, assume that there are no time-specific shocks, that is $e_t = 0$ for all t. Now the error components regression is

$$Y_{tj} = X_{tj}\beta + \nu_j + \varepsilon_{tj}$$

In matrix notation, the fixed effects version of the above regression is

$$Y = X\beta + D\nu + \varepsilon$$

where D represents n dummy variables corresponding to the n cross-sectional units in the sample. Provided ε_{tj} is *iid*, the model can be estimated via *OLS*. Or using *FWL*, the fixed effects estimator for β is

$$\widehat{\beta}^{WG} = \left(X^T M_D X\right)^{-1} X^T M_D Y$$

where $P_D = D\left(D^T D\right)^{-1} D^T$, projection into the columns of D, and $M_D = I - P_D$, the projection matrix that produces the deviations from cross-sectional group means. That is,

$$(M_D X)_{tj} = X_{tj} - \overline{X}_{\cdot j}$$

and

$$M_D Y_{tj} = Y_{tj} - \overline{Y}_{\cdot j}$$

where $\overline{X}_{\cdot j}$ and $\overline{Y}_{\cdot j}$ are the group (individual) j means for the regressors and regressand, respectively. Since this estimator only exploits the variation between the deviations of the regressand and the regressors from their respective group means, it is frequently referred to as a within-groups (*WG*) estimator.

Use of only the variation between deviations can be an advantage or a disadvantage. If the cross-sectional effects are correlated with the regressors, then the *OLS* estimator (without fixed effects) is inconsistent but the within-groups estimator is consistent. However, if the cross-sectional effects (i.e., the group means) are uncorrelated with the regressors then the within-groups (fixed effects) estimator is inefficient. In the extreme case in which there is an independent variable that has no variation between the deviations and only varies between group means, then the coefficient for this variable is not even identified by the within-groups estimator.

To see that *OLS* is inconsistent when the cross-sectional effects are correlated with the errors consider the complementary between-groups estimator. A between-groups estimator only utilizes the variation among group means.

$$\widehat{\beta}^{BG} = \left(X^T P_D X\right)^{-1} X^T P_D Y$$

The between-groups estimator is inconsistent if the (cross-sectional) group means are correlated with the regressors. Further, since the *OLS* estimator can be written as a matrix-weighted average of the within-groups and between-groups estimators, if the between-groups estimator is inconsistent, *OLS* (without fixed effects) is inconsistent as demonstrated below.

$$\widehat{\beta}^{OLS} = \left(X^T X\right)^{-1} X^T Y$$

Since $M_D + P_D = I$,

$$\widehat{\beta}^{OLS} = \left(X^T X\right)^{-1} \left(X^T M_D Y + X^T P_D Y\right)$$

Utilizing $\left(X^T X\right)^{-1} X^T X = \left(X^T X\right)^{-1} X^T \left(M_D + P_D\right) X = I$, we rewrite the *OLS* estimator as a matrix-weighted average of the within-groups and between-

3. Linear models

groups estimators

$$\begin{aligned}\widehat{\beta}^{OLS} &= \left(X^T X\right)^{-1} X^T M_D X \widehat{\beta}^{WG} + \left(X^T X\right)^{-1} X^T P_D X \widehat{\beta}^{BG} \\ &= \left(X^T X\right)^{-1} X^T M_D \left(X^T M_D X\right)^{-1} X^T M_D Y \\ &\quad + \left(X^T X\right)^{-1} X^T P_D X \left(X^T P_D X\right)^{-1} X^T P_D Y \\ &= \left(X^T X\right)^{-1} X^T M_D \left(X^T M_D X\right)^{-1} X^T M_D (X\beta + u) \\ &\quad + \left(X^T X\right)^{-1} X^T P_D X \left(X^T P_D X\right)^{-1} X^T P_D (X\beta + u) \end{aligned}$$

Now, if the group means are correlated with the regressors then

$$\begin{aligned} p\lim \widehat{\beta}^{BG} &= p\lim \left(X^T P_D X\right)^{-1} X^T P_D (X\beta + u) \\ &= \beta + p\lim \left(X^T P_D X\right)^{-1} X^T P_D u \\ &= \beta + \alpha \quad \alpha \neq 0 \end{aligned}$$

and

$$\begin{aligned} p\lim \widehat{\beta}^{OLS} &= \left(X^T X\right)^{-1} X^T X \beta + \left(X^T X\right)^{-1} X^T P_D X \alpha \\ &= \beta + \left(X^T X\right)^{-1} X^T P_D X \alpha \\ &\neq \beta \quad \text{if } \alpha \neq 0 \end{aligned}$$

Hence, *OLS* is inconsistent if the between-groups estimator is inconsistent, in other words, if the cross-sectional effects are correlated with the errors.

Random effects regressions are typically estimated via *GLS* or maximum likelihood (here we focus on *GLS* estimation of random effects models). If the individual error components are uncorrelated with the group means of the regressors, then *OLS* with fixed effects is consistent but inefficient. We may prefer to employ a random effects regression which is consistent and more efficient. *OLS* treats all observations equally but this is not an optimal usage of the data. On the other hand, a random effects regression treats ν_j as a component of the error rather than fixed. The variance of u_{jt} is $\sigma_\nu^2 + \sigma_\varepsilon^2$. The covariance of u_{ti} with u_{tj} is zero for $i \neq j$, under the conditions described above. But the covariance of u_{tj} with u_{sj} is σ_ν^2 for $s \neq t$. Thus, the $T \times T$ variance-covariance matrix is

$$\Sigma = \sigma_\varepsilon^2 I + \sigma_\nu^2 \iota \iota^T$$

where ι is a T-length vector of ones and the data are ordered first by individual unit and then by time. And the covariance matrix for the u_{tj} is

$$Var[u] = \begin{bmatrix} \Sigma & 0 & \cdots & 0 \\ 0 & \Sigma & \cdots & 0 \\ \vdots & \vdots & \ddots & \vdots \\ 0 & 0 & \cdots & \Sigma \end{bmatrix}$$

3.4 Fixed and random effects

GLS estimates can be computed directly or the data can be transformed and *OLS* applied. We'll briefly explore a transformation strategy. One transformation, derived via singular value decomposition (*SVD*), is

$$\Sigma^{-1/2} = \sigma_\varepsilon^{-1}(I - \alpha P_\iota)$$

where $P_\iota = \iota\left(\iota^T\iota\right)^{-1}\iota^T = \frac{1}{T}\iota\iota^T$ and α, between zero and one, is

$$\alpha = 1 - \sigma_\varepsilon\left(T\sigma_\nu^2 + \sigma_\varepsilon^2\right)^{-\frac{1}{2}}$$

The transformation is developed as follows. Since Σ is symmetric, *SVD* combined with the spectral theorem implies we can write

$$\Sigma = Q\Lambda Q^T$$

where Q is an orthogonal matrix ($QQ^T = Q^TQ = I$) with eigenvectors in its columns and Λ is a diagonal matrix with eigenvalues along its diagonal; $T - 1$ eigenvalues are equal to σ_ε^2 and one eigenvalue equals $T\sigma_\nu^2 + \sigma_\varepsilon^2$. To fix ideas, consider the $T = 2$ case,

$$\Sigma = \begin{bmatrix} \sigma_\nu^2 + \sigma_\varepsilon^2 & \sigma_\nu^2 \\ \sigma_\nu^2 & \sigma_\nu^2 + \sigma_\varepsilon^2 \end{bmatrix}$$

$$= Q\Lambda Q^T$$

where

$$\Lambda = \begin{bmatrix} \sigma_\varepsilon^2 & 0 \\ 0 & 2\sigma_\nu^2 + \sigma_\varepsilon^2 \end{bmatrix}$$

and

$$Q = \frac{1}{\sqrt{2}}\begin{bmatrix} 1 & 1 \\ -1 & 1 \end{bmatrix}$$

Since

$$\Sigma = Q \begin{bmatrix} \sigma_\varepsilon^2 & 0 \\ 0 & 2\sigma_\nu^2 + \sigma_\varepsilon^2 \end{bmatrix} Q^T$$

and

$$\begin{aligned}
\Sigma^{-1} &= Q \begin{bmatrix} \frac{1}{\sigma_\varepsilon^2} & 0 \\ 0 & \frac{1}{2\sigma_\nu^2 + \sigma_\varepsilon^2} \end{bmatrix} Q^T \\
&= Q \left(\begin{bmatrix} \frac{1}{\sigma_\varepsilon^2} & 0 \\ 0 & 0 \end{bmatrix} + \begin{bmatrix} 0 & 0 \\ 0 & \frac{1}{2\sigma_\nu^2 + \sigma_\varepsilon^2} \end{bmatrix} \right) Q^T \\
&= Q \begin{bmatrix} \frac{1}{\sigma_\varepsilon^2} & 0 \\ 0 & 0 \end{bmatrix} Q^T + Q \begin{bmatrix} 0 & 0 \\ 0 & \frac{1}{2\sigma_\nu^2 + \sigma_\varepsilon^2} \end{bmatrix} Q^T \\
&= \frac{1}{\sigma_\varepsilon^2} Q \begin{bmatrix} 1 & 0 \\ 0 & 0 \end{bmatrix} Q^T + \frac{1}{2\sigma_\nu^2 + \sigma_\varepsilon^2} Q \begin{bmatrix} 0 & 0 \\ 0 & 1 \end{bmatrix} Q^T \\
&= \frac{1}{\sigma_\varepsilon^2}(I - P_\iota) + \frac{1}{2\sigma_\nu^2 + \sigma_\varepsilon^2} P_\iota
\end{aligned}$$

3. Linear models

Note, the key to the general case is to construct Q such that

$$Q \begin{bmatrix} 0 & \cdots & 0 \\ \vdots & \ddots & \vdots \\ 0 & \cdots & 1 \end{bmatrix} Q^T = P_\iota$$

Since $I - P_\iota$ and P_ι are orthogonal projection matrices, we can write

$$\begin{aligned}
\Sigma^{-1} &= \Sigma^{-\frac{1}{2}} \Sigma^{-\frac{1}{2}} \\
&= \left[\frac{1}{\sigma_\varepsilon} (I - P_\iota) + \left(\frac{1}{T\sigma_\nu^2 + \sigma_\varepsilon^2} \right)^{\frac{1}{2}} P_\iota \right] \\
&\quad \times \left[\frac{1}{\sigma_\varepsilon} (I - P_\iota) + \left(\frac{1}{T\sigma_\nu^2 + \sigma_\varepsilon^2} \right)^{\frac{1}{2}} P_\iota \right]
\end{aligned}$$

and the above claim

$$\begin{aligned}
\Sigma^{-1/2} &= \sigma_\varepsilon^{-1} (I - \alpha P_\iota) \\
&= \sigma_\varepsilon^{-1} \left(I - \left[1 - \sigma_\varepsilon \left(T\sigma_\nu^2 + \sigma_\varepsilon^2 \right)^{-\frac{1}{2}} \right] P_\iota \right) \\
&= \frac{1}{\sigma_\varepsilon} (I - P_\iota) + \left(\frac{1}{T\sigma_\nu^2 + \sigma_\varepsilon^2} \right)^{\frac{1}{2}} P_\iota
\end{aligned}$$

is demonstrated.

A typical element of

$$\Sigma^{-1/2} Y_{\cdot j} = \sigma_\varepsilon^{-1} \left(Y_{tj} - \alpha \overline{Y}_{\cdot j} \right)$$

and for

$$\Sigma^{-1/2} X_{\cdot j} = \sigma_\varepsilon^{-1} \left(X_{tj} - \alpha \overline{X}_{\cdot j} \right)$$

GLS estimates then can be derived from the following *OLS* regression

$$\left(Y_{tj} - \alpha \overline{Y}_{\cdot j} \right) = \left(X_{tj} - \alpha \overline{X}_{\cdot j} \right) + residuals$$

Written in matrix terms this is

$$(I - \alpha P_\iota) Y = (I - \alpha P_\iota) X + (I - \alpha P_\iota) u$$

It is instructive to connect the *GLS* estimator to the *OLS* (without fixed effects) estimator and to the within-groups (fixed effects) estimator. When $\alpha = 0$, the *GLS* estimator is the same as the *OLS* (without fixed effects) estimator. Note $\alpha = 0$ when $\sigma_\nu = 0$ (i.e., the error term has only one component ε). When $\alpha = 1$, the *GLS* estimator equals the within-groups estimator. This is because $\alpha = 1$ when $\sigma_\varepsilon = 0$, or the between groups variation is zero. Hence, in this case the within-groups (fixed effects) estimator is fully efficient. In all other cases, α is between zero and one and the *GLS* estimator exploits both within-groups and between-groups variation. Finally, recall consistency of random effects estimators relies on there being no correlation between the error components and the regressors.

3.5 Random coefficients

Random effects can be generalized by random slopes as well as random intercepts in a random coefficients model. Then, individual-specific or heterogeneous response is more fully accommodated. Hence, for individual i, we have

$$Y_i = X_i \beta_i + \varepsilon_i$$

3.5.1 Nonstochastic regressors

Wald [1947], Hildreth and Houck [1968], and Swamy [1970] proposed standard identification conditions and (*OLS* and *GLS*) estimators for random coefficients. To fix ideas, we summarize Swamy's conditions. Suppose there are T observations on each of n individuals with observable outcomes Y_i and regressors X_i and unobservables β_i and ε_i.

$$\underset{(T\times 1)}{Y_i} = \underset{(T\times K)}{X_i} \underset{(K\times 1)}{\beta_i} + \underset{(T\times 1)}{\varepsilon_i} \quad (i=1,\ldots,n)$$

Condition 3.1 $E[\varepsilon_i] = 0 \quad E[\varepsilon_i \varepsilon_j^T] = \begin{matrix} \sigma_{ii} I & i=j \\ 0 & i\neq j \end{matrix}$

Condition 3.2 $E[\beta_i] = \beta$

Condition 3.3 $E\left[(\beta_i - \beta)(\beta_i - \beta)^T\right] = \begin{matrix} \Delta & i=j \\ 0 & i\neq j \end{matrix}$

Condition 3.4 β_i and ε_i are independent

Condition 3.5 β_i and β_j are independent for $i \neq j$

Condition 3.6 X_i $(i=1,\ldots,n)$ is a matrix of K nonstochastic regressors, x_{itk} $(t=1,\ldots,T; k=1,\ldots,K)$

It's convenient to define $\beta_i = \beta + \delta_i$ $(i=1,\ldots,n)$ where $E[\delta_i] = 0$ and

$$E\left[\delta_i \delta_i^T\right] = \begin{matrix} \Delta & i=j \\ 0 & i\neq j \end{matrix}$$

Now, we can write a stacked regression in error form

$$\begin{bmatrix} Y_1 \\ Y_2 \\ \vdots \\ Y_n \end{bmatrix} = \begin{bmatrix} X_1 \\ X_2 \\ \vdots \\ X_n \end{bmatrix} \beta + \begin{bmatrix} X_1 & 0 & \cdots & 0 \\ 0 & X_2 & \cdots & 0 \\ \vdots & \vdots & \ddots & \vdots \\ 0 & 0 & \cdots & X_n \end{bmatrix} \begin{bmatrix} \delta_1 \\ \delta_2 \\ \vdots \\ \delta_n \end{bmatrix} + \begin{bmatrix} \varepsilon_1 \\ \varepsilon_2 \\ \vdots \\ \varepsilon_n \end{bmatrix}$$

or in compact error form

$$Y = X\beta + H\delta + \varepsilon$$

where H is the $nT \times nT$ block matrix of regressors and the $nT \times 1$ disturbance vector, $H\delta + \varepsilon$, has variance

$$V \equiv Var[H\delta + \varepsilon]$$
$$= \begin{bmatrix} X_1 \Delta X_1^T + \sigma_{11}I & 0 & \cdots & 0 \\ 0 & X_2 \Delta X_2^T + \sigma_{22}I & \cdots & 0 \\ \vdots & \vdots & \ddots & \vdots \\ 0 & 0 & \cdots & X_n \Delta X_n^T + \sigma_{nn}I \end{bmatrix}$$

Therefore, while the parameters, β or $\beta + \delta$, can be consistently estimated via OLS, GLS is more efficient. Swamy [1970] demonstrates that β can be estimated directly via

$$b^{GLS} = (X^T V^{-1} X)^{-1} X^T V^{-1} Y$$
$$= \left[\sum_{j=1}^{n} X_j^T \left(X_j \Delta X_j^T + \sigma_{jj} I \right)^{-1} X_j \right]^{-1}$$
$$\times \sum_{i=1}^{n} X_i^T \left(X_i \Delta X_i^T + \sigma_{ii} I \right)^{-1} Y_i$$

or equivalently by a weighted average of the estimates for $\beta + \delta$

$$b^{GLS} = \sum_{i=1}^{n} W_i b_i$$

where, applying the matrix inverse result in Rao [1973, (2.9)],[3]

$$W_i = \left[\sum_{j=1}^{n} \left(\Delta + \sigma_{jj} \left(X_j^T X_j \right)^{-1} \right)^{-1} \right]^{-1} \left(\Delta + \sigma_{ii} \left(X_i^T X_i \right)^{-1} \right)^{-1}$$

and $b_i = (X_i^T X_i)^{-1} X_i^T Y_i$ is an *OLS* estimate for $\beta + \delta_i$.

3.5.2 Correlated random coefficients

As with random effects, a key weakness of random coefficients is the condition that the effects (coefficients) are independent of the regressors. When this

---

[3] Rao's inverse result follows. Let A and D be nonsingular matrices of orders m and n and B be an $m \times n$ matrix. Then

$$\left(A + BDB^T \right)^{-1} = A^{-1} - A^{-1} B \left(B^T A^{-1} B + D^{-1} \right)^{-1} B^T A^{-1}$$
$$= A^{-1} - A^{-1} B E B^T A^{-1} + A^{-1} B E (E + D)^{-1} E B^T A^{-1}$$

where $E = (B^T A^{-1} B)^{-1}$.

condition fails, *OLS* parameter estimation of β is likely inconsistent. However, Wooldridge [2002, ch. 18] suggests ignorability identification conditions. We briefly summarize a simple version of these conditions.[4] For a set of covariates W the following redundancy conditions apply:

Condition 3.7 $E[Y_i \mid X_i, \beta_i, W_i] = E[Y_i \mid X_i, \beta_i]$

Condition 3.8 $E[X_i \mid \beta_i, W_i] = E[X_i \mid W_i]$

Condition 3.9 $Var[X_i \mid \beta_i, W_i] = Var[X_i \mid W_i]$

and

Condition 3.10 $Var[X_i \mid W_i] > 0$ *for all* W_i

Then, β is identified as $\beta = E\left[\frac{Cov(X,Y|W)}{Var(X|W)}\right]$. Alternative ignorability conditions lead to a standard linear model.

Condition 3.11 $E[\beta_i \mid X_i, W_i] = E[\beta_i \mid W_i]$

Condition 3.12 *the regression of Y onto covariates W (as well as potentially correlated regressors, X) is linear*

Now, we can consistently estimate β via a linear panel data regression. For example, ignorable treatment allows identification of the average treatment effect[5] via the panel data regression

$$E[Y \mid D, W] = D\beta + H\delta + W\gamma_0 + D(W - E[W])\gamma_1$$

where D is (a vector of) treatments.

3.6 Ubiquity of the Gaussian distribution

Why is the Gaussian or normal distribution so ubiquitous? Jaynes [2003, ch. 7] argues probabilities are "states of knowledge" rather than long run frequencies. Further, probabilities as logic naturally draws attention to the Gaussian distribution. Before stating some general properties of this "central" distribution, we review it's development in Gauss [1809] as related by Jaynes [2003], p. 202. The Gaussian distribution is uniquely determined if we equate the error cancelling property of a maximum likelihood estimator (*MLE*; discussed in ch. 4) with the sample average. The argument proceeds as follows.

---

[4]Wooldridge [2002, ch. 18] discusses more general ignorable treatment (or conditional mean independence) conditions and also instrumental variables (*IV*) strategies. We defer *IV* approaches to chapter 10 when we consider average treatment effect identification strategies associated with continuous treatment.

[5]Average treatment effects for a continuum of treatments and their instrumental variable identification strategies are discussed in chapter 10.

3. Linear models

Suppose we have a sample of $n+1$ observations, $x_0, x_1, \ldots, x_n$, and the density function factors $f(x_0, x_1, \ldots, x_n \mid \theta) = f(x_0 \mid \theta) \cdots f(x_n \mid \theta)$. The log-likelihood is

$$\sum_{i=0}^{n} \log f(x_i \mid \theta) = \sum_{i=0}^{n} g(x_i - \theta)$$

so the MLE $\widehat{\theta}$ satisfies

$$\sum_{i=0}^{n} \frac{\partial g\left(\widehat{\theta} - x_i\right)}{\partial \widehat{\theta}} = \sum_{i=0}^{n} g'\left(\widehat{\theta} - x_i\right) = 0$$

Equating the MLE with the sample average we have

$$\widehat{\theta} = \bar{x} = \frac{1}{n+1} \sum_{i=0}^{n} x_i$$

In general, MLE and $\bar{x}$ are incompatible. However, consider a sample in which only x_0 is nonzero, that is, $x_1 = \cdots = x_n = 0$. Now, if we let $x_0 = (n+1)u$ and $\widehat{\theta} = u$ then

$$\widehat{\theta} - x_0 = u - (n+1)u = -nu$$

and

$$\sum_{i=0}^{n} g'\left(\widehat{\theta} - x_i\right) = 0$$

becomes

$$\sum_{i=0}^{n} g'(-nu) = 0$$

or since $u = \widehat{\theta} - 0$

$$g'(-nu) + n g'(u) = 0$$

The case $n = 1$ implies $g'(u)$ must be anti-symmetric, $g'(-u) = -g'(u)$. With this in mind, $g'(-nu) + n g'(u) = 0$ reduces to

$$g'(nu) = n g'(u)$$

Apparently, (and naturally if we consider the close connection between the Gaussian distribution and linearity)

$$g'(u) = au$$

that is, $g'(u)$ is a linear function and

$$g(u) = \frac{1}{2} au^2 + b$$

For this to be a normalizable function, a must be negative and b determines the normalization. Hence, we have

$$f(x \mid \theta) = \sqrt{\frac{\alpha}{2\pi}} \exp\left[-\frac{1}{2}\alpha(x-\theta)^2\right] \quad 0 < \alpha < \infty$$

and the "natural" way to think of error cancellation is the Gaussian distribution with only the scale parameter α unspecified. Since the maximum of the Gaussian likelihood function always equals the sample average and in the special case above this is true only for the Gaussian likelihood, the Gaussian distribution is necessary and sufficient.

Then, the ubiquity of the Gaussian distribution follows from its error cancellation properties described above, the central limit theorem (discussed in the appendix), and the following general properties (Jaynes [2003], pp. 220-221).

A. When any smooth function with a single mode is raised to higher and higher powers, it approaches a Gaussian function.

B. The product of two Gaussian functions is another Gaussian function.

C. The convolution of two Gaussian functions is another Gaussian function (see discussion below).

D. The Fourier transform of a Gaussian function is another Gaussian function.

E. A Gaussian probability distribution has higher entropy than any other distribution with equal variance.

Properties A and E suggest why various operations result in convergence toward the Gaussian distribution. Properties B, C, and D suggest why, once attained, a Gaussian distribution is preserved.

3.6.1 Convolution of Gaussians

Property C is pivotal as repeated convolutions lead to the central limit theorem. First, we discuss discrete convolutions (see Strang [1986],pp. 294-5). The convolution of f and g is written $f * g$. It is the sum (integral) of two functions after one has been reversed and shifted. Let $f = (f_0, f_1, \ldots, f_{n-1})$ and $g = (g_0, g_1, \ldots, g_{n-1})$ then

$$f * g = \begin{pmatrix} f_0 g_0 + f_1 g_{n-1} + f_2 g_{n-2} + \cdots + f_{n-1} g_1, f_0 g_1 + f_1 g_0 + f_2 g_{n-1} \\ + \cdots + f_{n-1} g_1, \ldots, f_0 g_{n-1} + f_1 g_{n-2} + f_2 g_{n-3} + \cdots + f_{n-1} g_0 \end{pmatrix}$$

For example, the convolution of $(1, 2, 3)$ and $(4, 5, 6)$ is $(1, 2, 3) * (4, 5, 6) = (1 \cdot 4 + 2 \cdot 6 + 3 \cdot 5, 1 \cdot 5 + 2 \cdot 4 + 3 \cdot 6, 1 \cdot 6 + 2 \cdot 5 + 3 \cdot 4) = (31, 31, 28)$.

Now, we discuss property C. The convolution property applied to Gaussians is

$$\int_{-\infty}^{\infty} \varphi(x - \mu_1 \mid \sigma_1) \, \varphi(y - x - \mu_2 \mid \sigma_2) \, dx = \varphi(y - \mu \mid \sigma)$$

where $\varphi(\cdot)$ is a Gaussian density function, $\mu = \mu_1 + \mu_2$ and $\sigma^2 = \sigma_1^2 + \sigma_2^2$. That is, two Gaussians convolve to make another Gaussian distribution with additive means and variances. For convenience let $w_i = \frac{1}{\sigma_i^2}$ and write

$$\left(\frac{x - \mu_1}{\sigma_1}\right)^2 + \left(\frac{y - x - \mu_2}{\sigma_2}\right)^2 = (w_1 + w_2)(x - \widehat{x})^2 + \frac{w_1 w_2}{w_1 + w_2}(y - \mu_1 - \mu_2)^2$$

where $\widehat{x} \equiv \frac{w_1 \mu_1 + w_2 y - w_2 \mu_2}{w_1 + w_2}$. Integrating out x produces the above result.

3.7 Interval estimation

Finite sampling distribution theory for the linear model $Y = X\beta + \varepsilon$ follows from assigning the errors independent, normal probability distributions with mean zero and constant variance σ^2.[6] Interval estimates for individual model parameters β_j are Student t distributed with $n-p$ degrees of freedom when σ^2 is unknown. This follows from

$$\frac{b_j - \beta_j}{Est.Var\,[b_j]} \sim t(n-p)$$

where $Est.Var\,[b_j] = s^2 \left(X_j^T M_{-j} X_j\right)^{-1}$, X_j is column j of X, $M_{-j} = I - P_{-j}$ and P_{-j} is the projection matrix onto all columns of X except j, $s^2 = \frac{e^T e}{n-p}$, and e is a vector of residuals. By *FWL*, the numerator is

$$\left(X_j^T M_{-j} X_j\right)^{-1} X_j^T M_{-j} Y$$

Rewriting yields

$$b_j = \left(X_j^T M_{-j} X_j\right)^{-1} X_j^T M_{-j} (X\beta + \varepsilon)$$

As M_{-j} annihilates all columns of X except X_j, we have

$$\begin{aligned} b_j &= \left(X_j^T M_{-j} X_j\right)^{-1} X_j^T M_{-j} \left(X_j \beta_j + \varepsilon\right) \\ &= \beta_j + \left(X_j^T M_{-j} X_j\right)^{-1} X_j^T M_{-j} \varepsilon \end{aligned}$$

Now,

$$b_j - \beta_j = \left(X_j^T M_{-j} X_j\right)^{-1} X_j^T M_{-j} \varepsilon$$

As this is a linear combination of independent, normal random variates, the transformed random variable also has a normal distribution with mean zero and variance $\sigma^2 \left(X_j^T M_{-j} X_j\right)^{-1}$. The estimated variance of b_j is $s^2 \left(X_j^T M_{-j} X_j\right)^{-1}$ and the t ratio is

$$\begin{aligned} \frac{b_j - \beta_j}{Est.Var\,[b_j]} &= \frac{b_j - \beta_j}{\sqrt{s^2 \left(X_j^T M_{-j} X_j\right)^{-1}}} \\ &= \sqrt{\frac{b_j - \beta_j}{\frac{e^T e \left(X_j^T M_{-j} X_j\right)^{-1}}{n-p}}} \end{aligned}$$

---

[6] It's instructive to recall the discussion of the ubiquity of the Gaussian distribution adapted from Jaynes [2003].

This can be rewritten as the ratio of a standard normal random variable to the square root of a chi-square random variable divided by its degrees of freedom.

$$\frac{b_j - \beta_j}{\sqrt{Est.Var[b_j]}} = \frac{\left(X_j^T M_{-j} X_j\right)^{-1} X_j^T M_{-j} \varepsilon}{\sqrt{\frac{\varepsilon^T M_X \varepsilon \left(X_j^T M_{-j} X_j\right)^{-1}}{n-p}}}$$

$$= \frac{\left(X_j^T M_{-j} X_j\right)^{-1} X_j^T M_{-j} (\varepsilon/\sigma)}{\sqrt{\frac{(\varepsilon/\sigma)^T M_X (\varepsilon/\sigma) \left(X_j^T M_{-j} X_j\right)^{-1}}{n-p}}}$$

In other words, a Student t distributed random variable with $n - p$ degrees of freedom which completes the demonstration.

Normal sampling distribution theory applied to joint parameter regions follow an F distribution. For example, the null hypothesis $H_0 : \beta_1 = ... = \beta_{p-1} = 0$ is tested via the F statistic $= \frac{MSR}{MSE} \sim F(p-1, n-p)$. As we observed above, the denominator is

$$MSE = \frac{e^T e}{n-p}$$

$$= \frac{\varepsilon^T M_X \varepsilon}{n-p} \sim \frac{\chi^2(n-p)}{n-p}$$

The numerator is $\frac{(Xb-\overline{Y})^T (Xb-\overline{Y})}{p-1}$. FWL indicates $Xb = P_\iota Y + M_\iota X_{-\iota} b_{-\iota}$ where ι refers to a vector of ones (for the intercept) and the subscript $-\iota$ refers to everything but the intercept (i.e., everything except the vector of ones in X). Therefore, $Xb - \overline{Y} = P_\iota Y + M_\iota X_{-\iota} b_{-\iota} - P_\iota Y = M_\iota X_{-\iota} b_{-\iota}$. Now,

$$MSR = \frac{b_{-\iota}^T X_{-\iota}^T M_\iota X_{-\iota} b_{-\iota}}{p-1}$$

$$= \frac{Y^T P_{M_\iota X_{-\iota}} Y}{p-1}$$

$$= \frac{(X\beta + \varepsilon)^T P_{M_\iota X_{-\iota}} (X\beta + \varepsilon)}{p-1}$$

under the null $\beta_{-\iota} = 0$ and β_0 is negated by M_ι. Hence,

$$MSR = \frac{\varepsilon^T P_{M_\iota X_{-\iota}} \varepsilon}{p-1} \sim \frac{\chi^2(p-1)}{p-1}$$

which completes the demonstration.

When the our understanding of the errors is weak, we frequently appeal to asymptotic or approximate sampling distributions. Asymptotic tests of restrictions are discussed next (also see the appendix).

3.8 Asymptotic tests of restrictions: Wald, *LM*, *LR* statistics

Tests of restrictions based on *Wald*, *LM* (Lagrange multiplier), and *LR* (likelihood ratio) statistics have a similar heritage. Asymptotically they are the same; only in finite samples do differences emerge. A brief sketch of each follows.

Intuition for these tests come from the *finite sample F statistic* (see Davidson and MacKinnon [1993], p. 82-83 and 452-6). The F statistic is valid if the errors have a normal probability assignment.

For the restriction H_0: $R\beta - r = 0$

$$F = \frac{\left(e_*^T e_* - e^T e\right)/J}{e^T e/(n-p)}$$

$$= \frac{(Rb-r)^T \left(Rs^2 \left(X^T X\right)^{-1} R^T\right)^{-1} (Rb-r)}{J} \sim F(J, n-p)$$

where R is $J \times p$, $e_* = (I - P_{X_*})Y = M_{X_*}Y$ and $e = (I - P_X)Y = M_X Y$ are the residuals from the restricted and unrestricted models, respectively, $P_x = X\left(X^T X\right)^{-1} X^T$, $P_{X_*} = X_*\left(X_*^T X_*\right)^{-1} X_*^T$ are the projection matrices, X_* is the restricted matrix of regressors, and $s^2 = e^T e/(n-p)$ is the sample variance. Recall the numerator and denominator of F are divided by σ^2 to yield the ratio of two chi-squared random variables. Since s^2 converges to σ^2 we have $p\lim\left(\frac{s^2}{\sigma^2}\right) = 1$ in the denominator. Hence, we have J squared standard normal random variables summed in the numerator or W converges in distribution to $\chi^2(J)$.

FWL provides another way to see the connection between the F statistic and the Wald statistic W,

$$F = \frac{\left(e_*^T e_* - e^T e\right)/J}{e^T e/(n-p)}$$

$$= \frac{(Rb-r)^T \left(Rs^2 \left(X^T X\right)^{-1} R^T\right)^{-1} (Rb-r)}{J}$$

$$= \frac{W}{J}$$

Consider the (partitioned) *DGP*:

$$Y = X\beta + \varepsilon = X_1\beta_1 + X_2\beta_2 + \varepsilon$$

3.8 Asymptotic tests of restrictions: Wald, LM, LR statistics

and restriction $H_0: \beta_2 = 0$. By *FWL*,[7]

$$e^T e = Y^T M_X Y = Y^T M_1 Y - Y^T M_1 X_2 \left(X_2^T M_1 X_2 \right)^{-1} X_2^T M_1 Y$$

and $e_*^T e_* = Y^T M_1 Y$. Hence, the numerator of F is

$$\left(e_*^T e_* - e^T e \right) / J = Y^T M_1 X_2 \left(X_2^T M_1 X_2 \right)^{-1} X_2^T M_1 Y / J$$

and the denominator is s^2. Since $R = \begin{bmatrix} 1 & 0 \end{bmatrix}$, rearrangement yields

$$W = \frac{b_2^T \left(\left(X^T X \right)^{-1} \right)_{22}^{-1} b_2 / J}{s^2}$$

where

$$\left(\left(X^T X \right)^{-1} \right)_{22} = \left(X_2^T X_2 - X_2^T X_1 \left(X_1^T X_1 \right)^{-1} X_1^T X_2 \right)^{-1}$$

is the lower right hand block of $\left(X^T X \right)^{-1}$. Both F and W are divided by J in the numerator and have s^2 in the numerator. Now, we show that

$$Y^T M_1 X_2 \left(X_2^T M_1 X_2 \right)^{-1} X_2^T M_1 Y = b_2^T \left(\left(X^T X \right)^{-1} \right)_{22}^{-1} b_2$$

by rewriting the right hand side

$$\begin{aligned} b_2^T \left(\left(X^T X \right)^{-1} \right)_{22}^{-1} b_2 &= b_2^T \left(X_2^T X_2 - X_2^T X_1 \left(X_1^T X_1 \right)^{-1} X_1^T X_2 \right) b_2 \\ &= b_2^T \left(X_2^T M_1 X_2 \right) b_2 \end{aligned}$$

by *FWL* for b_2

$$\begin{aligned} & b_2^T \left(X_2^T M_1 X_2 \right) b_2 \\ &= Y^T M_1 X_2 \left(X_2^T M_1 X_2 \right)^{-1} \left(X_2^T M_1 X_2 \right) \left(X_2^T M_1 X_2 \right)^{-1} X_2^T M_1 Y \\ &= Y^T M_1 X_2 \left(X_2^T M_1 X_2 \right)^{-1} X_2^T M_1 Y \end{aligned}$$

---

[7]From *FWL*,

$$\begin{aligned} Xb &= P_1 Y + M_1 X_2 b_2 \\ &= P_1 Y + M_1 X_2 \left(X_2^T M_1 X_2 \right)^{-1} X_2^T M_1 Y \\ &= P_1 Y + M_1 X_2 \left(X_2^T M_1 X_2 \right)^{-1} X_2^T M_1 Y \\ &= P_1 Y + P_{M_1 X_2} Y \end{aligned}$$

Since $P_1 P_{M_1} = 0$ (by orthogonality),

$$\begin{aligned} e^T e &= Y^T \left(I - P_1 Y - P_{M_1 X_2} \right) Y \\ &= Y^T M_1 Y - Y^T P_{M_1 X_2} Y \\ &= Y^T M_1 Y - Y^T M_1 X_2 \left(X_2^T M_1 X_2 \right)^{-1} X_2^T M_1 Y \end{aligned}$$

This completes the demonstration as the right hand side from W is the same as the left hand side from F.

If the errors do not have a normal probability assignment, the F statistic is invalid (even asymptotically) but the *Wald statistic* may be asymptotically valid

$$W = (Rb - r)^T \left(Rs^2 \left(X^T X\right)^{-1} R^T\right)^{-1} (Rb - r) \xrightarrow{d} \chi^2(J)$$

To see this apply the multivariate Lindberg-Feller version of the central limit theorem (see appendix on asymptotic theory) and recall if $x \sim N(\mu, \Sigma)$, then $(x - \mu)^T \Sigma^{-1} (x - \mu) \sim \chi^2(n)$. W is the quadratic form as under the null Rb has mean r and $Est.Var[Rb] = Rs^2 \left(X^T X\right)^{-1} R^T$.

The *LR statistic* is based on the log of the ratio of the restricted to unrestricted likelihoods

$$\begin{aligned} LR &= -2\left[LnL_* - LnL\right] \\ &= nLn\left[e_*^T e_* / e^T e\right] \xrightarrow{d} \chi^2(J) \end{aligned}$$

Asymptotically LR converges to W.

The *Lagrange Multiplier* (*LM*) *test* is based on the gradient of the log-likelihood. If the restrictions are valid then the derivatives of the log-likelihood evaluated at the restricted estimates should be close to zero.

Following manipulation of first order conditions we find

$$\lambda_* = \left(Rs_*^2 \left(X^T X\right)^{-1} R^T\right)^{-1} (Rb - r)$$

A Wald test of $\lambda_* = 0$ yields the statistic $LM = \lambda_*^T \left\{Est.Var[\lambda_*]\right\}^{-1} \lambda_*$ which simplifies to

$$LM = (Rb - r)^T \left(Rs_*^2 \left(X^T X\right)^{-1} R^T\right)^{-1} (Rb - r)$$

It is noteworthy that, unlike the *Wald* statistic above, the variance estimate is based on the restrictions.

In the classical regression model, the *LM* statistic can be simplified to an nR^2 test. Under the restrictions, $E\left[\frac{\partial LnL}{\partial \beta}\right] = E\left[\frac{1}{\sigma^2} X^T \varepsilon\right] = 0$ and $Asy.Var\left[\frac{\partial LnL}{\partial \beta}\right] = \left[\frac{\partial^2 LnL}{\partial \beta \partial \beta^T}\right]^{-1} = \sigma^2 \left(X^T X\right)^{-1}$. The *LM* statistic is

$$\frac{e_*^T X \left(X^T X\right)^{-1} X^T e_*}{e_*^T e_* / n} = nR_*^2 \xrightarrow{d} \chi^2(p - J)$$

LM is n times R^2 from a regression of the (restricted) residuals e_* on the full set of regressors.

From the above, we have $W > LR > LM$ in finite samples.

3.8.1 Nonlinear restrictions

More generally, suppose the restriction is nonlinear in β

$$H_0 : f(\beta) = 0$$

The corresponding *Wald* statistic is

$$W = f(b)^T \left[G(b) s^2 (X^T X)^{-1} G(b)^T \right]^{-1} f(b) \xrightarrow{d} \chi^2(J)$$

where $G(b) = \left[\frac{\partial f(b)}{\partial b^T} \right]$. This is an application of the *Delta method* (see the appendix on asymptotic theory). If $f(b)$ involves continuous functions of b such that $\Gamma = \left[\frac{\partial f(\beta)}{\partial \beta^T} \right]$, by the central limit theorem

$$f(b) \xrightarrow{d} N\left(f(\beta), \Gamma \left(\frac{\sigma^2}{n} Q^{-1} \right) \Gamma^T \right)$$

where $p \lim \left(\frac{X^T X}{n} \right)^{-1} = Q^{-1}$.

3.9 Misspecification and *IV* estimation

Misspecification arises from violation of $E\left[X^T \varepsilon\right] = 0$, or $E\left[\varepsilon \mid X\right] = 0$, or asymptotically, $p \lim \left(\frac{1}{n} X^T \varepsilon\right) = 0$. Omitted correlated regressors, measurement error in regressors, and endogeneity (including simultaneity and self-selection) produce such misspecification when not addressed.

Consider the *DGP*:

$$Y = X_1 \beta + X_2 \gamma + \varepsilon$$

where

$$\varepsilon \sim (0, \sigma^2 I), E\left[\begin{bmatrix} X_1 & X_2 \end{bmatrix}^T \varepsilon \right] = 0$$

and

$$p \lim \left(\frac{1}{n} \begin{bmatrix} X_1 & X_2 \end{bmatrix}^T \varepsilon \right) = 0$$

If X_2 is omitted then it effectively becomes part of the error term, say $\eta = X_2 \gamma + \varepsilon$. OLS yields

$$b = \left(X_1^T X_1\right)^{-1} X_1^T \left(X_1 \beta + X_2 \gamma + \varepsilon\right) = \beta + \left(X_1^T X_1\right)^{-1} X_1^T \left(X_2 \gamma + \varepsilon\right)$$

which is unbiased only if X_1 and X_2 are orthogonal (so the Gauss-Markov theorem likely doesn't apply). And, the estimator is asymptotically consistent only if $p \lim \left(\frac{1}{n} \left(X_1^T X_2 \right) \right) = 0$.

Instrumental variables (IV) estimation is a standard approach for addressing lack of independence between the regressors and the errors. A "good" set of instruments Z has two properties: (1) they are highly correlated with the (endogenous) regressors and (2) they are orthogonal to the errors (or $p \lim \left(\frac{1}{n} Z^T \varepsilon\right) = 0$).

3. Linear models

Consider the *DGP*:
$$Y = X\beta + \varepsilon$$
where $\varepsilon \sim (0, \sigma^2 I)$, but $E[X^T \varepsilon] \neq 0$, and $p\lim \left(\frac{1}{n} X^T \varepsilon\right) \neq 0$.

IV estimation proceeds as follows. Regress X onto Z to yield $\widehat{X} = P_Z X = Z(Z^T Z)^{-1} Z^T X$. Estimate β via b_{IV} by regressing Y onto $\widehat{X}$.

$$\begin{aligned} b_{IV} &= (X^T P_Z P_Z X)^{-1} X^T P_Z Y \\ &= (X^T P_Z X)^{-1} X^T P_Z Y \end{aligned}$$

Asymptotic consistency[8] follows as

$$\begin{aligned} p\lim(b_{IV}) &= p\lim\left((X^T P_Z X)^{-1} X^T P_Z Y\right) \\ &= p\lim\left((X^T P_Z X)^{-1} X^T P_Z (X\beta + \varepsilon)\right) \\ &= \beta + p\lim\left((X^T P_Z X)^{-1} X^T P_Z \varepsilon\right) \\ &= \beta + p\lim\left(\left(\frac{1}{n} X^T P_Z X\right)^{-1} 1/n X^T Z \left(\frac{1}{n} Z^T Z\right)^{-1} \frac{1}{n} Z^T \varepsilon\right) \\ &= \beta \end{aligned}$$

Note in the special case $Dim(Z) = Dim(X)$ (where Dim refers to the dimension or rank of the matrix), each regressor has one instrument associated with it, the instrumental variables estimator simplifies considerably as $(X^T Z)^{-1}$ and $(Z^T X)^{-1}$ exist. Hence,

$$\begin{aligned} b_{IV} &= (X^T P_Z X)^{-1} X^T P_z Y \\ &= \left(X^T Z (Z^T Z)^{-1} Z^T X\right)^{-1} X^T Z (Z^T Z)^{-1} Z^T Y \\ &= (Z^T X)^{-1} Z^T Y \end{aligned}$$

and
$$Asy.Var[b_{IV}] = \sigma^2 (Z^T X)^{-1} Z^T Z (X^T Z)^{-1}$$

There is a finite sample trade-off in choosing the number of instruments to employ. Asymptotic efficiency (inverse of variance) increases in the number of instruments but so does the finite-sample bias. Relatedly, if *OLS* is consistent the use of instruments inflates the variance of the estimates since $X^T P_Z X$ is smaller by a positive semidefinite matrix than $X^T X$ ($I = P_Z + (I - P_z)$, *IV* annihilates the left nullspace of Z).

---

[8] Slutsky's theorem is applied repeatedly below (see the appendix on asymptotic theory). The theorem indicates $plim(g(X)) = g(plim(X))$ and implies $plim(XY) = plim(X) plim(Y)$.

Importantly, if $Dim(Z) > Dim(X)$ then *over-identifying restrictions* can be used to test the instruments (Godfrey and Hutton, 1994). The procedure is regress the residuals from the second stage onto Z (all exogenous regressors). Provided there exists at least one exogenous regressor, then $nR^2 \sim \chi^2(K-L)$ where K is the number of exogenous regressors in the first stage and L is the number of endogenous regressors. Of course, under the null of exogenous instruments R^2 is near zero.

A *Hausman test* (based on a *Wald* statistic) can be applied to check the consistency of *OLS* (and is applied after the above exogeneity test and elimination of any offending instruments from the *IV* estimation).

$$W = (b - b_{IV})^T [V_1 - V_0]^{-1} (b - b_{IV}) \sim \chi^2(p)$$

where V_1 is the estimated asymptotic covariance for the *IV* estimator and $V_0 = s^2 (X^T X)^{-1}$ where s^2 is from the *IV* estimator (to ensure that $V_1 > V_0$).

3.10 Proxy variables

Frequently in accounting and business research we employ proxy variables as direct measures of constructs are not readily observable. Proxy variables can help to address potentially omitted, correlated variables. An important question is when do proxy variables aid the analysis and when is the cure worse than the disease.

Consider the *DGP*: $Y = \beta_0 + X\beta + Z\gamma + \varepsilon$. Let W be a set of proxy variables for Z (the omitted variables). Typically, there are two conditions to satisfy:
(1) $E[Y \mid X, Z, W] = E[Y \mid X, Z]$ This form of mean conditional independence is usually satisfied.
For example, suppose $W = Z + \nu$ and the variables are jointly normally distributed with ν independent of other variables. Then, the above condition is satisfied as follows. (For simplicity, we work with one-dimensional variables but the result

can be generalized to higher dimensions.[9])

$$E[Y \mid X, Z, W] = \mu_Y + \begin{bmatrix} \sigma_{YX} & \sigma_{YZ} & \sigma_{YZ} \end{bmatrix}$$
$$\times \begin{bmatrix} \sigma_{XX} & \sigma_{XZ} & \sigma_{XZ} \\ \sigma_{ZX} & \sigma_{ZZ} & \sigma_{ZZ} \\ \sigma_{ZX} & \sigma_{ZZ} & \sigma_{ZZ} + \sigma_{\nu\nu} \end{bmatrix}^{-1} \begin{bmatrix} x - \mu_X \\ z - \mu_Z \\ w - \mu_W \end{bmatrix}$$
$$= \mu_Y + \frac{\sigma_{YX}\sigma_{ZZ} - \sigma_{XZ}\sigma_{YZ}}{\sigma_{XX}\sigma_{ZZ} - \sigma_{XZ}^2}(x - \mu_X)$$
$$+ \frac{\sigma_{YZ}\sigma_{XX} - \sigma_{XZ}\sigma_{YX}}{\sigma_{XX}\sigma_{ZZ} - \sigma_{XZ}^2}(z - \mu_Z) + 0(w - \mu_W)$$

(2) $Cov[X_j, Z \mid W] = 0$ for all j. This condition is more difficult to satisfy. Again, consider proxy variables like $W = Z + \nu$ where $E[\nu] = 0$ and $Cov[Z, \nu] = 0$, then $Cov[X, Z \mid W] = \frac{\sigma_{XZ}\sigma_\nu^2}{\sigma_{ZZ} + \sigma_\nu^2}$. Hence, the smaller is σ_ν^2, the noise in the proxy variable, the better service provided by the proxy variable.

What is the impact of imperfect proxy variables on estimation? Consider proxy variables like $Z = \theta_0 + \theta_1 W + \nu$ where $E[\nu] = 0$ and $Cov[Z, \nu] = 0$. Let $Cov[X, \nu] = \rho \neq 0$, $Q = \begin{bmatrix} \iota & X & W \end{bmatrix}$, and

$$\omega^T = \begin{bmatrix} (\beta_0 + \gamma\theta_0) & \beta & \gamma\theta_1 \end{bmatrix}$$

The estimable equation is

$$Y = Q\omega + \epsilon = (\beta_0 + \gamma\theta_0) + \beta X + \gamma\theta_1 W + (\gamma\nu + \epsilon)$$

---

[9]A quick glimpse of the multivariate case can be found if we consider the simple case where the *DGP* omits X. If W doesn't contribute to $E[Y \mid Z, W]$, then it surely doesn't contribute to $E[Y \mid X, Z, W]$. It's readily apparent how the results generalize for the $E[Y \mid X, Z, W]$ case, though cumbersome. In block matrix form $E[Y \mid Z, W] =$

$$\mu_Y + \begin{bmatrix} \Sigma_{YZ} & \Sigma_{YZ} \end{bmatrix} \begin{bmatrix} \Sigma_{ZZ} & \Sigma_{ZZ} \\ \Sigma_{ZZ} & \Sigma_{ZZ} + \Sigma_{\nu\nu} \end{bmatrix}^{-1} \begin{bmatrix} z - \mu_Z \\ w - \mu_W \end{bmatrix}$$
$$= \mu_Y + \begin{bmatrix} \Sigma_{YZ} & \Sigma_{YZ} \end{bmatrix} \begin{bmatrix} \Sigma_{ZZ}^{-1} + \Sigma_{\nu\nu}^{-1} & -\Sigma_{\nu\nu}^{-1} \\ -\Sigma_{\nu\nu}^{-1} & \Sigma_{\nu\nu}^{-1} \end{bmatrix} \begin{bmatrix} z - \mu_Z \\ w - \mu_W \end{bmatrix}$$
$$= \mu_Y + \Sigma_{YZ}\Sigma_{ZZ}^{-1}(z - \mu_Z) + 0(w - \mu_W)$$

The key is recognizing that the partitioned inverse (following some rewriting of the off-diagonal blocks) for

$$\begin{bmatrix} \Sigma_{ZZ} & \Sigma_{ZZ} \\ \Sigma_{ZZ} & \Sigma_{ZZ} + \Sigma_{\nu\nu} \end{bmatrix}^{-1}$$
$$= \begin{bmatrix} [\Sigma_{ZZ} - \Sigma_{ZZ}(\Sigma_{ZZ} + \Sigma_{\nu\nu})^{-1}\Sigma_{ZZ}]^{-1} & -\Sigma_{ZZ}^{-1}\Sigma_{ZZ}\Sigma_{\nu\nu}^{-1} \\ -(\Sigma_{ZZ} + \Sigma_{\nu\nu})^{-1}\Sigma_{ZZ}\Sigma_{ZZ}^{-1}(\Sigma_{ZZ} + \Sigma_{\nu\nu})\Sigma_{\nu\nu}^{-1} & [\Sigma_{ZZ} + \Sigma_{\nu\nu} - \Sigma_{ZZ}\Sigma_{ZZ}^{-1}\Sigma_{ZZ}]^{-1} \end{bmatrix}$$
$$= \begin{bmatrix} [\Sigma_{ZZ} - \Sigma_{ZZ}(\Sigma_{ZZ} + \Sigma_{\nu\nu})^{-1}\Sigma_{ZZ}]^{-1} & -\Sigma_{ZZ}^{-1}\Sigma_{ZZ}\Sigma_{\nu\nu}^{-1} \\ -(\Sigma_{ZZ} + \Sigma_{\nu\nu})^{-1}\Sigma_{ZZ}\Sigma_{ZZ}^{-1}(\Sigma_{ZZ} + \Sigma_{\nu\nu})\Sigma_{\nu\nu}^{-1} & [\Sigma_{ZZ} + \Sigma_{\nu\nu} - \Sigma_{ZZ}\Sigma_{ZZ}^{-1}\Sigma_{ZZ}]^{-1} \end{bmatrix}$$
$$= \begin{bmatrix} \Sigma_{ZZ}^{-1} + \Sigma_{\nu\nu}^{-1} & -\Sigma_{\nu\nu}^{-1} \\ -\Sigma_{\nu\nu}^{-1} & \Sigma_{\nu\nu}^{-1} \end{bmatrix}$$

The *OLS* estimator of ω is $b = \left(Q^T Q\right)^{-1} Q^T Y$. Let $p \lim \left(1/n Q^T Q\right)^{-1} = \Omega$.

$$\begin{aligned} p \lim b &= \omega + \Omega \, p\lim 1/n \begin{bmatrix} \iota & X & W \end{bmatrix}^T (\gamma \nu + \epsilon) \\ &= \omega + \gamma \rho \begin{bmatrix} \Omega_{12} \\ \Omega_{22} \\ \Omega_{32} \end{bmatrix} = \begin{bmatrix} \beta_0 + \gamma \theta_0 + \Omega_{12} \gamma \rho \\ \beta + \Omega_{22} \gamma \rho \\ \gamma \theta_1 + \Omega_{32} \gamma \rho \end{bmatrix} \end{aligned}$$

Hence, b is asymptotically consistent when $\rho = 0$ and inconsistency ("bias") is increasing in the absolute value of $\rho = Cov[X, \nu]$.

3.10.1 Accounting and other information sources

Use of proxy variables in the study of information is even more delicate. Frequently we're interested in the information content of accounting in the midst of other information sources. As complementarity is the norm for information, we not only have the difficulty of identifying proxy variables for other information but also a functional form issue. Functional form is important as complementarity arises through joint information partitions. Failure to recognize these subtle interactions among information sources can yield spurious inferences regarding accounting information content.

A simple example (adapted from Antle, Demski, and Ryan[1994]) illustrates the idea. Suppose a nonaccounting information signal (x_1) precedes an accounting information signal (x_2). Both are informative of firm value (and possibly employ the language and algebra of valuation). The accounting signal however employs restricted recognition such that the nonaccounting signal is ignored by the accounting system. Table 3.1 identifies the joint probabilities associated with the information partitions and the firm's liquidating dividend (to be received at a future date and expressed in present value terms). Prior to any information reports,

Table 3.1: Multiple information sources case 1 setup

probabilities; payoffs		x_1		
		1	2	3
x_2	1	0.10;0	0.08;45	0.32;99
	2	0.32;1	0.08;55	0.10;100

firm value (expected present value of the liquidating dividend) is 50. The change in firm value at the time of the accounting report (following the second signal) as well as the valuation-scaled signals (recall accounting, the second signal, ignores the first signal) are reported in table 3.2. Due to the strong complementarity in the information and restricted recognition employed by accounting, response to earnings is negative. That is, the change in value moves in the opposite direction of the accounting earnings report x_2.

As it is difficult to identify other information sources (and their information partitions), often a proxy variable for x_1 is employed. Suppose our proxy variable

Table 3.2: Multiple information sources case 1 valuation implications

change in firm value		x_1		
		-49.238	0	49.238
x_2	20.56	-0.762	-5	-0.238
	-20.56	0.238	5	0.762

is added as a control variable and a linear model of the change in firm value as a function of the information sources is estimated. Even if we stack things in favor of the linear model by choosing $w = x_1$ we find

case 1: linear model with proxy variable
$$E[Y \mid w, x_2] = 0. + 0.0153w - 0.070x_2$$
$$R^2 = 0.618$$

While a saturated design matrix (an *ANOVA* with indicator variables associated with information partitions and interactions to capture potential complementarities between the signals) fully captures change in value

case 1: saturated *ANOVA*
$$E[Y \mid D_{12}, D_{13}, D_{22}, D_{12}D_{22}, D_{13}D_{22}] = -0.762 - 4.238D_{12}$$
$$+ 0.524D_{13} + 1.0D_{22} + 0.524D_{13} + 1.0D_{22}$$
$$+ 9.0D_{12}D_{22} + 0.0D_{13}D_{22}$$
$$R^2 = 1.0$$

where D_{ij} refers to information signal i and partition j, the linear model explains only slightly more than 60% of the variation in the response variable. Further, the linear model exaggerates responsiveness of firm value to earnings. This is a simple comparison of the estimated coefficient for γ (-0.070) compared with the mean effect scaled by reported earnings for the *ANOVA* design ($\frac{1.0}{-20.56} = -0.05$). Even if w effectively partitions x_1, without accommodating potential informational complementarity (via interactions), the linear model is prone to misspecification.

case 1: unsaturated *ANOVA*
$$E[Y \mid D_{12}, D_{13}, D_{22}] = -2.188 + 0.752D_{12} + 1.504D_{13} + 2.871D_{22}$$
$$R^2 = 0.618$$

The estimated earnings response for the discretized linear proxy model is $\frac{2.871}{-20.56} = -0.14$. In this case (call it case 1), it is even more overstated.

Of course, the linear model doesn't always overstate earnings response, it can also understate (case 2, tables 3.3 and 3.4) or produce opposite earnings response to the *DGP* (case 3, tables 3.5 and 3.6). Also, utilizing the discretized or partitioned proxy may yield earnings response that is closer or departs more from the *DGP* than the valuation-scaled proxy for x_1. The estimated results for case 2 are

case 2: linear model with proxy variable
$$E[Y \mid w, x_2] = 0. + 0.453w + 3.837x_2$$
$$R^2 = 0.941$$

Table 3.3: Multiple information sources case 2 setup

probabilities; payoffs		x_1		
		1	2	3
x_2	1	0.10;0	0.08;45	0.32;40
	2	0.32;60	0.08;55	0.10;100

Table 3.4: Multiple information sources case 2 valuation implications

change in firm value		x_1		
		-19.524	0.0	19.524
x_2	-4.400	-46.523	86.062	54.391
	4.400	61.000	-140.139	-94.107

case 2: saturated *ANOVA*
$$E[Y \mid D_{12}, D_{13}, D_{22}, D_{12}D_{22}, D_{13}D_{22}] = -30.476 + 25.476 D_{12}$$
$$+ 20.952 D_{13} + 40.0 D_{22} - 30.0 D_{12}D_{22} + 0.0 D_{13}D_{22}$$
$$R^2 = 1.0$$

case 2: unsaturated *ANOVA*
$$E[Y \mid D_{12}, D_{13}, D_{22}] = -25.724 + 8.842 D_{12} + 17.685 D_{13} + 33.762 D_{22}$$
$$R^2 = 0.941$$

Earnings response for the continuous proxy model is 3.837, for the partitioned proxy is $\frac{33.762}{4.4} = 7.673$, and for the *ANOVA* is $\frac{40.0}{4.4} = 9.091$. Hence, for case 2 the proxy variable models understate earnings response and the partitioned proxy is closer to the *DGP* earnings response than is the continuous proxy (unlike case 1).

For case 3, we have The estimated results for case 3 are

Table 3.5: Multiple information sources case 3 setup

probabilities; payoffs		x_1		
		1	2	3
x_2	1	0.10;4.802	0.08;105.927	0.32;50.299
	2	0.32;65.864	0.08;26.85	0.10;17.254

case 3: linear model with proxy variable
$$E[Y \mid w, x_2] = 0. + 0.063w + 1.766 x_2$$
$$R^2 = 0.007$$

case 3: saturated *ANOVA*
$$E[Y \mid D_{12}, D_{13}, D_{22}, D_{12}D_{22}, D_{13}D_{22}] = -46.523 + 86.062 D_{12}$$
$$+ 54.391 D_{13} + 61.062 D_{22} - 140.139 D_{12}D_{22} - 94.107 D_{13}D_{22}$$
$$R^2 = 1.0$$

48   3. Linear models

Table 3.6: Multiple information sources case 3 valuation implications

	change in firm value	x_1		
		1.326	16.389	-7.569
x_2	0.100	-46.523	86.062	54.391
	-0.100	61.000	-140.139	-94.107

case 3: unsaturated *ANOVA*
$$E[Y \mid D_{12}, D_{13}, D_{22}] = 4.073 - 1.400 D_{12} - 2.800 D_{13} - 5.346 D_{22}$$
$$R^2 = 0.009$$

Earnings response for the continuous proxy model is 1.766, for the partitioned proxy is $\frac{-5.346}{-0.100} = 53.373$, and for the *ANOVA* is $\frac{61.062}{-0.100} = -609.645$. Hence, for case 3 the proxy variable models yield earnings response opposite the *DGP*.

The above variety of misspecifications suggests that econometric analysis of information calls for nonlinear models. Various options may provide adequate summaries of complementary information sources. These choices include at least saturated *ANOVA* designs (when partitions are identifiable), polynomial regressions, and nonparametric and semiparametric regressions. Of course, the proxy variable problem still lurks. Next, we return to the equilibrium earnings management example discussed in chapter 2 and explore the (perhaps linear) relation between firm value and accounting accruals.

3.11 Equilibrium earnings management

The earnings management example in Demski [2004] provides a straightforward illustration of the econometric challenges faced when management's reporting behavior is endogenous and also the utility of the propensity score as an instrument. Suppose the objective is to track the relation between a firm's value P_t and its accruals z_t. To keep things simple, firm value equals the present value of expected future dividends, the market interest rate is zero, current period cash flows are fully paid out in dividends, and dividends $\widetilde{d}$ are normal *iid* with mean zero and variance σ^2. Firm managers have private information $\widetilde{y}_t^p$ about next period's dividend $\widetilde{y}_t^p = \widetilde{d}_{t+1} + \widetilde{\varepsilon}_t$ where $\widetilde{\varepsilon}$ are normal *iid* with mean zero and variance σ^2.[10] If the private information is revealed, ex dividend firm value at time t is

$$P_t \equiv E\left[\widetilde{d}_{t+1} \mid \widetilde{y}_t^p = y_t^p\right]$$
$$= \frac{1}{2} y_t^p$$

Suppose management reveals its private information through income I_t (cash flows plus change in accruals) where fair value accruals $z_t = E\left[\widetilde{d}_{t+1} \mid \widetilde{y}_t^p = y_t^p\right]$

---
[10]For simplicity, there is no other information.

$= \frac{1}{2}y_t^p$ are reported. Then, income is

$$\begin{aligned} I_t &= d_t + (z_t - z_{t-1}) \\ &= d_t + \frac{1}{2}\left(y_t^p - y_{t-1}^p\right) \end{aligned}$$

and

$$\begin{aligned} P_t &\equiv E\left[\tilde{d}_{t+1} \mid \tilde{d}_t = d_t, \tilde{I}_t = d_t + \frac{1}{2}\left(y_t^p - y_{t-1}^p\right)\right] \\ &= E\left[\tilde{d}_{t+1} \mid \tilde{z}_t = \frac{1}{2}y_t^p\right] \\ &= z_t \end{aligned}$$

There is a linear relation between price and fair value accruals.

Suppose the firm is owned and managed by an entrepreneur who, for intergenerational reasons, liquidates his holdings at the end of the period. The entrepreneur is able to misrepresent the fair value estimate by reporting, $z_t = \frac{1}{2}y_t^p + \theta$, where $\theta \geq 0$. Auditors are unable to detect any accrual overstatements below a threshold equal to $\frac{1}{2}\Delta$. Traders anticipate the firm reports $z_t = \frac{1}{2}y_t^p + \frac{1}{2}\Delta$ and the market price is

$$P_t = z_t - E[\theta] = z_t - \frac{1}{2}\Delta$$

Given this anticipated behavior, the entrepreneur's equilibrium behavior is to report as conjectured. Again, there is a linear relationship between firm value and reported "fair value" accruals.

Now, consider the case where the entrepreneur can misreport but with probability α. Investors process the entrepreneur's report with misreporting in mind. The probability of misreporting D, given an accrual report of z_t, is

$$\Pr(D \mid \tilde{z}_t = z_t) = \frac{\alpha\phi\left(\frac{z_t - 0.5\Delta}{\sqrt{0.5}\sigma}\right)}{\alpha\phi\left(\frac{z_t - 0.5\Delta}{\sqrt{0.5}\sigma}\right) + (1-\alpha)\phi\left(\frac{z_t}{\sqrt{0.5}\sigma}\right)}$$

where $\phi(\cdot)$ is the standard normal density function. In turn, the equilibrium price for the firm following the report is

$$\begin{aligned} P_t &= E\left[\tilde{d}_{t+1} \mid \tilde{z}_t = z_t\right] \\ &= \frac{\alpha(z_t - 0.5\Delta)\phi\left(\frac{z_t - 0.5\Delta}{\sqrt{0.5}\sigma}\right) + (1-\alpha)z_t\phi\left(\frac{z_t}{\sqrt{0.5}\sigma}\right)}{\alpha\phi\left(\frac{z_t - 0.5\Delta}{\sqrt{0.5}\sigma}\right) + (1-\alpha)\phi\left(\frac{z_t}{\sqrt{0.5}\sigma}\right)} \end{aligned}$$

Again, the entrepreneur's equilibrium reporting strategy is to misreport the maximum whenever possible and the accruals balance is $\alpha\left(\frac{1}{2}\Delta\right)$, on average. Price is no longer a linear function of reported "fair value".

Consider the following simulation to illustrate. Let $\sigma^2 = 2$, $\Delta = 4$, and $\alpha = \frac{1}{4}$. For sample size $n = 5,000$ and $1,000$ simulated samples, the regression is

$$P_t = \beta_0 + \beta_1 x_t$$

where

$$x_t = D_t \left(z_t^p + \frac{1}{2}\Delta \right) + (1 - D_t) z_t^p$$

$$D_t \sim Bernoulli\,(\alpha)$$

$$P_t = \frac{\alpha\,(x_t - 0.5\Delta)\,\phi\left(\frac{x_t - 0.5\Delta}{\sqrt{0.5}\sigma}\right) + (1-\alpha)\,x_t\phi\left(\frac{x_t}{\sqrt{0.5}\sigma}\right)}{\alpha\phi\left(\frac{x_t - 0.5\Delta}{\sqrt{0.5}\sigma}\right) + (1-\alpha)\,\phi\left(\frac{x_t}{\sqrt{0.5}\sigma}\right)}$$

and $z_t^p = \frac{1}{2}y_t^p$. A typical plot of the sampled data, price versus reported accruals is depicted in figure 3.1. There is a distinctly nonlinear pattern in the data.[11]

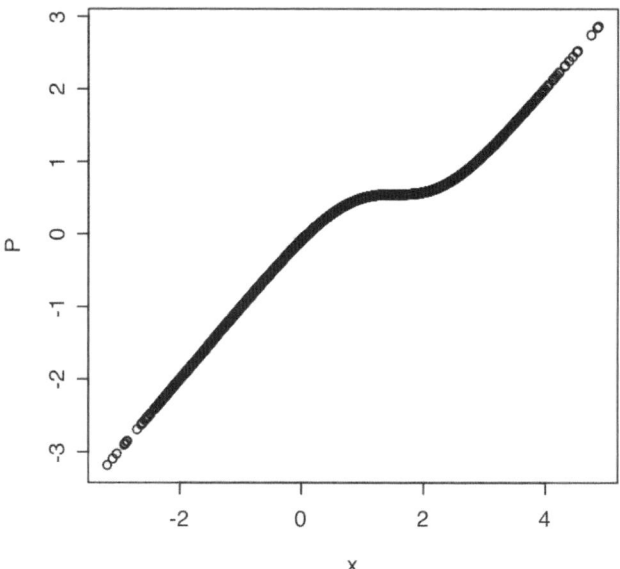

Figure 3.1: Price versus reported accruals

Sample statistics for the regression estimates are reported in table 3.7. The estimates of the slope are substantially biased downward. Recall the slope is one if there is no misreporting or if there is known misreporting. Suppose the analyst

---

[11] For larger (smaller) values of Δ, the nonlinearity is more (less) pronounced.

Table 3.7: Results for price on reported accruals regression

statistic	β_0	β_1
mean	−0.285	0.571
median	−0.285	0.571
standard deviation	0.00405	0.00379
minimum	−0.299	0.557
maximum	−0.269	0.584
$E[P_t \mid x_t] = \beta_0 + \beta_1 x_t$		

can ex post determine whether the firm misreported. Let $D_t = 1$ if the firm misreported in period t and 0 otherwise. Is price a linear function of reported accruals x_t conditional on D_t? Simulation results for the saturated regression

$$P_t = \beta_0 + \beta_1 x_t + \beta_2 D_t + \beta_3 x_t \times D_t$$

are reported in table 3.8. Perhaps surprisingly, the slope coefficient continues to

Table 3.8: Results for price on reported accruals saturated regression

statistic	β_0	β_1	β_2	β_3
mean	−0.244	0.701	0.117	−0.271
median	−0.244	0.701	0.117	−0.271
standard deviation	0.0032	0.0062	0.017	0.011
minimum	−0.255	0.680	0.061	−0.306
maximum	−0.233	0.720	0.170	−0.239
$E[P_t \mid x_t, D_t] = \beta_0 + \beta_1 x_t + \beta_2 D_t + \beta_3 x_t \times D_t$				

be biased toward zero.

Before we abandon hope for our econometric experiment, it is important to remember investors do not observe D_t but rather are left to infer any manipulation from reported accruals x_t. So what then is the omitted, correlated variable in this earnings management setting? Rather than D_t it's the propensity for misreporting inferred from the accruals report, in other words $\Pr(D_t \mid \tilde{x}_t = x_t) \equiv p(x_t)$. If the analyst knows what traders know, that is α, Δ, and σ, along with the observed report, then the regression for estimating the relation between price and fair value is

$$P_t = \beta_0 + \beta_1 x_t + \beta_2 p(x_t)$$

Simulation results are reported in table 3.9. Of course, this regression perfectly

52   3. Linear models

Table 3.9: Results for price on reported accruals and propensity score regression

statistic	β_0	β_1	β_2
mean	0.000	1.000	-2.000
median	0.000	1.000	-2.000
standard deviation	0.000	0.000	0.000
minimum	0.000	1.000	-2.000
maximum	0.000	1.000	-2.000
$E[P_t \mid x_t, p(x_t)] = \beta_0 + \beta_1 x_t + \beta_2 p(x_t)$			

fits the data as a little manipulation confirms.

$$\begin{aligned}
P_t &= \frac{\alpha(x_t - 0.5\Delta)\phi\left(\frac{x_t - 0.5\Delta}{\sqrt{0.5}\sigma}\right) + (1-\alpha)x_t\phi\left(\frac{x_t}{\sqrt{0.5}\sigma}\right)}{\alpha\phi\left(\frac{x_t - 0.5\Delta}{\sqrt{0.5}\sigma}\right) + (1-\alpha)\phi\left(\frac{x_t}{\sqrt{0.5}\sigma}\right)} \\
&= \beta_0 + \beta_1 x_t + \beta_2 p(x_t) \\
&= \beta_0 + \beta_1 x_t + \beta_2 \frac{\alpha\phi\left(\frac{x_t - 0.5\Delta}{\sqrt{0.5}\sigma}\right)}{\alpha\phi\left(\frac{x_t - 0.5\Delta}{\sqrt{0.5}\sigma}\right) + (1-\alpha)\phi\left(\frac{x_t}{\sqrt{0.5}\sigma}\right)} \\
&= \frac{(\beta_0 + \beta_1 x_t)\left[\alpha\phi\left(\frac{x_t - 0.5\Delta}{\sqrt{0.5}\sigma}\right) + (1-\alpha)\phi\left(\frac{x_t}{\sqrt{0.5}\sigma}\right)\right] + \beta_2\alpha\phi\left(\frac{x_t - 0.5\Delta}{\sqrt{0.5}\sigma}\right)}{\alpha\phi\left(\frac{x_t - 0.5\Delta}{\sqrt{0.5}\sigma}\right) + (1-\alpha)\phi\left(\frac{x_t}{\sqrt{0.5}\sigma}\right)}
\end{aligned}$$

For $\beta_1 = 1$, $P_t = \frac{\beta_1 x_t \left[\alpha\phi\left(\frac{x_t - 0.5\Delta}{\sqrt{0.5}\sigma}\right) + (1-\alpha)\phi\left(\frac{x_t}{\sqrt{0.5}\sigma}\right)\right] - \beta_1\alpha 0.5\Delta\phi\left(\frac{x_t - 0.5\Delta}{\sqrt{0.5}\sigma}\right)}{\alpha\phi\left(\frac{x_t - 0.5\Delta}{\sqrt{0.5}\sigma}\right) + (1-\alpha)\phi\left(\frac{x_t}{\sqrt{0.5}\sigma}\right)}$. Hence, $\beta_0 = 0$ and the above expression simplifies

$$\frac{(\beta_0 + \beta_1 x_t)\left[\alpha\phi\left(\frac{x_t - 0.5\Delta}{\sqrt{0.5}\sigma}\right) + (1-\alpha)\phi\left(\frac{x_t}{\sqrt{0.5}\sigma}\right)\right] + \beta_2\alpha\phi\left(\frac{x_t - 0.5\Delta}{\sqrt{0.5}\sigma}\right)}{\alpha\phi\left(\frac{x_t - 0.5\Delta}{\sqrt{0.5}\sigma}\right) + (1-\alpha)\phi\left(\frac{x_t}{\sqrt{0.5}\sigma}\right)}$$

$$= \frac{\beta_1\left[\alpha(x_t - 0.5\Delta)\phi\left(\frac{x_t - 0.5\Delta}{\sqrt{0.5}\sigma}\right) + (1-\alpha)x_t\phi\left(\frac{x_t}{\sqrt{0.5}\sigma}\right)\right]}{\alpha\phi\left(\frac{x_t - 0.5\Delta}{\sqrt{0.5}\sigma}\right) + (1-\alpha)\phi\left(\frac{x_t}{\sqrt{0.5}\sigma}\right)}$$

$$+ \frac{(\beta_1 0.5\Delta + \beta_2)\alpha\phi\left(\frac{x_t - 0.5\Delta}{\sqrt{0.5}\sigma}\right)}{\alpha\phi\left(\frac{x_t - 0.5\Delta}{\sqrt{0.5}\sigma}\right) + (1-\alpha)\phi\left(\frac{x_t}{\sqrt{0.5}\sigma}\right)}$$

Since the last term in the numerator must be zero and $\beta_1 = 1$, $\beta_2 = -\beta_1 0.5\Delta = -0.5\Delta$. In other words, reported accruals conditional on trader's perceptions of the propensity for misreporting map perfectly into price. The regression estimates the relation between price and fair value via β_1 and the magnitude of misreporting when the opportunity arises via β_2.

Of course, frequently the analyst (social scientist) suffers an informational disadvantage. Suppose the analyst ex post observes D_t (an information advantage

3.11 Equilibrium earnings management

relative to traders) but doesn't know α, Δ, and σ (an information disadvantage relative to traders). These parameters must be estimated from the data. An estimate of α is

$$\overline{D} = n^{-1} \sum_{t=1}^{n} D_t$$

An estimate of $\theta = \frac{1}{2}\Delta$ is

$$\widehat{\theta} = \frac{n^{-1} \sum_{t=1}^{n} x_t D_t}{\overline{D}} - \frac{n^{-1} \sum_{t=1}^{n} x_t (1 - D_t)}{(1 - \overline{D})}$$

An estimate of $\nu^2 = \frac{1}{2}\sigma^2$ is

$$\widehat{\nu}^2 = (n-1)^{-1} \sum_{t=1}^{n} (x_t - \overline{x})^2 - \widehat{\theta}^2 \overline{D} (1 - \overline{D})$$

Combining the above estimates[12] produces an estimate of $p(x_t)$

$$\widehat{p}(x_t) = \frac{\overline{D} \phi \left(\frac{x_t - \widehat{\theta}}{\widehat{\nu}} \right)}{\overline{D} \phi \left(\frac{x_t - \widehat{\theta}}{\widehat{\nu}} \right) + (1 - \overline{D}) \phi \left(\frac{x_t}{\widehat{\nu}} \right)}$$

And the regression now is

$$P_t = \beta_0 + \beta_1 x_t + \beta_2 \widehat{p}(x_t)$$

Simulation results reported in table 3.10 support the estimated propensity score $\widehat{p}(x_t)$.

Table 3.10: Results for price on reported accruals and estimated propensity score regression

statistic	β_0	β_1	β_2
mean	0.0001	0.9999	-2.0006
median	-0.0000	0.9998	-2.0002
standard deviation	0.0083	0.0057	0.0314
minimum	-0.025	0.981	-2.104
maximum	0.030	1.019	-1.906
$E[P_t \mid x_t, \widehat{p}(x_t)] = \beta_0 + \beta_1 x_t + \beta_2 \widehat{p}(x_t)$			

---

[12] If D_t is unobservable to the analyst then some other means of estimating $p(x_t)$ is needed (perhaps initial guesses for α and Δ followed by nonlinear refinement).

Rather than $\widehat{p}(x_t)$, the propensity score can be estimated via logit, $\widetilde{p}(x_t)$, (discussed in chapter 5) where D_t is regressed on x_t.[13] As expected, simulation results reported in table 3.11 are nearly identical to those reported above (the correlation between the two propensity score metrics is 0.999).

Table 3.11: Results for price on reported accruals and logit-estimated propensity score regression

statistic	β_0	β_1	β_2
mean	−0.000	1.000	−1.999
median	−0.000	1.000	−1.997
standard deviation	0.012	0.008	0.049
minimum	−0.035	0.974	−2.154
maximum	0.040	1.028	−1.863
$E[P_t \mid x_t, \widetilde{p}(x_t)] = \beta_0 + \beta_1 x_t + \beta_2 \widetilde{p}(x_t)$			

This stylized equilibrium earnings management example illustrates two points. First, it provides a setting in which the intuition behind the propensity score, a common econometric instrument, is clear. Second, it reinforces our theme concerning the importance of the union of theory, data, and model specification. Consistent analysis requires all three be carefully attended and the manner in which each is considered depends on the others.

3.12 Additional reading

Linear models have been extensively studied and accordingly there are many nice econometrics references. Some favorites include Davidson and MacKinnon [1993, 2003], Wooldridge [2002], Cameron and Trivedi [2005], Greene [1997], Amemiya [1985], Theil [1971], Rao [1973], and Graybill [1976]. Angrist and Pischke [2009] provide a provocative justification for the linear conditional expectation function (see the end of chapter appendix). Davidson and MacKinnon in particular offer excellent discussions of *FWL*. Bound, Brown, and Mathiowetz [2001] and Hausman [2001] provide extensive review of classical and nonclassical measurement error and their implications for proxy variables. Christensen and Demski [2003, ch. 9-10] provide a wealth of examples of accounting as an information source and the subtleties of multiple information sources. Their discussion of the correspondence (or lack thereof) between accounting metrics and firm value suggests that association studies are no less prone to challenging specification issues than are information content studies. Discussions in this chapter

---

[13]The posterior probability of manipulation given a normally distributed signal has a logistic distribution (see Kiefer [1980]). Probit results are very similar although the logit intervals are somewhat narrower. Of course, if D_t is unobservable (by the analyst) then discrete choice methods like logit or probit are not directly accessible.

refer specifically to information content. Finally, we reiterate Jayne's [2003] discussion regarding the ubiquity of the Gaussian distribution is provocative.

3.13 Appendix

Angrist and Pischke [2009, ch. 3] layout a foundation justifying regression analysis of economic data and building linkages to causal effects. The arguments begin with the population-level conditional expectation function (*CEF*)

$$E[Y_i \mid X_i = x] = \int t f_y (t \mid X_i = x) \, dt$$

where $f_y(t \mid X_i = x)$ is the conditional density function evaluated at $Y_i = t$ and the law of iterated expectations

$$E[Y_i] = E_X[E[Y_i \mid X_i]]$$

The law of iterated expectations allows us to separate the response variable into two components: the *CEF* and a residual.

Theorem 3.2 *CEF decomposition theorem.*

$$Y_i = E[Y_i \mid X_i] + \varepsilon_i$$

where (i) ε_i is mean independent of X_i, $E[\varepsilon_i \mid X_i] = 0$, and (ii) ε_i is uncorrelated with any function of X_i.

Proof. (i)

$$\begin{aligned} E[\varepsilon_i \mid X_i] &= E[Y_i - E[Y_i \mid X_i] \mid X_i] \\ &= E[Y_i \mid X_i] - E[Y_i \mid X_i] = 0 \end{aligned}$$

(ii) let $h(X_i)$ be some function of X_i. By the law of iterated expectations,

$$E[h(X_i)\varepsilon_i] = E_X[h(X_i) E[\varepsilon_i \mid X_i]]$$

and by mean independence $E[\varepsilon_i \mid X_i] = 0$. Hence, $E[h(X_i)\varepsilon_i] = 0$. ∎

The *CEF* optimally summarizes the relation between the response, Y_i, and explanatory variables, X_i, in a minimum mean square error (*MMSE*) sense.

Theorem 3.3 *CEF prediction theorem. Let $m(X_i)$ be any function of X_i. The CEF is the MMSE of Y_i given X_i in that it solves*

$$E[Y_i \mid X_i] = \arg\min_{m(X_i)} E\left[\{Y_i - m(X_i)\}^2\right]$$

Proof. Write

$$\begin{aligned}\{Y_i - m(X_i)\}^2 &= \{(Y_i - E[Y_i \mid X_i]) + (E[Y_i \mid X_i] - m(X_i))\}^2 \\ &= (Y_i - E[Y_i \mid X_i])^2 + 2(Y_i - E[Y_i \mid X_i]) \\ &\quad \times (E[Y_i \mid X_i] - m(X_i)) + (E[Y_i \mid X_i] - m(X_i))^2\end{aligned}$$

The first term can be ignored as it does not involve $m(X_i)$. By the *CEF* decomposition property, the second term is zero since we can think of $h(X_i) \equiv 2(Y_i - E[Y_i \mid X_i])$. Finally, the third term is minimized when $m(X_i)$ is the *CEF*. ∎

A closely related property involves decomposition of the variance. This property leads to the *ANOVA* table associated with many standard statistical analyses.

Theorem 3.4 *ANOVA theorem.*

$$Var[Y_i] = Var[E[Y_i \mid X_i]] + E_X[Var[Y_i \mid X_i]]$$

where $Var[\cdot]$ is the variance operator.

Proof. The *CEF* decomposition property implies the variance of Y_i equals the variance of the *CEF* plus the variance of the residual as the terms are uncorrelated.

$$Var[Y_i] = Var[E[Y_i \mid X_i]] + Var[\varepsilon_i \mid X_i]$$

Since $\varepsilon_i \equiv Y_i - E[Y_i \mid X_i]$ and $Var[\varepsilon_i \mid X_i] = Var[Y_i \mid X_i] = E[\varepsilon_i^2]$, by iterated expectations

$$\begin{aligned}E[\varepsilon_i^2] &= E_X[E[\varepsilon_i^2 \mid X_i]] \\ &= E_X[Var[Y_i \mid X_i]]\end{aligned}$$

∎

This background sets the stage for three linear regression justifications. Regression justification *I* is the linear *CEF* theorem which applies, for instance, when the data are jointly normally distributed (Galton [1886]).

Theorem 3.5 *Linear CEF theorem (regression justification I). Suppose the CEF is linear.*

$$E[Y_i \mid X_i] = X_i^T \beta$$

where

$$\begin{aligned}\beta &= \arg\min_b E\left[(Y_i - X_i^T b)^2\right] \\ &= E\left[(X_i X_i^T)^{-1}\right] E[X_i Y_i]\end{aligned}$$

Then the population regression function is linear.

Proof. Suppose $E[Y_i \mid X_i] = X_i^T \beta^*$ for some parameter vector β^*. By the *CEF* decomposition theorem,

$$E[X_i(Y_i - E[Y_i \mid X_i]) \mid X_i] = 0$$

Substitution yields

$$E[X_i(Y_i - X_i^T \beta^*) \mid X_i] = 0$$

Iterated expectations implies

$$E[X_i(Y_i - X_i^T \beta^*)] = 0$$

Rearrangement gives

$$\beta^* = E\left[(X_i X_i^T)^{-1}\right] E[X_i Y_i] = \beta$$

∎

Now, we explore approximate results associated with linear regression. First, we state the best linear predictor theorem (regression justification *II*). Then, we describe a linear approximation predictor result (regression justification *III*).

Theorem 3.6 *Best linear predictor theorem (regression justification II). The function $X_i^T \beta$ is the best linear predictor of Y_i given X_i in a MMSE sense.*

Proof. $\beta = E\left[(X_i X_i^T)^{-1}\right] E[X_i Y_i]$ is the solution to the population least squares problem as demonstrated in the proof to the linear CEF theorem. ∎

Theorem 3.7 *Regression CEF theorem (regression justification III). The function $X_i^T \beta$ provides the MMSE linear approximation to $E[Y_i \mid X_i]$. That is,*

$$\beta = \arg\min_b E\left[\left(E[Y_i \mid X_i] - X_i^T b\right)^2\right]$$

Proof. Recall β solves $\arg\min_b E\left[(Y_i - X_i^T b)^2\right]$. Write

$$
\begin{aligned}
(Y_i - X_i^T b)^2 &= \{(Y_i - E[Y_i \mid X_i]) + (E[Y_i \mid X_i] - X_i^T b)\}^2 \\
&= (Y_i - E[Y_i \mid X_i])^2 + (E[Y_i \mid X_i] - X_i^T b)^2 \\
&\quad + 2(Y_i - E[Y_i \mid X_i])(E[Y_i \mid X_i] - X_i^T b)
\end{aligned}
$$

The first term does not involve b and the last term has expected value equal to zero by the *CEF* decomposition theorem. Hence, the *CEF* approximation problem is the same as the population least squares problem (regression justification *II*). ∎

4
Loss functions and estimation

In the previous chapter we reviewed some results of linear (least squares) models without making the loss function explicit. In this chapter we remedy this and extend the discussion to various other (sometimes referred to as "robust") approaches. That the loss function determines the properties of estimators is common to classical and Bayesian statistics (whether made explicit or not). We'll review a few loss functions and the associated expected loss minimizing estimators. Then we briefly review maximum likelihood estimation (*MLE*) and nonlinear regression.

4.1 Loss functions

Let the loss function associated with the estimator $\hat{\theta}$ for θ be $C\left(\hat{\theta}, \theta\right)$ and the posterior distribution function be $f(\theta \mid y)$,[1] then minimum expected loss is

$$\min_{\hat{\theta}} E\left[C\left(\hat{\theta}, \theta\right)\right] = \int C\left(\hat{\theta}, \theta\right) f(\theta \mid y) \, d\theta$$

Briefly, a symmetric quadratic loss function results in an estimator equal to the posterior mean, a linear loss function results in an estimator equal to a quantile of the posterior distribution $f(\theta \mid y)$, and an all or nothing loss function results in an estimator for θ equal to the posterior mode.

---

[1] A source of controversy is whether the focus is the posterior distribution $f(\theta \mid y)$ or the likelihood function $f(y \mid \theta)$; see Poirier [1995]. We initially focus on the posterior distribution then review *MLE*.

4.1.1 Quadratic loss

The quadratic loss function is

$$C\left(\hat{\theta}, \theta\right) = \begin{array}{cc} c_1 \left(\hat{\theta} - \theta\right)^2 & \hat{\theta} \leq \theta \\ c_2 \left(\hat{\theta} - \theta\right)^2 & \hat{\theta} > \theta \end{array}$$

First order conditions are

$$\frac{d}{d\hat{\theta}} \left\{ \begin{array}{c} \int_{\hat{\theta}}^{\infty} c_1 \left(\hat{\theta} - \theta\right)^2 f\left(\theta \mid y\right) d\theta \\ + \int_{-\infty}^{\hat{\theta}} c_2 \left(\hat{\theta} - \theta\right)^2 f\left(\theta \mid y\right) d\theta \end{array} \right\} = 0$$

Rearrangement produces

$$\frac{d}{d\hat{\theta}} \left\{ \begin{array}{c} c_1 \left(1 - F\left(\hat{\theta}\right)\right) \hat{\theta}^2 - 2c_1 \hat{\theta} \int_{\hat{\theta}}^{\infty} \theta f\left(\theta \mid y\right) d\theta \\ + c_1 \int_{\hat{\theta}}^{\infty} \theta^2 f\left(\theta \mid y\right) d\theta + c_2 F\left(\hat{\theta}\right) \hat{\theta}^2 \\ - 2c_2 \int_{-\infty}^{\hat{\theta}} \theta f\left(\theta \mid y\right) d\theta + c_2 \int_{-\infty}^{\hat{\theta}} \theta^2 f\left(\theta \mid y\right) d\theta \end{array} \right\} = 0$$

where $F\left(\hat{\theta}\right)$ is the cumulative posterior distribution function for θ given the data y evaluated at $\hat{\theta}$. Differentiation reveals

$$\left\{ \begin{array}{c} c_1 \left[\begin{array}{c} 2\hat{\theta} \left(1 - F\left(\hat{\theta}\right)\right) - \hat{\theta}^2 f\left(\hat{\theta}\right) \\ -2 \int_{\hat{\theta}}^{\infty} \theta f\left(\theta \mid y\right) d\theta + 2\hat{\theta}^2 f\left(\hat{\theta}\right) - \hat{\theta}^2 f\left(\hat{\theta}\right) \end{array} \right] \\ + c_2 \left[\begin{array}{c} 2\hat{\theta} F\left(\hat{\theta}\right) + \hat{\theta}^2 f\left(\hat{\theta}\right) \\ -2 \int_{-\infty}^{\hat{\theta}} \theta f\left(\theta \mid y\right) d\theta - 2\hat{\theta}^2 f\left(\hat{\theta}\right) + \hat{\theta}^2 f\left(\hat{\theta}\right) \end{array} \right] \end{array} \right\} = 0$$

Simplification yields

$$\hat{\theta} \left[c_1 \left(1 - F\left(\hat{\theta}\right)\right) + c_2 F\left(\hat{\theta}\right) \right]$$
$$= c_1 \left(1 - F\left(\hat{\theta}\right)\right) E\left[\theta \mid y, \hat{\theta} \leq \theta\right] + c_{22} F\left(\hat{\theta}\right) E\left[\theta \mid y, \hat{\theta} > \theta\right]$$

Or,

$$\hat{\theta} = \frac{c_1 \left(1 - F\left(\hat{\theta}\right)\right) E\left[\theta \mid y, \theta \geq \hat{\theta}\right] + c_2 F\left(\hat{\theta}\right) E\left[\theta \mid y, \theta < \hat{\theta}\right]}{c_1 \left(1 - F\left(\hat{\theta}\right)\right) + c_2 F\left(\hat{\theta}\right)}$$

In other words, the quadratic expected loss minimizing estimator for θ is a cost-weighted average of truncated means of the posterior distribution. If $c_1 = c_2$ (symmetric loss), then $\hat{\theta} = E\left[\theta \mid y\right]$, the mean of the posterior distribution.

4.1.2 Linear loss

The linear loss function is

$$C\left(\hat{\theta}, \theta\right) = \begin{array}{ll} c_1 \left|\hat{\theta} - \theta\right| & \hat{\theta} \leq \theta \\ c_2 \left|\hat{\theta} - \theta\right| & \hat{\theta} > \theta \end{array}$$

First order conditions are

$$\frac{d}{d\hat{\theta}} \left\{ \begin{array}{l} -\int_{\hat{\theta}}^{\infty} c_1 \left(\hat{\theta} - \theta\right) f(\theta \mid y) \, d\theta \\ + \int_{-\infty}^{\hat{\theta}} c_2 \left(\hat{\theta} - \theta\right) f(\theta \mid y) \, d\theta \end{array} \right\} = 0$$

Rearranging yields

$$\frac{d}{d\hat{\theta}} \left\{ \begin{array}{l} -c_1 \hat{\theta} \left(1 - F\left(\hat{\theta}\right)\right) + c_1 \int_{\hat{\theta}}^{\infty} \theta f(\theta \mid y) \, d\theta \\ + c_2 \hat{\theta} F\left(\hat{\theta}\right) - c_2 \int_{-\infty}^{\hat{\theta}} \theta f(\theta \mid y) \, d\theta \end{array} \right\} = 0$$

Differentiation produces

$$\begin{aligned} 0 &= c_1 \left[-\left(1 - F\left(\hat{\theta}\right)\right) + \hat{\theta} f\left(\hat{\theta}\right) - \hat{\theta} f\left(\hat{\theta}\right)\right] \\ &+ c_2 \left[F\left(\hat{\theta}\right) + \hat{\theta} f\left(\hat{\theta}\right) - \hat{\theta} f\left(\hat{\theta}\right)\right] \end{aligned}$$

Simplification reveals

$$c_1 \left(1 - F\left(\hat{\theta}\right)\right) = c_2 F\left(\hat{\theta}\right)$$

Or

$$F\left(\hat{\theta}\right) = \frac{c_1}{c_1 + c_2}$$

The expected loss minimizing estimator is the quantile that corresponds to the relative cost $\frac{c_1}{c_1 + c_2}$. If $c_1 = c_2$, then the estimator is the median of the posterior distribution.

4.1.3 All or nothing loss

The all or nothing loss function is

$$C\left(\hat{\theta}, \theta\right) = \begin{array}{ll} c_1 & \hat{\theta} < \theta \\ 0 & \hat{\theta} = \theta \\ c_2 & \hat{\theta} > \theta \end{array}$$

If $c_1 > c_2$, then we want to choose $\hat{\theta} > \theta$, so $\hat{\theta}$ is the upper limit of support for $f(\theta \mid y)$. If $c_1 < c_2$, then we want to choose $\hat{\theta} < \theta$, so $\hat{\theta}$ is the lower limit of support for $f(\theta \mid y)$. If $c_1 = c_2$, then we want to choose $\hat{\theta}$ to maximize $f(\theta \mid y)$, so $\hat{\theta}$ is the mode of the posterior distribution.[2]

4.2 Nonlinear Regression

Many accounting and business settings call for analysis of data involving limited dependent variables (such as discrete choice models discussed in the next chapter).[3] Nonlinear regression frequently complements our understanding of standard maximum likelihood procedures employed for estimating such models as well as providing a means for addressing alternative functional forms. Here we review some basics of nonlinear least squares including Newton's method of optimization, Gauss-Newton regression (*GNR*), and artificial regressions.

Our discussion revolves around minimizing a smooth, twice continuously differentiable function, $Q(\beta)$. It's convenient to think $Q(\beta)$ equals $SSR(\beta)$, the residual sum of squares, but $-Q(\beta)$ might also refer to maximization of the log-likelihood.

4.2.1 Newton's method

A second order Taylor series approximation of $Q(\beta)$ around some initial values for β, say $\beta_{(0)}$ yields

$$Q^*(\beta) = Q\left(\beta_{(0)}\right) + g_{(0)}^T \left(\beta - \beta_{(0)}\right) + \frac{1}{2}\left(\beta - \beta_{(0)}\right)^T H_{(0)} \left(\beta - \beta_{(0)}\right)$$

where $g(\beta)$ is the $k \times 1$ gradient of $Q(\beta)$ with typical element $\frac{\partial Q(\beta)}{\partial \beta_j}$, $H(\beta)$ is the $k \times k$ Hessian of $Q(\beta)$ with typical element $\frac{\partial^2 Q(\beta)}{\partial \beta_j \partial \beta_i}$, and for notational simplicity, $g_{(0)} \equiv g\left(\beta_{(0)}\right)$ and $H_{(0)} \equiv H\left(\beta_{(0)}\right)$. The first order conditions for a minimum of $Q^*(\beta)$ with respect to β are

$$g_{(0)} + H_{(0)}\left(\beta - \beta_{(0)}\right) = 0$$

Solving for β yields a new value

$$\beta_{(1)} = \beta_{(0)} - H_{(0)}^{-1} g_{(0)}$$

This is the core of Newton's method. Successive values $\beta_{(1)}, \beta_{(2)}, \ldots$ lead to an approximation of the global minimum of $Q(\beta)$ at $\hat{\beta}$. If $Q(\beta)$ is approximately

---

[2] For a discrete probability mass distribution, the optimal estimator may be either the limit of support or the mode depending on the difference in cost. Clearly, large cost differentials are aligned with the limits and small cost differences are aligned with the mode.

[3] This section draws heavily from Davidson and MacKinnon [1993].

4.2 Nonlinear Regression

quadratic, as applies to sums of squares when sufficiently close to their minima, Newton's method usually converges quickly.[4]

4.2.2 Gauss-Newton regression

When minimizing a sum of squares function it is convenient to write the criterion as

$$Q(\beta) = \frac{1}{n} SSR(\beta) = \frac{1}{n} \sum_{t=1}^{n} (y_t - x_t(\beta))^2$$

Now, explicit expressions for the gradient and Hessian can be found. The gradient for the ith element is

$$g_i(\beta) = -\frac{2}{n} \sum_{t=1}^{n} X_{ti}(\beta)(y_t - x_t(\beta))$$

where $X_{ti}(\beta)$ is the partial derivative of $x_t(\beta)$ with respect to β_i. The more compact matrix notation is

$$\mathbf{g}(\beta) = -\frac{2}{n} \mathbf{X}^T(\beta)(\mathbf{y} - \mathbf{x}(\beta))$$

The Hessian $H(\beta)$ has typical element

$$H_{ij}(\beta) = -\frac{2}{n} \sum_{t=1}^{n} (y_t - x_t(\beta)) \frac{\partial X_{ti}(\beta)}{\partial \beta_j} - X_{ti}(\beta) X_{tj}(\beta)$$

Evaluated at β_0, this expression is asymptotically equivalent to[5]

$$\frac{2}{n} \sum_{t=1}^{n} X_{ti}(\beta) X_{tj}(\beta)$$

In matrix notation this is

$$D(\beta) = \frac{2}{n} \mathbf{X}^T(\beta) \mathbf{X}(\beta)$$

and $D(\beta)$ is positive definite when $X(\beta)$ is full rank. Now, writing Newton's method as

$$\beta_{(j+1)} = \beta_{(j)} - D_{(j)}^{-1} g_{(j)}$$

---

[4] If $Q^*(\beta)$ is strictly convex, as it is if and only if the Hessian is positive definite, then $\beta_{(1)}$ is the global minimum of $Q^*(\beta)$. Please consult other sources, such as Davidson and MacKinnon [2003, ch. 6] and references therein, for additional discussion of Newton's method including search direction, step size, and stopping rules.

[5] Since $y_t = x_t(\beta_0) + u_t$, the first term becomes $-\frac{2}{n} \sum_{t=1}^{n} \frac{\partial X_{ti}(\beta)}{\partial \beta_j} u_t$. By the law of large numbers this term tends to 0 as $n \to \infty$.

and substituting the above results we have the classic Gauss-Newton result

$$\beta_{(j+1)} = \beta_{(j)} - \left(\frac{2}{n} X_{(j)}^T X_{(j)}\right)^{-1} \left(-\frac{2}{n} \mathbf{X}_{(j)}^T (\mathbf{y} - \mathbf{x}_{(j)})\right)$$

$$= \beta_{(j)} + \left(\mathbf{X}_{(j)}^T \mathbf{X}_{(j)}\right)^{-1} \mathbf{X}_{(j)}^T (\mathbf{y} - \mathbf{x}_{(j)})$$

Artificial regression

The second term can be readily estimated by an artificial regression. It's called an artificial regression because functions of the variables and model parameters are employed. This artificial regression is referred to as a Gauss-Newton regression (*GNR*)

$$\mathbf{y} - \mathbf{x}(\beta) = \mathbf{X}(\beta) b + \ residuals$$

To be clear, Gaussian projection (*OLS*) produces the following estimate

$$\hat{\mathbf{b}} = \left(\mathbf{X}^T(\beta) \mathbf{X}(\beta)\right)^{-1} \mathbf{X}^T(\beta) (\mathbf{y} - \mathbf{x}(\beta))$$

To appreciate the *GNR*, consider a linear regression where X is the matrix of regressors. Then $\mathbf{X}(\beta)$ is simply replaced by X, the *GNR* is

$$\mathbf{y} - \mathbf{X}\beta_{(0)} = \mathbf{X}\mathbf{b} + \ residuals$$

and the artificial parameter estimates are

$$\hat{\mathbf{b}} = \left(\mathbf{X}^T \mathbf{X}\right)^{-1} \mathbf{X}^T \left(\mathbf{y} - \mathbf{X}\beta_{(0)}\right) = \hat{\beta} - \beta_{(0)}$$

where $\hat{\beta}$ is the *OLS* estimate. Rearranging we see that the Gauss-Newton estimate replicates *OLS*, $\beta_{(1)} = \beta_{(0)} + \hat{\mathbf{b}} = \beta_{(0)} + \hat{\beta} - \beta_{(0)} = \hat{\beta}$, as expected.

Covariance matrices

Return to the *GNR* above and substitute the nonlinear parameter estimates

$$\mathbf{y} - \mathbf{x}\left(\hat{\beta}\right) = \mathbf{X}\left(\hat{\beta}\right) \mathbf{b} + \ residuals$$

The artificial regression estimate is

$$\hat{\mathbf{b}} = \left(\mathbf{X}^T\left(\hat{\beta}\right) \mathbf{X}\left(\hat{\beta}\right)\right)^{-1} \mathbf{X}^T\left(\hat{\beta}\right) \left(\mathbf{y} - \mathbf{x}\left(\hat{\beta}\right)\right)$$

Since the first order or moment conditions require

$$\mathbf{X}^T\left(\hat{\beta}\right) \left(\mathbf{y} - \mathbf{x}\left(\hat{\beta}\right)\right) = 0$$

this regression cannot have any explanatory power, $\hat{\mathbf{b}} = 0$. Though this may not seem very interesting, it serves two useful functions. First, it provides a check on the consistency of the nonlinear optimization routine. Second, as it is the *GNR* variance estimate, it provides a quick estimator of the covariance matrix for the parameter estimates

$$\widehat{Var}\left[\hat{b}\right] = s^2 \left(\mathbf{X}^T\left(\hat{\beta}\right)\mathbf{X}\left(\hat{\beta}\right)\right)^{-1}$$

and it is readily available from the artificial regression.

Further, this same *GNR* readily supplies a heteroskedastic-consistent covariance matrix estimator. If $E\left[uu^T\right] = \Omega$, then a heteroskedastic-consistent covariance matrix estimator is

$$\widehat{Var}\left[\hat{b}\right] = \left(\mathbf{X}^T\left(\hat{\beta}\right)\mathbf{X}\left(\hat{\beta}\right)\right)^{-1} \mathbf{X}^T\left(\hat{\beta}\right) \hat{\Omega} \mathbf{X}\left(\hat{\beta}\right) \left(\mathbf{X}^T\left(\hat{\beta}\right)\mathbf{X}\left(\hat{\beta}\right)\right)^{-1}$$

where $\hat{\Omega}$ is a diagonal matrix with t^{th} element equal to the squared residual u_t^2. Next, we turn to maximum likelihood estimation and exploit some insights gained from nonlinear regression as they relate to typical *MLE* settings.

4.3 Maximum likelihood estimation (*MLE*)

Maximum likelihood estimation (*MLE*) applies to a wide variety of problems.[6] Since it is the most common method for estimating discrete choice models and discrete choice models are central to the discussion of accounting choice, we focus the discussion of *MLE* around discrete choice models.

4.3.1 Parameter estimation

The most common method for estimating the parameters of discrete choice models is maximum likelihood. Recall the likelihood is defined as the joint density for the parameters of interest β conditional on the data X_t. For binary choice models and $Y_t = 1$ the contribution to the likelihood is $F(X_t\beta)$, and for $Y_t = 0$ the contribution to the likelihood is $1 - F(X_t\beta)$ where these are combined as binomial draws. Hence,

$$L(\beta|X) = \prod_{t=1}^{n} F(X_t\beta)^{Y_t} \left[1 - F(X_t\beta)\right]^{1-Y_t}$$

The log-likelihood is

$$\ell(\beta|X) \equiv logL(\beta|X) = \sum_{t=1}^{n} Y_t log\left(F(X_t\beta)\right) + (1 - Y_t) log\left(1 - F(X_t\beta)\right)$$

---

[6]This section draws heavily from Davidson and MacKinnon [1993], chapter 8.

66   4. Loss functions and estimation

Since this function for binary response models like probit and logit is globally concave, numerical maximization is straightforward. The first order conditions for a maximum are

$$\sum_{t=1}^{n} \frac{Y_t f(X_t\beta) X_{it}}{F(X_t\beta)} - \frac{(1-Y_t)f(X_t\beta)X_{ti}}{1-F(X_t\beta)} = 0 \quad i=1,\ldots,k$$

where $f(\cdot)$ is the density function. Simplifying yields

$$\sum_{t=1}^{n} \frac{[Y_t - F(X_t\beta)]f(X_t\beta)X_{ti}}{F(X_t\beta)[1-F(X_t\beta)]} = 0 \quad i=1,\ldots,k$$

For the logit model the first order conditions simplify to

$$\sum_{t=1}^{n} [Y_t - \Lambda(X_{ti})] X_{ti} = 0 \quad i=1,\ldots,k$$

since the logit density is $\lambda(X_{ti}) = \Lambda(X_{ti})[1-\Lambda(X_{ti})]$ where $\Lambda(\cdot)$ is the logit (cumulative) distribution function.

Notice the above first order conditions look like the first order conditions for weighted nonlinear least squares with weights given by $[F(1-F)]^{-1/2}$. This is sensible because the error term in the nonlinear regression

$$Y_t = F(X_t\beta) + \varepsilon_t$$

has mean zero and variance

$$\begin{aligned} E\left[\varepsilon_t^2\right] &= E\left[\{Y_t - F(X_t\beta)\}^2\right] \\ &= Pr(Y_t = 1)[1 - F(X_t\beta)]^2 + Pr(Y_t = 0)[0 - F(X_t\beta)]^2 \\ &= F(X_t\beta)[1 - F(X_t\beta)]^2 + [1 - F(X_t\beta)]F(X_t\beta)^2 \\ &= F(X_t\beta)[1 - F(X_t\beta)] \end{aligned}$$

As *ML* is equivalent to weighted nonlinear least squares for binary response models, the asymptotic covariance matrix for $n^{1/2}\left(\hat{\beta} - \beta\right)$ is $\left(n^{-1}X^T\Psi X\right)^{-1}$ where Ψ is a diagonal matrix with elements $\frac{f(X_t\beta)^2}{F(X_t\beta)[1-F(X_t\beta)]}$. In the logit case, Ψ simplifies to λ (see Davidson and MacKinnon, p. 517-518).

4.3.2 *Estimated asymptotic covariance for MLE of $\hat{\theta}$*

There are (at least) three common estimators for the variance of $\hat{\theta}_{MLE}$:[7]

(i) $\left[-H\left(\hat{\theta}\right)\right]^{-1}$ the negative inverse of Hessian evaluated at $\hat{\theta}_{MLE}$,

(ii) $g\left(\hat{\theta}\right) g\left(\hat{\theta}\right)^T{}^{-1}$ the outer product of gradient (*OPG*) or Berndt, Hall, Hall, and Hausman (*BHHH*) estimator,

---

[7]This section draws heavily from Davidson and MacKinnon [1993], pp. 260-267.

(iii) $\left[\Im\left(\hat{\theta}\right)\right]^{-1}$ inverse of information matrix or negative expected value of Hessian, where the following definitions apply:

- *MLE* is defined as the solution to the first order conditions (*FOC*): $g\left(Y,\hat{\theta}\right) = 0$ where gradient or score vector g is defined by $g^T(Y,\theta) = D_\theta \ell(Y,\theta)$ (since $D_\theta \ell$ is row vector, g is column vector of partial derivatives of with respect to θ).

- Define $G(g,\theta)$ as the matrix of contributions to the gradient (*CG* matrix) with typical element $G_{ti}(g,\theta) \equiv \frac{\partial \ell_t(Y,\theta)}{\partial \theta_i}$.

- $H(Y,\theta)$ is the Hessian matrix for the log-likelihood with typical element $H_{ij}(Y,\theta) \equiv \frac{\partial^2 \ell_t(Y,\theta)}{\partial \theta_i \partial \theta_j}$.

- Define the expected average Hessian for sample of size n as $H_n(\theta) \equiv E_\theta\left[n^{-1} H(Y,\theta)\right]$.

- The limiting Hessian or asymptotic Hessian (if it exists) is $H(\theta) \equiv \lim_{n\to\infty} H_n(\theta)$ (the matrix is negative semidefinite).

- Define the information in observation t as $\Im_t(\theta)$ a $k \times k$ matrix with typical element $(\Im_t(\theta))_{ij} \equiv E_\theta[G_{ti}(\theta) G_{tj}(\theta)]$ (the information matrix is positive semidefinite).

- The average information matrix is $\Im_n(\theta) \equiv n^{-1} \sum_{t=1}^{n} \Im_t(\theta) = n^{-1} \Im_n$ and the limiting information matrix or asymptotic information matrix (if it exists) is $\Im(\theta) \equiv \lim_{n\to\infty} \Im_n(\theta)$.

The short explanation for these variance estimators is that *ML* estimators (under suitable regularity conditions) achieve the Cramer-Rao lower bound for consistent

68   4. Loss functions and estimation

estimators.[8] That is,

$$Asy.Var\left[\hat{\theta}\right] = \left\{-E\left[\frac{\partial^2 \ell(Y,\theta)}{\partial\theta\partial\theta^T}\right]\right\}^{-1} = \left\{E\left[\left(\frac{\partial \ell(Y,\theta)}{\partial\theta}\right)\left(\frac{\partial \ell(Y,\theta)}{\partial\theta^T}\right)\right]\right\}^{-1}$$

The expected outer product of the gradient (*OPG*) is an estimator of the inverse of the variance matrix for the gradient. Roughly speaking, the inverse of the gradient function yields *MLE* (type 2) parameter estimates and the inverse of expected *OPG* estimates the parameter variance matrix (see Berndt, Hall, Hall, and Hausman [1974]). Also, the expected value of the Hessian equals the negative of the information matrix.[9] In turn, the inverse of the information matrix is an estimator for the estimated parameter variance matrix.

Example: Consider the *MLE* of a standard linear regression model with *DGP*: $Y = X\beta + \varepsilon$ where $\varepsilon \sim N\left(0, \sigma^2 I\right)$ and $E\left[X^T \varepsilon\right] = 0$. Of course, the *MLE* for β is $b = \left(X^T X\right)^{-1} X^T Y$ as

$$g(\beta) \equiv \frac{\partial \ell(Y,\beta)}{\partial \beta} = -\frac{1}{\sigma^2}\begin{bmatrix} X_1^T(Y - X\beta) \\ \vdots \\ X_p^T(Y - X\beta) \end{bmatrix}$$

---

[8] See Theil [1971], pp. 384-385 and Amemiya [1985], pp. 14-17.

$$E\left[\frac{\partial^2 \ell}{\partial\theta\partial\theta^T}\right] = E\left[\frac{\partial}{\partial\theta}\left(\frac{1}{L}\frac{\partial L}{\partial\theta^T}\right)\right]$$

by the chain rule

$$= E\left[-\frac{1}{L^2}\frac{\partial L}{\partial\theta}\frac{\partial L}{\partial\theta^T} + \frac{1}{L}\frac{\partial^2 L}{\partial\theta\partial\theta^T}\right]$$

$$= E\left[-\left(\frac{1}{L}\frac{\partial L}{\partial\theta}\right)\left(\frac{\partial L}{\partial\theta^T}\frac{1}{L}\right)\right] + \int \frac{1}{L}\frac{\partial^2 L}{\partial\theta\partial\theta^T} L dx$$

$$= E\left[-\left(\frac{1}{L}\frac{\partial L}{\partial\theta}\right)\left(\frac{\partial L}{\partial\theta^T}\frac{1}{L}\right)\right] + \int \frac{\partial^2 L}{\partial\theta\partial\theta^T} dx$$

$$= -E\left[\frac{\partial \ell}{\partial\theta}\frac{\partial \ell}{\partial\theta^T}\right] + \int \frac{\partial^2 L}{\partial\theta\partial\theta^T} dx$$

since the regulatory conditions essentially make the order of integration and differentiation interchangeable the last term can be rewritten

$$\int \frac{\partial^2 L}{\partial\theta\partial\theta^T} dx = \frac{\partial}{\partial\theta}\int \frac{\partial L}{\partial\theta^T} dx = \frac{\partial}{\partial\theta}\frac{\partial}{\partial\theta^T}\int L dx = 0$$

Now we have

$$E\left[\frac{\partial^2 \ell}{\partial\theta\partial\theta^T}\right] = -E\left[\frac{\partial \ell}{\partial\theta}\frac{\partial \ell}{\partial\theta^T}\right]$$

[9] This is motivated by the fact that $plim \frac{1}{n} \sum_{i=1}^{n} g(y_i) = E[g(y)]$ for a random sample provided the first two moments of $g(y)$ are finite (see Greene [1997], ch. 4).

4.3 Maximum likelihood estimation (MLE)

where X_j refers to column j of X. Substituting $X\beta + \varepsilon$ for Y produces

$$\left(\frac{\partial \ell(Y,\beta)}{\partial \beta}\right)\left(\frac{\partial \ell(Y,\beta)}{\partial \beta^T}\right) = \left(\frac{1}{\sigma^2}\right)^2 \begin{bmatrix} X_1^T \varepsilon\varepsilon^T X_1 & \cdots & X_1^T \varepsilon\varepsilon^T X_p \\ \vdots & \ddots & \vdots \\ X_p^T \varepsilon\varepsilon^T X_1 & \cdots & X_p^T \varepsilon\varepsilon^T X_p \end{bmatrix}$$

Now,

$$E\left[\left(\frac{\partial \ell(Y,\beta)}{\partial \beta}\right)\left(\frac{\partial \ell(Y,\beta)}{\partial \beta^T}\right)\right] = \left(\frac{1}{\sigma^2}\right) \begin{bmatrix} X_1^T X_1 & \cdots & X_1^T X_p \\ \vdots & \ddots & \vdots \\ X_p^T X_1 & \cdots & X_p^T X_p \end{bmatrix}$$

Since

$$H(\beta) \equiv \frac{\partial^2 \ell(Y,\beta)}{\partial \beta \partial \beta^T} = -\left(\frac{1}{\sigma^2}\right) \begin{bmatrix} X_1^T X_1 & \cdots & X_1^T X_p \\ \vdots & \ddots & \vdots \\ X_p^T X_1 & \cdots & X_p^T X_p \end{bmatrix}$$

we have

$$E\left[\left(\frac{\partial \ell(Y,\beta)}{\partial \beta}\right)\left(\frac{\partial \ell(Y,\beta)}{\partial \beta^T}\right)\right] = -E\left[\frac{\partial^2 \ell(Y,\beta)}{\partial \beta \partial \beta^T}\right]$$

and the demonstration is complete as

$$\begin{aligned} Asy.Var[b] &= \left\{E\left[\left(\frac{\partial \ell(Y,\beta)}{\partial \beta}\right)\left(\frac{\partial \ell(Y,\beta)}{\partial \beta^T}\right)\right]\right\}^{-1} \\ &= -\left\{E\left[\frac{\partial^2 \ell(Y,\beta)}{\partial \beta \partial \beta^T}\right]\right\}^{-1} \\ &= \sigma^2 \left(X^T X\right)^{-1} \end{aligned}$$

A more complete explanation (utilizing results and notation developed in the appendix) starts with the *MLE* first order condition (*FOC*) $g\left(\hat{\theta}\right) = 0$. Now, a Taylor series expansion of the likelihood *FOC* around θ yields $0 = g\left(\hat{\theta}\right) \approx g(\theta) + H\left(\bar{\theta}\right)\left(\hat{\theta} - \theta\right)$ where $\bar{\theta}$ is convex combination (perhaps different for each row) of θ and $\hat{\theta}$. Solve for $\left(\hat{\theta} - \theta\right)$ and rewrite so every term is $O(1)$

$$n^{1/2}\left(\hat{\theta} - \theta\right) = -\left[n^{-1} H\left(\bar{\theta}\right)\right]^{-1}\left[n^{-1/2} g(\theta)\right]$$

By *WULLN* (weak uniform law of large numbers), the first term is asymptotically nonstochastic, by *CLT* (the central limit theorem) the second term is asymptotically normal, so $n^{1/2}\left(\hat{\theta} - \theta\right)$ is asymptotically normal. Hence, the asymptotic variance of $n^{1/2}\left(\hat{\theta} - \theta\right)$ is the asymptotic expectation of $n\left(\hat{\theta} - \theta\right)\left(\hat{\theta} - \theta\right)^T$.

Since $n^{1/2}\left(\hat{\theta} - \theta\right) \stackrel{a}{=} -\left[n^{-1}H\left(\theta\right)\right]^{-1}\left[n^{-1/2}g\left(\theta\right)\right]$, the asymptotic variance is $\left(-H^{-1}\left(\theta\right)\right)\left(n^{-1}E_\theta\left[g\left(\theta\right)g^T\left(\theta\right)\right]\right)\left[-H^{-1}\left(\theta\right)\right]$. Simplifying yields

$$Asym.Var\left[n^{1/2}\left(\hat{\theta} - \theta\right)\right] = H^{-1}\left(\theta\right)\Im\left(\theta\right)H^{-1}\left(\theta\right)$$

This can be simplified since $H\left(\theta\right) = -\Im\left(\theta\right)$ by *LLN*. Hence,

$$Asy.Var\left[n^{1/2}\left(\hat{\theta} - \theta\right)\right] = -H^{-1}\left(\theta\right) = \Im^{-1}\left(\theta\right)$$

And the statistic relies on estimation of $H^{-1}\left(\theta\right)$ or $\Im^{-1}\left(\theta\right)$.

- A common estimator of the empirical Hessian is

$$\hat{H} \equiv H_n\left(Y,\hat{\theta}\right) = n^{-1}D^2_{\theta\theta}\ell_t\left(Y,\hat{\theta}\right)$$

(*LLN* and consistency of $\hat{\theta}$ guarantee consistency of $\hat{H}$ for $H\left(\theta\right)$).

- The *OPG* or *BHHH* estimator is

$$\Im_{OPG} \equiv n^{-1}\sum_{t=1}^{n}D^T_\theta\ell_t\left(Y,\hat{\theta}\right)D_\theta\ell_t\left(Y,\hat{\theta}\right) = n^{-1}G^T\left(\hat{\theta}\right)G\left(\hat{\theta}\right)$$

(consistency is guaranteed by *CLT* and *LLN* for the sum).

- The third estimator evaluates the expected values of the second derivatives of the log-likelihood at $\hat{\theta}$. Since this form is not always known, this estimator may not be available. However, as this estimator does not depend on the realization of Y it is less noisy than the other estimators.

We round out this discussion of *MLE* by reviewing a surprising case where *MLE* is not the most efficient estimator. Next, we discuss James-Stein shrinkage estimators.

4.4 James-Stein shrinkage estimators

Stein [1955] showed that when estimating K parameters from independent normal observations with (for simplicity) unit variance, we can uniformly improve on the conventional maximum likelihood estimator in terms of expected squared error loss for $K > 2$. James and Stein [1961] determined such a shrinkage estimator can be written as a function of the maximum likelihood estimator $\hat{\theta}$

$$\theta^* = \hat{\theta}\left(1 - \frac{a}{\hat{\theta}^T\hat{\theta}}\right)$$

4.4 James-Stein shrinkage estimators

where $0 \leq a \leq 2(K-2)$. The expected squared error loss of the James-Stein estimator θ^* is

$$\begin{aligned}
\rho(\theta, \theta^*) &= E\left[(\theta - \theta^*)^T (\theta - \theta^*)\right] \\
&= E\left[\left((\widehat{\theta} - \theta) - \frac{a\widehat{\theta}}{\widehat{\theta}^T\widehat{\theta}}\right)^T \left((\widehat{\theta} - \theta) - \frac{a\widehat{\theta}}{\widehat{\theta}^T\widehat{\theta}}\right)\right] \\
&= E\left[(\widehat{\theta} - \theta)^T (\widehat{\theta} - \theta)\right] - 2aE\left[(\widehat{\theta} - \theta)^T \frac{\widehat{\theta}}{\widehat{\theta}^T\widehat{\theta}}\right] \\
&\quad + a^2 E\left[\frac{\widehat{\theta}^T\widehat{\theta}}{\left(\widehat{\theta}^T\widehat{\theta}\right)^2}\right] \\
&= E\left[(\widehat{\theta} - \theta)^T (\widehat{\theta} - \theta)\right] - 2aE\left[\frac{\widehat{\theta}^T\widehat{\theta}}{\widehat{\theta}^T\widehat{\theta}}\right] + 2a\theta^T E\left[\frac{\widehat{\theta}}{\widehat{\theta}^T\widehat{\theta}}\right] \\
&\quad + a^2 E\left[\frac{\widehat{\theta}^T\widehat{\theta}}{\left(\widehat{\theta}^T\widehat{\theta}\right)^2}\right] \\
&= E\left[(\widehat{\theta} - \theta)^T (\widehat{\theta} - \theta)\right] - 2a + 2a\theta^T E\left[\frac{\widehat{\theta}}{\widehat{\theta}^T\widehat{\theta}}\right] + a^2 E\left[\frac{1}{\widehat{\theta}^T\widehat{\theta}}\right]
\end{aligned}$$

This can be further simplified by exploiting the following theorems; we conclude this section with Judge and Bock's [1978, p. 322-3] proof following discussion of the James-Stein shrinkage estimator.

Theorem 4.1 $E\left[\frac{\widehat{\theta}}{\widehat{\theta}^T\widehat{\theta}}\right] = \theta E\left[\frac{1}{\chi^2_{(K+2,\lambda)}}\right]$ where $\widehat{\theta}^T\widehat{\theta} \sim \chi^2_{(K,\lambda)}$ and $\lambda = \theta^T\theta$ is a noncentrality parameter.[10]

Using

$$\begin{aligned}
E\left[(\widehat{\theta} - \theta)^T (\widehat{\theta} - \theta)\right] &= E\left[\widehat{\theta}^T\widehat{\theta}\right] - 2\theta^T E\left[\widehat{\theta}\right] + \theta^T\theta \\
&= K + \lambda - 2\lambda + \lambda = K
\end{aligned}$$

for the first term, a convenient substitution for one in the second term, and the above theorem for the third term, we rewrite the squared error loss (from above)

$$\rho(\theta, \theta^*) = E\left[(\widehat{\theta} - \theta)^T (\widehat{\theta} - \theta)\right] - 2a + 2a\theta^T E\left[\frac{\widehat{\theta}}{\widehat{\theta}^T\widehat{\theta}}\right] + a^2 E\left[\frac{1}{\widehat{\theta}^T\widehat{\theta}}\right]$$

---

[10] We adopt the convention the noncentrality parameter is the sum of squared means $\theta^T\theta$; others, including Judge and Bock [1978], employ $\frac{\theta^T\theta}{2}$.

4. Loss functions and estimation

as

$$\rho(\theta, \theta^*) = K - 2aE\left[\frac{\chi^2_{(K-2,\lambda)}}{\chi^2_{(K-2,\lambda)}}\right] + 2a\theta^T\theta E\left[\frac{1}{\chi^2_{(K+2,\lambda)}}\right] + a^2 E\left[\frac{1}{\chi^2_{(K,\lambda)}}\right]$$

Theorem 4.2 *For any real-valued function f and positive definite matrix A,*

$$E\left[f\left(\widehat{\theta}^T\widehat{\theta}\right)\left(\widehat{\theta}^T A\widehat{\theta}\right)\right] = E\left[f\left(\chi^2_{(K+2,\lambda)}\right) tr(A)\right]$$
$$+ E\left[f\left(\chi^2_{(K+4,\lambda)}\right)\right]\left(\theta^T A\theta\right)$$

where $tr(A)$ is trace of A.

Letting $f\left(\widehat{\theta}^T\widehat{\theta}\right) = \frac{1}{\chi^2_{(K-2,\lambda)}}$ and $A = I$ with rank $K - 2$,

$$-2aE\left[\frac{\chi^2_{(K-2,\lambda)}}{\chi^2_{(K-2,\lambda)}}\right] = -2aE\left[\frac{K-2}{\chi^2_{(K,\lambda)}}\right] - 2a\theta^T\theta E\left[\frac{1}{\chi^2_{(K+2,\lambda)}}\right]$$

and

$$\rho(\theta,\theta^*) = K - a\left[2(K-2) - a\right]E\left[\frac{1}{\chi^2_{(K,\lambda)}}\right]$$

Hence, $\rho(\theta,\theta^*) = K - a\left[2(K-2) - a\right]E\left[\frac{1}{\chi^2_{(K,\lambda)}}\right] \le \rho\left(\theta,\widehat{\theta}\right) = K$ for all θ if $0 < a < 2(K-2)$ with strict inequality for some $\theta^T\theta$.

Now, we can find the optimal James-Stein shrinkage estimator. Solving the first order condition

$$\frac{\partial \rho(\theta,\theta^*)}{\partial a} = 0$$

$$(-2(K-2) - a + 2a)E\left[\frac{1}{\chi^2_{(K,\lambda)}}\right] = 0$$

leads to $a^* = K - 2$; hence, $\theta^* = \widehat{\theta}\left(1 - \frac{K-2}{\widehat{\theta}^T\widehat{\theta}}\right)$. As $E\left[\frac{1}{\chi^2_{(K,\lambda)}}\right] = \frac{1}{K-2}$, the James-Stein estimator has minimum expected squared error loss when $\theta = 0$,

$$\rho(\theta,\theta^*) = K - (K-2)^2 E\left[\frac{1}{\chi^2_{(K,\lambda)}}\right]$$
$$= K - (K-2) = 2$$

and its *MSE* approaches that for the *MLE* as $\lambda = \theta^T\theta$ approaches infinity. Next, we sketch proofs of the theorems.

Stein [1966] identified a key idea used in the proofs. Suppose a $J \times 1$ random vector w is distributed as $N(\theta, I)$, then its quadratic form $w^T w$ has a noncentral

$\chi^2_{(J,\lambda)}$ where $\lambda = \theta^T \theta$. This quadratic form can be regarded as having a central $\chi^2_{(J+2H)}$ where H is a Poisson random variable with parameter $\frac{\lambda}{2}$. Hence,

$$E\left[f\left(\chi^2_{(J,\lambda)}\right)\right] = E_H\left[E\left[f\left(\chi^2_{(J+2H)}\right)\right]\right]$$
$$= \sum_{t=0}^{\infty} \left(\frac{\lambda}{2}\right)^t \frac{\exp\left[-\frac{\lambda}{2}\right]}{t!} E\left[f\left(\chi^2_{(J+2t)}\right)\right]$$

Now, we proceed with proofs to the above theorems.

Theorem 4.3 $E\left[\frac{\hat{\theta}}{\hat{\theta}^T\hat{\theta}}\right] = \theta E\left[\frac{1}{\chi^2_{(K+2,\lambda)}}\right]$.

Proof. Write

$$E\left[f\left(w^2\right)w\right] = \frac{1}{\sqrt{2\pi}} \int_{-\infty}^{\infty} f\left(w^2\right) w \exp\left[-\frac{(w-\theta)^2}{2}\right] dw$$
$$= \exp\left[-\frac{\theta^2}{2}\right] \frac{1}{\sqrt{2\pi}} \int_{-\infty}^{\infty} f\left(w^2\right) w \exp\left[-\frac{w^2}{2} + \theta w\right] dw$$

Rewrite as

$$\exp\left[-\frac{\theta^2}{2}\right] \frac{1}{\sqrt{2\pi}} \frac{\partial}{\partial \theta} \int_{-\infty}^{\infty} f\left(w^2\right) \exp\left[-\frac{w^2}{2} + \theta w\right] dw$$

complete the square and write

$$\exp\left[-\frac{\theta^2}{2}\right] \frac{\partial}{\partial \theta} \left\{ E\left[f\left(w^2\right)\right] \exp\left[\frac{\theta^2}{2}\right] \right\}$$

Since $w \sim N(\theta, 1)$, $w^2 \sim \chi^2_{(1,\theta^2)}$. Now, apply Stein's observation

$$\exp\left[-\frac{\theta^2}{2}\right] \frac{\partial}{\partial \theta} \left\{ E\left[f\left(w^2\right)\right] \exp\left[\frac{\theta^2}{2}\right] \right\}$$
$$= \exp\left[-\frac{\theta^2}{2}\right] \frac{\partial}{\partial \theta} \left\{ \exp\left[\frac{\theta^2}{2}\right] \sum_{j=0}^{\infty} \left(\frac{\theta^2}{2}\right)^j \frac{\exp\left[-\frac{\theta^2}{2}\right]}{j!} E\left[f\left(\chi^2_{(1+2j)}\right)\right] \right\}$$
$$= \exp\left[-\frac{\theta^2}{2}\right] \frac{\partial}{\partial \theta} \left\{ \sum_{j=0}^{\infty} \left(\frac{\theta^2}{2}\right)^j \frac{1}{j!} E\left[f\left(\chi^2_{(1+2j)}\right)\right] \right\}$$

Taking the partial derivative yields

$$\exp\left[-\frac{\theta^2}{2}\right] \theta \left\{ \sum_{j=1}^{\infty} \left(\frac{\theta^2}{2}\right)^{j-1} \frac{1}{(j-1)!} E\left[f\left(\chi^2_{(3+2(j-1))}\right)\right] \right\}$$

or
$$E\left[f\left(w^2\right)w\right] = \theta E\left[f\left(\chi^2_{(3,\theta^2)}\right)\right]$$

For the multivariate case at hand, this implies

$$E\left[\frac{\widehat{\theta}}{\widehat{\theta}^T\widehat{\theta}}\right] = \theta E\left[\frac{1}{\chi^2_{(K+2,\lambda)}}\right]$$

∎

Lemma 4.1 $E\left[f\left(w^2\right)w^2\right] = E\left[f\left(\chi^2_{(3,\theta^2)}\right)\right] + \theta^2 E\left[f\left(\chi^2_{(5,\theta^2)}\right)\right].$

Proof. Let $z \sim \chi^2_{(1,\theta^2)}$.

$$\begin{aligned}E\left[f\left(w^2\right)w^2\right] &= E\left[f(z)z\right] \\ &= \sum_{j=0}^{\infty}\left(\frac{\theta^2}{2}\right)^j \frac{1}{j!}\exp\left[-\frac{\theta^2}{2}\right] E\left[f\left(\chi^2_{(1+2j)}\right)\right]\end{aligned}$$

Since

$$E\left[f\left(\chi^2_{(n)}\right)\chi^2_{(n)}\right] = \int_0^{\infty} \frac{f(s) s \exp\left[-\frac{s}{2}\right] s^{\frac{n}{2}-1}}{\Gamma\left(\frac{n}{2}\right) 2^{\frac{n}{2}}} ds$$

combining terms involving powers of s and rewriting

$$\Gamma(t) = \int_0^{\infty} x^{t-1}\exp[-x]\,dx$$

for $t > 0$, as $\Gamma(t=1) = t\Gamma(t)$, leads to

$$\begin{aligned}&n\int_0^{\infty} \frac{f(s)\exp\left[-\frac{s}{2}\right] s^{\frac{n+2}{2}-1}}{\Gamma\left(\frac{n+2}{2}\right) 2^{\frac{n+2}{2}}} ds \\ &= nE\left[f\left(\chi^2_{(n+2)}\right)\right]\end{aligned}$$

Now,

$$\begin{aligned}&\sum_{j=0}^{\infty}\left(\frac{\theta^2}{2}\right)^j \frac{1}{j!}\exp\left[-\frac{\theta^2}{2}\right] E\left[f\left(\chi^2_{(1+2j)}\right)\right] \\ &= \sum_{j=0}^{\infty}\left(\frac{\theta^2}{2}\right)^j \frac{1}{j!}\exp\left[-\frac{\theta^2}{2}\right](1+2j) E\left[f\left(\chi^2_{(3+2j)}\right)\right]\end{aligned}$$

Rewrite as

$$\begin{aligned}&\sum_{j=0}^{\infty}\left(\frac{\theta^2}{2}\right)^j \frac{1}{j!}\exp\left[-\frac{\theta^2}{2}\right] E\left[f\left(\chi^2_{(3+2j)}\right)\right] \\ &+ 2\frac{\theta^2}{2}\sum_{j=0}^{\infty}\left(\frac{\theta^2}{2}\right)^{j-1}\frac{1}{(j-1)!}\exp\left[-\frac{\theta^2}{2}\right] E\left[f\left(\chi^2_{(5+2(j-1))}\right)\right]\end{aligned}$$

Again, apply Stein's observation to produce

$$E\left[f\left(w^2\right)w^2\right] = E\left[f\left(\chi^2_{(3,\theta^2)}\right)\right] + \theta^2 E\left[f\left(\chi^2_{(5,\theta^2)}\right)\right]$$

∎

Theorem 4.4

$$E\left[f\left(\widehat{\theta}^T\widehat{\theta}\right)\left(\widehat{\theta}^T A\widehat{\theta}\right)\right] = E\left[f\left(\chi^2_{(K+2,\lambda)}\right) tr(A)\right]$$
$$+ E\left[f\left(\chi^2_{(K+4,\lambda)}\right)\right]\left(\theta^T A\theta\right)$$

Proof. Let P be an orthogonal matrix such that $PAP^T = D$, a diagonal matrix with eigenvalues of A, $d_j > 0$, along the diagonal. Define vector $\omega = Pw \sim N(P\theta, I)$. Since

$$\omega^T\omega = w^T P^T P w = w^T w$$

and

$$\omega^T D\omega = w^T P^T A P w = w^T A w$$

$$E\left[f\left(\omega^T\omega\right)\omega^T D\omega\right] = \sum_{i=1}^J d_i E\left[E\left[f\left(\omega_i^2 + \sum_{j\neq i}\omega_j^2\right)\omega_i^2 \mid \omega_j, i\neq j\right]\right]$$

Using the lemma, this can be expressed as

$$\sum_{i=1}^J d_i \left\{ \begin{array}{l} E\ f\left(\chi^2_{(3,(p_i^T\theta)^2)} + \sum_{j\neq i}\omega_j^2\right) \\ + \left(p_i^T\theta\right)^2 E\ f\left(\chi^2_{(5,(p_i^T\theta)^2)}\right) + \sum_{j\neq i}\omega_j^2 \end{array} \right\}$$

where p_i^T is the ith row of P. Since $\sum_{i=1}^J d_i \left(p_i^T\theta\right)^2 = \theta^T A\theta$ and $\sum_{i=1}^J d_i = tr(A)$,

$$E\left[f\left(\widehat{\theta}^T\widehat{\theta}\right)\left(\widehat{\theta}^T A\widehat{\theta}\right)\right] = E\left[f\left(\chi^2_{(K+2,\lambda)}\right) tr(A)\right]$$
$$+ E\left[f\left(\chi^2_{(K+4,\lambda)}\right)\right]\left(\theta^T A\theta\right)$$

∎

4.5 Summary

This chapter has briefly reviewed loss functions, nonlinear regression, maximum likelihood estimation, and some alternative estimation methods (including James-Stein shrinkage estimators). It is instructive to revisit nonlinear regression (especially, *GNR*) in the next chapter when we address specification and estimation of discrete choice models.

4.6 Additional reading

Poirier [1995] provides a nice discussion of loss functions. Conditional linear loss functions lead to quantile regression (see Koenker and Bassett [1978], Koenker [2005], and Koenker [2009] for an **R** computational package). Shugan and Mitra [2008] offer an intriguing discussion of when and why non-averaging statistics (e.g., maximum and variance) explain more variance than averaging metrics. Maximum likelihood estimation is discussed by a broad range of authors including Davidson and MacKinnon [1993], Greene [1997], Amemiya [1985], Rao [1973], and Theil [1971]. Stigler [2007] provides a fascinating account of the history of maximum likelihood estimation including the pioneering contributions of Gauss and Fisher as well as their detractors. The nonlinear regression section draws heavily from a favorite reference, Davidson and MacKinnon [2003]. Their chapter 6 and references therein provide a wealth of ideas related to estimation and specification of nonlinear models.

5
Discrete choice models

Choice models attempt to analyze decision marker's preferences amongst alternatives. We'll primarily address the binary case to simplify the illustrations though in principle any number of discrete choices can be analyzed. A key is choices are mutually exclusive and exhaustive. This framing exercise impacts the interpretation of the data.

5.1 Latent utility index models

Maximization of expected utility representation implies that two choices a and b involve comparison of expected utilities such that $U_a > U_b$ (the reverse, or the decision maker is indifferent). However, the analyst typically cannot observe all attributes that affect preferences. The functional representation of observable attributes affecting preferences, Z_i, is often called representative utility. Typically $U_i \neq Z_i$ and Z_i is linear in the parameters $X_i \beta$,[1]

$$U_a = Z_a + \varepsilon_a = X_a \beta + \varepsilon_a$$

$$U_b = Z_b + \varepsilon_b = X_b \beta + \varepsilon_b$$

---

[1] Discrete response models are of the form

$$P_i \equiv E[Y_i | \Omega_i] = F(h(X_i, \beta))$$

This is a general specification. $F(X_i \beta)$ is more common. The key is to employ a transformation (link) function $F(X)$ that has the properties

$F(-\infty) = 0$, $F(\infty) = 1$, and $f(X) \equiv \frac{\partial F(X)}{\partial X} > 0$.

where X is observable attributes, characteristics of decision maker, etc. and ε represents unobservable (to the analyst) features. Since utility is ordinal, addition of a constant to all Z or scaling by $\lambda > 0$ has no substantive impact and is not identified for probability models.

Consequently, β_0 for either a or b is fixed (at zero) and the scale is chosen ($\sigma = 1$, for probit). Hence, the estimated parameters are effectively β/σ and reflect the differences in contribution to preference $X_a - X_b$. Even the error distribution is based on differences $\varepsilon = \varepsilon_a - \varepsilon_b$ (the difference in preferences related to the unobservable attributes). Of course, this is why this is a probability model as some attributes are not observed by the analyst. Hence, only probabilistic statements of choice can be offered.

$$\begin{aligned} & Pr\left(U_a > U_b\right) \\ = & Pr\left(Z_a + \varepsilon_a > Z_b + \varepsilon_b\right) \\ = & Pr\left(X_a \beta + \varepsilon_a > X_b \beta + \varepsilon_b\right) \\ = & Pr\left(\varepsilon_b - \varepsilon_a < Z_a - Z_b\right) \\ = & F_\varepsilon\left(X\beta\right) \end{aligned}$$

where $\varepsilon = \varepsilon_b - \varepsilon_a$ and $X = X_a - X_b$. This reinforces why sometimes the latent utility index model is written $Y^* = U_a - U_b = X\beta - V$, where $V = \varepsilon_a - \varepsilon_b$.

We're often interested in the effect of a regressor on choice. Since the model is a probability model this translates into the *marginal probability effect*. The marginal probability effect is $\frac{\partial F(X\beta)}{\partial x} = f(X\beta)\beta$ where x is a row of X. Hence, the marginal effect is proportional (not equal) to the parameter with changing proportionality over the sample. This is often summarized for the population by reference to the sample mean or some other population level reference (see Greene [1997], ch. 21 for more details).

5.2 Linear probability models

Linear probability models

$$Y = X\beta + \varepsilon$$

where $Y \in \{0, 1\}$ (in the binary case), are not really probability models as the predicted values are not bounded between 0 and 1. But they are sometimes employed for exploratory analysis or to identify relative starting values for *MLE*.

5.3 Logit (logistic regression) models

As the first demonstrated random utility model (*RUM*) — consistent with expected utility maximizing behavior (Marschak [1960]) — logit is the most popular discrete choice model. Standard logit models assume independence of irrelevant alternatives (*IIA*) (Luce [1959]) which can simplify experimentation but can also be unduly restrictive and produce perverse interpretations.[2] We'll explore the emergence of this property when we discuss multinomial and conditional logit.

A variety of closely related logit models are employed in practice. Interpretation of these models can be subtle and holds the key to their distinctions. A few popular variations are discussed below.

5.3.1 Binary logit

The logit model employs the latent utility index model where ε is extreme value distributed.

$$Y^* = U_a - U_b = X\beta + \varepsilon$$

where $X = X_a - X_b, \varepsilon = \varepsilon_b - \varepsilon_a$. The logit model can be derived by assuming the log of the odds ratio equals an index function $X_t\beta$. That is, $log\frac{P_t}{1-P_t} = X_t\beta$ where $P_t = E[Y_t|\Omega_t] = F(X_t\beta)$ and Ω_t is the information set available at t.

First, the logistic (cumulative distribution) function is

$$\begin{aligned} \Lambda(X) &\equiv \left(1 + e^{-X}\right)^{-1} \\ &= \frac{e^X}{1 + e^X} \end{aligned}$$

It has first derivative or density function

$$\begin{aligned} \lambda(X) &\equiv \frac{e^X}{(1+e^X)^2} \\ &= \Lambda(X)\Lambda(-X) \end{aligned}$$

Solving the log-odds ratio for P_t yields

$$\begin{aligned} \frac{P_t}{1 - P_t} &= exp(X_t\beta) \\ P_t &= \frac{exp(X_t\beta)}{1 + exp(X_t\beta)} \\ &= [1 + exp(-X_t\beta)]^{-1} \\ &= \Lambda(X_t\beta) \end{aligned}$$

Notice if the regressors are all binary the log-odds ratio provides a particularly straightforward and simple method for estimating the parameters. This also points out a difficulty with estimation that sometimes occurs if we encounter a perfect classifier. For a perfect classifier, there is some range of the regressor(s) for which Y_t is always 1 or 0 (a separating hyperplane exists). Since β is not identifiable (over a compact parameter space), we cannot obtain sensible estimates for β as any sensible optimization approach will try to choose β arbitrarily large or small.

---

[2]The connection between discrete choice models and *RUM* is reviewed in McFadden [1981,2001].

5.3.2 Multinomial logit

Multinomial logit is a natural extension of binary logit designed to handle $J+1$ alternatives ($J = 1$ produces binary logit).[3] The probability of observing alternative k is

$$\Pr(Y_t = k) = P_{kt} = \frac{exp\left(X_t\beta^k\right)}{\sum_{j=0}^{J} exp\left(X_t\beta^j\right)}$$

with parameter vector $\beta^0 = 0$.[4] Notice the vector of regressors remains constant but additional parameter vectors are added for each additional alternative. It may be useful to think of the regressors as individual-specific characteristics rather than attributes of specific-alternatives. This is a key difference between multinomial and conditional logit (see the discussion below). For multinomial logit, we have

$$\frac{P_{kt}}{P_{lt}} = \frac{exp\left(X_t\beta^k\right)}{exp\left(X_t\beta^l\right)} = exp\left(X_t\left(\beta^k - \beta^l\right)\right)$$

That is, the odds of two alternatives depend on the regressors and the difference in their parameter vectors. Notice the odds ratio does not depend on other alternatives; hence *IIA* applies.

5.3.3 Conditional logit

The conditional logit model deals with J alternatives where utility for alternative k is

$$Y_k^* = X_k\beta + \varepsilon_k$$

where ε_k is *iid* Gumbel distributed. The Gumbel distribution has density function $f(\varepsilon) = \exp(-\varepsilon)\exp(-e^{-\varepsilon})$ and distribution function $\exp(-e^{-\varepsilon})$. The probability that alternative k is chosen is the $Pr(U_k > U_j)$ for $j \neq k$ which is[5]

$$\Pr(Y_t = k) = P_{kt} = \frac{exp\left(X_{kt}\beta\right)}{\sum_{j=1}^{J} exp\left(X_{jt}\beta\right)}$$

---

[3] Multinomial choice models can represent unordered or ordered choices. For simplicity, we focus on the unordered variety.

[4] Hence, $P_{0t} = \dfrac{1}{1+\sum_{j=1}^{J} exp(x_t\beta^j)}$.

[5] See Train [2003, section 3.10] for a derivation. The key is to rewrite $Pr(U_k > U_j)$ as $Pr(\varepsilon_j < \varepsilon_k + V_k - V_j)$. Then recall that

$$Pr(\varepsilon_j < \varepsilon_k + V_k - V_j) = \int Pr(\varepsilon_j < \varepsilon_k + V_k - V_j \mid \varepsilon_k) f(\varepsilon_k) d(\varepsilon_k)$$

Notice a vector of regressors is associated with each alternative (J vectors of regressors) and one parameter vector for the model. It may be useful to think of the conditional logit regressors as attributes associated with specific-alternatives. The odds ratio for conditional logit

$$\frac{P_{kt}}{P_{lt}} = \frac{exp(X_{kt}\beta)}{exp(X_{lt}\beta)} = exp(X_{kt} - X_{lt})\beta$$

is again independent of other alternatives; hence *IIA*. *IIA* arises as a result of probability assignment to the unobservable component of utility. Next, we explore a variety of models that relax these restrictions in some manner.

5.3.4 GEV (generalized extreme value) models

GEV models are extreme valued choice models that seek to relax the *IIA* assumption of conditional logit. McFadden [1978] developed a process to generate *GEV* models.[6] This process allows researchers to develop new *GEV* models that best fit the choice situation at hand.

Let $Y_j \equiv exp(Z_j)$, where Z_j is the observable part of utility associated with choice j. $G = G(Y_j, \ldots, Y_J)$ is a function that depends on Y_j for all j. If G satisfies the properties below, then

$$P_i = \frac{Y_i G_i}{G} = \frac{exp(Z_j) \frac{\partial G}{\partial Y_i}}{G}$$

where $G_i = \frac{\partial G}{\partial Y_i}$.

Condition 5.1 $G \geq 0$ *for all positive values of* Y_j.

Condition 5.2 G *is homogeneous of degree one.*[7]

Condition 5.3 $G \to \infty$ *as* $Y_j \to \infty$ *for any* j.

Condition 5.4 *The cross partial derivatives alternate in signs as follows:* $G_i = \frac{\partial G}{\partial Y_i} \geq 0$, $G_{ij} = \frac{\partial^2 G}{\partial Y_i \partial Y_j} \leq 0$, $G_{ijk} = \frac{\partial^3 G}{\partial Y_i \partial Y_j \partial Y_k} \geq 0$ *for i, j, and k distinct, and so on.*

These conditions are not economically intuitive but it's straightforward to connect the ideas to some standard logit and *GEV* models as depicted in table 5.1 (person n is suppressed in the probability descriptions).

5.3.5 Nested logit models

Nested logit models relax *IIA* in a particular way. Suppose a decision maker faces a set of alternatives that can be partitioned into subsets or *nests* such that

---

[6] See Train [2003] section 4.6.
[7] $G(\rho Y_1, \ldots \rho Y_J) = \rho G(Y_1, \ldots Y_J)$. Ben-Akiva and Francois [1983] show this condition can be relaxed. For simplicity, it's maintained for purposes of the present discussion.

5. Discrete choice models

Table 5.1: Variations of multinomial logits

model	G	P_i
ML	$\sum_{j=0}^{J} Y_j,\ Y_0 = 1$	$\dfrac{\exp(Z_i)}{\sum_{j=0}^{J} \exp(Z_j)}$
CL	$\sum_{j=1}^{J} Y_j$	$\dfrac{\exp(Z_i)}{\sum_{j=1}^{J} \exp(Z_j)}$
NL	$\sum_{k=1}^{K}\left(\sum_{j \in B_k} Y_j^{1/\lambda_k}\right)^{\lambda_k}$ where $0 \leq \lambda_k \leq 1$	$\dfrac{\exp(Z_i/\lambda_k)\left(\sum_{j \in B_k} \exp[Z_j/\lambda_k]\right)^{\lambda_k - 1}}{\sum_{\ell=1}^{K}\left(\sum_{j \in B_\ell} \exp[Z_j/\lambda_\ell]\right)^{\lambda_\ell}}$
GNL	$\sum_{k=1}^{K}\left(\sum_{j \in B_k}(\alpha_{jk} Y_j)^{1/\lambda_k}\right)^{\lambda_k}$ $\alpha_{jk} \geq 0$ and $\sum_k \alpha_{jk} = 1\ \forall j$	$\dfrac{\sum_k (\alpha_{ik} e^{Z_i})^{1/\lambda_k}\left(\sum_{j \in B_k}(\alpha_{jk} e^{Z_j})^{1/\lambda_k}\right)^{\lambda_k - 1}}{\sum_{\ell=1}^{K}\left(\sum_{j \in B_\ell}(\alpha_{j\ell} \exp[Z_j])^{1/\lambda_\ell}\right)^{\lambda_\ell}}$
	$Z_j = X\beta_j$ for multinomial logit ML but $Z_j = X_j\beta$ for conditional logit CL NL refers to nested logit with J alternatives in K nests $B_1,\ldots,B_K$ GNL refers to generalized nested logit	

Condition 5.5 *IIA holds within each nest. That is, the ratio of probabilities for any two alternatives in the same nest is independent of other alternatives.*

Condition 5.6 *IIA does not hold for alternatives in different nests. That is, the ratio of probabilities for any two alternatives in different nests can depend on attributes of other alternatives.*

The nested logit probability can be decomposed into a marginal and conditional probability

$$P_i = P_{i|B_k} P_{B_k}$$

where the conditional probability of choosing alternative i given that an alternative in nest B_k is chosen is

$$P_{i|B_k} = \frac{\exp\left(\dfrac{Z_i}{\lambda_k}\right)}{\sum_{j \in B_k} \exp\left(\dfrac{Z_j}{\lambda_k}\right)}$$

and the marginal probability of choosing an alternative in nest B_k is

$$P_{B_k} = \frac{\left(\sum_{j \in B_k} \exp\left[\frac{Z_j}{\lambda_k}\right]\right)^{\lambda_k}}{\sum_{\ell=1}^{K}\left(\sum_{j \in B_\ell} \exp\left[\frac{Z_j}{\lambda_\ell}\right]\right)^{\lambda_\ell}}$$

which can be rewritten (see Train [2003, ch. 4, p. 90])

$$= \frac{e^{W_k}\left(\sum_{j \in B_k} \exp\left[\frac{Z_j}{\lambda_k}\right]\right)^{\lambda_k}}{\sum_{\ell=1}^{K} e^{W_\ell}\left(\sum_{j \in B_\ell} \exp\left[\frac{Z_j}{\lambda_\ell}\right]\right)^{\lambda_\ell}}$$

where observable utility is decomposed as $V_{nj} = W_{nk} + Z_{nj}$, W_{nk} depends only on variables that describe nest k (variation over nests but not over alternatives within each nest) and Z_{nj} depends on variables that describe alternative j (variation over alternatives within nests) for individual n.

The parameter λ_k indicates the degree of independence in unobserved utility among alternatives in nest B_k. The level of independence or correlation can vary across nests. If $\lambda_k = 1$ for all nests, there is independence among all alternatives in all nests and the nested logit reduces to a standard logit model. An example seems appropriate.

Example 5.1 *For simplicity we consider two nests $k \in \{A, B\}$ and two alternatives within each nest $j \in \{a, b\}$ and a single variable to differentiate each of the various choices.*[8] *The latent utility index is*

$$U_{kj} = W_k \beta_1 + X_{kj} \beta_2 + \varepsilon$$

where $\varepsilon = \sqrt{\rho_k}\eta_k + \sqrt{1-\rho_k}\eta_j$, $k \neq j$, and η has a Gumbel (type I extreme value) distribution. This implies the lead term captures dependence within a nest. Samples of $1,000$ observations are drawn with parameters $\beta_1 = 1$, $\beta_2 = -1$, and $\rho_A = \rho_B = 0.5$ for $1,000$ simulations with W_k and X_{kj} drawn from independent uniform$(0, 1)$ distributions. Observables are defined as $Y_{ka} = 1$ if $U_{ka} > U_{kb}$ and 0 otherwise, and $Y_A = 1$ if $max\{U_{Aa}, U_{Ab}\} > max\{U_{Ba}, U_{Bb}\}$ and 0 otherwise. Now, the log-likelihood can be written as

$$\begin{aligned} L &= \sum Y_A Y_{Aa} \log(P_{Aa}) + Y_A(1 - Y_{Aa})\log(P_{Ab}) \\ &+ Y_B Y_{Ba} \log(P_{Ba}) + Y_B Y_{Ba} \log(P_{Bb}) \end{aligned}$$

where P_{kj} is as defined the table above. Results are reported in tables 5.2 and 5.3.

---

[8]There is no intercept in this model as the intercept is unidentified.

The parameters are effectively recovered with nested logit. However, for conditional logit recovery is poorer. Suppose we vary the correlation in the within nest errors (unobserved components). Tables 5.4 and 5.5 report comparative results with low correlation ($\rho_A = \rho_B = 0.01$) within the unobservable portions of the nests. As expected, conditional logit performs well in this setting.
Suppose we try high correlation ($\rho_A = \rho_B = 0.99$) within the unobservable portions of the nests. Table 5.6 reports nested logit results. As indicated in table 5.7 conditional logit performs poorly in this setting as the proportional relation between the parameters is substantially distorted.

5.3.6 Generalizations

Generalized nested logit models involve nests of alternatives where each alternative can be a member of more than one nest. Their membership is determined by an allocation parameter α_{jk} which is non-negative and sums to one over the nests for any alternative. The degree of independence among alternatives is determined, as in nested logit, by parameter λ_k. Higher λ_k means greater independence and less correlation. Interpretation of *GNL* models is facilitated by decomposition of the probability.

$$P_i = P_{i|B_k} P_k$$

where P_k is the marginal probability of nest k

$$\frac{\sum_{j \in B_k} (\alpha_{jk} \exp[Z_j])^{\frac{1}{\lambda_k}}}{\sum_{\ell=1}^{K} \left(\sum_{j \in B_\ell} (\alpha_{j\ell} \exp[Z_j])^{\frac{1}{\lambda_\ell}} \right)^{\lambda_\ell}}$$

and $P_{i|B_k}$ is the conditional probability of alternative i given nest k

$$\frac{(\alpha_{jk} \exp[Z_j])^{\frac{1}{\lambda_k}}}{\sum_{j \in B_\ell} (\alpha_{jk} \exp[Z_j])^{\frac{1}{\lambda_k}}}$$

Table 5.2: Nested logit with moderate correlation

	$\hat{\beta}_1$	$\hat{\beta}_2$	$\hat{\lambda}_A$	$\hat{\lambda}_B$
mean	0.952	−0.949	0.683	0.677
std. dev.	0.166	0.220	0.182	0.185
(.01, .99) quantiles	(0.58, 1.33)	(−1.47, −0.46)	(0.29, 1.13)	(0.30, 1.17)
		$\beta_1 = 1, \beta_2 = -1, \rho_A = \rho_B = 0.5$		

5.3 Logit (logistic regression) models

Table 5.3: Conditional logit with moderate correlation

	$\hat{\beta}_1$	$\hat{\beta}_2$
mean	0.964	−1.253
std. dev.	0.168	0.131
(.01, .99) quantiles	(0.58, 1.36)	(−1.57, −0.94)
$\beta_1 = 1, \beta_2 = -1, \rho_A = \rho_B = 0.5$		

Table 5.4: Nested logit with low correlation

	$\hat{\beta}_1$	$\hat{\beta}_2$	$\hat{\lambda}_A$	$\hat{\lambda}_B$
mean	0.994	−0.995	1.015	1.014
std. dev.	0.167	0.236	0.193	0.195
(.01, .99) quantiles	(0.56, 1.33)	(−1.52, −0.40)	(0.27, 1.16)	(0.27, 1.13)
$\beta_1 = 1, \beta_2 = -1, \rho_A = \rho_B = 0.01$				

Table 5.5: Conditional logit with low correlation

	$\hat{\beta}_1$	$\hat{\beta}_2$
mean	0.993	−1.004
std. dev.	0.168	0.132
(.01, .99) quantiles	(0.58, 1.40)	(−1.30, −0.72)
$\beta_1 = 1, \beta_2 = -1, \rho_A = \rho_B = 0.01$		

Table 5.6: Nested logit with high correlation

	$\hat{\beta}_1$	$\hat{\beta}_2$	$\hat{\lambda}_A$	$\hat{\lambda}_B$
mean	0.998	−1.006	0.100	0.101
std. dev.	0.167	0.206	0.023	0.023
(.01, .99) quantiles	(0.62, 1.40)	(−1.51, −0.54)	(0.05, 0.16)	(0.05, 0.16)
$\beta_1 = 1, \beta_2 = -1, \rho_A = \rho_B = 0.99$				

Table 5.7: Conditional logit with high correlation

	$\hat{\beta}_1$	$\hat{\beta}_2$
mean	1.210	−3.582
std. dev.	0.212	0.172
(.01, .99) quantiles	(0.73, 1.71)	(−4.00, −3.21)
$\beta_1 = 1, \beta_2 = -1, \rho_A = \rho_B = 0.99$		

5.4 Probit models

Probit models involve weaker restrictions from a utility interpretation perspective (no *IIA* conditions) than logit. Probit models assume the same sort of latent utility index form except that V is assigned a normal or Gaussian probability distribution. Some circumstances might argue that normality is an unduly restrictive or logically inconsistent mapping of unobservables into preferences.

A derivation of the latent utility probability model is as follows.

$$\begin{aligned} Pr\left(Y_t=1\right) &= Pr\left(Y_t^* > 0\right) \\ &= Pr\left(X_t\beta + V_t > 0\right) \\ &= 1 - Pr\left(V_t \leq -X_t\beta\right) \end{aligned}$$

For symmetric distributions, like the normal (and logistic), $F(-X) = 1 - F(X)$ where $F(\cdot)$ refers to the cumulative distribution function. Hence,

$$\begin{aligned} Pr\left(Y_t=1\right) &= 1 - Pr\left(V_t \leq -X_t\beta\right) \\ &= 1 - F\left(-X_t\beta\right) = F\left(X_t\beta\right) \end{aligned}$$

Briefly, first order conditions associated with maximization of the log-likelihood (L) for the binary case are

$$\frac{\partial L}{\partial \beta_j} = \sum_{t=1}^{n} Y_t \frac{\phi\left(X_t\beta\right)}{\Phi\left(X_t\beta\right)} X_{jt} + (1-Y_t) \frac{-\phi\left(X_t\beta\right)}{1-\Phi\left(X_t\beta\right)} X_{jt}$$

where $\phi(\cdot)$ and $\Phi(\cdot)$ refer to the standard normal density and cumulative distribution functions, respectively, and scale is normalized to unity.[9] Also, the marginal probability effects associated with the regressors are

$$\frac{\partial p_t}{\partial X_{jt}} = \phi\left(X_t\beta\right)\beta_j$$

5.4.1 Conditionally-heteroskedastic probit

Discrete choice model specification may be sensitive to changes in variance of the unobservable component of expected utility (see Horowitz [1991] and Greene [1997]). Even though choice models are normalized as scale cannot be identified, parameter estimates (and marginal probability effects) can be sensitive to changes in the variability of the stochastic component as a function of the level of regressors. In other words, parameter estimates (and marginal probability effects) can be sensitive to conditional-heteroskedasticity. Hence, it may be useful to consider a model specification check for conditional-heteroskedasticity. Davidson and

---

[9] See chapter 4 for a more detailed discussion of maximum likelihood estimation.

MacKinnon [1993] suggest a standard (restricted vs. unrestricted) likelihood ratio test where the restricted model assumes homoskedasticity and the unrestricted assumes conditional-heteroskedasticity.

Suppose we relax the latent utility specification of a standard probit model by allowing conditional heteroskedasticity.

$$Y^* = Z + \varepsilon$$

where $\varepsilon \sim N(0, \exp(2W\gamma))$. In a probit frame, the model involves rescaling the index function by the conditional standard deviation

$$p_t = \Pr(Y_t = 1) = \Phi\left(\frac{X_t\beta}{\exp[W_t\gamma]}\right)$$

where the conditional standard deviation is given by $\exp[W_t\gamma]$ and W refers to some rank q subset of the regressors X (for notational convenience, subscripts are matched so that $X_j = W_j$) and of course, cannot include an intercept (recall the scale or variance is not identifiable in discrete choice models).[10]

Estimation and identification of marginal probability effects of regressors proceed as usual but the expressions are more complex and convergence of the likelihood function is more delicate. Briefly, first order conditions associated with maximization of the log-likelihood (L) for the binary case are

$$\frac{\partial L}{\partial \beta_j} = \sum_{t=1}^{n} Y_t \frac{\phi\left(\frac{X_t\beta}{\exp[W_t\gamma]}\right)}{\Phi\left(\frac{X_t\beta}{\exp[W_t\gamma]}\right)} \frac{X_{jt}}{\exp[W_t\gamma]} + (1 - Y_t) \frac{-\phi\left(\frac{X_t\beta}{\exp[W_t\gamma]}\right)}{1 - \Phi\left(\frac{X_t\beta}{\exp[W_t\gamma]}\right)} \frac{X_{jt}}{\exp[W_t\gamma]}$$

where $\phi(\cdot)$ and $\Phi(\cdot)$ refer to the standard normal density and cumulative distribution functions, respectively. Also, the marginal probability effects are[11]

$$\frac{\partial p_t}{\partial W_{jt}} = \phi\left(\frac{X_t\beta}{\exp[W_t\gamma]}\right)\left(\frac{\beta_j - X_t\beta\gamma_j}{\exp[W_t\gamma]}\right)$$

5.4.2 Artificial regression specification test

Davidson and MacKinnon [2003] suggest a simple specification test for conditional-heteroskedasticity. As this is not restricted to a probit model, we'll explore a general link function $F(\cdot)$. In particular, a test of $\gamma = 0$, implying homoskedasticity

---

[10] Discrete choice models are inherently conditionally-heteroskedastic as a function of the regressors (MacKinnon and Davidson [1993]). Consider the binary case, the binomial setup produces variance equal to $p_j(1-p_j)$ where p_j is a function of the regressors X. Hence, the heteroskedastic probit model enriches the error (unobserved utility component).

[11] Because of the second term, the marginal effects are not proportional to the parameter estimates as in the standard discrete choice model. Rather, the sign of the marginal effect may be opposite that of the parameter estimate. Of course, if heteroskedasticity is a function of some variable not included as a regressor the marginal effects are simpler

$$\frac{\partial p_t}{\partial W_{jt}} = \phi\left(\frac{X_t\beta}{\exp[W_t\gamma]}\right)\left(\frac{\beta_j}{\exp[W_t\gamma]}\right)$$

5. Discrete choice models

with $exp\left[W_t\gamma\right] = 1$, in the following artificial regression (see chapter 4 to review artificial regression)

$$\tilde{V}_t^{-\frac{1}{2}}\left(Y_t - \tilde{F}_t\right) = \tilde{V}_t^{-\frac{1}{2}}\tilde{f}_t X_t b - \tilde{V}_t^{-\frac{1}{2}}\tilde{f}_t X_t\tilde{\beta}W_t c + residual$$

where $\tilde{V}_t = \tilde{F}_t\left(1 - \tilde{F}_t\right)$, $\tilde{F}_t = F\left(X_t\tilde{\beta}\right)$, $\tilde{f}_t = f\left(X_t\tilde{\beta}\right)$, and $\tilde{\beta}$ is the *ML* estimate under the hypothesis that $\gamma = 0$. The test for heteroskedasticity is based on the explained sum of squares (*ESS*) for the above *BRMR* which is asymptotically distributed $\chi^2(q)$ under $\gamma = 0$. That is, under the null, neither term offers any explanatory power.

Let's explore the foundations of this test. The nonlinear regression for a discrete choice model is

$$Y_t = F\left(X_t\beta\right) + v_t$$

where v_t have zero mean, by construction, and variance

$$E\left[v_t^2\right] = E\left[\left(Y_t - F\left(X_t\beta\right)\right)^2\right]$$

$$= F\left(X_t\beta\right)\left(1 - F\left(X_t\beta\right)\right)^2 + \left(1 - F\left(X_t\beta\right)\right)\left(0 - F\left(X_t\beta\right)\right)^2$$

$$= F\left(X_t\beta\right)\left(1 - F\left(X_t\beta\right)\right)$$

The simplicity, of course, is due to the binary nature of Y_t. Hence, even though the latent utility index representation here is homoskedastic, the nonlinear regression is heteroskedastic.[12]

The Gauss-Newton regression (*GNR*) that corresponds to the above nonlinear regression is

$$Y_t - F\left(X_t\beta\right) = f\left(X_t\beta\right)X_t b + residual$$

as the estimate corresponds to the updating term, $-H_{(j-1)}g_{(j-1)}$, in Newton's method (see chapter 4)

$$\hat{b} = \left(X^T\mathbf{f}^2\left(X\beta\right)X\right)^{-1}X^T\mathbf{f}\left(X\beta\right)\left(y - F\left(X\beta\right)\right)$$

where $\mathbf{f}\left(X\beta\right)$ is a diagonal matrix. The artificial regression for binary response models (*BRMR*) is the above *GNR* after accounting for heteroskedasticity noted above. That is,

$$V_t^{-\frac{1}{2}}\left(Y_t - F\left(X_t\beta\right)\right) = V_t^{-\frac{1}{2}}f\left(X_t\beta\right)X_t b + residual$$

where $V_t = F\left(X_t\beta\right)\left(1 - F\left(X_t\beta\right)\right)$. The artificial regression used for specification testing reduces to this *BRMR* when $\gamma = 0$ and therefore $c = 0$.

---

[12] The nonlinear regression for the binary choice problem could be estimated via iteratively reweighted nonlinear least squares using Newton's method (see chapter 4). Below we explore an alternative approach that is usually simpler and computationally faster.

The artificial regression used for specification testing follows simply from using the development in chapter 4 which says asymptotically the change in estimate via Newton's method is

$$-H_{(j)}g_{(j)} \stackrel{a}{=} \left(X_{(j)}^T X_{(j)}\right)^{-1} X_{(j)}^T \left(y - x_{(j)}\right)$$

Now, for the binary response model recall $x_{(j)} = F\left(\frac{X\beta_{(j)}}{\exp(W\gamma)}\right)$ and, after conveniently partitioning, the matrix of partial derivatives

$$X_{(j)}^T = \left[\begin{array}{cc} \frac{\partial F\left(\frac{X\beta_{(j)}}{\exp(W\gamma)}\right)}{\partial \beta_{(j)}} & \frac{F\left(\frac{X\beta_{(j)}}{\exp(W\gamma)}\right)}{\partial \gamma} \end{array} \right]$$

Under the hypothesis $\gamma = 0$,

$$X_{(j)}^T = \left[\begin{array}{cc} f\left(X\beta_{(j)}\right) X & -f\left(X\beta_{(j)}\right) X\beta_{(j)} Z \end{array} \right]$$

Replace $\beta_{(j)}$ by the *ML* estimate for β and recognize a simple way to compute

$$\left(X_{(j)}^T X_{(j)}\right)^{-1} X_{(j)}^T \left(Y - x_{(j)}\right)$$

is via artificial regression. After rescaling for heteroskedasticity, we have the artificial regression used for specification testing

$$\tilde{V}_t^{-\frac{1}{2}} \left(Y_t - \tilde{F}_t\right) = \tilde{V}_t^{-\frac{1}{2}} \tilde{f}_t X_t b - \tilde{V}_t^{-\frac{1}{2}} \tilde{f}_t X_t \tilde{\beta} W_t c + residual$$

See Davidson and MacKinnon [2003, ch. 11] for additional discrete choice model specification tests. It's time to explore an example.

Example 5.2 *Suppose the DGP is heteroskedastic*

$$Y^* = X_1 - X_2 + \varepsilon, \ \varepsilon \sim N\left(0, (\exp(X_1))^2\right)$$ [13]

Y^* is unobservable but $Y = 1$ if $Y^* > 0$ and $Y = 0$ otherwise is observed. Further, x_1 and x_2 are both uniformly distributed between $(0, 1)$ and the sample size $n = 1,000$. First, we report standard (assumed homoskedastic) binary probit results based on $1,000$ simulations in table 5.8. Though the parameter estimates remain proportional, they are biased towards zero.

Let's explore some variations of the above BRMR specification test. Hence, as reported in table 5.9, the appropriately specified heteroskedastic test has reasonable power. Better than 50% of the simulations produce evidence of misspecification at the 80% confidence level.[14]

Now, leave the DGP unaltered but suppose we suspect the variance changes due to another variable, say X_2. Misidentification of the source of heteroskedasticity

---

[13] To be clear, the second parameter is the variance so that the standard deviation is $\exp(X_1)$.
[14] As this is a specification test, a conservative approach is to consider a lower level (say, 80% vs. 95%) for our confidence intervals.

Table 5.8: Homoskedastic probit results with heteroskedastic DGP

	$\hat{\beta}_0$	$\hat{\beta}_1$	$\hat{\beta}_2$
mean	-0.055	0.647	-0.634
std. dev.	0.103	0.139	0.140
$(0.01, 0.99)$ quantiles	$(-0.29, 0.19)$	$(0.33, 0.98)$	$(-0.95, -0.31)$

Table 5.9: BRMR specification test 1 with heteroskedastic DGP

	b_0	b_1	b_2
mean	0.045	0.261	-0.300
std. dev.	0.050	0.165	0.179
$(0.01, 0.99)$ quantiles	$(-0.05, 0.18)$	$(-0.08, 0.69)$	$(-0.75, 0.08)$
	c_1	ESS ($\chi^2(1) prob$)	
mean	0.972	3.721 (0.946)	
std. dev.	0.592	3.467	
$(0.01, 0.99)$ quantiles	$(-0.35, 2.53)$	$(0.0, 14.8)$	

$$\tilde{V}_t^{-\frac{1}{2}}\left(Y_t - \tilde{F}_t\right) = \tilde{V}_t^{-\frac{1}{2}}\tilde{f}_t X_t b - \tilde{V}_t^{-\frac{1}{2}}\tilde{f}_t X_t \tilde{\beta} X_{1t} c_1 + residual$$

substantially reduces the power of the test as depicted in table 5.10. Only between 25% and 50% of the simulations produce evidence of misspecification at the 80% confidence level.

Next, we explore changing variance as a function of both regressors. As demonstrated in table 5.11, the power is comprised relative to the proper specification. Although better than 50% of the simulations produce evidence of misspecification at the 80% confidence level. However, one might be inclined to drop $X_2 c_2$ and re-estimate.

Assuming the evidence is against homoskedasticity, we next report simulations for the heteroskedastic probit with standard deviation $exp(X_1\gamma)$. As reported in table 5.12, on average, ML recovers the parameters of a properly-specified heteroskedastic probit quite effectively. Of course, we may incorrectly conclude the data are homoskedastic.

Now, we explore the extent to which the specification test is inclined to indicate heteroskedasticity when the model is homoskedastic. Everything remains the same except the DGP is homoskedastic

$$Y^* = X_1 - X_2 + \varepsilon, \; \varepsilon \sim N(0,1)$$

As before, we first report standard (assumed homoskedastic) binary probit results based on $1,000$ simulations in table 5.13. On average, the parameter estimates are recovered effectively.

Table 5.10: BRMR specification test 2 with heteroskedastic DGP

	b_0	b_1	b_2
mean	0.029	−0.161	0.177
std. dev.	0.044	0.186	0.212
$(0.01, 0.99)$ quantiles	$(-0.05, 0.16)$	$(-0.65, 0.23)$	$(-0.32, 0.73)$
	c_2	ESS $(\chi^2(1) prob)$	
mean	−0.502	1.757 (0.815)	
std. dev.	0.587	2.334	
$(0.01, 0.99)$ quantiles	$(-1.86, 0.85)$	$(0.0, 10.8)$	
$\tilde{V}_t^{-\frac{1}{2}} \left(y_t - \tilde{F}_t \right) = \tilde{V}_t^{-\frac{1}{2}} \tilde{f}_t X_t b - \tilde{V}_t^{-\frac{1}{2}} \tilde{f}_t X_t \tilde{\beta} X_{2t} c_2 + residual$			

Table 5.11: BRMR specification test 3 with heteroskedastic DGP

	b_0	b_1	b_2
mean	0.050	0.255	−0.306
std. dev.	0.064	0.347	0.416
$(0.01, 0.99)$ quantiles	$(-0.11, 0.23)$	$(-0.56, 1.04)$	$(-1.33, 0.71)$
	c_1	c_2	ESS $(\chi^2(2) prob)$
mean	0.973	0.001	4.754 (0.907)
std. dev.	0.703	0.696	3.780
$(0.01, 0.99)$ quantiles	$(-0.55, 2.87)$	$(-1.69, 1.59)$	$(0.06, 15.6)$
$\tilde{V}_t^{-\frac{1}{2}} \left(y_t - \tilde{F}_t \right) = \tilde{V}_t^{-\frac{1}{2}} \tilde{f}_t X_t b - \tilde{V}_t^{-\frac{1}{2}} \tilde{f}_t X_t \tilde{\beta} \begin{bmatrix} X_{1t} & X_{2t} \end{bmatrix} \begin{bmatrix} c_1 \\ c_2 \end{bmatrix} + residual$			

Next, we explore some variations of the BRMR specification test. Based on table 5.14, the appropriately specified heteroskedastic test seems, on average, resistant to rejecting the null (when it should not reject). Fewer than 25% of the simulations produce evidence of misspecification at the 80% confidence level.

Now, leave the DGP unaltered but suppose we suspect the variance changes due to another variable, say X_2. Even though the source of heteroskedasticity is misidentified, the test reported in table 5.15 produces similar results, on average. Again, fewer than 25% of the simulations produce evidence of misspecification at the 80% confidence level.

Next, we explore changing variance as a function of both regressors. Again, we find similar specification results, on average, as reported in table 5.16. That is, fewer than 25% of the simulations produce evidence of misspecification at the 80% confidence level. Finally, assuming the evidence is against homoskedasticity, we next report simulations for the heteroskedastic probit with standard devia-

Table 5.12: Heteroskedastic probit results with heteroskedastic DGP

	β_0	β_1	$\hat{\beta}_2$	$\hat{\gamma}$
mean	0.012	1.048	-1.048	1.0756
std. dev.	0.161	0.394	0.358	0.922
$(0.01, 0.99)$ quantiles	$(-0.33, 0.41)$	$(0.44, 2.14)$	$(-2.0, -0.40)$	$(-0.30, 2.82)$

Table 5.13: Homoskedastic probit results with homoskedastic DGP

	β_0	β_1	$\hat{\beta}_2$
mean	-0.002	1.005	-0.999
std. dev.	0.110	0.147	0.148
$(0.01, 0.99)$ quantiles	$(-0.24, 0.25)$	$(0.67, 1.32)$	$(-1.35, -0.67)$

tion $exp(X_1 \gamma)$. On average, table 5.17 results support the homoskedastic choice model as $\hat{\gamma}$ is near zero. Of course, the risk remains that we may incorrectly conclude that the data are heteroskedastic.

5.5 Robust choice models

A few robust (relaxed distribution or link function) discrete choice models are briefly discussed next.

5.5.1 Mixed logit models

Mixed logit is a highly flexible model that can approximate any random utility. Unlike logit, it allows random taste variation, unrestricted substitution patterns, and correlation in unobserved factors (over time). Unlike probit, it is not restricted to normal distributions.

Mixed logit probabilities are integrals of standard logit probabilities over a density of parameters. $P = \int L(\beta) f(\beta) d\beta$ where $L(\beta)$ is the logit probability evaluated at β and $f(\beta)$ is a density function. In other words, the mixed logit is a weighted average of the logit formula evaluated at different values of β with weights given by the density $f(\beta)$. The mixing distribution $f(\beta)$ can be discrete or continuous.

5.5.2 Semiparametric single index discrete choice models

Another "robust" choice model draws on kernel density-based regression (see chapter 6 for more details). In particular, the density-weighted average derivative estimator from an index function yields $E[Y_t | X_t] = G(X_t b)$ where $G(\cdot)$ is some general nonparametric function. As Stoker [1991] suggests the bandwidth

Table 5.14: BRMR specification test 1 with homoskedastic DGP

	b_0	b_1	b_2
mean	-0.001	-0.007	0.007
std. dev.	0.032	0.193	0.190
$(0.01, 0.99)$ quantiles	$(-0.09, 0.08)$	$(-0.46, 0.45)$	$(-0.44, 0.45)$
	c_1	ESS ($\chi^2(1)prob$)	
mean	-0.013	0.983 (0.679)	
std. dev.	0.382	1.338	
$(0.01, 0.99)$ quantiles	$(-0.87, 0.86)$	$(0.0, 6.4)$	
$\tilde{V}_t^{-\frac{1}{2}}\left(Y_t - \tilde{F}_t\right) = \tilde{V}_t^{-\frac{1}{2}}\tilde{f}_t X_t b - \tilde{V}_t^{-\frac{1}{2}}\tilde{f}_t X_t \tilde{\beta} X_{1t} c_1 + residual$			

Table 5.15: BRMR specification test 2 with homoskedastic DGP

	b_0	b_1	b_2
mean	-0.001	0.009	-0.010
std. dev.	0.032	0.187	0.187
$(0.01, 0.99)$ quantiles	$(-0.08, 0.09)$	$(-0.42, 0.44)$	$(-0.46, 0.40)$
	c_2	ESS ($\chi^2(1)prob$)	
mean	0.018	0.947 (0.669)	
std. dev.	0.377	1.362	
$(0.01, 0.99)$ quantiles	$(-0.85, 0.91)$	$(0.0, 6.6)$	
$\tilde{V}_t^{-\frac{1}{2}}\left(y_t - \tilde{F}_t\right) = \tilde{V}_t^{-\frac{1}{2}}\tilde{f}_t X_t b - \tilde{V}_t^{-\frac{1}{2}}\tilde{f}_t X_t \tilde{\beta} X_{2t} c_2 + residual$			

is chosen based on "critical smoothing." Critical smoothing refers to selecting the bandwidth as near the "optimal" bandwidth as possible such that monotonicity of probability in the index function is satisfied. Otherwise, the estimated "density" function involves negative values.

5.5.3 Nonparametric discrete choice models

The nonparametric kernel density regression model can be employed to estimate very general (without index restrictions) probability models (see chapter 6 for more details).

$$E[Y_t|X_t] = m(X_t)$$

Table 5.16: BRMR specification test 3 with homoskedastic DGP

	b_0	b_1	b_2
mean	-0.000	0.004	-0.006
std. dev.	0.045	0.391	0.382
$(0.01, 0.99)$ quantiles	$(-0.11, 0.12)$	$(-0.91, 0.93)$	$(-0.91, 0.86)$
	c_1	c_2	ESS $(\chi^2(2) prob)$
mean	-0.006	0.015	1.943 (0.621)
std. dev.	0.450	0.696	3.443
$(0.01, 0.99)$ quantiles	$(-1.10, 1.10)$	$(-0.98, 1.06)$	$(0.03, 8.2)$

$$\tilde{V}_t^{-\frac{1}{2}} \left(y_t - \tilde{F}_t \right) = \tilde{V}_t^{-\frac{1}{2}} \tilde{f}_t X_t b - \tilde{V}_t^{-\frac{1}{2}} \tilde{f}_t X_t \tilde{\beta} \begin{bmatrix} X_{1t} & X_{2t} \end{bmatrix} \begin{bmatrix} c_1 \\ c_2 \end{bmatrix} + residual$$

Table 5.17: Heteroskedastic probit results with homoskedastic DGP

	$\hat{\beta}_0$	$\hat{\beta}_1$	$\hat{\beta}_2$	$\hat{\gamma}$
mean	0.001	1.008	-1.009	-0.003
std. dev.	0.124	0.454	0.259	0.521
$(0.01, 0.99)$ quantiles	$(-0.24, 0.29)$	$(0.55, 1.67)$	$(-1.69, -0.50)$	$(-0.97, 0.90)$

5.6 Tobit (censored regression) models

Sometimes the dependent variable is censored at a value (we assume zero for simplicity).

$$Y_t^* = X_t \beta + \varepsilon_t$$

where $\varepsilon \sim N\left(0, \sigma^2 I\right)$ and $Y_t = Y_t^*$ if $Y_t^* > 0$ and $Y_t = 0$ otherwise. Then we have a mixture of discrete and continuous outcomes. Tobin [1958] proposed writing the likelihood function as a combination of a discrete choice model (binomial likelihood) and standard regression (normal likelihood), then estimating β and σ via maximum likelihood. The log-likelihood is

$$\sum_{y_t=0} \log \Phi \left(-\frac{X_t \beta}{\sigma} \right) + \sum_{y_t>0} \log \left[\frac{1}{\sigma} \phi \left(\frac{Y_t - X_t \beta}{\sigma} \right) \right]$$

where, as usual, $\phi(\cdot), \Phi(\cdot)$ are the unit normal density and distribution functions, respectively.

5.7 Bayesian data augmentation

Albert and Chib's [1993] idea is to treat the latent variable Y^* as missing data and use Bayesian analysis to estimate the missing data. Typically, Bayesian analysis draws inferences by sampling from the posterior distribution $p(\theta|Y)$. However, the marginal posterior distribution in the discrete choice setting is often not recognizable though the conditional posterior distributions may be. In this case, Markov chain Monte Carlo (*McMC*) methods, in particular a Gibbs sampler, can be employed (see chapter 7 for more details on *McMC* and the Gibbs sampler).

Albert and Chib use a Gibbs sampler to estimate a Bayesian probit.

$$p(\beta|Y, X, Y^*) \sim N\left(b_1, \left(Q^{-1} + X^T X\right)^{-1}\right)$$

where

$$\begin{aligned} b_1 &= \left(Q^{-1} + X^T X\right)\left(^{-1}Q^{-1}b_0 + X^T X b\right) \\ b &= \left(X^T X\right)^{-1} X^T Y \end{aligned}$$

b_0 = prior means for β and $Q = \left(X_0^T X_0\right)^{-1}$ is the prior on the covariance.[15] The conditional posterior distributions for the latent utility index are

$$\begin{aligned} p(Y^*|Y=1, X, \beta) &\sim N\left(X^T \beta, I | Y^* > 0\right) \\ p(Y^*|Y=0, X, \beta) &\sim N\left(X^T \beta, I | Y^* \leq 0\right) \end{aligned}$$

For the latter two, random draws from a truncated normal (truncated from below for the first and truncated from above for the second) are employed.

5.8 Additional reading

There is an extensive literature addressing discrete choice models. Some favorite references are Train [2003], and McFadden's [2001] Nobel lecture. Connections between discrete choice models, nonlinear regression, and related specification tests are developed by Davidson and MacKinnon [1993, 2003]. Coslett [1981] discusses efficient estimation of discrete choice models with emphasis on choice-based sampling. Mullahy [1997] discusses instrumental variable estimation of count data models.

---

[15] Bayesian inference works as if we have data from the prior period $\{Y_0, X_0\}$ as well as from the sample period $\{Y, X\}$ from which β is estimated ($b_0 = \left(X_0^T X_0\right)^{-1} X_0^T Y_0$ as if taken from prior sample $\{X_0, Y_0\}$; see Poirier [1995], p. 527). Applying *OLS* yields

$$\begin{aligned} b_1 &= (X_0^T X_0 + X^T X)^{-1}(X_0^T Y_0 + X^T Y) \\ &= (Q^{-1} + X^T X)^{-1}(Q^{-1} b_0 + X^T X b) \end{aligned}$$

since $Q^{-1} = (X_0^T X_0)$, $X_0^T Y_0 = Q^{-1} b_0$, and $X^T Y = X^T X b$.

6
Nonparametric regression

Frequently in econometric analysis of accounting data, one is concerned with departures from standard parametric model probability assignments. Semi- and nonparametric methods provide an alternative means to characterize data and assess parametric model robustness or logical consistency. Here, we focus on regression. That is, we examine the conditional relation between Y and X. The most flexible fit of this conditional relation is nonparametric regression where flexible fit refers to the degree of distributional or structural form restrictions imposed on the data in estimating the relationship.

6.1 Nonparametric (kernel) regression

Nonparametric regression is motivated by at least the following four objectives: (1) it provides a versatile method for exploring a general relation between variables, (2) it give predictions without reference to a fixed parametric model, (3) it provides a tool for identifying spurious observations, and (4) it provides a method for 'fixing' missing values or interpolating between regressor values (see Hardle [1990, p 6-7]).

A nonparametric (kernel) regression can be represented as follows (Hardle [1990]).

$$E[Y|X] = m(X)$$

where $m(X) = \dfrac{n^{-1}h^{-d} \sum_{i=1}^{n} K\left(\frac{X-x_i}{h}\right) y_i}{n^{-1}h^{-d} \sum_{i=1}^{n} K\left(\frac{X-x_i}{h}\right)}$, $y_i(x_i)$ is the ith observation for $Y(X)$, n is the number of observations, d is the dimension (number of regressors) of X,

$K\left(\cdot\right)$ is any well-defined (multivariate) kernel, and h is the smoothing parameter or bandwidth (see *GCV* below for bandwidth estimation). Notice as is the case with linear regression each predictor is constructed by regressor-based weights of each observed value of the response variable $M\left(h\right)Y$ where

$$M\left(h\right) = \begin{bmatrix} \frac{K\left(\frac{X_1-x_1}{h}\right)}{\sum_{i=1}^{n}K\left(\frac{X_i-x_1}{h}\right)} & \frac{K\left(\frac{X_2-x_1}{h}\right)}{\sum_{i=1}^{n}K\left(\frac{X_i-x_1}{h}\right)} & \cdots & \frac{K\left(\frac{X_n-x_1}{h}\right)}{\sum_{i=1}^{n}K\left(\frac{X_i-x_1}{h}\right)} \\ \frac{K\left(\frac{X_1-x_2}{h}\right)}{\sum_{i=1}^{n}K\left(\frac{X_i-x_2}{h}\right)} & \frac{K\left(\frac{X_2-x_2}{h}\right)}{\sum_{i=1}^{n}K\left(\frac{X_i-x_2}{h}\right)} & \cdots & \frac{K\left(\frac{X_n-x_2}{h}\right)}{\sum_{i=1}^{n}K\left(\frac{X_i-x_2}{h}\right)} \\ \vdots & \vdots & \ddots & \vdots \\ \frac{K\left(\frac{X_1-x_n}{h}\right)}{\sum_{i=1}^{n}K\left(\frac{X_i-x_n}{h}\right)} & \frac{K\left(\frac{X_2-x_n}{h}\right)}{\sum_{i=1}^{n}K\left(\frac{X_i-x_n}{h}\right)} & \cdots & \frac{K\left(\frac{X_n-x_n}{h}\right)}{\sum_{i=1}^{n}K\left(\frac{X_i-x_n}{h}\right)} \end{bmatrix}$$

To fix ideas, compare this with liner regression. For linear regression, the predictions are $\widehat{Y} = P_X Y$, where

$$P_X = X\left(X^T X\right)^{-1} X^T$$

the projection matrix (into the columns of X), again a linear combination (based on the regressors) of the response variable.

A multivariate kernel is constructed, row by row, by computing the product of marginal densities for each variable in the matrix of regressors X.[1] That is, $h^{-d}K\left(\frac{X-x_i}{h}\right) = \prod_{j=1}^{d} h^{-1}K\left(\frac{x_j - x_{ji}}{h}\right)$, where x_j is the jth column vector in the regressors matrix. Typically, we employ leave-one-out kernels. That is, the current observation is excluded in the kernel construction to avoid overfitting — the principal diagonal in $M\left(h\right)$ is zeroes. Since nonparametric regression simply exploits the explanatory variables to devise a weighting scheme for Y, assigning no weight to the current observation of Y is an intuitively appealing means of avoiding overfitting.

Nonparametric (kernel) regression is the most flexible model that we employ and forms the basis for many other kernel density estimators. While nonparametric regression models provide a very flexible fit of the relation between Y and X, this does not come at zero cost. In particular, it is more difficult to succinctly describe this relation, especially when X is a high dimension matrix. Also, nonparametric regressions typically do not achieve parametric rates of convergence (i.e., they converge more slowly than square root n).[2] Next, we turn to models that retain

---

[1] As we typically estimate one bandwidth for all regressors, the variables are first scaled by their estimated standard deviation.

[2] It can be shown that optimal rates of convergence for nonparametric models are $n - r$, $0 < r < 1/2$. More specifically, $r = (\rho + \beta - k)/(2[\rho + \beta] - d)$, where ρ is the number of times the smoothing

some of the flexibility of nonparametric regression but enhance interpretability (i.e., semiparametric models).

6.2 Semiparametric regression models

6.2.1 Partial linear regression

Frequently, we are concerned about the relation between Y and X but troubled that the analysis is plagued by omitted, correlated variables. One difficulty is that we do not know the functional form of the relation between our variables of interest and these other control variables. That is, we envision a *DGP* where

$$E[Y \mid X, Z] = X\beta + \theta(Z)$$

Provided that we can observe these control variables, Robinson [1988] suggests a two-stage approach analogous to *FWL* (see chapter 3) which is called partial linear regression. Partial linear regression models nonparametrically fit the relation between the dependent variable, Y, and the control variables, Z, and also the experimental regressors of interest, X, and the control variables Z. The residuals from each nonparametric regression are retained, $e_Y = Y - E[Y|Z]$ and $e_X = X - E[X|Z]$, in standard double residual regression fashion.

Next, we simply employ no-intercept *OLS* regression of the dependent variable residuals on the regressor residuals, $e_Y = e_X \beta$. The parameter estimator for β fully captures the influence of the otherwise omitted, control variables and is accordingly, asymptotically consistent. Of course, we now have parameters to succinctly describe the relation between Y and X conditional on Z. Robinson demonstrates that this estimator converges at the parametric (square-root n) rate.

6.2.2 Single-index regression

The partial linear model discussed above imposes distributional restrictions on the relation between Y and X in the second stage. One (semiparametric) approach for relaxing this restriction and retaining ease of interpretability is single-index regression. Single-index regression follows from the idea that the average derivative of a general function with respect to the regressor is proportional to the parameters of the index. Suppose the DGP is

$$E[Y|X] = G(X\beta)$$

then define $\delta = \partial E[Y|X]/\partial X = dG/d(X\beta)\beta = \gamma\beta$. Thus, the derivative with respect to X is proportional to β for all X, and likewise the average derivative $E[dG/d(X\beta)]\beta = \gamma\beta$, for $\gamma \neq 0$, is proportional to β.

---

function is differentiable, k is the order of the derivative of the particular estimate of interest ($k \leq \rho$), β is the characteristic or exponent for the smoothness class, and d is the order of the regressors (Hardle [1990, p. 93]).

6. Nonparametric regression

Our applications employ the density-weighted average derivative single-index model of Powell, Stock, and Stoker [1989].[3] That is,

$$\hat{\delta} = -2n^{-1} \sum_{i=1}^{n} \frac{\partial \hat{f}_i(X_i)}{\partial X} Y_i$$

exploiting the U statistic structure (see Hoeffding [1948])

$$= -2\left[n(n-1)\right]^{-1} \sum_{i=1}^{n-1} \sum_{j=i+1}^{n} h^{-(d+1)} K'\left(\frac{X_i - X_j}{h}\right)(Y_i - Y_j)$$

For a Gaussian kernel, K, notice that $K\prime(u) = -uK(u)$. Thus,

$$\hat{\delta} = 2\left[n(n-1)\right]^{-1} \sum_{i=1}^{n-1} \sum_{j=i+1}^{n} h^{-(d+1)} K\left(\frac{X_i - X_j}{h}\right)\left(\frac{X_i - X_j}{h}\right)(Y_i - Y_j)$$

where $K(u) = (2\pi)^{-1/2} exp\left\{-\frac{u^2}{2}\right\}$. The asymptotic covariance matrix for the parameters $\Sigma_{\hat{\delta}}$ is estimated as

$$\widehat{\Sigma}_{\hat{\delta}} = 4n^{-1} \sum_{i=1}^{n} \hat{r}(Z_i) \hat{r}(Z_i)^T - 4\widetilde{\delta}\widetilde{\delta}^T$$

where

$$\hat{r}(Z_i) = (-n-1)^{-1} \sum_{\substack{j=1 \\ i \neq j}}^{n} h^{-(d+1)} K\left(\frac{X_i - X_j}{h}\right)\left(\frac{X_i - X_j}{h}\right)(Y_i - Y_j)$$

The above estimator is proportional to the index parameters. Powell, et al also proposed a properly-scaled instrumental variable version of the density-weighted average derivative. We refer to this estimator as $\hat{d} = \hat{\delta}_X^{-1} \hat{\delta}$, where

$$\hat{\delta}_X = -2n^{-1} \sum_{i=1}^{n} \frac{\partial \hat{f}_i(X_i)}{\partial X} X_i^T$$

$$= \frac{2 \sum_{i=1}^{n-1} \sum_{j=i+1}^{n} h^{-(d+1)} K\left(\frac{X_i - X_j}{h}\right) \left(\frac{X_i - X_j}{h}\right) (X_i - X_j)^T}{n(n-1)}$$

---

[3]Powell et al's description of the asymptotic properties of their average derivative estimator exploits a 'leave-one-out' approach, as discussed for nonparametric regression above. This estimator also achieves the parametric (square-root n) rate of convergence.

rescales $\hat{\delta}$. The asymptotic covariance estimator for the parameters $\Sigma_{\hat{d}}$ is estimated as $\hat{\Sigma}_{\hat{d}} = 4n^{-1} \sum_{i=1}^{n} \hat{r}_d(Z_i) \hat{r}_d(Z_i)^T$, where

$$\hat{r}_d(Z_i) = \hat{\delta}_x^{-1} \frac{\sum_{\substack{j=1 \\ i \neq j}}^{n} h^{-(d+1)} K\left(\frac{X_i - X_j}{h}\right)\left(\frac{X_i - X_j}{h}\right)\left(\hat{U}_i - \hat{U}_j\right)}{-n-1}$$

$$\hat{U}_i = Y_i - X_i \hat{d}$$

The optimal bandwidth is estimated similarly to that described for nonparametric regression. First, $\hat{d}$ (and its covariance matrix) is estimated (for various bandwidths). Then, the bandwidth that produces minimum mean squared error is identified from the leave-one out nonparametric regression of Y on the index $X\hat{d}$ (the analog to regressing Y on X in fully nonparametric regression). This yields a readily interpretable, flexibly fit set of index parameters, the counterpart to the slope parameter in *OLS* (linear) regression.

6.2.3 *Partial index regression models*

Now, we put together the last two sets of ideas. That is, nonparametric estimates for potentially omitted, correlated (control) variables as in the partial linear model are combined with single index model parameter estimates for the experimental regressors. That is, we envision a *DGP* where

$$E[Y \mid X, Z] = G(X\beta) + \theta(Z)$$

Following Stoker [1991], these are called partial index models. As with partial linear models, the relation between Y and Z (the control variables) and the relation between X and Z are estimated via nonparametric regression. As before, separate bandwidths are employed for the regression of Y on Z and X on Z. Again, residuals are computed, e_Y and e_X. Now, single index regression of e_Y on e_X completes the partial index regression. Notice, that a third round of bandwidth selection is involved in the second stage.

6.3 Specification testing against a general nonparametric benchmark

Specification or logical consistency testing lies at the heart of econometric analysis. Borrowing from conditional moment tests (Ruud [1984], Newey [1985], Pagan and Vella [1989]) and the U statistic structure employed by Powell et al, Zheng [1996] proposed a specification test of any parametric model $f(X, \theta)$ against a general nonparametric benchmark $g(X)$.

Let $\varepsilon_i \equiv Y_i - f(X_i, \theta)$ and $p(\bullet)$ denote the density function of X_i. The null hypothesis is that the parametric model is correct (adequate for summarizing the data)

$$H_0 : \Pr E[Y_i|X_i] = f(X_i, \theta_0) = 1 \text{ for some } \theta_0 \in \Theta$$

where $\theta_0 = \arg\min_{\theta \in \Theta} E[Y_i - f(X_i, \theta_0)]^2$. The alternative is the null is false, but there is no specific alternative model

$$H_0 : \Pr E[Y_i|X_i] = f(X_i, \theta) < 1 \text{ for all } \theta \in \Theta$$

The idea is under the null, $E[\varepsilon_i|X_i] = 0$. Therefore, we have

$$E[\varepsilon_i E[\varepsilon_i|X_i]p(X_i)] = 0$$

while under the alternative we have

$$E[\varepsilon_i E[\varepsilon_i|X_i]p(X_i)] = E\left[\{E[\varepsilon_i|X_i]\}^2 p(X_i)\right]$$

since $E[\varepsilon_i|X_i] = g(X_i) - f(X_i, \theta)$

$$E[\varepsilon_i E[\varepsilon_i|X_i]p(X_i)] = E\left[\{g(X_i) - f(X_i, \theta)\}^2 p(X_i)\right] > 0$$

The sample analog of $E[\varepsilon_i E[\varepsilon_i|X_i]p(X_i)]$ is used to form a test statistic. In particular, kernel estimators of the components are employed. A kernel estimator of the density function p is

$$\widehat{p}(x_i) = (-n-1)^{-1} \sum_{\substack{j=1 \\ i \neq j}}^{n} h^{-d} K\left(\frac{X_i - X_j}{h}\right)$$

and a kernel estimator of the regression function $E[\varepsilon_i|X_i]$ is

$$E[\varepsilon_i|X_i] = (-n-1)^{-1} \sum_{\substack{j=1 \\ i \neq j}}^{n} h^{-d} \frac{K\left(\frac{X_i - X_j}{h}\right) \varepsilon_i}{\widehat{p}(X_i)}$$

The sample analog to $E[\varepsilon_i E[\varepsilon_i|X_i]p(X_i)]$ is completed by replacing ε_i with $e_i \equiv Y_i - f\left(X_i, \widehat{\theta}\right)$ and we have

$$V_n \equiv (-n-1)^{-1} \sum_{i=1}^{n} \sum_{\substack{j=1 \\ i \neq j}}^{n} h^{-d} K\left(\frac{X_i - X_j}{h}\right) e_i e_j$$

Under the null, Zheng shows that the statistic $nh^{d/2}V_n$ is consistent asymptotic normal (*CAN*; see appendix) with mean zero and variance Σ. Also, the variance can be consistently estimated by

$$\widehat{\Sigma} = 2(n(-n-1))^{-1} \sum_{i=1}^{n} \sum_{\substack{j=1 \\ i \neq j}}^{n} h^{-d} K^2\left(\frac{X_i - X_j}{h}\right) e_i^2 e_j^2$$

Consequently, a standardized test statistic is

$$T_n \equiv \sqrt{\frac{n-1}{n}} \frac{nh^{d/2}V_n}{\sqrt{\hat{\Sigma}}}$$

$$= \frac{\sum_{i=1}^{n}\sum_{\substack{j=1 \\ i\neq j}}^{n} h^{-d}K\left(\frac{X_i-X_j}{h}\right) e_i e_j}{\left\{2\sum_{i=1}^{n}\sum_{\substack{j=1 \\ i\neq j}}^{n} h^{-d}K^2\left(\frac{X_i-X_j}{h}\right) e_i^2 e_j^2\right\}^{1/2}}$$

Since V_n is *CAN* under the null, the standardized test statistic converges in distribution to a standard normal, $Tn \xrightarrow{d} N(0,1)$ (see the appendix for discussion of convergence in distribution).

6.4 Locally linear regression

Another local method, *locally linear regression*, produces smaller bias (especially at the boundaries of X) and no greater variance than regular kernel regression.[4] Hence, it produces smaller *MSE*.

Regular kernel regression solves

$$\min_{g} \sum_{i=1}^{n} (y_i - g)^2 h^{-d} K\left(\left|\frac{X-x_i}{h}\right|\right)$$

while locally linear regression solves

$$\min_{g,\beta} \sum_{i=1}^{n} \left(y_i - g - (X-x_i)^T \beta\right)^2 h^{-d} K\left(\left|\frac{X-x_i}{h}\right|\right)$$

Effectively, kernel regression is a constrained version of locally linear regression with $\beta = 0$. Both are regressor-based weighted averages of Y.

Newey [2007] shows the asymptotic *MSE* for locally linear regression is

$$MSE_{LLR} = \frac{1}{nh}\nu_0 \frac{\sigma^2(X)}{f_0(X)} + \frac{h^4}{4} g_0''(X) \mu_2^2$$

while for kernel regression we have

$$MSE_{KR} = \frac{1}{nh}\nu_0 \frac{\sigma^2(X)}{f_0(X)} + \frac{h^4}{4}\left[g_0''(X) + 2g_0'(X)\frac{f_0'(X)}{f_0(X)}\right] \mu_2^2$$

---

[4]This section draws heavily from Newey [2007].

where $f_0(X)$ is the density function for $X = [x_1, \ldots, x_n]^T$ with variance $\sigma^2(X)$, $g_0(X) = E[Y|X]$,

$$u = \frac{X - X_i}{h}, \mu_2 = \left| \int K(u) u^2 du, \nu_0 = \left| \int K(u)^2 du \right.\right.$$

and kernel regression bias is

$$bias_{KR} = \left| \frac{1}{2} g_0''(X) + g_0'(X) \frac{f_0'(X)}{f_0(X)} \right| \mu_2 h^2$$

Hence, locally linear regression has smaller bias and smaller *MSE* everywhere.

6.5 Generalized cross-validation (*GCV*)

The bandwidth h is frequently chosen via generalized cross validation (*GCV*) (Craven and Wahba [1979]). *GCV* utilizes principles developed in ridge regression for addressing computational instability problems in a regression context.

$$GCV(h) = \frac{n^{-1} ||y - \hat{m}(h)||^2}{[1 - n^{-1} tr(M(h))]^2}$$

where $\hat{m}(h) = M(h)Y$ is the nonparametric regression of Y on X given bandwidth h, $||\cdot||^2$ is the squared norm or vector inner product, and $tr(\cdot)$ is the trace of the matrix.

Since the properties of this statistic are data specific and convergence at a uniform rate cannot be assured, we evaluate a dense grid of values for h to numerically find the minimum MSE. Optimal bandwidths are determined by trading off a 'good approximation' to the regression function (reduction in bias) and a 'good reduction' of observational noise (reduction in noise). The former (latter) is increasing (decreasing) in the bandwidth (Hardle [1990, p. 29-30, 149]).

For leave-one-out nonparametric regression estimator, *GCV* chooses the bandwidth h that minimizes the mean squared errors

$$\min_h n^{-1} ||Y - \hat{m}_{-t}(h)||^2$$

That is, the penalty function in *GCV* is avoided (as $tr(M_{-t}(h)) = 0$, the denominator is 1) and *GCV* effectively chooses the bandwidth to minimize the model mean square error.

$$M_{-t}(h) = \begin{bmatrix} \frac{0}{n} & \frac{K\left(\frac{x_2-x_1}{h}\right)}{n} & \cdots & \frac{K\left(\frac{x_n-x_1}{h}\right)}{n} \\ \sum_{i=2}K\left(\frac{x_i-x_1}{h}\right) & \sum_{i=2}K\left(\frac{x_i-x_1}{h}\right) & & \sum_{i=2}K\left(\frac{x_i-x_1}{h}\right) \\ \frac{K\left(\frac{x_1-x_2}{h}\right)}{n} & \frac{0}{n} & \cdots & \frac{K\left(\frac{x_n-x_2}{h}\right)}{n} \\ \sum_{i=1,3}K\left(\frac{x_i-x_2}{h}\right) & \sum_{i=1,3}K\left(\frac{x_i-x_2}{h}\right) & & \sum_{i=1,3}K\left(\frac{x_i-x_2}{h}\right) \\ \vdots & \vdots & \ddots & \vdots \\ \frac{K\left(\frac{x_1-x_n}{h}\right)}{n-1} & \frac{K\left(\frac{x_2-x_n}{h}\right)}{n-1} & \cdots & \frac{0}{n-1} \\ \sum_{i=1}K\left(\frac{x_i-x_n}{h}\right) & \sum_{i=1}K\left(\frac{x_i-x_n}{h}\right) & & \sum_{i=1}K\left(\frac{x_i-x_n}{h}\right) \end{bmatrix}$$

As usual, the mean squared error is composed of squared bias and variance.

$$\begin{aligned} MSE\left(\hat{\theta}\right) &= E\left[\left(\hat{\theta}-\theta\right)^2\right] \\ &= E\left[\hat{\theta}^2\right] - 2E\left[\hat{\theta}\right]\theta + \theta^2 \\ &= \left\{E\left[\hat{\theta}^2\right] - E\left[\hat{\theta}\right]^2\right\} + \left\{E\left[\hat{\theta}\right]^2 - 2E\left[\hat{\theta}\right]\theta + \theta^2\right\} \end{aligned}$$

The leading term is the variance of $\hat{\theta}$ and the trailing term is the squared bias.

6.6 Additional reading

There is a burgeoning literature on nonparametric regression and its semiparametric cousins. Hardle [1990] and Stoker [1991] offer eloquent overviews. Newey and Powell [2003] discuss instrumental variable estimation of nonparametric models. Powell et al's average derivative estimator assumes the regressors are continuous. Horowitz and Hardle [1996] proposed a semiparametric model that accommodates some discrete as well as continuous regressors.

When estimating causal effects in a selection setting, the above semiparametric methods are lacking as the intercept is suppressed by nonparametric regression. Andrews and Schafgans [1998] suggested a semiparametric selection model to remedy this deficiency. Variations on these ideas are discussed in later chapters.

7
Repeated-sampling inference

Much of the discussion regarding econometric analysis of endogenous relations centers around identification issues. In this chapter we review the complementary matter of inference. Exchangeability or symmetric dependence and de Finetti's theorem lie at the heart of most (perhaps all) statistical inference. A simple binomial example illustrates. Exchangeability says that a sequence of coin flips has the property

$$Pr(X_1 = 1, X_2 = 0, X_3 = 1, X_4 = 1)$$
$$= Pr(X_3 = 1, X_4 = 0, X_2 = 1, X_1 = 1)$$

and so on for all permutations of the random variable index. de Finetti's theorem [1937, reprinted in 1964] provides justification for typical statistical sampling from a population with unknown distribution based on a large number of *iid* draws from the unknown distribution. That is, if ex ante the analyst assesses that samples are exchangeable (and from a large population), then the samples can be viewed as independent and identically distributed from an unknown distribution function. Perhaps it is instructive to consider whether (most) specification issues can be thought of as questions of the validity of some exchangeability conditions. While we ponder this, we review repeated-sampling based inference with particular attention to bootstrapping and Bayesian simulation.[1]

---

[1] MacKinnon [2002] suggests three fruitful avenues for exploiting abundant computing capacity: (1) structural models at the individual level that frequently draw on simulation, (2) Markov chain Monte Carlo (*McMC*) analysis, and (3) bootstrap inference.

7. Repeated-sampling inference

7.1 Monte Carlo simulation

Monte Carlo simulation can be applied when the statistic of interest is pivotal.

Definition 7.1 *A pivotal statistic is one that depends only on the data and no unknown parameters.*

Monte Carlo simulation of pivotal statistics produces exact tests.

Definition 7.2 *Exact tests are tests for which a true null hypothesis is rejected with probability precisely equal to α, the nominal size of the test.*

However, if the test statistic is not pivotal (for instance, the distribution is unknown), a Monte Carlo test doesn't apply.

7.2 Bootstrap

Inference based on bootstrapping is simply an application of the *fundamental theorem of statistics*. That is, when randomly sampled with replacement the empirical distribution function is consistent for the population distribution function (see appendix).

To bootstrap a single parameter such as the correlation between two random variables, say x and y, we simply sample randomly with replacement from the pair (x, y). Then, utilize the empirical distribution of the statistic (say, sample correlation) to draw inferences, for instance, about the mean, etc. (see Efron [1979, 2000]).

7.2.1 Bootstrap regression

For a regression that satisfies standard *OLS* (spherical) conditions, bootstrapping involves first estimating the regression via *OLS* $X_i\hat{\beta}$ and calculating the residuals.[2] The second step involves randomly sampling with replacement a residual for each estimated regression observation $X_i\hat{\beta}$. Pseudo responses $\widehat{Y}$ are constructed by adding the sampled residual to the estimated regression $X_i\hat{\beta}$ for each draw desired (often this is simply n, the original sample size). Next, b_k is estimated via *OLS* regression of $\widehat{Y}$ on the matrix of regressors. Steps two and three are repeated B times to produce an empirical sample of b_k, $k = 1, \ldots, B$. Davidson and MacKinnon [2003] recommend choosing B such that $\alpha(B+1)$ is an integer where α is the proposed size of the test. Inferences (such as interval estimates) are then based on this empirical sample.

---

[2] The current and next section draw heavily from Freedman [1981] and Freedman and Peters [1984].

7.2.2 Bootstrap panel data regression

If the errors are heteroskedastic and/or correlated, then the bootstrapping procedure above is modified to accommodate these features. The key is we bootstrap exchangeable partitions of the data. Suppose we have panel data stacked by time series of length T by J cross-sectional individuals in the sample (the sample size is $n = T * J$).

Heteroskedasticity

If we suppose the errors are independent but the variance depends on the cross-sectional unit,

$$\Sigma = \begin{bmatrix} \sigma_1^2 I_T & 0 & \cdots & 0 \\ 0 & \sigma_2^2 I_T & \cdots & 0 \\ \vdots & \vdots & \ddots & \vdots \\ 0 & 0 & \cdots & \sigma_J^2 I_T \end{bmatrix}$$

then random draws with replacement of the first step residuals (whether estimated by *OLS* or *WLS*, weighted least squares) are taken from the size T sample of residuals for each cross-sectional unit or group of cross-sectional individuals with the same variance. As these partitions are exchangeable, this preserves the differences in variances across cross-sectional units. The remainder of the process remains as described above for bootstrapping regression.

When the nature of the heteroskedasticity is unknown, Freedman [1981] suggests a *paired bootstrap* where $[Y_i, X_i]$ are sampled simultaneously. MacKinnon [2002, p. 629-631] also discusses a *wild bootstrap* to deal with unknown heteroskedasticity.

Correlated errors

If the errors are serially correlated but the variance is constant across cross-sectional units,

$$\Sigma = \begin{bmatrix} V & 0 & \cdots & 0 \\ 0 & V & \cdots & 0 \\ \vdots & \vdots & \ddots & \vdots \\ 0 & 0 & \cdots & V \end{bmatrix}$$

where

$$V = \sigma^2 \begin{bmatrix} 1 & \rho_1 & \cdots & \rho_T \\ \rho_1 & 1 & \cdots & \rho_{T-1} \\ \vdots & \vdots & \ddots & \vdots \\ \rho_T & \rho_{t-1} & \cdots & 1 \end{bmatrix}$$

then random vector (of length T) draws with replacement of the first step residuals (whether estimated by *OLS* or *GLS*, generalized least squares) are taken from the cross-sectional units.[3] As these partitions are exchangeable, this preserves the serial correlation inherent in the data. The remainder of the process is as described above for bootstrapping regression.[4]

Heteroskedasticity and serial correlation

If the errors are serially correlated and the variance is nonconstant across cross-sectional units,

$$\Sigma = \begin{bmatrix} V_1 & 0 & \cdots & 0 \\ 0 & V_2 & \cdots & 0 \\ \vdots & \vdots & \ddots & \vdots \\ 0 & 0 & \cdots & V_J \end{bmatrix}$$

where

$$V_j = \sigma_j^2 \begin{bmatrix} 1 & \rho_1 & \cdots & \rho_T \\ \rho_1 & 1 & \cdots & \rho_{T-1} \\ \vdots & \vdots & \ddots & \vdots \\ \rho_T & \rho_{t-1} & \cdots & 1 \end{bmatrix}$$

then a combination of the above two sampling procedures is employed.[5] That is, groups of cross-section units with the same variance-covariance structure are identified and random vector (of length T) draws with replacement of the first step residuals (whether estimated by *OLS* or *GLS*) are taken from the groups of cross-sectional units. As these partitions are exchangeable, this preserves the heteroskedasticity and serial correlation inherent in the data. The remainder of the process is as described above for bootstrapping regression.

---

[3] For cross-sectional correlation (but independent errors through time)

$$\Sigma = \sigma^2 \begin{bmatrix} I_T & \rho_{12}I_T & \cdots & \rho_{1J}I_T \\ \rho_{12}I_T & I_T & \cdots & \rho_{2J}I_T \\ \vdots & \vdots & \ddots & \vdots \\ \rho_{1J}I_T & \rho_{2J}I_T & \cdots & I_T \end{bmatrix}$$

simply apply the same ideas to the length J vector of residuals over cross-sectional units in place of the length T vector of residuals through time.

[4] When the nature of the serial correlation is unknown, as expected the challenge is greater. MacKinnon [2002] discusses two approaches: *sieve bootstrap* and *block bootstrap*. Not surprisingly, when the nature of the correlation or heteroskedasticity is unknown the bootstrap performs more poorly than otherwise.

[5] Cross-sectional correlation and heteroskedasticity

$$\Sigma = \begin{bmatrix} \sigma_1^2 I_T & \rho_{12}\sigma_1\sigma_2 I_T & \cdots & \rho_{1J}\sigma_1\sigma_J I_T \\ \rho_{12}\sigma_1\sigma_2 I_T & \sigma_2^2 I_T & \cdots & \rho_{2J}\sigma_2\sigma_J I_T \\ \vdots & \vdots & \ddots & \vdots \\ \rho_{1J}\sigma_1\sigma_J I_T & \rho_{2J}\sigma_2\sigma_J I_T & \cdots & \sigma_J^2 I_T \end{bmatrix}$$

again calls for sampling from like variance-covariance groups.

7.2.3 Bootstrap summary

Horowitz [2001] relates the bootstrap to asymptotically pivotal statistics in discussing effective usage of the bootstrap.

Definition 7.3 *An asymptotically pivotal statistic is a statistic whose asymptotic distribution does not depend on unknown population parameters.*

Horowitz concludes

- If an asymptotically pivotal statistic is available, use the bootstrap to estimate the probability distribution of the asymptotically pivotal statistic or a critical test value based on the asymptotically pivotal statistic.

- Use an asymptotically pivotal statistic if available rather than bootstrapping a non-asymptotically pivotal statistic such as a regression slope coefficient or standard error to estimate the probability distribution of the statistic.

- Recenter the residuals of an overidentified model before applying the bootstrap.

- Extra care is called for when bootstrapping models for dependent data, semi- or non-parametric estimators, or non-smooth estimators.

7.3 Bayesian simulation

Like bootstrapping, Bayesian simulation employs repeated sampling with replacement to draw inferences. Bayesian sampling in its simplest form utilizes Bayes' theorem to identify the posterior distribution of interest $p(\theta \mid Y)$ from the likelihood function $p(Y \mid \theta)$ and prior distribution for the parameters of interest $p(\theta)$.

$$p(\theta \mid Y) = \frac{p(Y \mid \theta) p(\theta)}{p(Y)}$$

The marginal distribution of the data $p(Y)$ is a normalizing adjustment. Since it does not affect the kernel of the distribution it is typically suppressed and the posterior is written

$$p(\theta \mid Y) \propto p(Y \mid \theta) p(\theta)$$

7.3.1 Conjugate families

It is straightforward to sample from the posterior distribution when its kernel (the portion of the density function or probability mass function that depends on the parameters of interest) is readily recognized. For a number of prior distributions (and likelihood functions), the posterior distribution is readily recognized as a standard distribution. This is referred to as *conjugacy* and the matching prior distribution is called the *conjugate prior*. A formal definition follows.

Definition 7.4 *If $\mathcal{F}$ is a class of sampling distributions $p(Y \mid \theta)$ and $\wp$ is a class of prior distributions for θ, then class $\wp$ is conjugate to $\mathcal{F}$ class if $p(\theta \mid Y) \in \wp$ for all $p(\cdot \mid \theta) \in \mathcal{F}$ and $p(\cdot) \in \wp$.*

For example, a binomial likelihood

$$\ell(\theta \mid s; n) = \binom{n}{s} \theta^s (1-\theta)^{n-s}$$

$$s = \sum_{i=1}^{n} y_i, \quad y_i = \{0, 1\}$$

combines with a beta$(\theta; \alpha, \beta)$ prior

$$p(\theta) = \frac{\Gamma(\alpha+\beta)}{\Gamma(\alpha)\Gamma(\beta)} \theta^{\alpha-1} (1-\theta)^{\beta-1}$$

to yield

$$\begin{aligned} p(\theta \mid y) &\propto \theta^s (1-\theta)^{n-s} \theta^{\alpha-1} (1-\theta)^{\beta-1} \\ &= \theta^{s+\alpha-1} (1-\theta)^{n-s+\beta-1} \end{aligned}$$

which is the kernel of a beta$(\theta \mid y; \alpha + s, \beta + n - s)$ distribution.

Also, a single draw from a Gaussian likelihood with known standard deviation, σ

$$\ell(\theta \mid y, \sigma) \propto \exp\left[-\frac{1}{2}\frac{(y-\theta)^2}{\sigma^2}\right]$$

combines with a Gaussian or normal prior

$$p(\theta \mid \mu_0, \tau_0) \propto \exp\left[-\frac{1}{2}\frac{(\theta-\mu_0)^2}{\tau_0^2}\right]$$

to yield[6]

$$p(\theta \mid y, \sigma, \mu_0, \tau_0) \propto \exp\left[-\frac{1}{2}\frac{(\theta-\mu_1)^2}{\tau_1^2}\right]$$

where $\mu_1 = \frac{\frac{1}{\tau_0^2}\mu_0 + \frac{1}{\sigma^2}y}{\frac{1}{\tau_0^2} + \frac{1}{\sigma^2}}$ and $\tau_1^2 = \frac{1}{\frac{1}{\tau_0^2} + \frac{1}{\sigma^2}}$. The posterior distribution of the mean given the data and priors is Gaussian. And, for a sample of n exchangeable draws, the likelihood is

$$\ell(\theta \mid y, \sigma) \propto \prod_{i=1}^{n} \exp\left[-\frac{1}{2}\frac{(y_i-\theta)^2}{\sigma^2}\right]$$

---

[6]The product gives

$$\exp\left[-\frac{1}{2}\left(\frac{(y-\theta)^2}{\sigma^2} + \frac{(\theta-\mu_0)^2}{\tau_0^2}\right)\right]$$

Then, expand the exponent and complete the square. Any constants are ignored in the identification of the kernel as they're absorbed through normalization of the posterior kernel.

combined with the above prior yields

$$p(\theta \mid y, \sigma, \mu_0, \tau_0) \propto \exp\left[-\frac{1}{2}\frac{(\theta - \mu_n)^2}{\tau_n^2}\right]$$

where $\mu_1 = \frac{\frac{1}{\tau_0^2}\mu_0 + \frac{n}{\sigma^2}\bar{y}}{\frac{1}{\tau_0^2} + \frac{n}{\sigma^2}}$, $\bar{y}$ is the sample mean, and $\tau_1^2 = \frac{1}{\frac{1}{\tau_0^2} + \frac{n}{\sigma^2}}$. The posterior distribution of the mean given the data and priors is again Gaussian.

These and some other well-known and widely used conjugate family distributions are summarized in tables 7.1, 7.2, 7.3, and 7.4 (see Bernardo and Smith [1994] and Gelman et al [2003]).

Table 7.1: Conjugate families for univariate discrete distributions

likelihood $p(Y \mid \theta)$	**conjugate prior** $p(\theta)$	**posterior** $p(\theta \mid Y)$
Binomial $(s \mid n, \theta)$ where $s = \sum_{i=1}^{n} y_i, y_i \in \{0, 1\}$	Beta $(\theta; \alpha, \beta)$ $\propto \theta^{\alpha-1}(1-\theta)^{\beta-1}$	Beta $(\theta \mid \alpha + s, \beta + n - s)$
Poisson $(s \mid n\lambda)$ where $s = \sum_{i=1}^{n} y_i, y_i = 0, 1, 2, \ldots$	Gamma $(\theta; \alpha, \beta)$ $\propto \theta^{\alpha-1}e^{-\beta\theta}$	Gamma $(\theta \mid \alpha + s, \beta + n)$
Exponential $(t \mid n, \theta)$ where $t = \sum_{i=1}^{n} y_i, y_i = 0, 1, 2, \ldots$	Gamma $(\theta; \alpha, \beta)$ $\propto \theta^{\alpha-1}e^{-\beta\theta}$	Gamma $(\theta \mid \alpha + n, \beta + t)$
Negative-binomial $(s \mid \theta, nr)$ where $s = \sum_{i=1}^{n} y_i, y_i = 0, 1, 2, \ldots$	Beta $(\theta; \alpha, \beta)$ $\propto \theta^{\alpha-1}(1-\theta)^{\beta-1}$	Beta $(\theta \mid \alpha + nr, \beta + s)$
Beta and gamma are continuous distributions		

A few words regarding the multi-parameter Gaussian case with unknown mean and variance seem appropriate. The joint prior combines a Gaussian prior for the mean conditional on the variance and an inverse-gamma or inverse-chi square prior for the variance.[7] The joint posterior distribution is the same form as the prior

---

[7] The inverse-gamma(α, β) distribution

$$p(\sigma^2; \alpha, \beta) \propto (\sigma^2)^{-(\alpha+1)} \exp\left[-\frac{\beta}{\sigma^2}\right]$$

Table 7.2: Conjugate families for univariate continuous distributions

likelihood $p(Y \mid \theta)$	conjugate prior $p(\theta)$	marginal posterior $p(\theta \mid Y)$
Uniform $(Y_i \mid 0, \theta)$ where $0 < Y_i < \theta$, $t = \max\{Y_1, \ldots, Y_n\}$	Pareto $(\theta; \alpha, \beta)$ $\propto \theta^{-(\alpha+1)}$	Pareto $(\theta; \alpha + n, \max\{\beta, t\})$
Normal $(Y \mid \theta, \sigma^2)$ variance known	Normal $(\theta \mid \sigma^2; \theta_0, \tau_0^2)$ $\propto \tau_0^{-1} e^{-\frac{(y-\theta_0)^2}{2\tau_0^2}}$	Normal $\left(\mu \mid \sigma^2; \dfrac{\frac{\theta_0}{\tau_0^2} + \frac{n\overline{Y}}{\sigma^2}}{\frac{1}{\tau_0^2} + \frac{n}{\sigma^2}}, \dfrac{1}{\frac{1}{\tau_0^2} + \frac{n}{\sigma^2}}\right)$
Normal $(Y \mid \mu, \theta)$ mean known, $\sigma^2 = \theta$	Inverse-gamma $(\theta; \alpha, \beta)$ $\propto \theta^{-(\alpha+1)} e^{-\beta/\theta}$	Inverse-gamma $(\theta; \frac{n+2\alpha}{2}, \beta + \frac{1}{2}t)$ where $t = \sum_{i=1}^{n}(Y_i - \mu)^2$
Normal $(Y \mid \theta, \sigma^2)$ both unknown	Normal$(\theta \mid \sigma^2; \theta_0, n_0)$ $*Inverse-$ $gamma\,(\sigma^2; \alpha, \beta)$	Student t $(\theta; \theta_n, \gamma, 2\alpha + n)$; Inverse-gamma $(\sigma^2; \alpha + \frac{1}{2}n, \beta_n)$

For the normal-inverse gamma posterior the parameters are
$$\theta_n = (n_0 + n)^{-1}(n_0\theta_0 + n\overline{Y})$$
$$\gamma = (n + n_0)\left(\alpha + \tfrac{1}{2}n\right)\beta_n^{-1}$$
$$\beta_n = \beta + \tfrac{1}{2}(n-1)s^2 + \tfrac{1}{2}(n_0 + n)^{-1} n_0 n \left(\theta_0 - \overline{Y}\right)^2$$
$$s^2 = (n-1)^{-1}\sum_{i=1}^{n}(Y_i - \overline{Y})^2$$

— Gaussian $\left(\theta \mid \sigma^2; \theta_n, \sigma_n^2\right) \times$ inverse-gamma$\left(\sigma^2 \mid \alpha + \tfrac{1}{2}n, \beta_n\right)$. Hence, the conditional distribution for the mean given the variance is Gaussian$\left(\theta \mid \sigma^2; \theta_n, \sigma_n^2\right)$ where $\sigma_n^2 = \dfrac{\sigma^2}{n_0+n}$. On integrating out the variance from the joint posterior the marginal posterior for the mean is noncentral, scaled Student t$(\theta \mid \theta_n, \gamma, \nu)$ distributed.

A scaled Student t$\left(X \mid \mu, \lambda = \tfrac{1}{\sigma^2}, \nu\right)$ is symmetric with mean μ, variance $\tfrac{1}{\lambda}\tfrac{\nu}{\nu-2} = \sigma^2 \tfrac{\nu}{\nu-2}$, ν degrees of freedom, and the density function kernel is

$$\left[1 + \nu^{-1}\lambda(X-\mu)^2\right]^{-(\nu+1)/2} = \left[1 + \nu^{-1}\left(\frac{X-\mu}{\sigma}\right)^2\right]^{-(\nu+1)/2}$$

---

can be reparameterized as an inverse-χ^2 distribution(ν, σ_0^2)

$$p\left(\sigma^2; \nu, \sigma_0^2\right) \propto \left(\sigma^2\right)^{-(\nu/2+1)} \exp\left[-\frac{\nu\sigma_0^2}{2\sigma^2}\right]$$

(see Gelman et al [2003], p. 50). Hence, $\alpha = \tfrac{\nu}{2}$ or $\nu = 2\alpha$ and $\beta = \tfrac{\nu\sigma_0^2}{2}$ or $\nu\sigma_0^2 = 2\beta$.

Hence, the *standard* t distribution is Student t$(Z \mid 0, 1, \nu)$ where $Z = \frac{X-\mu}{\sigma}$. Marginalization of the mean follows Gelman et al [2003] p. 76. For uninformative priors, $p(\theta, \sigma^2) \propto \sigma^{-2}$

$$p(\theta \mid y) = \int_0^\infty p(\theta, \sigma^2 \mid y)[d\sigma^2]$$

$$= \int_0^\infty \sigma^{-n-2} \exp\left[-\frac{A}{2\sigma^2}\right] d\sigma^2$$

where $A = (n-1)s^2 + n(\theta - \bar{y})^2$. Let $z = \frac{A}{2\sigma^2}$, then transformation of variables yields

$$p(\theta \mid y) \propto A^{-n/2} \int_0^\infty z^{(n-2)/2} \exp[-z] dz$$

Since the integral involves the kernel for a gamma, it integrates to a constant and can be ignored for identifying the marginal posterior kernel. Hence, we recognize

$$p(\theta \mid y) \propto A^{-n/2} = \left[(n-1)s^2 + n(\theta - \bar{y})^2\right]^{-\frac{n}{2}}$$

$$\propto \left[1 + \frac{n(\theta - \bar{y})^2}{(n-1)s^2}\right]^{-\frac{n}{2}}$$

is the kernel for a noncentral, scaled Student t$\left(\theta; \bar{y}, \frac{s^2}{n}, n-1\right)$. Marginalization with informed conjugate priors works in analogous fashion.

Table 7.3: Conjugate families for multivariate discrete distributions

likelihood $p(Y \mid \theta)$	conjugate prior $p(\theta)$	posterior $p(\theta \mid Y)$
Multinomial$_k$ $(r; \theta, n)$ where $r_i = 0, 1, 2, \ldots$	Dirichlet$_k$ $(\theta; \alpha)$ where $\alpha = \{\alpha_1, \ldots, \alpha_{k+1}\}$	Dirichlet$_k$ $\left(\theta; \begin{array}{c} \alpha_1 + r_1, \ldots, \\ \alpha_{k+1} + r_{k+1} \end{array}\right)$

The Dirichlet distribution is a multivariate analog to the beta distribution and has continuous support where $r_{k+1} = n - \sum_{\ell=1}^{k} r_\ell$. Ferguson [1973] proposed the Dirichlet process as a Bayesian nonparametric approach. Some properties of the Dirichlet distribution include

$$E[\theta_i \mid \alpha] = \frac{\alpha_i}{\alpha_0}$$

$$Var[\theta_i \mid \alpha] = \frac{\alpha_i(\alpha_0 - \alpha_i)}{\alpha_0^2(\alpha_0 + 1)}$$

$$Cov\left[\theta_i, \theta_j \mid \alpha\right] = \frac{-\alpha_i \alpha_j}{\alpha_0^2 (\alpha_0 + 1)}$$

where $\alpha_0 = \sum_{i=1}^{k+1} \alpha_i$

Table 7.4: Conjugate families for multivariate continuous distributions

likelihood $p(Y \mid \theta)$	conjugate prior $p(\theta)$	marginal posterior $p(\theta \mid Y)$
Normal $(Y \mid \theta, \Sigma)$ parameters unknown	Normal$(\theta \mid \Sigma; \theta_0, n_0)$ *Inverse- Wishart $(\Sigma; \alpha, \beta)$	Student $t_k\left(\theta; \theta_n, \Gamma, 2\alpha_n\right)$; Inverse- Wishart $\left(\Sigma; \alpha + \tfrac{1}{2}n, \beta_n\right)$
Linear regression Normal$(Y \mid X\theta, \sigma^2)$ parameters unknown	Normal$\left(\theta \mid \sigma^2; \theta_0, n_0^{-1}\sigma^2\right)$ *Inverse- gamma $\left(\sigma^2; \alpha, \beta\right)$	Student $t_k\left(\theta; \theta_n, \Gamma, 2\alpha + n\right)$; Inverse- gamma $\left(\sigma^2; \alpha + \tfrac{1}{2}n, \beta_n\right)$

The multivariate Student $t_k(X \mid \mu, \Gamma, \nu)$ is analogous to the univariate Student $t(X \mid \mu, \gamma, \nu)$ as it is symmetric with mean vector (length k) μ, $k \times k$ symmetric, positive definite variance matrix $\Gamma^{-1}\frac{\nu}{\nu-2}$, and ν degrees of freedom. For the Student t and inverse-Wishart marginal posteriors associated with multivariate normal likelihood function, the parameters are

$$\theta_n = (n_0 + n)^{-1}\left(n_0 \theta_0 + n\overline{Y}\right)$$

$$\Gamma = (n + n_0)\alpha_n \beta_n^{-1}$$

$$\beta_n = \beta + \frac{1}{2}S + \frac{1}{2}(n_0 + n)^{-1} n_0 n \left(\theta_0 - \overline{Y}\right)\left(\theta_0 - \overline{Y}\right)^T$$

$$S = \sum_{i=1}^{n}\left(Y_i - \overline{Y}\right)\left(Y_i - \overline{Y}\right)^T$$

$$\alpha_n = \alpha + \frac{1}{2}n - \frac{1}{2}(k-1)$$

For the Student t and inverse-gamma marginal posteriors associated with linear regression, the parameters are[8]

$$\theta_n = \left(n_0 + X^T X\right)^{-1}\left(n_0 \theta_0 + X^T Y\right)$$

$$n_0 = X_0^T X_0$$

$$\Gamma = \left(n_0 + X^T X\right)\left(\alpha + \frac{1}{2}n\right)\beta_n^{-1}$$

---

[8] Notice, linear regression subsumes the univariate, multi-parameter Gaussian case. If we let $X = \iota$ (a vector of ones), then linear regression becomes the univariate Gaussian case.

$$\beta_n = \beta + \frac{1}{2}(Y - X\theta_n)^T Y + \frac{1}{2}(\theta_0 - \theta_n)^T n_0 \theta_0$$

Bayesian regression with conjugate priors works as if we have data from a prior period $\{Y_0, X_0\}$ and the current period $\{Y, X\}$ from which to estimate θ_n. Applying *OLS* to the stack of equations $\begin{array}{c} Y_0 \\ Y \end{array} = \begin{array}{c} X_0 \\ X \end{array} \theta_n + \begin{array}{c} \varepsilon_0 \\ \varepsilon \end{array}$ yields[9]

$$\begin{aligned}\theta_n &= \left(X_0^T X_0 + X^T X\right)^{-1}\left(X_0^T Y_0 + X^T Y\right) \\ &= \left(n_0 + X^T X\right)^{-1}\left(n_0 \theta_0 + X^T Y\right)\end{aligned}$$

The inverse-Wishart and multivariate Student t distributions are multivariate analogs to the inverse-gamma and (noncentral, scaled) univariate Student t distributions, respectively.

7.3.2 McMC simulations

Markov chain Monte Carlo (*McMC*) simulations are employed when the marginal posterior distributions cannot be derived or are extremely cumbersome to derive. McMC approaches draw from the set of conditional posterior distributions instead of the marginal posterior distributions. The Hammersley-Clifford theorem (Hammersley and Clifford [1971] and Besag [1974]) provides regulatory conditions

---

[9] This perspective of Bayesian regression is consistent with recursive least squares where the previous estimate θ_{t-1} based on data $\{Y_{t-1}, X_{t-1}\}$ is updated for data $\{Y_t, X_t\}$ as $\theta_t = \theta_{t-1} + \Im_t^{-1} X_t^T (Y_t - X_t \theta_{t-1})$, where $\theta_{t-1} = \left(X_{t-1}^T X_{t-1}\right)^{-1} X_{t-1}^T Y_{t-1}$ and the information matrix is updated as $\Im_t = \Im_{t-1} + X_t^T X_t$. To see this, note

$$\theta_t = \Im_t^{-1} X_t^T Y_t + \left(I - \Im_t^{-1} X_t^T X_t\right) \theta_{t-1}$$

but

$$\begin{aligned}\left(I - \Im_t^{-1} X_t^T X_t\right) \theta_{t-1} &= \Im_t^{-1}\left(\Im_t - X_t^T X_t\right) \theta_{t-1} \\ &= \Im_t^{-1}\left(X_{t-1}^T X_{t-1} + X_t^T X_t - X_t^T X_t\right)\theta_{t-1} \\ &= \Im_t^{-1} X_{t-1}^T X_{t-1} \theta_{t-1} \\ &= \Im_t^{-1} X_{t-1}^T Y_{t-1}\end{aligned}$$

since $\theta_{t-1} = \left(X_{t-1}^T X_{t-1}\right)^{-1} X_{t-1}^T Y_{t-1}$. Hence,

$$\begin{aligned}\theta_t &= \Im_t^{-1} X_t^T Y_t + \Im_t^{-1} X_{t-1}^T Y_{t-1} \\ &= \Im_t^{-1}\left(X_t^T Y_t + X_{t-1}^T Y_{t-1}\right) \\ &= \left(X_{t-1}^T X_{t-1} + X_t^T X_t\right)^{-1}\left(X_{t-1}^T Y_{t-1} + X_t^T Y_t\right)\end{aligned}$$

or, in the notation above

$$\theta = \left(X_0^T X_0 + X^T X\right)^{-1}\left(X_0^T Y_0 + X^T Y\right)$$

as indicated above.

for when a set of conditional distributions characterizes a unique joint distribution. The regulatory conditions are essentially that every point in the marginal and conditional distributions have positive mass. Common *McMC* approaches (Gibbs sampler and Metropolis-Hastings algorithm) are supported by the Hammersley-Clifford theorem. The utility of *McMC* simulation has evolved along with the **R** Foundation for Statistical Computing.

Gibbs sampler

Suppose we cannot derive $p(\theta \mid Y)$ in closed form (it does not have a standard probability distribution) but we can identify the conditional posterior distributions. We can utilize the full conditional posterior distributions to draw dependent samples for parameters of interest via *McMC* simulation.

For full conditional posterior distributions

$$p(\theta_1 \mid \theta_{-1}, Y)$$
$$\vdots$$
$$p(\theta_k \mid \theta_{-k}, Y)$$

draws are made for θ_1 conditional on starting values for parameters other than θ_1, that is θ_{-1}. Then, θ_2 is drawn conditional on the θ_1 draw and the starting value for the remaining θ. Next, θ_3 is drawn conditional on the draws for θ_1 and θ_2 and the remaining θ. This continues until all θ have been sampled. Then the sampling is repeated for a large number of draws with parameters updated each iteration by the most recent draw.

The samples are dependent. Not all samples will be from the posterior; only after a finite (but unknown) number of iterations are draws from the marginal posterior distribution (see Gelfand and Smith [1990]). (Note, in general, $p(\theta_1, \theta_2 \mid Y) \neq p(\theta_1 \mid \theta_2, Y) p(\theta_1 \mid \theta_2, Y)$.) Convergence is usually checked using trace plots, burn-in iterations, and other convergence diagnostics. Model specification includes convergence checks, sensitivity to starting values and possibly prior distribution and likelihood assignments, comparison of draws from the posterior predictive distribution with the observed sample, and various goodness of fit statistics.

Albert and Chib's Gibbs sampler Bayes' probit

The challenge with discrete choice models (like probit) is that latent utility is unobservable, rather the analyst observes only discrete (usually binary) choices (see chapter 5). Albert & Chib [1993] employ Bayesian data augmentation to "supply" the latent variable. Hence, parameters of a probit model are estimated via normal Bayesian regression (see earlier discussion in this chapter). Consider the latent utility model

$$U_D = W\theta - V$$

The conditional posterior distribution for θ is

$$p(\theta \mid D, W, U_D) \sim N\left(b_1, \left(Q^{-1} + W^T W\right)^{-1}\right)$$

7.3 Bayesian simulation

where
$$b_1 = \left(Q^{-1} + W^T W\right)^{-1} \left(Q^{-1} b_0 + W^T W b\right)$$
$$b = \left(W^T W\right)^{-1} W^T U_D$$

b_0 = prior means for θ and $Q = \left(W_0^T W_0\right)^{-1}$ is the prior for the covariance. The conditional posterior distribution for the latent variables are

$$p(U_D | D = 1, W, \theta) \sim N(W\theta, I | U_D > 0) \text{ or } TN_{(0,\infty)}(W\theta, I)$$
$$p(U_D | D = 0, W, \theta) \sim N(W\theta, I | U_D \leq 0) \text{ or } TN_{(-\infty,0)}(W\theta, I)$$

where $TN(\cdot)$ refers to random draws from a truncated normal (truncated below for the first and truncated above for the second). Iterative draws for $(U_D | D, W, \theta)$ and $(\theta | D, W, U_D)$ form the Gibbs sampler. Interval estimates of θ are supplied by post-convergence draws of $(\theta | D, W, U_D)$. For simulated normal draws of the unobservable portion of utility, V, this Bayes' augmented data probit produces remarkably similar inferences to *MLE*.[10]

Metropolis-Hastings algorithm

If neither some conditional posterior, $p(\theta_j | Y, \theta_{-j})$, or the marginal posterior, $p(\theta | Y)$, is recognizable, then we can employ the Metropolis-Hastings (*MH*) algorithm. The Gibbs sampler is a special case of the *MH* algorithm. The random walk Metropolis algorithm is most common and outlined next.

The random walk Metropolis algorithm is as follows. We wish to draw from $p(\theta | \cdot)$ but we only know $p(\theta | \cdot)$ up to constant of proportionality, $p(\theta | \cdot) = cf(\theta | \cdot)$ where c is unknown.

- Let $\theta^{(k-1)}$ be a draw from $p(\theta | \cdot)$.[11]

- Draw θ^* from $N\left(\theta^{(k-1)}, s^2\right)$ where s^2 is fixed.

---

[10] An efficient algorithm for this Gibbs sampler probit, rbprobitGibbs, is available in the bayesm package of R (http://www.r-project.org/), the open source statistical computing project. Bayesm is a package written to complement Rossi, Allenby, and McCulloch [2005].

[11] The procedure describes the algorithm for a single parameter. A general K parameter algorithm works similarly (see Train [2002], p. 305):
(a) Start with a value β_n^0.
(b) Draw K independent values from a standard normal density, and stack the draws into a vector labeled η^1.
(c) Create a trial value of $\beta_n^1 = \beta_n^0 + \sigma \Gamma \eta^1$ where σ is the researcher-chosen jump size parameter, Γ is the Cholesky factor of W such that $\Gamma \Gamma^T = W$. Note the proposal distribution is specified to be normal with zero mean and variance $\sigma^2 W$.
(d) Draw a standard uniform variable μ_1.
(e) Calculate the ratio $F = \frac{L(y_n | \beta_n^1) \phi(\beta_n^1 | b, W)}{L(y_n | \beta_n^0) \phi(\beta_n^0 | b, W)}$ where $L(y_n | \beta_n^1)$ is a product of logits, and $\phi(\beta_n^1 | b, W)$ is the normal density.
(f) If $\mu_1 \leq F$, accept β_n^1; if $\mu_1 > F$, reject β_n^1 and let $\beta_n^1 = \beta_n^0$.
(g) Repeat the process many times. For sufficiently large t, β_n^t is a draw from the marginal posterior.

7. Repeated-sampling inference

- Let $\alpha = min\left\{1, \frac{p(\theta^*|\cdot)}{p(\theta^{(k-1)}|\cdot)} = \frac{cf(\theta^*|\cdot)}{cf(\theta^{(k-1)}|\cdot)}\right\}$.

- Draw z^* from $U(0, 1)$.

- If $z^* < \alpha$ then $\theta^{(k)} = \theta^*$, otherwise $\theta^{(k)} = \theta^{(k-1)}$. In other words, with probability α set $\theta^{(k)} = \theta^*$, and otherwise set $\theta^{(k)} = \theta^{(k-1)}$.[12]

These draws converge to random draws from the marginal posterior distribution after a burn-in interval if properly tuned.

Tuning the Metropolis algorithm involves selecting s^2 (jump size) so that the parameter space is explored appropriately (see Halton sequences discussion below). Usually, smaller jump size results in more accepts and larger jump size results in fewer accepts. If s^2 is too small, the Markov chain will not converge quickly, has more serial correlation in the draws, and may get stuck at a local mode (multi-modality can be a problem). If s^2 is too large, the Markov chain will move around too much and not be able to thoroughly explore areas of high posterior probability. Of course, we desire concentrated samples from the posterior distribution. A commonly-employed rule of thumb is to target an acceptance rate for θ^* around 30% (20 − 80% is usually considered "reasonable").[13]

Some other *McMC* methods

Other acceptance sampling procedures such as WinBUGs (see Spiegelhalter, et al. [2003]) are self-tuned. That is, the algorithm adaptively tunes the jump size in generating random post convergence joint posterior draws. A difficulty with WinBUGs is that it can mysteriously crash with little diagnostic aid.

Halton sequences

Random sampling can be slow to provide good coverage and hence prove to be a costly way to simulate data. An alternative that provides better coverage with fewer draws involves *Halton sequences* (see Train [2002], ch. 9, p. 224-238). Unlike other methods discussed above, Halton draws tend to be negatively correlated. Importantly, Bhat [2001] finds that 100 Halton draws provided lower simulation error for his mixed logit than $1,000$ random draws, for discrete choice models. Further, the error rate with 125 Halton draws was half as large as with $1,000$ random draws and somewhat smaller than with $2,000$ random draws.

A Halton sequence builds around a pre-determined number k (usually a prime number). The Halton sequence is

$$s_{t+1} = \left\{s_t, s_t + \frac{1}{k^t}, s_t + \frac{2}{k^t}, \ldots, s_t + \frac{k-1}{k^t}\right\}$$

---

[12] A modification of the *RW* Metropolis algorithm sets $\theta^{(k)} = \theta^*$ with $log(\alpha)$ probability where $\alpha = min\{0, log[f(\theta^*|\cdot)] - log[f(\theta^{(k-1)}|\cdot)]\}$.

[13] Gelman, et al [2004] report the optimal acceptance rate is 0.44 when the number of parameters $K = 1$ and drops toward 0.23 as K increases.

7.3 Bayesian simulation

starting with $s_0 = 0$ (even though zero is ignored). An example helps to fix ideas.

Example 7.1 *Consider the prime $k = 3$. The sequence through two iterations is*

$$\left\{\begin{array}{c} 0 + 1/3 = 1/3, 0 + 2/3 = 2/3, \\ 0 + 1/9 = 1/9, 1/3 + 1/9 = 4/9, 2/3 + 1/9 = 7/9, \\ 0 + 2/9 = 2/9, 1/3 + 2/9 = 5/9, 2/3 + 2/9 = 8/9, \ldots \end{array}\right\}$$

This procedure describes uniform Halton draws. Other distributions are accommodated in the usual way — by inverse distribution functions.

Example 7.2 *For example, normal draws are found by $\Phi^{-1}(s_t)$. Continuing with the above Halton sequence, standard normal draws are*

$$\left\{\begin{array}{c} \Phi^{-1}(1/3) \approx -0.43, \Phi^{-1}(2/3) \approx 0.43, \\ \Phi^{-1}(1/9) \approx -1.22, \Phi^{-1}(4/9) \approx -0.14, \Phi^{-1}(7/9) \approx 0.76, \\ \Phi^{-1}(2/9) \approx -0.76, \Phi^{-1}(5/9) \approx 0.14, \Phi^{-1}(8/9) \approx 1.22, \ldots \end{array}\right\}$$

Example 7.3 *For two independent standard normal unobservables we create Halton sequences for each from different primes and transform. Suppose we use $k = 2$ and $k = 3$. The first few draws are*

$$\left\{\begin{array}{c} \varepsilon_1 = \left\langle \Phi^{-1}\left(\tfrac{1}{2}\right) = 0, \Phi^{-1}\left(\tfrac{1}{3}\right) = -0.43 \right\rangle, \\ \varepsilon_2 = \left\langle \Phi^{-1}\left(\tfrac{1}{4}\right) = -.67, \Phi^{-1}\left(\tfrac{2}{3}\right) = 0.43 \right\rangle, \\ \varepsilon_3 = \left\langle \Phi^{-1}\left(\tfrac{3}{4}\right) = 0.67, \Phi^{-1}\left(\tfrac{1}{9}\right) = -1.22 \right\rangle, \\ \varepsilon_4 = \left\langle \Phi^{-1}\left(\tfrac{1}{8}\right) = -1.15, \Phi^{-1}\left(\tfrac{4}{9}\right) = -0.14 \right\rangle, \\ \varepsilon_5 = \left\langle \Phi^{-1}\left(\tfrac{5}{8}\right) = 0.32, \Phi^{-1}\left(\tfrac{7}{9}\right) = 0.76 \right\rangle, \\ \varepsilon_6 = \left\langle \Phi^{-1}\left(\tfrac{3}{8}\right) = -0.32, \Phi^{-1}\left(\tfrac{2}{9}\right) = -0.76 \right\rangle, \\ \varepsilon_7 = \left\langle \Phi^{-1}\left(\tfrac{7}{8}\right) = 1.15, \Phi^{-1}\left(\tfrac{5}{9}\right) = 0.14 \right\rangle, \ldots \end{array}\right\}$$

As the initial cycle of elements (from near zero to near one) for multiple dimension sequences are highly correlated, the initial elements are usually discarded (treated as burn-in). The number of elements discarded is at least as large as the largest prime used in creating the sequences. Since primes cycle at different rates after the first cycle, primes are more effective bases (they have smaller correlation) for Halton sequences.

Randomized Halton draws

Halton sequences are systematic, not random, while asymptotic properties of estimators assume random (or at least pseudo-random) draws of unobservables. Halton sequences can be transformed in a way that makes draws pseudo-random (as is the case for all computer-based randomizations). Bhat [2003] suggests the following procedure:
1. Take a draw μ from a standard uniform distribution.
2. Add μ to each element of the Halton sequence. If the resulting element exceeds one, subtract 1 from it. That is, $s_n = \text{mod}(s_0 + \mu)$ where s_0 (s_n) is the original (transformed) element of the Halton sequence and $\text{mod}(\cdot)$ returns the fractional

part of the argument.

Suppose $\mu = 0.4$ for the above Halton sequence (again through two iterations), the pseudo-random sequence is

$$\{0.4, 0.733, 0.067, 0.511, 0.844, 0.178, 0.622, 0.956, 0.289, \ldots\}$$

The spacing remains the same so we achieve the same coverage but draws are random. In a sense, this "blocking" approach is similar to bootstrapping regressions with heteroskedastic and/or correlated errors. A different draw for μ is taken for each unobservable.

Bhat [2003] also proposes scrambled Halton draws to deal with high dimension issues. Halton sequences for high dimension problems utilize larger prime numbers. For large prime numbers, correlation in the sequences may persist for much longer than the first cycle as discussed above. Bhat proposes scrambling the sequence so that if we think of the above sequence as BC then the sequence is reversed to be CB where $B = \frac{1}{3}$ and $C = \frac{2}{3}$. Different permutations are employed for different primes. Continuing with the above Halton sequence for $k = 3$, the original and scrambled sequences are tabulated below.

Original	Scrambled
1/3	2/3
2/3	1/3
1/9	2/9
4/9	8/9
7/9	5/9
2/9	1/9
5/9	7/9
8/9	4/9

7.4 Additional reading

Kreps [1988, ch. 11] and McCall [1991] discuss exchangeability and de Finetti's theorem as well as implications for economics. Davidson and MacKinnon [2003], MacKinnon [2002], and Cameron and Trivedi [2005] discuss bootstrapping, pivotal statistics, etc., and Horowitz [2001] provides an extensive discussion of bootstrapping. Casella and George [1992] and Chib and Hamilton [1995] offer basic introductions to the Gibbs sampler and Metropolis-Hastings algorithm, respectively. Tanner and Wong [1987] discuss calculating posterior distributions by data augmentation. Train [2002, ch. 9] discusses various Halton sequence approaches and other remaining open questions associated with this relatively new, but promising technique.

8
Overview of endogeneity

"A government study today revealed that 83% of statistics are misleading."
- Ziggy by Tom Wilson

As discussed in chapter 2, managers actively make production-investment, financing, and accounting choices. These choices are intertwined and far from innocuous. Design of accounting (like other information systems) is highly dependent on the implications and responses to accounting information in combination with other information. As these decisions are interrelated, their analysis is inherently endogenous (Demski [2004]). Endogeneity presents substantial challenges for econometric analysis. The behavior of unobservable (to the analyst) components and omitted, correlated variables are continuing themes.

In this chapter, we briefly overview econometric analysis of endogeneity, explore some highly stylized examples that motivate its importance, and lay some ground work for exploring treatment effects in the following chapters. A theme for this discussion is that econometric analysis of endogeneity is a three-legged problem: theory, data, and model specification (or logically consistent discovery of the *DGP*). Failure to support any leg and the entire inquiry is likely to collapse. Progress is impeded when authors fail to explicitly define the causal effects of interest or state what conditions are perceived for identification of the estimand of interest. As Heckman and Vytlacil [2007] argue in regards to the economics literature, this makes it difficult to build upon past literature and amass a coherent body of evidence. We explore various identifying conditions in the ensuing discussions of endogenous causal effects.

8.1 Overview

Many, perhaps all, endogeneity concerns can be expressed in the form of an omitted, correlated variable problem. We remind the reader (see chapter 3) that standard parameter estimators (such as *OLS*) are not asymptotically consistent in the face of omitted, correlated variables.

8.1.1 Simultaneous equations

When many of us think of endogeneity, simultaneous equations is one of the first settings that comes to mind. That is, when we have multiple variables whose behavior are interrelated such that they are effectively simultaneously determined, endogeneity is a first-order consideration. For instance, consider a simple example where the *DGP* is expressed as the following structural equations[1]

$$Y_1 = \beta_1 X_1 + \beta_2 Y_2 + \varepsilon_1$$
$$Y_2 = \gamma_1 X_2 + \gamma_2 Y_1 + \varepsilon_2$$

Clearly, little can be said about either Y_1 or Y_2 without including the other (a form of omitted variable). It is not possible to speak of manipulation of only Y_1 or Y_2. Perhaps, this is most readily apparent if we rewrite the equations in reduced form:

$$\begin{bmatrix} 1 & -\beta_2 \\ -\gamma_2 & 1 \end{bmatrix} \begin{bmatrix} Y_1 \\ Y_2 \end{bmatrix} = \begin{bmatrix} \beta_1 & 0 \\ 0 & \gamma_1 \end{bmatrix} \begin{bmatrix} X_1 \\ X_2 \end{bmatrix} + \begin{bmatrix} \varepsilon_1 \\ \varepsilon_2 \end{bmatrix}$$

assuming $\beta_2 \gamma_2 \neq 1$

$$\begin{bmatrix} Y_1 \\ Y_2 \end{bmatrix} = \begin{bmatrix} 1 & -\beta_2 \\ -\gamma_2 & 1 \end{bmatrix}^{-1} \left\{ \begin{bmatrix} \beta_1 & 0 \\ 0 & \gamma_1 \end{bmatrix} \begin{bmatrix} X_1 \\ X_2 \end{bmatrix} + \begin{bmatrix} \varepsilon_1 \\ \varepsilon_2 \end{bmatrix} \right\}$$

$$Y_1 = \frac{\beta_1}{1 - \beta_2 \gamma_2} X_1 + \frac{\beta_2 \gamma_1}{1 - \beta_2 \gamma_2} X_2 + \frac{1}{1 - \beta_2 \gamma_2} \varepsilon_1 + \frac{\beta_2}{1 - \beta_2 \gamma_2} \varepsilon_2$$

$$Y_2 = \frac{\beta_1 \gamma_2}{1 - \beta_2 \gamma_2} X_1 + \frac{\gamma_1}{1 - \beta_2 \gamma_2} X_2 + \frac{\gamma_2}{1 - \beta_2 \gamma_2} \varepsilon_1 + \frac{1}{1 - \beta_2 \gamma_2} \varepsilon_2$$

which can be rewritten as

$$Y_1 = \omega_{11} X_1 + \omega_{12} X_2 + \eta_1$$
$$Y_2 = \omega_{21} X_1 + \omega_{22} X_2 + \eta_2$$

where $Var \begin{bmatrix} \eta_1 \\ \eta_2 \end{bmatrix} = \begin{bmatrix} v_{11} & v_{12} \\ v_{12} & v_{22} \end{bmatrix}$. Since rank and order conditions are satisfied (assuming $\beta_2 \gamma_2 \neq 1$), the structural parameters can be recovered from the reduced

---

[1]Goldberger [1972, p. 979] defines structural equations as an approach that employs "stochastic models in which each equation represents a causal link, rather than a mere empirical association."

form parameters as follows.

$$\beta_1 = \omega_{11} - \frac{\omega_{12}\omega_{21}}{\omega_{22}}$$

$$\beta_2 = \frac{\omega_{12}}{\omega_{22}}$$

$$\gamma_1 = \omega_{22} - \frac{\omega_{12}\omega_{21}}{\omega_{11}}$$

$$\gamma_2 = \frac{\omega_{21}}{\omega_{11}}$$

$$Var\left[\varepsilon_1\right] = v_{11} + \frac{\omega_{12}\left(v_{22}\omega_{12} - 2v_{12}\omega_{22}\right)}{\omega_{22}^2}$$

$$Var\left[\varepsilon_2\right] = v_{22} + \frac{\omega_{21}\left(v_{11}\omega_{21} - 2v_{12}\omega_{11}\right)}{\omega_{11}^2}$$

$$Cov\left[\varepsilon_1, \varepsilon_2\right] = \frac{v_{12}\left(\omega_{12}\omega_{21} + \omega_{11}\omega_{22}\right) - v_{11}\omega_{21}\omega_{22} - v_{22}\omega_{11}\omega_{12}}{\omega_{11}\omega_{22}}$$

Suppose the causal effects of interest are β_1 and γ_1. Examination of the reduced form equations reveals that ignoring simultaneity produces inconsistent estimates of β_1 and γ_1 even if X_1 and X_2 are uncorrelated (unless β_2 or $\gamma_2 = 0$).

More naively, suppose we attempt to estimate the structural equations directly (say, via *OLS*). Since the response variables are each a function of the other response variable, the regressors are correlated with the errors and the fundamental condition of regression $E\left[X^T \varepsilon\right] = 0$ is violated and *OLS* parameter estimates are inconsistent. A couple of recursive substitutions highlight the point. For illustrative purposes, we work with Y_1 but the same ideas obviously apply to Y_2.

$$\begin{aligned} Y_1 &= \beta_1 X_1 + \beta_2 Y_2 + \varepsilon_1 \\ &= \beta_1 X_1 + \beta_2 \left(\gamma_1 X_2 + \gamma_2 Y_1 + \varepsilon_2\right) + \varepsilon_1 \end{aligned}$$

Of course, if $E\left[\varepsilon_2^T \varepsilon_1\right] \neq 0$ then we've demonstrated the point; notice this is a standard endogenous regressor problem. Simultaneity bias (inconsistency) is illustrated with one more substitution.

$$Y_1 = \beta_1 X_1 + \beta_2 \left(\gamma_1 X_2 + \gamma_2 \left(\beta_1 X_1 + \beta_2 Y_2 + \varepsilon_1\right) + \varepsilon_2\right) + \varepsilon_1$$

Since Y_2 is a function of Y_1, inclusion of Y_2 as a regressor produces a clear violation of $E\left[X^T \varepsilon\right] = 0$ as we have $E\left[\varepsilon_1^T \varepsilon_1\right] \neq 0$.

Notice, we can think of simultaneity problems as arising from omitted, correlated unobservable variables. Hence, this simple example effectively identifies the basis — *omitted, correlated unobservable variables* — for most (perhaps all) endogeneity concerns. Further, this simple structural example readily connects to estimation of causal effects.

Definition 8.1 *Causal effects are the ceteris paribus response to a change in variable or parameter (Marshall [1961] and Heckman [2000]).*

As the simultaneity setting illustrates, endogeneity often makes it infeasible to "turn one dial at a time."

8.1.2 Endogenous regressors

Linear models with endogenous regressors are commonplace (see Larcker and Rusticus [2004] for an extensive review of the accounting literature). Suppose the *DGP* is

$$Y_1 = X_1\beta_1 + Y_2\beta_2 + \varepsilon_1$$

$$Y_2 = \gamma_1 X_2 + \varepsilon_2$$

where $E\left[X^T\varepsilon_1\right] = 0$ and $E\left[X_2^T\varepsilon_2\right] = 0$ but $E\left[\varepsilon_2^T\varepsilon_1\right] \neq 0$. In other words, $Y_1 = \beta_1 X_1 + \beta_2(\gamma_1 X_2 + \varepsilon_2) + \varepsilon_1$. Of course, *OLS* produces inconsistent estimates. Instrumental variables (*IV*) are a standard remedy. Suppose we observe variables X_2. Variables X_2 are clearly instruments as they are unrelated to ε_1 but highly correlated with the endogenous regressors Y_2 (assuming $\gamma_1 \neq 0$).

Two-stage least squares instrumental variable (*2SLS-IV*) estimation is a standard approach for dealing with endogenous regressors. In the first stage, project all of the regressors (endogenous plus exogenous) onto the instruments plus all other exogenous regressors (see chapter 3 on overidentifying restrictions and *IV*). Let $X = \begin{bmatrix} X_1 & Y_2 \end{bmatrix}$ and $Z = \begin{bmatrix} X_1 & X_2 \end{bmatrix}$

$$\hat{X} = Z\left(Z^TZ\right)^{-1}Z^TX = P_ZX = \begin{bmatrix} P_ZX_1 & P_ZY_2 \end{bmatrix}$$

In the second stage, replace the regressors with the predicted values from the first stage regression.

$$Y_1 = P_ZX_1\beta_1 + P_ZY_2\beta_2 + \varepsilon_1'$$

The *IV* estimator for β (for convenience, we have reversed the order of the variables) is

$$\left(\begin{bmatrix} Y_2^T P_Z \\ X_1^T P_Z \end{bmatrix} \begin{bmatrix} P_ZY_2 & P_ZX_1 \end{bmatrix}\right)^{-1} \begin{bmatrix} Y_2^T P_Z \\ X_1^T P_Z \end{bmatrix} Y_1$$

The probability limit of the estimator is

$$\text{plim} \begin{bmatrix} Y_2^T P_Z Y_2 & Y_2^T P_Z X_1 \\ X_1^T P_Z Y_2 & X_1^T P_Z X_1 \end{bmatrix}^{-1} \begin{bmatrix} Y_2^T P_Z \\ X_1^T P_Z \end{bmatrix} (X_1\beta_1 + Y_2\beta_2 + \varepsilon_1)$$

$$= \begin{matrix} \beta_2 \\ \beta_1 \end{matrix}$$

To see this, recall the inverse of the partitioned matrix

$$\begin{bmatrix} Y_2^T P_Z Y_2 & Y_2^T P_Z X_1 \\ X_1^T P_Z Y_2 & X_1^T P_Z X_1 \end{bmatrix}^{-1}$$

via block "rank-one" LDL^T representation (see *FWL* in chapter 3) is

$$\left(\begin{bmatrix} I & 0 \\ A & I \end{bmatrix} \begin{bmatrix} Y_2^T P_Z Y_2 & 0 \\ 0 & X_1^T P_Z M_{P_Z Y_2} P_Z X_1 \end{bmatrix} \begin{bmatrix} I & A^T \\ 0 & I \end{bmatrix}\right)^{-1}$$

where $A = X_1^T P_Z Y_2 \left(Y_2^T P_Z Y_2\right)^{-1}$. Simplification gives

$$\begin{bmatrix} I & -A^T \\ 0 & I \end{bmatrix} \begin{bmatrix} \left(Y_2^T P_Z Y_2\right)^{-1} & 0 \\ 0 & B \end{bmatrix} \begin{bmatrix} I & 0 \\ -A & I \end{bmatrix}$$

$$= \begin{bmatrix} \left(Y_2^T P_Z Y_2\right)^{-1} + A^T B A & -A^T B \\ -BA & B \end{bmatrix}$$

where

$$B = \left(X_1^T P_Z M_{P_Z Y_2} P_Z X_1\right)^{-1}$$

and

$$M_{P_Z Y_2} = I - P_Z Y_2 \left(Y_2^T P_Z Y_2\right)^{-1} Y_2^T P_Z$$

Now, focus on the second equation.

$$\begin{aligned}
&= \left(-BAY_2^T P_Z + BX_1^T P_Z\right) Y_1 \\
&= \begin{Bmatrix} -\left(X_1^T P_Z M_{P_Z Y_2} P_Z X_1\right)^{-1} X_1^T P_Z Y_2 \left(Y_2^T P_Z Y_2\right)^{-1} Y_2^T P_Z \\ + \left(X_1^T P_Z M_{P_Z Y_2} P_Z X_1\right)^{-1} X_1^T P_Z \end{Bmatrix} Y_1 \\
&= \left(X_1^T P_Z M_{P_Z Y_2} P_Z X_1\right)^{-1} X_1^T P_Z \left(I - P_Z Y_2 \left(Y_2^T P_Z Y_2\right)^{-1} Y_2^T P_Z\right) \\
&\quad (X_1 \beta_1 + Y_2 \beta_2 + \varepsilon_1) \\
&= \left(X_1^T P_Z M_{P_Z Y_2} P_Z X_1\right)^{-1} X_1^T P_Z M_{P_Z Y_2} (X_1 \beta_1 + Y_2 \beta_2 + \varepsilon_1)
\end{aligned}$$

Since $P_Z M_{P_Z Y_2} = P_Z M_{P_Z Y_2} P_Z$, the second equation can be rewritten as

$$\begin{aligned}
&\left(X_1^T P_Z M_{P_Z Y_2} P_Z X_1\right)^{-1} X_1^T P_Z M_{P_Z Y_2} P_Z (X_1 \beta_1 + Y_2 \beta_2 + \varepsilon_1) \\
&= \beta_1 + \left(X_1^T P_Z M_{P_Z Y_2} P_Z X_1\right)^{-1} X_1^T P_Z M_{P_Z Y_2} P_Z (Y_2 \beta_2 + \varepsilon_1)
\end{aligned}$$

Since $M_{P_Z Y_2} P_Z Y_2 = 0$ (by orthogonality) and $p \lim \frac{1}{n} P_Z \varepsilon_1 = 0$, the estimator for β_1 is consistent. The derivation is completed by reversing the order of the variables in the equations again to show that β_2 is consistent.[2]

8.1.3 Fixed effects

Fixed effects models allow for time and/or individual differences in panel data. That is, separate regressions, say for m firms in the sample, are estimated with differences in intercepts but pooled slopes as illustrated in figure 8.1.

$$Y = X\beta + Z\gamma + \sum_{j=1}^{m} \alpha_j D_j + \varepsilon$$

---

[2] Of course, we could simplify the first equation but it seems very messy so why not exploit the effort we've already undertaken.

128   8. Overview of endogeneity

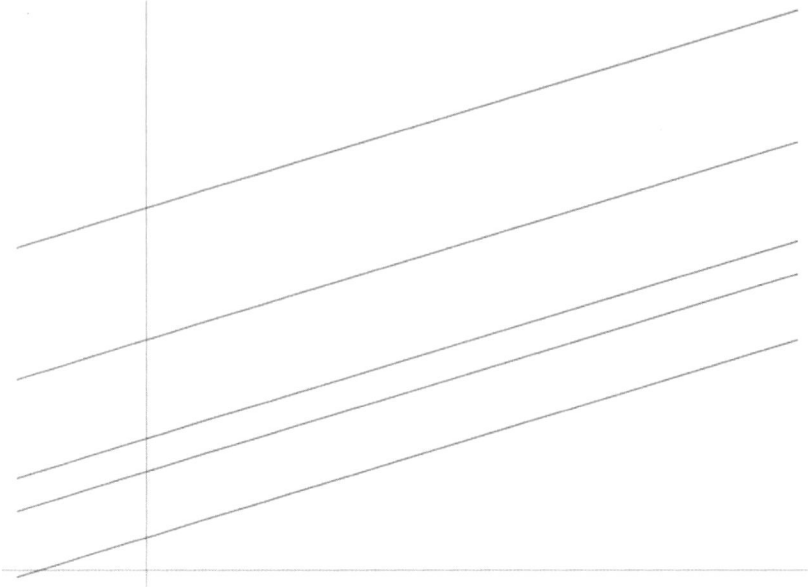

Figure 8.1: Fixed effects regression curves

where D_j is a firm indicator variable, X represents the experimental regressors and Z the control variables.[3] Geometrically, it's instructive to think of *FWL* (see chapter 3) where we condition on all control variables, then the experimental explanatory variables of interest are evaluated conditional on the control variables.[4]

$$M_Z Y = M_Z X \beta + \sum_{j=1}^{m} \alpha_j M_Z D_j + \epsilon$$

Of course, we can also consider semi- and non-parametric fixed effects regressions as well if we think of the nonparametric analog to *FWL* initiated by Robinson [1988] in the form of partial linear models and Stoker's [1991] partial index models (see chapter 6).

Causal effects are identified via a fixed effects model when there are constant, unobserved (otherwise they could be included as covariates) individual characteristics that because they are related to both outcomes and causing variables would be omitted, correlated variables if ignored. Differencing approaches such as fixed effects are simple and effective so long as individual fixed effects do not vary across periods and any correlation between treatment and unobserved outcome potential is described by an additive time-invariant covariate. Since this condition

---

[3]Clearly, time fixed effects can be accommodated in analogous fashion with time subscripts and indicator variables replacing the firm or individual variables.

[4]Of course, if an intercept is included in the fixed effects regression then the summation index is over $m - 1$ firms or individuals instead of m.

doesn't usually follow from economic theory or institutionally-relevant information, the utility of the fixed effects approach for identifying causal effects is limited.[5]

Nikolaev and Van Lent [2005] study variation through time in a firm's disclosure quality and its impact on the marginal cost of debt. In their setting, unobservable cross-firm heterogeneity, presumed largely constant through time, is accommodated via firm fixed effects. That is, firm-by-firm regressions that vary in intercept but have the same slope are estimated. Nikolaev and Van Lent argue that omitted variables and endogeneity plague evaluation of the impact of disclosure quality on cost of debt capital and the problem is mitigated by fixed effects.

Robinson [1989] concludes that fixed effects analysis more effectively copes with endogeneity than longitudinal, control function, or *IV* approaches in the analysis of the differential effects of union wages. In his setting, endogeneity is primarily related to worker behavior and measurement error. Robinson suggests that while there is wide agreement that union status is not exogenous, there is little consistency in teasing out the effect of union status on wages. While longitudinal analysis typically reports smaller effects than *OLS*, cross-sectional approaches such as *IV* or control function approaches (inverse Mills ratio) typically report larger effects than *OLS*. Robinson concludes that a simple fixed effects analysis of union status is a good compromise. (Also, see Wooldridge [2002], p. 581-590.)

On the other hand, Lalonde [1986] finds that regression approaches (including fixed effects) perform poorly compared with "experimental" methods in the analysis of the National Supported Work (NSW) training program. Dehejia and Wahba [1995] reanalyze the NSW data via propensity score matching and find similar results to Lalonde's experimental evidence. Once again we find no single approach works in all settings and the appropriate method depends on the context.

8.1.4   *Differences-in-differences*

Differences-in-differences (*DID*) is a close cousin to fixed effects. *DID* is a panel data approach that identifies causal effects when certain groups are treated and other groups are not. The treated are exposed to sharp changes in the causing variable due to shifts in the economic environment or changes in (government) policy. Typically, potential outcomes, in the absence of the change, are composed of the sum of a time effect that is common to all groups and a time invariant individual fixed effect, say,

$$E\left[Y_0 \mid t, i\right] = \beta_t + \gamma_i$$

Then, the causal effect δ is simply the difference between expected outcomes with treatment and expected outcomes without treatment

$$E\left[Y_1 \mid t, i\right] = E\left[Y_0 \mid t, i\right] + \delta$$

---

[5]It is well-known that fixed effects yield inconsistent parameter estimates when the model involves lagged dependent variables (see Chamberlain [1984] and Angrist and Krueger [1998]).

The key identifying condition for *DID* is the parameters associated with (treatment time and treatment group) interaction terms are zero in the absence of intervention.

Sometimes apparent interventions are themselves terminated and provide opportunities to explore the absence of intervention. Relatedly, R. A. Fisher (quoted in Cochran [1965]) suggested the case for causality is stronger when the model has many implications supported by the evidence. This emerges in terms of robustness checks, exploration of sub-populations in which treatment effects should not be observed (because the subpopulation is insensitive or immune to treatment or did not receive treatment), and comparison of experimental and non-experimental research methods (Lalonde [1986]). However, general equilibrium forces may confound direct evidence from such absence of intervention analyses. As Angrist and Krueger [1998, p. 56] point out, "Tests of refutability may have flaws. It is possible, for example, that a subpopulation that is believed unaffected by the intervention is indirectly affected by it."

8.1.5 Bivariate probit

A variation on a standard self-selection theme is when both selection and outcome equations are observed as discrete responses. If the unobservables are jointly normally distributed a bivariate probit accommodates endogeneity in the same way that a standard Heckman (inverse Mills ratio) control function approach works with continuous outcome response. Endogeneity is reflected in nonzero correlation among the unobservables. Dubin and Rivers [1989] provide a straightforward overview of this approach.

$$U_D = Z\theta + V, D = \begin{cases} 1 & \text{if } U_D > 0 \\ 0 & \text{otherwise} \end{cases}$$

$$\begin{bmatrix} Y^* \end{bmatrix} = X\beta + \varepsilon, Y = \begin{cases} 1 & \text{if } Y^* > 0 \\ 0 & \text{otherwise} \end{cases}$$

$$E \begin{bmatrix} V \\ \varepsilon \end{bmatrix} = 0, Var \begin{bmatrix} V \\ \varepsilon \end{bmatrix} = \Sigma = \begin{bmatrix} 1 & \rho \\ \rho & 1 \end{bmatrix}$$

Following the shorthand of Greene [1997], let $q_{i1} = 2U_{iD} - 1$ and $q_{i2} = 2Y_i - 1$, so that $q_{ij} = 1$ or -1. The bivariate normal cumulative distribution function is

$$\Pr(X_1 < x_1, X_2 < x_2) = \Phi_2(z_1, z_2, \rho) = \int_{-\infty}^{x_2} \int_{-\infty}^{x_1} \phi_2(z_1, z_2, \rho) \, dz_1 dz_2$$

where

$$\phi_2(z_1, z_2, \rho) = \frac{1}{2\pi(1-\rho^2)^{\frac{1}{2}}} \exp\left[-\frac{x_1^2 + x_2^2 - 2\rho x_1 x_2}{2(1-\rho^2)}\right]$$

denotes the bivariate normal (unit variance) density. Now let

$$\begin{aligned} z_{i1} &= \theta^T Z_i & w_{i1} &= q_{i1} z_{i1} \\ z_{i2} &= \beta^T X_i & w_{i2} &= q_{i2} z_{i2} \end{aligned}$$

$$\rho_{i*} = q_{i1}q_{i2}\rho$$

With this setup, the log-likelihood function can be written in a simple form where all the sign changes associated with D and Y equal to 0 and 1 are accounted for

$$lnL = \sum_{i=1}^{n} \Phi_2(w_{i1}, w_{i2}, \rho_{i*})$$

and maximization proceeds in the usual manner (see, for example, Greene [1997] for details).[6]

8.1.6 Simultaneous probit

Suppose we're investigating a discrete choice setting where an experimental variable (regressor) is endogenously determined. An example is Bagnoli, Liu, and Watts [2006] (BLW). BLW are interested in the effect of family ownership on the inclusion of covenants in debt contracts. Terms of debt contracts, such as covenants, are likely influenced by interest rates and interest rates are likely determined simultaneously with terms such as covenants. A variety of limited information approaches[7] have been proposed for estimating these models - broadly referred to as simultaneous probit models (see Rivers and Vuong [1988]). BLW adopted two stage conditional maximum likelihood estimation (*2SCML*; discussed below).

The base model involves a structural equation

$$y^* = Y\gamma + X_1\beta + u$$

where discrete D is observed

$$D_i = \begin{array}{ll} 1 & \text{if } y_i^* > 0 \\ 0 & \text{if } y_i^* \leq 0 \end{array}$$

The endogenous explanatory variables have reduced form

$$Y = \Pi X + V$$

where exogenous variable X and X_1 are related via matrix J, $X_1 = JX$, Y is an $n \times m$ matrix of endogenous variables, X_1 is $n \times k$, and X is $n \times p$. The following conditions are applied to all variations:

Condition 8.1 *(X_i, u_i, V_i) is iid with X_i having finite positive definite variance matrix Σ_{XX}, and $(u_i, V_i \mid X_i)$ are jointly normally distributed with mean zero and finite positive definite variance matrix* $\Omega = \begin{array}{cc} \sigma_{uu} & \Sigma_{uV} \\ \Sigma_{Vu} & \Sigma_{VV} \end{array}$.

---

[6] Evans and Schwab [1995] employ bivariate probit to empirically estimate causal effects of schooling.

[7] They are called limited information approaches in that they typically focus on one equation at a time and hence ignore information in other equations.

8. Overview of endogeneity

Condition 8.2 $rank(\Pi, J) = m + k$.

Condition 8.3 $(\gamma, \beta, \Pi, \Omega)$ *lie in the interior of a compact parameter space* Θ.

Identification of the parameters in the structural equation involves normalization. A convenient normalization is

$$Var\left[y_i^* \mid X_i, Y_i\right] = \sigma_{uu} - \lambda^T \Sigma_{VV} \lambda = 1,$$

where $\lambda = \Sigma_{VV}^{-1} \Sigma_{Vu}$, the structural equation is rewritten as

$$y^* = Y\gamma + X_1\beta + V\lambda + \eta$$

and $\eta_i = u_i - V_i^T \lambda \sim N\left(0, \sigma_{uu} - \lambda^T \Sigma_{VV} \lambda = 1\right)$.

Limited information maximum likelihood (*LIML*)

A limited information maximum likelihood (*LIML*) approach was adopted by Godfrey and Wickens [1982]. The likelihood function is

$$\prod_{i=1}^{n} (2\pi)^{-\frac{(m+1)}{2}} |\Omega|^{-\frac{1}{2}} \left[\int_{c_i}^{\infty} \exp\left\{-\frac{1}{2} \left(u, V_i^T\right) \Omega^{-1} \left(u, V_i^T\right)^T\right\} du\right]^{D_i}$$

$$\times \left[\int_{-\infty}^{c_i} \exp\left\{-\frac{1}{2} \left(u, V_i^T\right) \Omega^{-1} \left(u, V_i^T\right)^T\right\} du\right]^{1-D_i}$$

where $c_i = -\left(Y_i^T \gamma - X_{1i}^T \beta\right)$. Following some manipulation, estimation involves maximizing the log-likelihood with respect to $(\gamma, \beta, \lambda, \Pi, \Sigma_{VV})$. As *LIML* is computationally difficult in large models, it has received little attention except as a benchmark case.

Instrumental variables probit (*IVP*)

Lee [1981] proposed an instrumental variables probit (*IVP*). Lee rewrites the structural equation in reduced form

$$y_i^* = \left(\Pi^T X_i\right) \gamma + X_{1i}\beta + V_i\lambda + \eta_i$$

The log-likelihood for D given X is

$$\sum_{i=1}^{n} D_i \log \Phi\left[\left(\Pi^T X_i\right) \gamma_* + X_{1i}^T \beta_*\right]$$

$$+ (1 - D_i) \log \left\{1 - \Phi\left[\left(\Pi^T X_i\right) \gamma_* + X_{1i}^T \beta_*\right]\right\}$$

where $\Phi(\cdot)$ denotes a standard normal *cdf* and

$$\gamma_* = \frac{\gamma}{\omega} \quad \beta_* = \frac{\beta}{\omega}$$

8.1 Overview

$$\begin{aligned}
\omega^2 &= Var\left[u_i + V_i^T\gamma\right] = \sigma_{uu}^2 + \gamma^T\Sigma_{VV}\gamma + \gamma^T\Sigma_{VV}\lambda + \lambda^T\Sigma_{VV}\gamma \\
&= 1 + \lambda^T\Sigma_{VV}\lambda + \gamma^T\Sigma_{VV}\gamma + \gamma^T\Sigma_{VV}\lambda + \lambda^T\Sigma_{VV}\gamma \\
&= 1 + (\gamma + \lambda)^T \Sigma_{VV} (\gamma + \lambda)
\end{aligned}$$

Consistent estimates for Π, $\hat{\Pi}$, are obtained via *OLS*. Then, utilizing $\hat{\Pi}$ in place of Π, maximization of the log-likelihood with respect to γ_* and β_* is computed via m regressions followed by a probit estimation.

Generalized two-stage simultaneous probit (*G2SP*)

Amemiya [1978] suggested a general method for obtaining structural parameter estimates from reduced form estimates (*G2SP*). Heckman's [1978] two-stage endogenous dummy variable model is a special case of *G2SP*. Amemiya's proposal is a variation on *IVP* where the unconstrained log-likelihood is maximized with respect to τ_*

$$\sum_{i=1}^n D_i \log \Phi\left(X_i^T\tau_*\right) + (1 - D_i) \log\left[1 - \Phi\left(X_i^T\tau_*\right)\right]$$

$$\tau_* = \Pi\gamma_* + J\beta_*$$

In terms of the sample estimates we have the regression problem

$$\begin{aligned}
\hat{\tau}_* &= \begin{bmatrix} \hat{\Pi} & J \end{bmatrix} \begin{bmatrix} \gamma_* \\ \beta_* \end{bmatrix} + (\hat{\tau}_* - \tau_*) - \left(\hat{\Pi} - \Pi\right)\gamma_* \\
&= \hat{H} \begin{bmatrix} \gamma_* \\ \beta_* \end{bmatrix} + e
\end{aligned}$$

where $e = (\hat{\tau}_* - \tau_*) - \left(\hat{\Pi} - \Pi\right)\gamma_*$. *OLS* provides consistent estimates of γ_* and β_* but *GLS* is more efficient. Let $\hat{V}$ denote an asymptotic consistent estimator for the variance e. Then Amemiya's *G2SP* estimator is

$$\begin{bmatrix} \hat{\gamma}_* \\ \hat{\beta}_* \end{bmatrix} = \left(\hat{H}^T\hat{V}^{-1}\hat{H}\right)^{-1} \hat{H}^T\hat{V}^{-1}\hat{\tau}_*$$

This last step constitutes one more computational step (in addition to the m reduced form regressions and one probit) than required for *IVP* (and *2SCML* described below).

Two-stage conditional maximum likelihood (*2SCML*)

Rivers and Vuong [1988] proposed two-stage conditional maximum likelihood (*2SCML*). Vuong [1984] notes when the joint density for a set of endogenous variables can be factored into a conditional distribution for one variable and a marginal distribution for the remaining variables, estimation can often be simplified by using conditional maximum likelihood methods. In the simultaneous

8. Overview of endogeneity

probit setting, the joint density for D_i and Y_i factors into a probit likelihood and a normal density.

$$h(D_i, Y_i \mid X_i; \gamma, \beta, \lambda, \Pi, \Sigma_{VV})$$
$$= f(D_i \mid Y_i, X_i; \gamma, \beta, \lambda, \Pi) g(Y_i \mid X_i; \Pi, \Sigma_{VV})$$

where

$$f(D_i \mid Y_i, X_i; \gamma, \beta, \lambda, \Pi)$$
$$= \Phi\left(Y_i^T \gamma + X_{1i}^T \beta + V_i^T \lambda\right)^{D_i} \left[1 - \Phi\left(Y_i^T \gamma + X_{1i}^T \beta + V_i^T \lambda\right)\right]^{(1-D_i)}$$

$$g(Y_i \mid X_i; \Pi, \Sigma_{VV})$$
$$= (2\pi)^{-\frac{m}{2}} |\Sigma_{VV}|^{-\frac{1}{2}} \exp\left\{-\frac{1}{2} (Y_i - \Pi^T X_i)^T \Sigma_{VV}^{-1} (Y_i - \Pi^T X_i)\right\}$$

Two steps are utilized to compute the *2SCML* estimator. First, the marginal log-likelihood for Y_i is maximized with respect to $\hat{\Pi}$ and $\hat{\Sigma}_{VV}$. This is computed by m reduced form regressions of Y on X to obtain $\hat{\Pi}$. Let the residuals be $\hat{V}_i = Y_i - \hat{\Pi} X_i$, then the standard variance estimator is $\hat{\Sigma}_{VV} = n^{-1} \sum_{i=1}^{n} \hat{V}_i \hat{V}_i^T$. Second, replacing Π with $\hat{\Pi}$, the conditional log-likelihood for D_i is maximized with respect to $\left(\hat{\gamma}, \hat{\beta}, \hat{\lambda}\right)$. This is computed via a probit analysis of D_i with regressors Y_i, X_{1i}, and $\hat{V}_i$.

2SCML provides several convenient tests of endogeneity. When Y_i and u_i are correlated, standard probit produces inconsistent estimators for γ and β. However, if $\Sigma_{Vu} = 0$, or equivalently, $\lambda = 0$, the Y_is are effectively exogenous. A modified *Wald* statistic is

$$MW = n \, \hat{\lambda}^T \hat{V}_0\left(\hat{\lambda}\right)^{-1} \hat{\lambda}$$

where $\hat{V}_0\left(\hat{\lambda}\right)$ is a consistent estimator for the lower right-hand block (corresponding to λ) of $V_0(\theta) = \left(\tilde{H}^T \tilde{\Sigma} \tilde{H}\right)^{-1}$ where

$$\tilde{H} = \begin{bmatrix} \Pi & J & 0 \\ I_m & 0 & I_m \end{bmatrix}$$

and

$$\tilde{\Sigma} = \begin{bmatrix} \tilde{\Sigma}_{XX} & \tilde{\Sigma}_{XV} \\ \tilde{\Sigma}_{VX} & \tilde{\Sigma}_{VV} \end{bmatrix}$$
$$= E \frac{\phi\left(Z_i^T \delta + V_i^T \lambda\right)^2}{\Phi\left(Z_i^T \delta + V_i^T \lambda\right)\left[1 - \Phi\left(Z_i^T \delta + V_i^T \lambda\right)\right]} \begin{bmatrix} X_i \\ V_i \end{bmatrix} \begin{bmatrix} X_i \\ V_i \end{bmatrix}^T$$

with $Z_i = \begin{bmatrix} Y_i \\ X_{1i} \end{bmatrix}$, $\delta = \begin{bmatrix} \gamma \\ \beta \end{bmatrix}$, and $\phi(\cdot)$ is the standard normal density. Notice the modified *Wald* statistic draws from the variance estimator under the null. The conditional score statistic is

$$CS = \frac{1}{n} \frac{\partial L\left(\tilde{\gamma}, \tilde{\beta}, 0, \hat{\Pi}\right)}{\partial \lambda^T} \hat{V}_0\left(\hat{\lambda}\right) \frac{\partial L\left(\tilde{\gamma}, \tilde{\beta}, 0, \hat{\Pi}\right)}{\partial \lambda}$$

where $\left(\tilde{\gamma}, \tilde{\beta}\right)$ are the standard probit maximum likelihood estimators. The conditional likelihood ratio statistic is

$$CLR = 2\left[L\left(\hat{\gamma}, \hat{\beta}, \hat{\lambda}, \hat{\Pi}\right) - L\left(\hat{\gamma}, \hat{\beta}, 0, \hat{\Pi}\right)\right]$$

As is typical (see chapter 3), the modified *Wald*, conditional score, and conditional likelihood ratio statistics have the same asymptotic properties.[8]

8.1.7 Strategic choice model

Amemiya [1974] and Heckman [1978] suggest resolving identification problems in simultaneous probit models by making the model recursive. Bresnahan and Reiss [1990] show that this approach rules out interesting interactions in strategic choice models. Alternatively, they propose modifying the error structure to identify unique equilibria in strategic, multi-person choice models.

Statistical analysis of strategic choice extends random utility analysis by adding game structure and Nash equilibrium strategies (Bresnahan and Reiss [1990, 1991] and Berry [1992]). McKelvey and Palfrey [1995] proposed quantal response equilibrium analysis by assigning extreme value (logistic) distributed random errors to players' strategies. Strategic error by the players makes the model amenable to statistical analysis as the likelihood function does not degenerate. Signorino [2003] extends the idea to political science by replacing extreme value errors with assignment of normally distributed errors associated with analyst uncertainty and/or private information regarding the players' utility for outcomes. Since analyst error due to unobservable components is ubiquitous in business and economic data and private information problems are typical in settings where accounting plays an important role, we focus on the game setting with analyst error and private information.

A simple two player, sequential game with analyst error and private information (combined as π) is depicted in figure 8.2. Player A moves first by playing either left (l) or right (r). Player B moves next but player A's choice depends on the anticipated response of player B to player A's move. For simplicity, assume $\pi_i \sim N\left(0, \sigma^2 I\right)$ where

$$\pi_i^T = \begin{bmatrix} \pi_{lLi}^A & \pi_{lLi}^B & \pi_{lRi}^A & \pi_{lRi}^B & \pi_{rLi}^A & \pi_{rLi}^B & \pi_{rRi}^A & \pi_{rRi}^B \end{bmatrix}$$

---

[8]Rivers and Vuong also identify three Hausman-type test statistics for endogeneity but their simulations suggest the modified *Wald*, conditional score, and conditional likelihood ratio statistics perform at least as well and in most cases better.

8. Overview of endogeneity

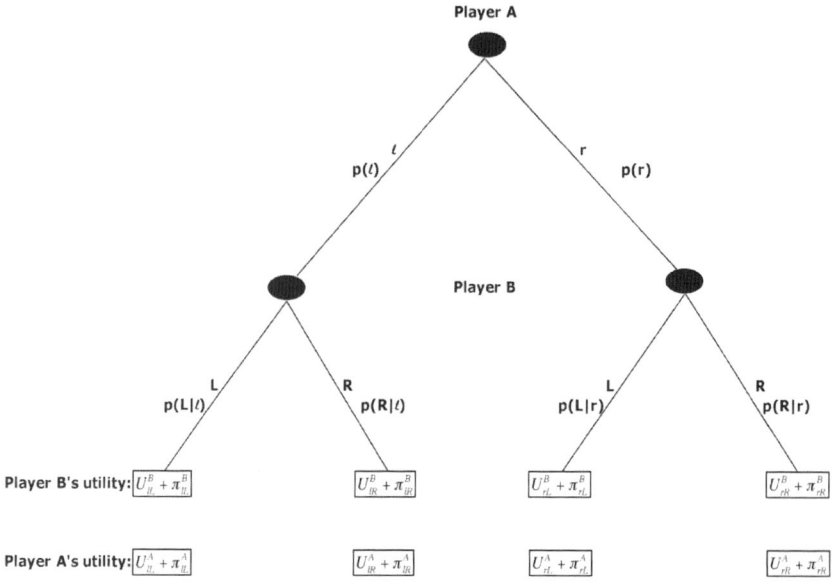

Figure 8.2: Strategic choice game tree

Since choice is scale-free (see chapter 5) maximum likelihood estimation proceeds with σ^2 normalized to 1.

The log-likelihood is

$$\sum_{i=1}^{n} Y_{lLi} \log\left(P_{lLi}\right) + Y_{lRi} \log\left(P_{lRi}\right) + Y_{rLi} \log\left(P_{rLi}\right) + Y_{rRi} \log\left(P_{rRi}\right)$$

where $Y_{jki} = 1$ if strategy j is played by A and k is played by B for sample i, and P_{jki} is the probability that strategy j is played by A and k is played by B for sample i. The latter requires some elaboration. Sequential play yields $P_{jk} = P_{(k|j)} P_j$. Now, only the conditional and marginal probabilities remain to be identified. Player B's strategy depends on player A's observed move. Hence,

$$P_{(L|l)} = \Phi\left(\frac{U_{lL} - U_{lR}}{\sqrt{2\sigma^2}}\right)$$
$$P_{(R|l)} = 1 - P_{(L|l)}$$
$$P_{(R|r)} = 1 - P_{(L|r)}$$
$$P_{(L|r)} = \Phi\left(\frac{U_{rL} - U_{rR}}{\sqrt{2\sigma^2}}\right)$$

Player A's strategy however depends on B's response to A's move. Therefore,

$$P_l = \Phi\left(\frac{P_{(L|l)} U_{lL} - P_{(L|r)} U_{rL} + P_{(R|l)} U_{lR} - P_{(R|r)} U_{rR}}{\sqrt{\left(P_{(L|l)}^2 + P_{(L|r)}^2 + P_{(R|l)}^2 + P_{(R|r)}^2\right) \sigma^2}}\right)$$

and
$$P_r = 1 - P_l$$

Usually, the observable portion of expected utility is modeled as an index function; for Player B we have

$$U_{jk} - U_{jk'} = U_j^B = \left(X_{jk} - X_{jk'}\right)\beta_{jk}^B = X_j^B \beta_j^B$$

Since Player B moves following Player A, stochastic analysis of Player B's utility is analogous to the simple binary discrete choice problem. That is,

$$P_{(L|l)} = \Phi\left(\frac{U_{lL} - U_{lR}}{\sqrt{2\sigma^2}}\right)$$
$$= \Phi\left(\frac{X_l^B \beta_l^B}{\sqrt{2}}\right)$$

and
$$P_{(L|r)} = \Phi\left(\frac{X_r^B \beta_r^B}{\sqrt{2}}\right)$$

However, stochastic analysis of Player A's utility is a little more subtle. Player A's expected utility depends on Player B's response to Player A's move. Hence, Player A's utilities are weighted by the conditional probabilities associated with Player B's strategies. That is, from an estimation perspective the regressors X interact with the conditional probabilities to determine the coefficients in Player A's index function.

$$U_{jk} - U_{j'k} = X_{jk}\beta_{jk}^A - X_{j'k}\beta_{j'k}^A$$

Consequently, Player A's contribution to the likelihood function is a bit more complex than that representing Player B's utilities.[9] Stochastic analysis of Player A's strategy is

$$P_l = \Phi\left(\frac{P_{(L|l)}U_{lL} - P_{(L|r)}U_{rL} + P_{(R|l)}U_{lR} - P_{(R|r)}U_{rR}}{\sqrt{\left(P_{(L|l)}^2 + P_{(L|r)}^2 + P_{(R|l)}^2 + P_{(R|r)}^2\right)\sigma^2}}\right)$$

$$= \Phi\left(\frac{P_{(L|l)}X_{lL}\beta_{lL}^A - P_{(L|r)}X_{rL}\beta_{rL}^A + P_{(R|l)}X_{lR}\beta_{lR}^A - P_{(R|r)}X_{rR}\beta_{rR}^A}{\sqrt{\left(P_{(L|l)}^2 + P_{(L|r)}^2 + P_{(R|l)}^2 + P_{(R|r)}^2\right)}}\right)$$

---

[9] Recall the analysis is stochastic because the analyst doesn't observe part of the agents' utilities. Likewise, private information produces agent uncertainty regarding the other player's utility. Hence, private information produces a similar stochastic analysis. This probabilistic nature ensures that the likelihood doesn't degenerate even in a game of pure strategies.

Example 8.1 *Consider a simple experiment comparing a sequential strategic choice model with standard binary choice models for each player. We generated* 200 *simulated samples of size* $n = 2,000$ *with uniformly distributed regressors and standard normal errors. In particular,*

$$X_l^B \sim U(-2, 2)$$
$$X_r^B \sim U(-5, 5)$$
$$X_{lL}^A, X_{lR}^A, X_{rL}^A, X_{rR}^A \sim U(-3, 3)$$

and

$$\beta_l^B = \begin{bmatrix} -0.5 & 1 \end{bmatrix}^T$$
$$\beta_r^B = \begin{bmatrix} 0.5 & -1 \end{bmatrix}^T$$
$$\beta^A = \begin{bmatrix} 0.5 & 1 & 1 & -1 & -1 \end{bmatrix}^T$$

where the leading element of each vector is an intercept β_0.[10] *Results (means, standard deviations, and the* 0.01 *and* 0.99 *quantiles) are reported in tables 8.1 and 8.2. The standard discrete choice (DC) estimates seem to be more systemati-*

Table 8.1: Strategic choice analysis for player B

parameter	β_{l0}^B	β_l^B	β_{r0}^B	β_r^B
parameter	−0.5	1	0.5	−1
SC mean	−0.482	0.932	0.460	−0.953
DC mean	−0.357	0.711	0.354	−0.713
SC std dev	0.061	0.057	0.101	0.059
DC std dev	0.035	0.033	0.050	0.030
SC $\begin{pmatrix} 0.01, \\ 0.99 \end{pmatrix}$ quantiles	$\begin{pmatrix} -0.62, \\ -0.34 \end{pmatrix}$	$\begin{pmatrix} 0.80, \\ 1.10 \end{pmatrix}$	$\begin{pmatrix} 0.22, \\ 0.69 \end{pmatrix}$	$\begin{pmatrix} -1.10, \\ 0.82 \end{pmatrix}$
DC $\begin{pmatrix} 0.01, \\ 0.99 \end{pmatrix}$ quantiles	$\begin{pmatrix} -0.43, \\ -0.29 \end{pmatrix}$	$\begin{pmatrix} 0.65, \\ 0.80 \end{pmatrix}$	$\begin{pmatrix} 0.23, \\ 0.47 \end{pmatrix}$	$\begin{pmatrix} -0.79, \\ -0.64 \end{pmatrix}$

cally biased towards zero. Tables 8.3 and 8.4 expressly compare the parameter estimate differences between the strategic choice model (SC) and the discrete choice models (DC). Hence, not only are the standard discrete choice parameter estimates biased toward zero but also there is almost no overlap with the (0.01, 0.99) *interval estimates for the strategic choice model.*

As in the case of conditionally-heteroskedastic probit (see chapter 5), marginal probability effects of regressors are likely to be nonmonotonic due to cross agent

---

[10] The elements of β^A correspond to $\begin{bmatrix} intercept & \beta_{lL}^A & \beta_{lR}^A & \beta_{rL}^A & \beta_{rR}^A \end{bmatrix}$ where the intercept is the mean difference in observed utility (conditional on the regressors) between strategies l and r.

8.1 Overview 139

Table 8.2: Strategic choice analysis for player A

	β_0^A	β_{lL}^A	β_{lR}^A
parameter	0.5	1	1
SC mean	0.462	0.921	0.891
DC mean	0.304	0.265	0.360
SC std dev	0.044	0.067	0.053
DC std dev	0.032	0.022	0.021
SC $\binom{0.01,}{0.99}$ quantiles	$\binom{0.34,}{0.56}$	$\binom{0.78,}{1.08}$	$\binom{0.78,}{1.01}$
DC $\binom{0.01,}{0.99}$ quantiles	$\binom{0.23,}{0.38}$	$\binom{0.23,}{0.32}$	$\binom{0.31,}{0.41}$

	β_{rL}^A	β_{rR}^A
parameter	-1	-1
SC mean	-0.911	-0.897
DC mean	-0.352	-0.297
SC std dev	0.053	0.058
DC std dev	0.022	0.023
SC $\binom{0.01,}{0.99}$ quantiles	$\binom{-1.04,}{-0.79}$	$\binom{-1.05,}{-0.78}$
DC $\binom{0.01,}{0.99}$ quantiles	$\binom{-0.40,}{-0.30}$	$\binom{-0.34,}{-0.25}$

probability interactions. Indeed, comparison of marginal effects for strategic probit with those of standard binary probit helps illustrate the contrast between statistical analysis of strategic and single person decisions. For the sequential strategic game above, the marginal probabilities for player A's regressors include

$$\frac{\partial P_{lLj}}{\partial X_{ikj}^A} = P_{(L|l)j} f_{lj} (sign_j) P_{(k|i)j} \beta_{ik}^A Den^{-\frac{1}{2}}$$

$$\frac{\partial P_{lRj}}{\partial X_{ikj}^A} = P_{(R|l)j} f_{lj} (sign_j) P_{(k|i)j} \beta_{ik}^A Den^{-\frac{1}{2}}$$

$$\frac{\partial P_{rLj}}{\partial X_{ikj}^A} = P_{(L|r)j} f_{rj} (sign_j) P_{(k|i)j} \beta_{ik}^A Den^{-\frac{1}{2}}$$

$$\frac{\partial P_{rRj}}{\partial X_{ikj}^A} = P_{(R|r)j} f_{rj} (sign_j) P_{(k|i)j} \beta_{ik}^A Den^{-\frac{1}{2}}$$

where $sign_j$ is the sign of the X_{ikj} term in P_{mnj}, f_{ij} and $f_{(k|i)j}$ is the standard normal density function evaluated at the same arguments as P_{ij} and $P_{(k|i)j}$,

$$Den = \left(P_{(L|l)j}^2 + P_{(L|r)j}^2 + P_{(R|l)j}^2 + P_{(R|r)j}^2\right)$$

Table 8.3: Parameter differences in strategic choice analysis for player B

SC-DC	β_{l0}^B	β_l^B	β_{r0}^B	β_r^B
parameter	-0.5	1	0.5	-1
mean	-0.125	0.221	0.106	-0.241
std dev	0.039	0.041	0.079	0.049
$\begin{pmatrix} 0.01, \\ 0.99 \end{pmatrix}$ quantiles	$\begin{pmatrix} -0.22, \\ -0.03 \end{pmatrix}$	$\begin{pmatrix} 0.13, \\ 0.33 \end{pmatrix}$	$\begin{pmatrix} -0.06, \\ 0.29 \end{pmatrix}$	$\begin{pmatrix} -0.36, \\ -0.14 \end{pmatrix}$

Table 8.4: Parameter differences in strategic choice analysis for player A

SC-DC	β_0^A	β_{lL}^A	β_{lR}^A
parameter	0.5	1	1
mean	0.158	0.656	0.531
std dev	0.027	0.056	0.044
$(0.01, 0.99)$ quantiles	$(0.10, 0.22)$	$(0.54, 0.80)$	$(0.43, 0.62)$
SC-DC	β_{rL}^A	β_{rR}^A	
parameter	-1	-1	
mean	-0.559	-0.600	
std dev	0.045	0.050	
$(0.01, 0.99)$ quantiles	$(-0.67, -0.46)$	$(-0.73, -0.49)$	

and

$$Num = \left\{ \begin{array}{l} P_{(L|l)j} X_{iLj}^A \beta_{iL}^A - P_{(L|r)j} X_{rLj}^A \beta_{rL}^A \\ + P_{(R|l)j} X_{lRj}^A \beta_{lR}^A - P_{(R|r)j} X_{rRj}^A \beta_{rR}^A \end{array} \right\}$$

Similarly, the marginal probabilities with respect to player B's regressors include

$$\frac{\partial P_{lLj}}{\partial X_{lj}^B} = f_{(L|l)j} \frac{\beta_l^B}{\sqrt{2}} P_{lj} + P_{(L|l)j} f_{lj} f_{(L|l)j} \frac{\beta_l^B}{\sqrt{2}}$$
$$\times \left\{ \begin{array}{l} \left(X_{lLj}^A \beta_{lL}^A - X_{lRj}^A \beta_{lR}^A \right) Den^{-\frac{1}{2}} \\ - Num\, Den^{-\frac{3}{2}} \left(P_{(L|l)j} - P_{(R|l)j} \right) \end{array} \right\}$$

$$\frac{\partial P_{lLj}}{\partial X_{rj}^B} = P_{(l|l)j} f_{lj} f_{(L|r)j} \frac{\beta_r^B}{\sqrt{2}}$$
$$\times \left\{ \begin{array}{l} -\left(X_{rLj}^A \beta_{rL}^A - X_{rRj}^A \beta_{rR}^A \right) Den^{-\frac{1}{2}} \\ - Num\, Den^{-\frac{3}{2}} \left(P_{(L|r)j} - P_{(R|r)j} \right) \end{array} \right\}$$

$$\frac{\partial P_{lRj}}{\partial X_{lj}^{B}} = f_{(R|l)j}\frac{-\beta_{l}^{B}}{\sqrt{2}}P_{lj} + P_{(R|l)j}f_{lj}f_{(R|l)j}\frac{\beta_{l}^{B}}{\sqrt{2}}$$
$$\times \left\{ \begin{array}{l} \left(X_{iLj}^{A}\beta_{iL}^{A} - X_{iRj}^{A}\beta_{iR}^{A}\right)Den^{-\frac{1}{2}} \\ -NumDen^{-\frac{3}{2}}\left(P_{(L|l)j} - P_{(R|l)j}\right) \end{array} \right\}$$

$$\frac{\partial P_{lRj}}{\partial X_{rj}^{B}} = P_{(R|l)j}f_{lj}f_{(R|r)j}\frac{-\beta_{l}^{B}}{\sqrt{2}}$$
$$\times \left\{ \begin{array}{l} \left(X_{rLj}^{A}\beta_{rL}^{A} - X_{rRj}^{A}\beta_{rR}^{A}\right)Den^{-\frac{1}{2}} \\ +NumDen^{-\frac{3}{2}}\left(P_{(L|r)j} - P_{(R|r)j}\right) \end{array} \right\}$$

$$\frac{\partial P_{rLj}}{\partial X_{lj}^{B}} = P_{(L|r)j}f_{rj}f_{(L|l)j}\frac{\beta_{l}^{B}}{\sqrt{2}}$$
$$\times \left\{ \begin{array}{l} -\left(X_{iLj}^{A}\beta_{iL}^{A} - X_{iRj}^{A}\beta_{iR}^{A}\right)Den^{-\frac{1}{2}} \\ +NumDen^{-\frac{3}{2}}\left(P_{(L|l)j} - P_{(R|l)j}\right) \end{array} \right\}$$

$$\frac{\partial P_{rLj}}{\partial X_{rj}^{B}} = f_{(L|r)j}\frac{\beta_{r}^{B}}{\sqrt{2}}P_{rj} + P_{(L|r)j}f_{rj}f_{(L|r)j}\frac{\beta_{r}^{B}}{\sqrt{2}}$$
$$\times \left\{ \begin{array}{l} \left(X_{rLj}^{A}\beta_{rL}^{A} - X_{rRj}^{A}\beta_{rR}^{A}\right)Den^{-\frac{1}{2}} \\ +NumDen^{-\frac{3}{2}}\left(P_{(L|r)j} - P_{(R|r)j}\right) \end{array} \right\}$$

$$\frac{\partial P_{rRj}}{\partial X_{lj}^{B}} = P_{(R|r)j}f_{rj}f_{(R|l)j}\frac{-\beta_{l}^{B}}{\sqrt{2}}$$
$$\times \left\{ \begin{array}{l} \left(X_{iLj}^{A}\beta_{iL}^{A} - X_{iRj}^{A}\beta_{iR}^{A}\right)Den^{-\frac{1}{2}} \\ -NumDen^{-\frac{3}{2}}\left(P_{(L|l)j} - P_{(R|l)j}\right) \end{array} \right\}$$

$$\frac{\partial P_{rRj}}{\partial X_{rj}^{B}} = f_{(R|r)j}\frac{-\beta_{r}^{B}}{\sqrt{2}}P_{rj} + P_{(R|r)j}f_{rj}f_{(R|r)j}\frac{\beta_{r}^{B}}{\sqrt{2}}$$
$$\times \left\{ \begin{array}{l} \left(X_{rLj}^{A}\beta_{rL}^{A} - X_{rRj}^{A}\beta_{rR}^{A}\right)Den^{-\frac{1}{2}} \\ +NumDen^{-\frac{3}{2}}\left(P_{(L|r)j} - P_{(R|r)j}\right) \end{array} \right\}$$

Clearly, analyzing responses to anticipated moves by other agents who themselves are anticipating responses changes the game. In other words, endogeneity is fundamental to the analysis of strategic play.

Multi-person strategic choice models can be extended in a variety of ways including simultaneous move games, games with learning, games with private information, games with multiple equilibria, etc. (Bresnahan and Reiss [1990], Tamer [2003]). The key point is that strategic interaction is endogenous and standard (single-person) discrete choice models (as well as simultaneous probit models) ignore this source of endogeneity.

8.1.8 Sample selection

A common problem involves estimation of β for the model

$$Y^* = X\beta + \varepsilon$$

however sample selection results in Y being observed only for individuals receiving treatment (when $D = 1$). The data are censored but not at a fixed value (as in a Tobit problem; see chapter 5). Treating sample selection D as an exogenous variable is inappropriate if the unobservable portion of the selection equation, say V_D, is correlated with unobservables in the outcome equation ε.

Heckman [1974, 1976, 1979] addressed this problem and proposed the classic two stage approach. In the first stage, estimate the selection equation via probit. Identification in this model does not depend on an exclusion restriction (Z need not include variables appropriately excluded from X) but if instruments are available they're likely to reduce collinearity issues.

To fix ideas, identification conditions include

Condition 8.4 (X, D) *are always observed,* Y_1 *is observed when* $D = 1$ $(D^* > 1)$,

Condition 8.5 (ε, V_D) *are independent of X with mean zero,*

Condition 8.6 $V_D \sim N(0, 1)$,

Condition 8.7 $E[\varepsilon \mid V_D] = \gamma_1 V_D$.[11]

The two-stage procedure estimates θ from a first stage probit.

$$D^* = Z\theta - V_D$$

These estimates $\widehat{\theta}$ are used to construct the inverse Mills ratio $\lambda_i = \frac{\phi(Z_i\widehat{\theta})}{\Phi(Z_i\widehat{\theta})}$ which is utilized as a covariate in the second stage regression.

$$Y_1 = X\beta + \gamma\lambda + \eta$$

where $E[\eta \mid X, \lambda] = 0$. Given proper specification of the selection equation (including normality of V_D), Heckman shows that the two-step estimator is asymptotically consistent (if not efficient) for β, the focal parameter of the analysis.[12]

---

[11] Bivariate normality of (ε, V_D) is often posed, but strictly speaking is not required for identification.

[12] It should be noted that even though Heckman's two stage approach is commonly employed to estimate treatment effects (discussed later), treatment effects are not the object of the sample selection model. In fact, since treatment effects involve counterfactuals and we have no data from which to identify population parameters for the counterfactuals, treatment effects in this setting are unassailable.

A semi-nonparametric alternative

Concern over reliance on normal probability assignment to unobservables in the selection equation as well as the functional form of the outcome equation, has resulted in numerous proposals to relax these conditions. Ahn and Powell [1993] provide an alternative via their semi-nonparametric two stage approach. However, nonparametric identification involves an exclusion restriction or, in other words, at least one instrument. That is, (at least) one variable included in the selection equation is properly omitted from the outcome equation. Intuitively, this is because the selection equation could be linear and the second stage would then involve colinear regressors. Ahn and Powell propose a nonparametric selection model coupled with a partial index outcome (second stage) model. The first stage selection index is estimated via nonparametric regression

$$\widehat{g}_i = \frac{\sum_{j=1}^{n} K\left(\frac{w_i - w_j}{h_1}\right) D_j}{\sum_{j=1}^{n} K\left(\frac{w_i - w_j}{h_1}\right)}$$

The second stage uses instruments Z, which are functions of W, and the estimated selection index.

$$\widehat{\beta} = \left[\widehat{S}_{XX}\right]^{-1} \widehat{S}_{XY}$$

where

$$\widehat{S}_{XX} = \binom{n}{2}^{-1} \sum_{i=1}^{n-1} \sum_{j=i+1}^{n} \widehat{\omega}_{ij} (z_i - z_j)(x_i - x_j)^T$$

$$\widehat{S}_{XY} = \binom{n}{2}^{-1} \sum_{i=1}^{n-1} \sum_{j=i+1}^{n} \widehat{\omega}_{ij} (z_i - z_j)(y_i - y_j)$$

and

$$\widehat{\omega}_{ij} = \frac{1}{h_2} K\left(\frac{\widehat{g}_i - \widehat{g}_j}{h_2}\right) D_i D_j$$

Ahn and Powell show the instrumental variable density-weighted average derivative estimator for β achieves root-n convergence (see the discussion of nonparametric regression and Powell, Stock, and Stoker's [1989] instrumental variable density-weighted average derivative estimator in chapter 6).

8.1.9 Duration models

Sometimes the question involves how long to complete a task. For instance, how long to complete an audit (internal or external), how long to turn around a distressed business unit or firm, how long to complete custom projects, how long will a recession last, and so on. Such questions can be addressed via duration models.

8. Overview of endogeneity

The most popular duration models are proportional hazard models. Analysis of such questions can be plagued by the same challenges of endogeneity and unobservable heterogeneity as other regression models.

We'll explore a standard version of the model and a couple of relaxations. Namely, we'll look at Horowitz's [1999] semiparametric proportional hazard (classical) model with unobserved heterogeneity and Campolieti's [2001] Bayesian semiparametric duration model with unobserved heterogeneity.

Unconditional hazard rate

The probability that an individual leaves a state during a specified interval given the individual was previously in the particular state is

$$\Pr(t < T < t+h \mid T > t)$$

The hazard function, then is $\lambda(t) = \lim_{h \to 0} \frac{\Pr(t<T<t+h|T>t)}{h}$, the instantaneous rate of leaving per unit of time. To relate this to the hazard function write

$$\Pr(t < T < t+h \mid T > t) = \frac{\Pr(t < T < t+h)}{\Pr(T > t)}$$
$$= \frac{F(t+h) - F(t)}{1 - F(t)}$$

where F is the probability distribution function and f is the density function for T. When F is differentiable, the hazard rate is seen as the limit of the right hand side divided by h as h approaches 0 (from above)

$$\lambda(t) = \lim_{h \to 0} \frac{F(t+h) - F(t)}{h} \frac{1}{1 - F(t)}$$
$$= \frac{f(t)}{1 - F(t)}$$

To move this closer to a version of the model that is frequently employed define the integrated hazard function as[13]

$$\Lambda(t) \equiv \int_0^t \lambda(s)\, ds$$

Now,

$$\frac{d \int_0^t \lambda(s)\, ds}{dt} = \lambda(t)$$

---

[13] The lower limit of integration is due to $F(0) = 0$.

and
$$\lambda(t) = \frac{f(t)}{1 - F(t)} = \frac{f(t)}{S(t)} = -\frac{d \ln S(t)}{dt}$$

Hence, $-\ln S(t) = \int_0^t \lambda(s)\, ds$ and the survivor function is

$$S(t) = \exp\left[-\int_0^t \lambda(s)\, ds\right]$$

Since $S(t) = 1 - F(t)$, the distribution function can be written

$$F(t) = 1 - \exp\left[-\int_0^t \lambda(s)\, ds\right]$$

and the density function (following differentiation) can be written

$$f(t) = \lambda(t) \exp\left[-\int_0^t \lambda(s)\, ds\right]$$

And all probabilities can conveniently be expressed in terms of the hazard function. For instance,

$$\Pr(T \geq t_2 \mid T \geq t_1) = \frac{1 - F(t_2)}{1 - F(t_1)}$$
$$= \exp\left[-\int_{t_1}^{t_2} \lambda(s)\, ds\right]$$

for $t_2 > t_1$. The above discussion focuses on unconditional hazard rates but frequently we're interested in conditional hazard rates.

Regression (conditional hazard rate) models

Conditional hazard rate models may be parametric or essentially nonparametric (Cox [1972]). Parametric models focus on $\lambda(t \mid x)$ where the conditional distribution is known (typically, Weibull, exponential, or lognormal). Much conditional duration analysis is based on the proportional hazard model. The proportional hazard model relates the hazard rate for an individual with characteristics x to some (perhaps unspecified) baseline hazard rate by some positive function of x. Since, as seen above, the probability of change is an exponential function it is convenient to also express this positive function as an exponential function. The proportional hazard model then is

$$\lambda(t \mid x, u) = \lambda_0(t) \exp\left[-(x\beta + u)\right]$$

where λ is the hazard that $T = t$ conditional on observables $X = x$ and unobservables $U = u$, λ_0 is the baseline hazard function, and β is a vector of (constant) parameters.

A common parameterization follows from a Weibull (α, γ) distribution. Then, the baseline hazard rate is

$$\lambda_0(t) = \frac{\alpha}{\gamma} \left(\frac{t}{\gamma}\right)^{\alpha-1}$$

and the hazard rate is

$$\lambda(t \mid x_1) = \frac{\alpha}{\gamma} \left(\frac{t}{\gamma}\right)^{\alpha-1} \exp[-x_1\beta_1]$$

The latter is frequently rewritten by adding a vector of ones to x_1 (denote this x) and absorbing γ (denote the augmented parameter vector β) so that

$$\lambda(t \mid x) = \alpha t^{\alpha-1} \exp[-x\beta]$$

This model can be estimated in standard fashion via maximization of the log-likelihood.

Since Cox's [1972] method doesn't require the baseline hazard function to be estimated, the method is essentially nonparametric in nature. Heterogeneity stems from observable and unobservable components of

$$\exp[-(x\beta + u)]$$

Cox's method accommodates observed heterogeneity but assumes unobserved homogeneity. As usual, unobservable heterogeneity can be problematic as conditional exchangeability is difficult to satisfy. Therefore, we look to alternative approaches to address unobservable heterogeneity.

Horowitz [1999] describes an approach for nonparametrically estimating the baseline hazard rate λ_0 and the integrated hazard rate Λ. In addition, the distribution function F and density function f for U, the unobserved source of heterogeneity with time-invariant covariates x, are nonparametrically estimated. The approach employs kernel density estimation methods similar to those discussed in chapter 6. As the estimators for F and f are slow to converge, the approach calls for large samples.

Campolieti [2001] addresses unobservable heterogeneity and the unknown error distribution via an alternative tack - Bayesian data augmentation (similar to that discussed in chapter 7). Discrete duration is modeled as a sequence of multi-period probits where duration dependence is accounted for via nonparametric estimation of the baseline hazard. A Dirichlet process prior supplies the nonparametric nature to the baseline hazard estimation.

8.1.10 Latent IV

Sometimes (perhaps frequently) it is difficult to identify instruments. Of course, this makes instrumental variable (*IV*) estimation unattractive. However, latent *IV* methods may help to overcome this deficiency. If the endogenous data are nonnormal (exhibit skewness and/or multi-modality) then it may be possible to decompose the data into parts that are unrelated to the regressor error and the part that is related. This is referred to as latent *IV*. Ebbes [2004] reviews the history of latent *IV* related primarily to measurement error and extends latent *IV* via analysis and simulation to various endogeneity concerns, including self selection.

8.2 Selectivity and treatment effects

This chapter is already much too long so next we only briefly introduce our main thesis - analysis of treatment effects in the face of potential endogeneity. Treatment effects are a special case of causal effects which we can under suitable conditions address without a fully structural model. As such treatment effects are both simple and challenging at the same time. Discussion of treatment effects occupies much of our focus in chapters 9 through 12.

First, we describe a prototypical setting. Then, we identify some typical treatment effects followed by a brief review of various identification conditions.

Suppose the *DGP* is
outcome equations:
$$Y_j = \mu_j(X) + V_j, j = 0, 1$$

selection equation:[14]
$$D^* = \mu_D(Z) - V_D$$

observable response:
$$Y = DY_1 + (1 - D)Y_0$$

where
$$D = \begin{matrix} 1 & D^* > 0 \\ 0 & otherwise \end{matrix}$$

In the binary case, the treatment effect is the effect on outcome of treatment compared with no treatment, $\Delta = Y_1 - Y_0$. Typical average treatment effects include *ATE*, *ATT*, and *ATUT*.[15] *ATE* refers to the average treatment effect,

$$ATE = E[\Delta] = E[Y_1 - Y_0]$$

In other words, the average effect on outcome of treatment for a random draw from the population. *ATT* refers to the average treatment effect on the treated,

$$ATT = E[\Delta \mid D = 1] = E[Y_1 - Y_0 \mid D = 1]$$

---

[14] We'll stick with binary choice for simplicity, though this can be readily generalized to the multinomial case.

[15] Additional treatment effects are discussed in subsequent chapters.

In other words, the average effect on outcome of treatment for a random draw from the subpopulation selecting (or assigned) treatment. *ATUT* refers to the average treatment effect on the untreated,

$$ATUT = E\left[\Delta \mid D = 0\right] = E\left[Y_1 - Y_0 \mid D = 0\right]$$

In other words, the average effect on outcome of treatment for a random draw from the subpopulation selecting (or assigned) no treatment.

The simplest approaches (strongest data conditions) involve ignorable treatment (sometimes referred to as selection on observables). These approaches include exogenous dummy variable regression, nonparametric regression, propensity score, propensity score matching, and control function methods. Various conditions and relaxations are discussed in the next chapter.

Instrumental variables (*IV*) are a common treatment effect identification strategy when ignorability is ill-suited to the data at hand. *IV* strategies accommodate homogeneous response at their simplest (strongest conditions) or unobservable heterogeneity at their most challenging (weakest conditions). Various *IV* approaches including standard *IV*, propensity score *IV*, control function *IV*, local *IV*, and Bayesian data augmentation are discussed in subsequent chapters. Heckman and Vytlacil [2005] argue that each of these strategies potentially estimate different treatment effects under varying conditions including continuous treatment and general equilibrium treatment effects.

8.3 Why bother with endogeneity?

Despite great effort by analysts, experiments frequently fail to identify substantive endogenous effects (Heckman [2000, 2001]). Why then do we bother? In this section we present a couple of stylized examples that depict some of our concerns regarding ignoring endogeneity. A theme of these examples is that failing to adequately attend to the *DGP* may produce a Simpson's paradox result.

8.3.1 Sample selection example

Suppose a firm has two production facilities, *A* and *B*. Facility *A* is perceived to be more efficient (produces a higher proportion of non-defectives). Consequently, production has historically been skewed in favor of facility *A*. The firm is interested in improving production efficiency, and particularly, improving facility *B*. Management has identified new production technology and is interested in whether the new technology improves production efficiency. Production using the new technology is skewed toward facility *B*. This "experiment" generates the data depicted in table 8.5.

Is the new technology more effective than the old technology? What is the technology treatment effect? As management knows, the choice of facility is important. The facility is a sufficiently important variable that its inclusion illuminates

Table 8.5: Production data: Simpson's paradox

	Facility A		Facility B		Total	
Technology	New	Old	New	Old	New	Old
Successes	10	120	133	25	143	145
Trials	10	150	190	50	200	200
% successes	100	80	70	50	71.5	72.5

the production technology treatment effect but its exclusion obfuscates the effect.[16] Aggregate results reported under the "Total" columns are misleading. For facility A, on average, there is a 20% improvement from the new technology. Likewise, for facility B, there is an average 20% improvement from the new technology.

Now, suppose an analyst collects the data but is unaware that there are two different facilities (the analyst only has the last two columns of data). What conclusion regarding the technology treatment effect is likely to be reached? This level of aggregation results in a serious omitted variable problem that leads to inferences opposite what the data suggest. This, of course, is a classic Simpson's paradox result produced via a sample selection problem. The data are not generated randomly but rather reflect management's selective "experimentation" on production technology.

8.3.2 Tuebingen-style treatment effect examples

Treatment effects are the focus of much economic self-selection analyses. When we ask what is the potential outcome response (Y) to treatment? — we pose a treatment effect question. A variety of treatment effects may be of interest. To setup the next example we define a few of the more standard treatment effects that may be of interest.

Suppose treatment is binary ($D = 1$ for treatment, $D = 0$ for untreated), for simplicity. As each individual is only observed either with treatment or without treatment, the observed outcome is

$$Y = DY_1 + (1 - D) Y_0$$

where

$$Y_1 = \mu_1 + V_1$$

is outcome response with treatment,

$$Y_0 = \mu_0 + V_0$$

is outcome response without treatment, μ_j is observed outcome for treatment $j = 0$ or 1, and V_j is unobserved (by the analyst) outcome for treatment j. Now, the

---

[16] This is an example of ignorable treatment (see ch. 9 for additional details).

treatment effect is

$$\begin{aligned}\Delta &= Y_1 - Y_0 \\ &= \mu_1 + V_1 - \mu_0 - V_0 \\ &= (\mu_1 - \mu_0) + (V_1 - V_0)\end{aligned}$$

an individual's (potential) outcome response to a change in treatment from regime 0 to regime 1. Note $(\mu_1 - \mu_0)$ is the population level effect (based on observables) and $(V_1 - V_0)$ is the individual-specific gain. That is, while treatment effects focus on potential gains for an individual, the unobservable nature of counterfactuals often lead analysts to focus on population level parameters.

The average treatment effect

$$ATE = E[\Delta] = E[Y_1 - Y_0]$$

is the average response to treatment for a random sample from the population. Even though seemingly cumbersome, we can rewrite ATE in a manner that illuminates connections with other treatment effects,

$$\begin{aligned}E[Y_1 - Y_0] &= E[Y_1 - Y_0 | D = 1]\Pr(D = 1) \\ &+ E[Y_1 - Y_0 | D = 0]\Pr(D = 0)\end{aligned}$$

The average treatment effect on the treated

$$ATT = E[\Delta | D = 1] = E[Y_1 - Y_0 | D = 1]$$

is the average response to treatment for a sample of individuals that choose (or are assigned) treatment. Selection (or treatment) is assumed to follow some *RUM* (random utility model; see chapter 5), $D^* = Z - V_D$ where D^* is latent utility index associated with treatment, Z is the observed portion, V_D is the part unobserved by the analyst, and $D = 1$ if $D^* > 0$ or $D = 0$ otherwise.

The average treatment effect on the untreated

$$ATUT = E[\Delta | D = 0] = E[Y_1 - Y_0 | D = 0]$$

is the average response to treatment for a sample of individuals that choose (or are assigned) no treatment. Again, selection (or treatment) is assumed to follow some *RUM*, $D^* = Z - V_D$.

To focus attention on endogeneity, it's helpful to identify what is estimated by *OLS* (exogenous treatment). Exogenous dummy variable regression estimates

$$OLS = E[Y_1 | D = 1] - E[Y_0 | D = 0]$$

An important question is when and to what extent is OLS a biased measure of the treatment effect.

Bias in the OLS estimate for ATT is

$$\begin{aligned}OLS &= ATT + bias_{ATT} \\ &= E[Y_1 | D = 1] - E[Y_0 | D = 0] \\ &= E[Y_1 | D = 1] - E[Y_0 | D = 1] + \{E[Y_0 | D = 1] - E[Y_0 | D = 0]\}\end{aligned}$$

8.3 Why bother with endogeneity?

Hence,
$$bias_{ATT} = \{E[Y_0|D=1]] - E[Y_0|D=0]\}$$

Bias in the OLS estimate for $ATUT$ is

$$\begin{aligned} OLS &= ATUT + bias_{ATUT} \\ &= E[Y_1|D=1] - E[Y_0|D=0] \\ &= E[Y_1|D=0] - E[Y_0|D=0] + \{E[Y_1|D=1] - E[Y_1|D=0]\} \end{aligned}$$

Hence,
$$bias_{ATUT} = \{E[Y_1|D=1] - E[Y_1|D=0]\}$$

Since

$$\begin{aligned} ATE &= \Pr(D=1) E[Y_1 - Y_0|D=1] + \Pr(D=0) E[Y_1 - Y_0|D=0] \\ &= \Pr(D=1) ATT + \Pr(D=0) ATUT \end{aligned}$$

bias in the OLS estimate for ATE can be written as a function of the bias in other treatment effects

$$bias_{ATE} = \Pr(D=1) bias_{ATT} + \Pr(D=0) bias_{ATUT}$$

Now we explore some examples.

Case 1

The setup involves simple (no regressors), discrete probability and outcome structure. It is important for identification of counterfactuals that outcome distributions are not affected by treatment selection. Hence, outcomes Y_0 and Y_1 vary only between states (and not by D within a state) as described, for instance, in table 8.6. Key components, the treatment effects, and any bias for case 1 are reported in table 8.7. Case 1 exhibits no endogeneity bias. This, in part, can be attributed to the idea that Y_1 is constant. However, even with Y_1 constant, this is a knife-edge result as the next cases illustrate.

Case 2

Case 2, depicted in table 8.8, perturbs the state two conditional probabilities only. Key components, the treatment effects, and any bias for case 2 are reported in table 8.9. Hence, a modest perturbation of the probability structure produces endogeneity bias in both ATT and ATE (but of course not $ATUT$ as Y_1 is constant).

Table 8.6: Tuebingen example case 1: ignorable treatment

State (s)	one		two		three	
$\Pr(Y,D,s)$	0.0272	0.0128	0.32	0.0	0.5888	0.0512
D	0	1	0	1	0	1
Y	0	1	1	1	2	1
Y_0	0	0	1	1	2	2
Y_1	1	1	1	1	1	1

152   8. Overview of endogeneity

Table 8.7: Tuebingen example case 1 results: ignorable treatment

Results	Key components
$ATE = E[Y_1 - Y_0]$ $= -0.6$	$p = \Pr(D = 1) = 0.064$
$ATT = E[Y_1 - Y_0 \mid D = 1]$ $= -0.6$	$E[Y_1 \mid D = 1] = 1.0$
$ATUT = E[Y_1 - Y_0 \mid D = 0]$ $= -0.6$	$E[Y_1 \mid D = 0] = 1.0$
$OLS = E[Y_1 \mid D = 1]$ $-E[Y_0 \mid D = 0] = -0.6$	$E[Y_1] = 1.0$
$bias_{ATT} = E[Y_0 \mid D = 1]$ $-E[Y_0 \mid D = 0] = 0.0$	$E[Y_0 \mid D = 1] = 1.6$
$bias_{ATUT} = E[Y_1 \mid D = 1]$ $-E[Y_1 \mid D = 0] = 0.0$	$E[Y_0 \mid D = 0] = 1.6$
$bias_{ATE} = p\, bias_{ATT}$ $+ (1-p)\, bias_{ATUT} = 0.0$	$E[Y_0] = 1.6$

Table 8.8: Tuebingen example case 2: heterogeneous response

State (s)	one		two		three	
$\Pr(Y, D, s)$	0.0272	0.0128	0.224	0.096	0.5888	0.0512
D	0	1	0	1	0	1
Y	0	1	1	1	2	1
Y_0	0	0	1	1	2	2
Y_1	1	1	1	1	1	1

Table 8.9: Tuebingen example case 2 results: heterogeneous response

Results	Key components
$ATE = E[Y_1 - Y_0]$ $= -0.6$	$p = \Pr(D = 1) = 0.16$
$ATT = E[Y_1 - Y_0 \mid D = 1]$ $= -0.24$	$E[Y_1 \mid D = 1] = 1.0$
$ATUT = E[Y_1 - Y_0 \mid D = 0]$ $= -0.669$	$E[Y_1 \mid D = 0] = 1.0$
$OLS = E[Y_1 \mid D = 1]$ $-E[Y_0 \mid D = 0] = -0.669$	$E[Y_1] = 1.0$
$bias_{ATT} = E[Y_0 \mid D = 1]$ $-E[Y_0 \mid D = 0] = -0.429$	$E[Y_0 \mid D = 1] = 1.24$
$bias_{ATUT} = E[Y_1 \mid D = 1]$ $-E[Y_1 \mid D = 0] = 0.0$	$E[Y_0 \mid D = 0] = 1.669$
$bias_{ATE} = p\, bias_{ATT}$ $+ (1-p)\, bias_{ATUT} = -0.069$	$E[Y_0] = 1.6$

Table 8.10: Tuebingen example case 3: more heterogeneity

State (s)	one		two		three	
$\Pr(Y, D, s)$	0.0272	0.0128	0.224	0.096	0.5888	0.0512
D	0	1	0	1	0	1
Y	0	1	1	1	2	0
Y_0	0	0	1	1	2	2
Y_1	1	1	1	1	0	0

Table 8.11: Tuebingen example case 3 results: more heterogeneity

Results	Key components
$ATE = E[Y_1 - Y_0]$ $= -1.24$	$p = \Pr(D = 1) = 0.16$
$ATT = E[Y_1 - Y_0 \mid D = 1]$ $= -0.56$	$E[Y_1 \mid D = 1] = 0.68$
$ATUT = E[Y_1 - Y_0 \mid D = 0]$ $= -1.370$	$E[Y_1 \mid D = 0] = 0.299$
$OLS = E[Y_1 \mid D = 1]$ $-E[Y_0 \mid D = 0] = -0.989$	$E[Y_1] = 0.36$
$bias_{ATT} = E[Y_0 \mid D = 1]$ $-E[Y_0 \mid D = 0] = -0.429$	$E[Y_0 \mid D = 1] = 1.24$
$bias_{ATUT} = E[Y_1 \mid D = 1]$ $-E[Y_1 \mid D = 0] = 0.381$	$E[Y_0 \mid D = 0] = 1.669$
$bias_{ATE} = p \, bias_{ATT}$ $+ (1-p) \, bias_{ATUT} = 0.251$	$E[Y_0] = 1.6$

Case 3

Case 3, described in table 8.10, maintains the probability structure of case 2 but alters the outcomes with treatment Y_1. Key components, the treatment effects, and any bias for case 3 are reported in table 8.11. A modest change in the outcomes with treatment produces endogeneity bias in all three average treatment effects (ATT, ATE, and $ATUT$).

Case 4

Case 4 maintains the probability structure of case 3 but alters the outcomes with treatment Y_1 as described in table 8.12. Key components, the treatment effects, and any bias for case 4 are reported in table 8.13. Case 4 is particularly noteworthy as OLS indicates a negative treatment effect, while all standard treatment effects, ATE, ATT, and $ATUT$ are positive. The endogeneity bias is so severe that it produces a Simpson's paradox result. Failure to accommodate endogeneity results in inferences opposite the *DGP*. Could this *DGP* represent earnings management? While these examples may not be as rich and deep as Lucas' [1976] critique of econometric policy evaluation, the message is similar — *endogeneity matters*!

Table 8.12: Tuebingen example case 4: Simpson's paradox

State (s)	one		two		three	
$\Pr(Y, D, s)$	0.0272	0.0128	0.224	0.096	0.5888	0.0512
D	0	1	0	1	0	1
Y	0	1	1	1	2	2.3
Y_0	0	0	1	1	2	2
Y_1	1	1	1	1	2.3	2.3

Table 8.13: Tuebingen example case 4 results: Simpson's paradox

Results	Key components
$ATE = E[Y_1 - Y_0]$ $= 0.232$	$p = \Pr(D = 1) = 0.16$
$ATT = E[Y_1 - Y_0 \mid D = 1]$ $= 0.176$	$E[Y_1 \mid D = 1] = 1.416$
$ATUT = E[Y_1 - Y_0 \mid D = 0]$ $= 0.243$	$E[Y_1 \mid D = 0] = 1.911$
$OLS = E[Y_1 \mid D = 1]$ $-E[Y_0 \mid D = 0] = -0.253$	$E[Y_1] = 1.832$
$bias_{ATT} = E[Y_0 \mid D = 1]$ $-E[Y_0 \mid D = 0] = -0.429$	$E[Y_0 \mid D = 1] = 1.24$
$bias_{ATUT} = E[Y_1 \mid D = 1]$ $-E[Y_1 \mid D = 0] = -0.495$	$E[Y_0 \mid D = 0] = 1.669$
$bias_{ATE} = p\,bias_{ATT}$ $+ (1-p)\,bias_{ATUT} = -0.485$	$E[Y_0] = 1.6$

8.4 Discussion and concluding remarks

> "All models are wrong but some are useful."
> - G. E. P. Box

It's time to return to our theme. Identifying causal effects suggests close attention to the interplay between theory, data, and model specification. Theory frames the problem so that economically meaningful effects can be deduced. Data supplies the evidence from which inference is drawn. Model specification attempts to consistently identify properties of the *DGP*. These elements are interdependent and iteratively divined.

Heckman [2000,2001] criticizes the selection literature for periods of preoccupation with devising estimators with nice statistical properties (e.g., consistency) but little economic import. Heckman's work juxtaposes policy evaluation implications of the treatment effects literature with the more ambitious structural modeling of the Cowles commission. It is clear for policy evaluation that theory or framing is of paramount importance.

> "Every econometric study is incomplete."
> - Zvi Griliches

In his discussion of economic data issues, Griliches [1986] reminds us that the quality of the data depends on both its source and its use. This suggests that creativity is needed to embrace the data issue. Presently, it seems that creativity in the address of omitted correlated variables, unobservable heterogeneity, and identification of instruments is in short supply in the accounting and business literature.

Model specification receives more attention in these pages but there is little to offer if theory and data are not carefully and creatively attended. With our current understanding of econometrics it seems we can't say much about a potential specification issue (including endogeneity) unless we accommodate it in the analysis. Even so, it is typically quite challenging to assess the nature and extent of the problem. If there is a mismatch with the theory or data, then discovery of (properties of) the *DGP* is likely hopelessly confounded. Logical consistency has been compromised.

8.5 Additional reading

The accounting literature gives increasing attention to endogeneity issues. Larcker and Rusticus [2004] review much of this work. Thought-provoking discussions of accounting and endogeneity are reported in an issue of *The European Accounting Review* including Chenhall and Moers. [2007a,2007b], Larcker and Rusticus [2007], and Van Lent [2007].

156   8. Overview of endogeneity

Amemiya [1985], Wooldridge [2002], Cameron and Trivedi [2005], Angrist and Krueger [1998], and the volumes of *Handbook of Econometrics* (especially volumes 5 and 6b) offer extensive reviews of econometric analysis of endogeneity. Latent *IV* traces back to Madansky [1959] and is resurrected by Lewbel [1997]. Heckman and Singer [1985,1986] discuss endogeneity challenges in longitudinal studies or duration models. The treatment effect examples are adapted from Joel Demski's seminars at the University of Florida and Eberhard Karls University of Tuebingen, Germany.

9
Treatment effects: ignorability

First, we describe a prototypical selection setting. Then, we identify some typical average treatment effects followed by a review of various identification conditions assuming ignorable treatment (sometimes called selection on observables). Ignorable treatment approaches are the simplest to implement but pose the strongest conditions for the data. That is, when the data don't satisfy the conditions it makes it more likely that inferences regarding properties of the *DGP* are erroneous.

9.1 A prototypical selection setting

Suppose the *DGP* is
outcome equations:[1]
$$Y_j = \mu_j(X) + V_j, j = 0, 1$$
selection equation:[2]
$$D^* = \mu_D(Z) - V_D$$
observable response:
$$Y = DY_1 + (1-D)Y_0$$

---

[1] Sometimes we'll find it convenient to write the outcome equations as a linear response
$$Y_j = \mu_j + X\beta_j + V_j$$

[2] We'll stick with binary choice for simplicity, though this can be readily generalized to the multinomial case (as discussed in the marginal treatment effects chapter).

where

$$D = \begin{cases} 1 & D^* > 0 \\ 0 & otherwise \end{cases}$$

and Y_1 is (potential) outcome with treatment and Y_0 is outcome without treatment.

In the binary case, the treatment effect is the effect on outcome of treatment compared with no treatment, $\Delta = Y_1 - Y_0$. Some typical treatment effects include: *ATE, ATT*, and *ATUT*. *ATE* refers to the average treatment effect, by iterated expectations

$$\begin{aligned} ATE &= E_X\left[ATE\left(X\right)\right] \\ &= E_X\left[E\left[\Delta \mid X = x\right]\right] = E\left[Y_1 - Y_0\right] \end{aligned}$$

In other words, the average effect on outcome of treatment for a random draw from the population. *ATT* refers to the average treatment effect on the treated,

$$\begin{aligned} ATT &= E_X\left[ATT\left(X\right)\right] \\ &= E_X\left[E\left[\Delta \mid X = x, D = 1\right]\right] = E\left[Y_1 - Y_0 \mid D = 1\right] \end{aligned}$$

In other words, the average effect on outcome of treatment for a random draw from the subpopulation selecting (or assigned) treatment. *ATUT* refers to the average treatment effect on the untreated,

$$\begin{aligned} ATUT &= E_X\left[ATUT\left(X\right)\right] \\ &= E_X\left[E\left[\Delta \mid X = x, D = 0\right]\right] = E\left[Y_1 - Y_0 \mid D = 0\right] \end{aligned}$$

In other words, the average effect on outcome of treatment for a random draw from the subpopulation selecting (or assigned) no treatment.

The remainder of this chapter is devoted to simple identification and estimation strategies. These simple strategies pose strong conditions for the data that may lead to logically inconsistent inferences.

9.2 Exogenous dummy variable regression

The simplest strategy (strongest data conditions) is exogenous dummy variable regression. Suppose D is independent of (Y_1, Y_0) conditional on X, response is linear, and errors are normally distributed, then *ATE* is identified via exogenous dummy variable (*OLS*) regression.[3] For instance, suppose the *DGP* is

$$Y = \delta + \varsigma D + X\beta_0 + DX\left(\beta_1 - \beta_0\right) + \varepsilon$$

Since Y_1 and Y_0 are conditionally mean independent of D given X

$$\begin{aligned} E\left[Y_1 \mid X, D = 1\right] &= E\left[Y_1 \mid X\right] \\ &= \delta + \varsigma + X\beta_0 + X\left(\beta_1 - \beta_0\right) \end{aligned}$$

---

[3] These conditions are stronger than necessary as we can get by with conditional mean independence in place of conditional stochastic independence.

and
$$E[Y_0 \mid X, D = 0] = E[Y_0 \mid X]$$
$$= \delta + X\beta_0$$
then
$$ATE(X) = E[Y_1 \mid X] - E[Y_0 \mid X]$$
$$= \varsigma + X(\beta_1 - \beta_0)$$

Then, by iterated expectations, $ATE = \varsigma + E[X](\beta_1 - \beta_0)$. ATE can be directly estimated via α if we rewrite the response equation as

$$Y = \delta + \alpha D + X\beta_0 + D(X - E[X])(\beta_1 - \beta_0) + \varepsilon$$

which follows from rewriting the *DGP* as

$$Y = \delta + (\varsigma + E[X](\beta_1 - \beta_0)) D + X\beta_0$$
$$+ D[X(\beta_1 - \beta_0) - E[X](\beta_1 - \beta_0)] + \varepsilon$$

9.3 Tuebingen-style examples

To illustrate ignorable treatment, we return to the Tuebingen-style examples of chapter 8 and add regressors to the mix. For each case, we compare treatment effect analyses when the analyst observes the states with when the analyst observes only the regressor, X. The setup involves simple discrete probability and outcome structure. Identification of counterfactuals is feasible if outcome distributions are not affected by treatment selection. Hence, outcomes Y_0 and Y_1 vary only between states (and not by D within a state).

Case 1

The first case depicted in table 9.1 involves extreme homogeneity (no variation in Y_0 and Y_1). Suppose the states are observable to the analyst. Then, we have

Table 9.1: Tuebingen example case 1: extreme homogeneity

State (s)	one		two		three	
$\Pr(Y, D, s)$	0.0272	0.0128	0.224	0.096	0.5888	0.0512
D	0	1	0	1	0	1
Y	0	1	0	1	0	1
Y_0	0	0	0	0	0	0
Y_1	1	1	1	1	1	1
X	1	1	1	1	0	0

a case of perfect regressors and no residual uncertainty. Consequently, we can identify treatments effects by states. The treatment effect for all three states is homogeneously one.

Now, suppose the states are unobservable but the analyst observes X. Then, conditional average treatment effects are

$$E[Y_1 - Y_0 \mid X = 1] = E[Y_1 - Y_0 \mid X = 0] = 1$$

Key components, unconditional average (integrating out X) treatment effects, and any bias for case 1 are reported in table 9.2. Case 1 exhibits no endogeneity bias.

Table 9.2: Tuebingen example case 1 results: extreme homogeneity

Results	Key components
$ATE = E[Y_1 - Y_0]$ $= 1.0$	$p = \Pr(D = 1) = 0.16$
$ATT = E[Y_1 - Y_0 \mid D = 1]$ $= 1.0$	$E[Y_1 \mid D = 1] = 1.0$
$ATUT = E[Y_1 - Y_0 \mid D = 0]$ $= 1.0$	$E[Y_1 \mid D = 0] = 1.0$
$OLS = E[Y_1 \mid D = 1]$ $-E[Y_0 \mid D = 0] = 1.0$	$E[Y_1] = 1.0$
$bias_{ATT} = E[Y_0 \mid D = 1]$ $-E[Y_0 \mid D = 0] = 0.0$	$E[Y_0 \mid D = 1] = 0.0$
$bias_{ATUT} = E[Y_1 \mid D = 1]$ $-E[Y_1 \mid D = 0] = 0.0$	$E[Y_0 \mid D = 0] = 0.0$
$bias_{ATE} = p\, bias_{ATT}$ $+ (1 - p)\, bias_{ATUT} = 0.0$	$E[Y_0] = 0.0$

Extreme homogeneity implies stochastic independence of (Y_0, Y_1) and D conditional on X.

Case 2

Case 2 adds variation in outcomes but maintains treatment effect homogeneity as displayed in table 9.3. Suppose the states are observable to the analyst. Then, we

Table 9.3: Tuebingen example case 2: homogeneity

State (s)	one		two		three	
$\Pr(Y, D, s)$	0.0272	0.0128	0.224	0.096	0.5888	0.0512
D	0	1	0	1	0	1
Y	0	1	1	2	2	3
Y_0	0	0	1	1	2	2
Y_1	1	1	2	2	3	3
X	1	1	1	1	0	0

can identify treatments effects by states. The treatment effect for all three states is homogeneously one.

Now, suppose the states are unobservable but the analyst observes X. Then, conditional average treatment effects are

$$E[Y_1 - Y_0 \mid X = 1] = E[Y_1 - Y_0 \mid X = 0] = 1$$

which follows from

$$E_X[E[Y_1 \mid X]] = 0.36(1.889) + 0.64(3) = 2.6$$

$$E_X[E[Y_0 \mid X]] = 0.36(0.889) + 0.64(2) = 1.6$$

but *OLS* (or, for that matter, nonparametric regression) estimates

$$E_X[E[Y_1 \mid X, D = 1]] = 0.68(1.882) + 0.32(3) = 2.24$$

and

$$E_X[E[Y_0 \mid X, D = 0]] = 0.299(0.892) + 0.701(2) = 1.669$$

Clearly, outcomes are not conditionally mean independent of treatment given X ($2.6 \neq 2.24$ for Y_1 and $1.6 \neq 1.669$ for Y_0). Key components, unconditional average (integrating out X) treatment effects, and any bias for case 2 are summarized in table 9.4. Hence, homogeneity does not ensure exogenous dummy variable (or

Table 9.4: Tuebingen example case 2 results: homogeneity

Results	Key components
$ATE = E[Y_1 - Y_0]$ $= 1.0$	$p = \Pr(D = 1) = 0.16$
$ATT = E[Y_1 - Y_0 \mid D = 1]$ $= 1.0$	$E[Y_1 \mid D = 1] = 2.24$
$ATUT = E[Y_1 - Y_0 \mid D = 0]$ $= 1.0$	$E[Y_1 \mid D = 0] = 2.669$
$OLS = E[Y_1 \mid D = 1]$ $-E[Y_0 \mid D = 0] = 0.571$	$E[Y_1] = 2.6$
$bias_{ATT} = E[Y_0 \mid D = 1]$ $-E[Y_0 \mid D = 0] = -0.429$	$E[Y_0 \mid D = 1] = 1.24$
$bias_{ATUT} = E[Y_1 \mid D = 1]$ $-E[Y_1 \mid D = 0] = -0.429$	$E[Y_0 \mid D = 0] = 1.669$
$bias_{ATE} = p \, bias_{ATT}$ $+ (1 - p) \, bias_{ATUT} = -0.429$	$E[Y_0] = 1.6$

nonparametric) identification of average treatment effects.

Case 3

Case 3 slightly perturbs outcomes with treatment, Y_1, to create heterogeneous response as depicted in table 9.5. Suppose the states are observable to the analyst. Then, we can identify treatments effects by states. The treatment effect for all three states is homogeneously one.

162   9. Treatment effects: ignorability

Table 9.5: Tuebingen example case 3: heterogeneity

State (s)	one		two		three	
$\Pr(Y, D, s)$	0.0272	0.0128	0.224	0.096	0.5888	0.0512
D	0	1	0	1	0	1
Y	0	1	1	1	2	0
Y_0	0	0	1	1	2	2
Y_1	1	1	2	2	2	2
X	1	1	1	1	0	0

But, suppose the states are unobservable and the analyst observes X. Then, conditional average treatment effects are heterogeneous

$$E[Y_1 - Y_0 \mid X = 1] = 1$$
$$E[Y_1 - Y_0 \mid X = 0] = 0$$

This follows from

$$E_X[E[Y_1 \mid X]] = 0.36(1.889) + 0.64(2) = 1.96$$

$$E_X[E[Y_0 \mid X]] = 0.36(0.889) + 0.64(2) = 1.6$$

but *OLS* (or nonparametric regression) estimates

$$E_X[E[Y_1 \mid X, D = 1]] = 0.68(1.882) + 0.32(2) = 1.92$$

and

$$E_X[E[Y_0 \mid X, D = 0]] = 0.299(0.892) + 0.701(2) = 1.669$$

Clearly, outcomes are not conditionally mean independent of treatment given X (1.96 ≠ 1.92 for Y_1 and 1.6 ≠ 1.669 for Y_0). Key components, unconditional average (integrating out X) treatment effects, and any bias for case 3 are summarized in table 9.6. A modest change in outcomes with treatment produces endogeneity bias in all three average treatment effects (ATT, ATE, and $ATUT$). Average treatment effects are not identified by dummy variable regression (or nonparametric regression) in case 3.

Case 4

Case 4, described in table 9.7, maintains the probability structure of case 3 but alters outcomes with treatment, Y_1, to produce a Simpson's paradox result. Suppose the states are observable to the analyst. Then, we can identify treatments effects by states. The treatment effect for all three states is homogeneously one. But, suppose the states are unobservable and the analyst observes X. Then, conditional average treatment effects are heterogeneous

$$E[Y_1 - Y_0 \mid X = 1] = 0.111$$
$$E[Y_1 - Y_0 \mid X = 0] = 0.3$$

9.3 Tuebingen-style examples 163

Table 9.6: Tuebingen example case 3 results: heterogeneity

Results	Key components
$ATE = E[Y_1 - Y_0]$ $= 0.36$	$p = \Pr(D = 1) = 0.16$
$ATT = E[Y_1 - Y_0 \mid D = 1]$ $= 0.68$	$E[Y_1 \mid D = 1] = 1.92$
$ATUT = E[Y_1 - Y_0 \mid D = 0]$ $= 0.299$	$E[Y_1 \mid D = 0] = 1.968$
$OLS = E[Y_1 \mid D = 1]$ $- E[Y_0 \mid D = 0] = 0.251$	$E[Y_1] = 1.96$
$bias_{ATT} = E[Y_0 \mid D = 1]$ $- E[Y_0 \mid D = 0] = -0.429$	$E[Y_0 \mid D = 1] = 1.24$
$bias_{ATUT} = E[Y_1 \mid D = 1]$ $- E[Y_1 \mid D = 0] = -0.048$	$E[Y_0 \mid D = 0] = 1.669$
$bias_{ATE} = p\, bias_{ATT}$ $+ (1-p)\, bias_{ATUT} = -0.109$	$E[Y_0] = 1.6$

Table 9.7: Tuebingen example case 4: Simpson's paradox

State (s)	one		two		three	
$\Pr(Y, D, s)$	0.0272	0.0128	0.224	0.096	0.5888	0.0512
D	0	1	0	1	0	1
Y	0	1	1	1	2	2.3
Y_0	0	0	1	1	2	2
Y_1	1	1	1	1	2.3	2.3
X	1	1	1	1	0	0

This follows from

$$E_X[E[Y_1 \mid X]] = 0.36(1.0) + 0.64(2.3) = 1.832$$

$$E_X[E[Y_0 \mid X]] = 0.36(0.889) + 0.64(2) = 1.6$$

but *OLS* (or nonparametric regression) estimates

$$E_X[E[Y_1 \mid X, D = 1]] = 0.68(1.0) + 0.32(2.3) = 1.416$$

and

$$E_X[E[Y_0 \mid X, D = 0]] = 0.299(0.892) + 0.701(2) = 1.669$$

Clearly, outcomes are not conditionally mean independent of treatment given X (1.932 $\neq$ 1.416 for Y_1 and 1.6 $\neq$ 1.669 for Y_0). Key components, unconditional average (integrating out X) treatment effects, and any bias for case 4 are summarized in table 9.8. Case 4 is particularly noteworthy as dummy variable regression (or nonparametric regression) indicates a negative treatment effect, while all three standard average treatment effects, ATE, ATT, and $ATUT$, are positive. Hence,

164 9. Treatment effects: ignorability

Table 9.8: Tuebingen example case 4 results: Simpson's paradox

Results	Key components
$ATE = E[Y_1 - Y_0]$ $= 0.232$	$p = \Pr(D = 1) = 0.16$
$ATT = E[Y_1 - Y_0 \mid D = 1]$ $= 0.176$	$E[Y_1 \mid D = 1] = 1.416$
$ATUT = E[Y_1 - Y_0 \mid D = 0]$ $= 0.243$	$E[Y_1 \mid D = 0] = 1.911$
$OLS = E[Y_1 \mid D = 1]$ $-E[Y_0 \mid D = 0] = -0.253$	$E[Y_1] = 1.832$
$bias_{ATT} = E[Y_0 \mid D = 1]$ $-E[Y_0 \mid D = 0] = -0.429$	$E[Y_0 \mid D = 1] = 1.24$
$bias_{ATUT} = E[Y_1 \mid D = 1]$ $-E[Y_1 \mid D = 0] = -0.495$	$E[Y_0 \mid D = 0] = 1.669$
$bias_{ATE} = p\, bias_{ATT}$ $+ (1 - p)\, bias_{ATUT} = -0.485$	$E[Y_0] = 1.6$

average treatment effects are not identified by exogenous dummy variable regression (or nonparametric regression) for case 4.

How do we proceed when ignorable treatment (conditional mean independence) fails? A common response is to look for instruments and apply *IV* approaches to identify average treatment effects. Chapter 10 explores instrumental variable approaches. The remainder of this chapter surveys some other ignorable treatment approaches and applies them to the asset revaluation regulation problem introduced in chapter 2.

9.4 Nonparametric identification

Suppose treatment is ignorable or, in other words, treatment is conditionally mean independent of outcome,

$$E[Y_1 \mid X, D] = E[Y_1 \mid X]$$

and

$$E[Y_0 \mid X, D] = E[Y_0 \mid X]$$

This is also called "selection on observables" as the regressors are so powerful that we can ignore choice D. For binary treatment, this implies

$$E[Y_1 \mid X, D = 1] = E[Y_1 \mid X, D = 0]$$

and

$$E[Y_0 \mid X, D = 1] = E[Y_0 \mid X, D = 0]$$

9.4 Nonparametric identification

The condition is difficult to test directly as it involves $E[Y_1 \mid X, D = 0]$ and $E[Y_0 \mid X, D = 1]$, the counterfactuals. Let $p(X) = \Pr(D = 1 \mid X)$. Ignorable treatment implies the average treatment effect is nonparametrically identified.

$$\begin{aligned} ATE(X) &= E[\Delta \mid X] = E[Y_1 - Y_0 \mid X] \\ &= E[Y_1 \mid X] - E[Y_0 \mid X] \end{aligned}$$

By Bayes' theorem we can rewrite the expression as

$$p(X) E[Y_1 \mid X, D = 1] + (1 - p(X)) E[Y_1 \mid X, D = 0]$$
$$-p(X) E[Y_0 \mid X, D = 1] - (1 - p(X)) E[Y_0 \mid X, D = 0]$$

conditional mean independence allows simplification to

$$E[Y_1 \mid X] - E[Y_0 \mid X] = ATE(X)$$

Consider a couple of ignorable treatment examples which distinguish between exogenous dummy variable and nonparametric identification.

Example 9.1 *The first example posits a simple case of stochastic independence between treatment D and response (Y_1, Y_0) conditional on X. The DGP is depicted in table 9.9 (values of D, Y_1, and Y_0 vary randomly at each level of X).[4] Clearly, if the response variables are stochastically independent of D conditional*

Table 9.9: Exogenous dummy variable regression example

probability	$\frac{1}{6}$	$\frac{1}{6}$	$\frac{1}{6}$	$\frac{1}{6}$	$\frac{1}{6}$	$\frac{1}{6}$	$E[\cdot]$
$(Y_1 \mid X, D = 1)$	0	1	0	2	0	3	1
$(Y_1 \mid X, D = 0)$	0	1	0	2	0	3	1
$(Y_0 \mid X, D = 1)$	-1	0	-2	0	-3	0	-1
$(Y_0 \mid X, D = 0)$	-1	0	-2	0	-3	0	-1
X	1	1	2	2	3	3	2
$(D \mid X)$	0	1	0	1	0	1	0.5

on X

$$\Pr(Y_1 = y_1 \mid X = x, D = 1) = \Pr(Y_1 = y_1 \mid X = x, D = 0)$$

and

$$\Pr(Y_0 = y_0 \mid X = x, D = 1) = \Pr(Y_0 = y_0 \mid X = x, D = 0)$$

---

[4]The columns in the table are not states of nature but merely indicate the values the response Y_j and treatment D variables are allowed to take and their likelihoods. Conditional on X, the likelihoods for Y_j and D and independent.

9. Treatment effects: ignorability

then they are also conditionally mean independent

$$E[Y_1 \mid X = 1, D = 1] = E[Y_1 \mid X = 1, D = 0] = 0.5$$
$$E[Y_1 \mid X = 2, D = 1] = E[Y_1 \mid X = 2, D = 0] = 1$$
$$E[Y_1 \mid X = 3, D = 1] = E[Y_1 \mid X = 3, D = 0] = 1.5$$

and

$$E[Y_0 \mid X = 1, D = 1] = E[Y_0 \mid X = 1, D = 0] = -0.5$$
$$E[Y_0 \mid X = 2, D = 1] = E[Y_0 \mid X = 2, D = 0] = -1$$
$$E[Y_0 \mid X = 3, D = 1] = E[Y_0 \mid X = 3, D = 0] = -1.5$$

Conditional average treatment effects are

$$ATE(X = 1) = 0.5 - (-0.5) = 1$$
$$ATE(X = 2) = 1 - (-1) = 2$$
$$ATE(X = 3) = 1.5 - (-1.5) = 3$$

and unconditional average treatment effects are

$$ATE = E[Y_1 - Y_0] = 1 - (-1) = 2$$
$$ATT = E[Y_1 - Y_0 \mid D = 1] = 1 - (-1) = 2$$
$$ATUT = E[Y_1 - Y_0 \mid D = 0] = 1 - (-1) = 2$$

Exogenous dummy variable regression

$$Y = \delta + \alpha D + X\beta_0 + D(X - E[X])(\beta_1 - \beta_0) + \varepsilon$$

consistently estimates ATE via α. Based on a saturated "sample" of size 384 reflecting the DGP, dummy variable regression results are reported in table 9.10.

Table 9.10: Exogenous dummy variable regression results

parameter	coefficient	se	t-statistic
δ	0.000	0.207	0.000
α	2.000	0.110	18.119
β_0	−0.500	0.096	−5.230
$\beta_1 - \beta_0$	1.000	0.135	7.397

The conditional regression estimates of average treatment effects

$$ATE(X = 1) = 2 + 1(1 - 2) = 1$$
$$ATE(X = 2) = 2 + 1(2 - 2) = 2$$
$$ATE(X = 3) = 2 + 1(3 - 2) = 3$$

correspond well with the DGP. In this case, exogenous dummy variable regression identifies the average treatment effects.

9.4 Nonparametric identification   167

Example 9.2 *The second example relaxes the DGP such that responses are conditionally mean independent but not stochastically independent and, importantly, the relations between outcomes and X are nonlinear. The DGP is depicted in table 9.11 (values of D, Y_1, and Y_0 vary randomly at each level of X).[5] Again,*

Table 9.11: Nonparametric treatment effect regression

probability	$\frac{1}{6}$	$\frac{1}{6}$	$\frac{1}{6}$	$\frac{1}{6}$	$\frac{1}{6}$	$\frac{1}{6}$	$E[\cdot]$
$(Y_1 \mid X, D = 1)$	0	1	0	2	0	3	1
$(Y_1 \mid X, D = 0)$	0.5	0.5	1	1	1.5	1.5	1
$(Y_0 \mid X, D = 1)$	-1	0	-2	0	-3	0	-1
$(Y_0 \mid X, D = 0)$	-0.5	-0.5	-1	-1	-1.5	-1.5	-1
X	-1	-1	-2	-2	3	3	0
$(D \mid X)$	0	1	0	1	0	1	0.5

population average treatment effects are

$$ATE = E[Y_1 - Y_0] = 1 - (-1) = 2$$

$$ATT = E[Y_1 - Y_0 \mid D = 1] = 1 - (-1) = 2$$

$$ATUT = E[Y_1 - Y_0 \mid D = 0] = 1 - (-1) = 2$$

Further, the average treatment effects conditional on X are

$$ATE(X = -1) = 0.5 - (-0.5) = 1$$

$$ATE(X = -2) = 1 - (-1) = 2$$

$$ATE(X = 3) = 1.5 - (-1.5) = 3$$

Average treatment effects are estimated in two ways. First, exogenous dummy variable regression

$$Y = \delta + \alpha D + X\beta_0 + D(X - E[X])(\beta_1 - \beta_0) + \varepsilon$$

consistently estimates ATE via α. A saturated "sample" of 48 observations reflecting the DGP produces the results reported in table 9.12. However, the regression-estimated average treatment effects conditional on X are

$$ATE(X = -1) = 1.714$$

$$ATE(X = -2) = 1.429$$

$$ATE(X = 3) = 2.857$$

---

[5]Again, the columns of the table are not states of nature but merely indicate the values the variables can take conditional on X.

Table 9.12: Nonparametrically identified treatment effect: exogenous dummy variable regression results

parameter	coefficient	se	t-statistic
δ	-1.000	0.167	-5.991
α	2.000	0.236	8.472
β_0	-0.143	0.077	-1.849
$\beta_1 - \beta_0$	0.286	0.109	2.615

Hence, the conditional average treatment effects are not identified by exogenous dummy variable regression for this case. Second, let $\Im_x$ be an indicator variable for $X = x$. ANOVA is equivalent to nonparametric regression since X is sparse.

$$Y = \alpha D + \gamma_1 \Im_{-1} + \gamma_2 \Im_{-2} + \gamma_3 \Im_3 + \gamma_4 D\Im_{-1} + \gamma_5 D\Im_{-2} + \varepsilon$$

ANOVA results are reported in table 9.13. The ANOVA-estimated conditional av-

Table 9.13: Nonparametric treatment effect regression results

parameter	coefficient	se	t-statistic
α	3.000	0.386	7.774
γ_1	-0.500	0.273	-1.832
γ_2	-1.000	0.273	-3.665
γ_3	-1.500	0.273	-5.497
γ_4	-2.000	0.546	-3.665
γ_5	-1.000	0.546	-1.832

erage treatment effects are

$$ATE\,(X = -1) = 3 - 2 = 1$$
$$ATE\,(X = -2) = 3 - 1 = 2$$
$$ATE\,(X = 3) = 3$$

and the unconditional average treatment effect is

$$ATE = \frac{1}{3}(1 + 2 + 3) = 2$$

Therefore, even though the estimated average treatment effects for exogenous dummy variable regression are consistent with the DGP, the average treatment effects conditional on X do not correspond well with the DGP. Further, the treatment effects are not even monotonic in X. However, the ANOVA results properly account for the nonlinearity in the data and correspond nicely with the DGP for both unconditional and conditional average treatment effects. Hence, average treatment effects are nonparametrically identified for this case but not identified by exogenous dummy variable regression.

9.5 Propensity score approaches

Suppose the data are conditionally mean independent

$$E[Y_1 \mid X, D] = E[Y_1 \mid X]$$
$$E[Y_0 \mid X, D] = E[Y_0 \mid X]$$

so treatment is ignorable, and common X support leads to nondegenerate propensity scores

$$0 < p(X) = \Pr(D = 1 \mid X) < 1 \text{ for all } X$$

then average treatment effect estimands are

$$ATE = E\left[\frac{(D - p(X))Y}{p(X)(1 - p(X))}\right]$$
$$ATT = E\left[\frac{(D - p(X))Y}{(1 - p(X))}\right] / \Pr(D = 1)$$
$$ATUT = E\left[\frac{(D - p(X))Y}{p(X)}\right] / \Pr(D = 0)$$

The econometric procedure is to first estimate the propensity for treatment or propensity score, $p(X)$, via some flexible model (e.g., nonparametric regression; see chapter 6), then *ATE*, *ATT*, and *ATUT* are consistently estimated via sample analogs to the above.

9.5.1 ATE and propensity score

$ATE = E\left[\frac{(D - p(X))Y}{p(X)(1 - p(X))}\right]$ is identified as follows. Observed outcome is

$$Y = DY_1 + (1 - D)Y_0$$

Substitution for Y and evaluation of the conditional expectation produces

$$E[(D - p(X))Y \mid X]$$
$$= E[DDY_1 + D(1 - D)Y_0 - p(X)DY_1 - p(X)(1 - D)Y_0 \mid X]$$
$$= E[DY_1 + 0 - p(X)DY_1 - p(X)(1 - D)Y_0 \mid X]$$

Letting $m_j(X) \equiv E[Y_j \mid X]$ and recognizing

$$p(X) \equiv \Pr(D = 1 \mid X)$$
$$= E[D = 1 \mid X]$$

gives

$$E[DY_1 - p(X)DY_1 - p(X)(1 - D)Y_0 \mid X]$$
$$= p(X)m_1(X) - p^2(X)m_1(X) - p(X)(1 - p(X))m_0(X)$$
$$= p(X)(1 - p(X))(m_1(X) - m_0(X))$$

This leads to the conditional average treatment effect

$$E\left[\frac{p(X)(1-p(X))(m_1(X)-m_0(X))}{p(X)(1-p(X))}\;\Big|\;X\right] = m_1(X) - m_0(X)$$
$$= E[Y_1 - Y_0 \mid X]$$

The final connection to the estimand is made by iterated expectations,

$$ATE = E[Y_1 - Y_0]$$
$$= E_X[Y_1 - Y_0 \mid X]$$

9.5.2 ATT, ATUT, and propensity score

Similar logic identifies the estimand for the average treatment effect on the treated

$$ATT = E\left[\frac{(D-p(X))Y}{(1-p(X))}\right] / \Pr(D=1)$$

Utilize

$$E[(D-p(X))Y \mid X] = p(X)(1-p(X))(m_1(X) - m_0(X))$$

from the propensity score identification of ATE. Eliminating $(1-p(X))$ and rewriting gives

$$\frac{p(X)(1-p(X))(m_1(X)-m_0(X))}{(1-p(X))}$$
$$= p(X)(m_1(X) - m_0(X))$$
$$= \Pr(D=1 \mid X)(E[Y_1 \mid X] - E[Y_0 \mid X])$$

Conditional mean independence implies

$$\Pr(D=1 \mid X)(E[Y_1 \mid X] - E[Y_0 \mid X])$$
$$= \Pr(D=1 \mid X)(E[Y_1 \mid D=1, X] - E[Y_0 \mid D=1, X])$$
$$= \Pr(D=1 \mid X) E[Y_1 - Y_0 \mid D=1, X]$$

Then, by iterated expectations, we have

$$E_X[\Pr(D=1 \mid X) E[Y_1 - Y_0 \mid D=1, X]]$$
$$= \Pr(D=1) E[Y_1 - Y_0 \mid D=1]$$

Putting it all together produces the estimand

$$ATT = E_X\left[\frac{(D-p(X))Y}{(1-p(X))}\right] / \Pr(D=1)$$
$$= E[Y_1 - Y_0 \mid D=1]$$

9.5 Propensity score approaches

For the average treatment effect on the untreated estimand

$$ATUT = E\left[\frac{(D-p(X))Y}{p(X)}\right]/\Pr(D=0)$$

identification is analogous to that for *ATT*. Eliminating $p(X)$ from

$$E\left[(D-p(X))Y \mid X\right] = p(X)(1-p(X))(m_1(X) - m_0(X))$$

and rewriting gives

$$\frac{p(X)(1-p(X))(m_1(X) - m_0(X))}{p(X)}$$
$$= (1-p(X))(m_1(X) - m_0(X))$$
$$= \Pr(D=0 \mid X)(E[Y_1 \mid X] - E[Y_0 \mid X])$$

Conditional mean independence implies

$$\Pr(D=0 \mid X)(E[Y_1 \mid X] - E[Y_0 \mid X])$$
$$= \Pr(D=0 \mid X)(E[Y_1 \mid D=0, X] - E[Y_0 \mid D=0, X])$$
$$= \Pr(D=0 \mid X)E[Y_1 - Y_0 \mid D=0, X]$$

Iterated expectations yields

$$E_X\left[\Pr(D=0 \mid X)E[Y_1 - Y_0 \mid D=0, X]\right]$$
$$= \Pr(D=0)E[Y_1 - Y_0 \mid D=0]$$

Putting everything together produces the estimand

$$ATUT = E\left[\frac{(D-p(X))Y}{p(X)}\right]/\Pr(D=0)$$
$$= E[Y_1 - Y_0 \mid D=0]$$

Finally, the average treatment effects are connected as follows.

$$ATE = \Pr(D=1) ATT + \Pr(D=0) ATUT$$
$$= \Pr(D=1) E\left[\frac{(D-p(X))Y}{(1-p(X))}\right]/\Pr(D=1)$$
$$\quad + \Pr(D=0) E\left[\frac{(D-p(X))Y}{p(X)}\right]/\Pr(D=0)$$
$$= E\left[\frac{(D-p(X))Y}{(1-p(X))}\right] + E\left[\frac{(D-p(X))Y}{p(X)}\right]$$
$$= E_X\left[\Pr(D=1 \mid X)(E[Y_1 \mid X] - E[Y_0 \mid X])\right]$$
$$\quad + E_X\left[\Pr(D=0 \mid X)(E[Y_1 \mid X] - E[Y_0 \mid X])\right]$$
$$= \Pr(D=1) E[Y_1 - Y_0] + \Pr(D=0) E[Y_1 - Y_0]$$
$$= E[Y_1 - Y_0]$$

9.5.3 Linearity and propensity score

If we add the condition $E[Y_0 \mid p(X)]$ and $E[Y_1 \mid p(X)]$ are linear in $p(X)$ then α in the expression below consistently estimates *ATE*

$$E[Y \mid X, D] = \varsigma_0 + \alpha D + \varsigma_1 \hat{p} + \varsigma_2 D (\hat{p} - \hat{\mu}_p)$$

where $\hat{\mu}_p$ is the sample average of the estimated propensity score $\hat{p}$.

9.6 Propensity score matching

Rosenbaum and Rubin's [1983] propensity score matching is a popular propensity score approach. Rosenbaum and Rubin suggest selecting a propensity score at random from the sample, then matching two individuals with this propensity score — one treated and one untreated. The expected outcome difference $E[Y_1 - Y_0 \mid p(X)]$ is ATE conditional on $p(X)$. Hence, by iterated expectations

$$ATE = E_{p(X)} [E[Y_1 - Y_0 \mid p(X)]]$$

ATE identification by propensity score matching poses strong ignorability. That is, outcome (Y_1, Y_0) independence of treatment D given X (a stronger condition than conditional mean independence) and, as before, common X support leads to nondegenerate propensity scores $p(X) \equiv \Pr(D = 1 \mid X)$

$$0 < \Pr(D = 1 \mid X) < 1 \text{ for all } X$$

As demonstrated by Rosenbaum and Rubin, strong ignorability implies index sufficiency. In other words, outcome (Y_1, Y_0) independence of treatment D given $p(X)$ and

$$0 < \Pr(D = 1 \mid p(X)) < 1 \text{ for all } p(X)$$

The latter (inequality) condition is straightforward. Since X is finer than $p(X)$, the first inequality (for X) implies the second (for $p(X)$). The key is conditional stochastic independence given the propensity score

$$\Pr(D = 1 \mid Y_1, Y_0, p(X)) = \Pr(D = 1 \mid p(X))$$

This follows from

$$\begin{aligned} \Pr(D = 1 \mid Y_1, Y_0, p(X)) &= E[\Pr(D = 1 \mid Y_1, Y_0, X) \mid Y_1, Y_0, p(X)] \\ &= E[p(X) \mid Y_1, Y_0, p(X)] = p(X) \\ &= E[D \mid p(X)] \\ &= \Pr(D = 1 \mid p(X)) \end{aligned}$$

For a general matching strategy on X, Heckman, Ichimura, and Todd [1998] point out that for *ATT*, strong ignorability can be relaxed to conditional mean

9.6 Propensity score matching

independence for outcomes without treatment and full support S for the treated subsample. This allows counterfactuals to be related to observables

$$E[Y_0 \mid D = 1, X] = E[Y_0 \mid D = 0, X] \text{ for } X \in S$$

so that *ATT(X)* can be expressed in terms of observables only

$$\begin{aligned} ATT(X) &= E[Y_1 \mid D = 1, X] - E[Y_0 \mid D = 1, X] \\ &= E[Y_1 \mid D = 1, X] - E[Y_0 \mid D = 0, X] \end{aligned}$$

Iterated expectations gives the unconditional estimand

$$\begin{aligned} ATT &= E_{X \in S}[E[Y_1 \mid D = 1, X] - E[Y_0 \mid D = 1, X]] \\ &= E[Y_1 - Y_0 \mid D = 1] \end{aligned}$$

For *ATUT* the analogous condition applies to outcomes with treatment

$$E[Y_1 \mid D = 0, X] = E[Y_1 \mid D = 1, X] \text{ for } X \in S'$$

so that the counterfactual mean can be identified from observables.

$$\begin{aligned} ATUT(X) &= E[Y_1 \mid D = 0, X] - E[Y_0 \mid D = 0, X] \\ &= E[Y_1 \mid D = 1, X] - E[Y_0 \mid D = 0, X] \end{aligned}$$

Again, iterated expectations gives

$$\begin{aligned} ATUT &= E_{X \in S'}[E[Y_1 \mid D = 0, X] - E[Y_0 \mid D = 0, X]] \\ &= E[Y_1 - Y_0 \mid D = 0] \end{aligned}$$

Heckman, et al relate this general matching strategy to propensity score matching by the following arguments.[6] Partition X into two (not necessarily mutually exclusive) sets of variables, (T, Z), where the T variables determine outcomes and outcomes are additively separable

$$Y_0 = g_0(T) + U_0$$

$$Y_1 = g_1(T) + U_1$$

and the Z variables determine selection.

$$P(X) \equiv \Pr(D = 1 \mid X) = \Pr(D = 1 \mid Z) \equiv P(Z)$$

ATT is identified via propensity score matching if the following conditional mean independence condition for outcomes without treatment is satisfied

$$E[U_0 \mid D = 1, P(Z)] = E[U_0 \mid D = 0, P(Z)]$$

---

[6]Heckman, Ichimura, and Todd [1998] also discuss trade-offs between general matching on X and propensity score matching.

Then, the counterfactual $E[Y_0 \mid D = 1, P(Z)]$ can be replaced with the mean of the observable

$$\begin{aligned}
ATT(P(Z)) &= E[Y_1 - Y_0 \mid D = 1, P(Z)] \\
&= g_1(T) + E[U_1 \mid D = 1, P(Z)] \\
&\quad - \{g_0(T) + E[U_0 \mid D = 1, P(Z)]\} \\
&= g_1(T) + E[U_1 \mid D = 1, P(Z)] \\
&\quad - \{g_0(T) + E[U_0 \mid D = 0, P(Z)]\}
\end{aligned}$$

Iterated expectations over $P(Z)$ produces the unconditional estimand

$$ATT = E_{P(Z)}[ATT(P(Z))]$$

Also, *ATUT* is identified if

$$E[U_1 \mid D = 1, P(Z)] = E[U_1 \mid D = 0, P(Z)]$$

is satisfied for outcomes with treatment. Analogous to *ATT*, the counterfactual $E[Y_1 \mid D = 0, P(Z)]$ can be replaced with the mean of the observable

$$\begin{aligned}
ATUT(P(Z)) &= E[Y_1 - Y_0 \mid D = 0, P(Z)] \\
&= g_1(T) + E[U_1 \mid D = 0, P(Z)] \\
&\quad - \{g_0(T) + E[U_0 \mid D = 0, P(Z)]\} \\
&= g_1(T) + E[U_1 \mid D = 1, P(Z)] \\
&\quad - \{g_0(T) + E[U_0 \mid D = 0, P(Z)]\}
\end{aligned}$$

Iterated expectations over $P(Z)$ produces the unconditional estimand

$$ATUT = E_{P(Z)}[ATUT(P(Z))]$$

Interestingly, the original strategy of Rosenbaum and Rubin implies homogeneous response while the relaxed approach of Heckman, et al allows for heterogeneous response. To see this, notice the above conditions say nothing about

$$E[U_0 \mid D, P(Z)] = E[U_0 \mid P(Z)] = 0$$

or

$$E[U_1 \mid D, P(Z)] = E[U_1 \mid P(Z)] = 0$$

so individual effects (heterogeneity) are identified by conditional mean independence along with additive separability.

A strength of propensity score matching is that it makes the importance of overlaps clear. However, finding matches can be difficult. Heckman, Ichimura, and Todd [1997] discuss trimming strategies in a nonparametric context and derive asymptotically-valid standard errors. Next, we revisit our second example from chapter 2 to explore ignorable treatment implications in a richer accounting setting.

9.7 Asset revaluation regulation example

Our second example from chapter 2 explores the ex ante impact of accounting asset revaluation policies on owners' welfare through their investment decisions (a treatment effect) in an economy of, on average, price protected buyers.[7] Prior to investment, an owner evaluates both investment prospects from asset retention and the market for resale in the event the owner becomes liquidity stressed. The payoff from investment I is distributed uniformly and centered at $\hat{x} = \frac{\beta}{\alpha} I^\alpha$ where $\alpha, \beta > 0$ and $\alpha < 1$. That is, support for investment payoff is $x : \hat{x} \pm f = [\underline{x}, \overline{x}]$. A potential problem with the resale market is the owner will have private information — knowledge of the asset value. However, since there is some positive probability the owner becomes distressed π (as in Dye [1985]) the market will not collapse. The equilibrium price is based on distressed sellers marketing potentially healthy assets combined with non-distressed sellers opportunistically marketing impaired assets. Regulators may choose to prop-up the price to support distressed sellers by requiring certification of assets at cost k [8] with values below some cutoff x_c.[9] The owner's ex ante expected payoff from investment I and certification cutoff x_c is

$$E[V \mid I, x_c] = \pi \frac{1}{2f} \left[\frac{1}{2} \left(x_c^2 - \underline{x}^2 \right) - k \left(x_c - \underline{x} \right) + P \left(\overline{x} - x_c \right) \right]$$
$$+ (1 - \pi) \frac{1}{2f} \left[\frac{1}{2} \left(x_c^2 - \underline{x}^2 \right) + P \left(P - x_c \right) + \frac{1}{2} \left(\overline{x}^2 - P^2 \right) \right]$$
$$- I$$

The equilibrium uncertified asset price is

$$P = \frac{x_c + \sqrt{\pi} \overline{x}}{1 + \sqrt{\pi}}$$

This follows from the equilibrium condition

$$P = \frac{1}{4fq} \left[\pi \left(\overline{x}^2 - x_c^2 \right) + (1 - \pi) \left(P^2 - x_c^2 \right) \right]$$

where

$$q = \frac{1}{2f} \left[\pi \left(\overline{x} - x_c \right) + (1 - \pi) \left(P - x_c \right) \right]$$

is the probability that an uncertified asset is marketed. When evaluating the welfare effects of their policies, regulators may differentially weight the welfare,

---

[7] This example draws heavily from Demski, Lin, and Sappington [2008].
[8] This cost is incremental to normal audit cost. As such, even if audit fee data is available, k may be difficult for the analyst to observe.
[9] Owners never find it ex ante beneficial to commit to any certified revaluation because of the certification cost. We restrict attention to targeted certification but certification could be proportional rather than targeted (see Demski, et al [2008] for details). For simplicity, we explore only targeted certification.

$W(I, x_c)$, of distressed sellers and non-distressed sellers. Specifically, regulators may value distressed seller's net gains dollar-for-dollar but value non-distressed seller's gains at a fraction w on the dollar.

$$W(I, x_c) = \pi \frac{1}{2f} \frac{1}{2}\left(x_c^2 - \underline{x}^2\right) - k(x_c - \underline{x}) + P(\overline{x} - x_c)$$
$$+ w(1-\pi) \frac{1}{2f} \frac{1}{2}\left(x_c^2 - \underline{x}^2\right) + P(P - x_c) + \frac{1}{2}\left(\overline{x}^2 - P^2\right)$$
$$- I[\pi + (1-\pi)w]$$

9.7.1 Numerical example

Consider the following parameters

$$\left\{\alpha = \frac{1}{2}, \beta = 10, \pi = 0.7, k = 2, f = 100\right\}$$

Owners will choose to never certify asset values. No certification ($x_c = \underline{x}$) results in investment $I = 100$, owner's expected payoff $E[V \mid I, x_c] = 100$, and equilibrium uncertified asset price $P \approx 191.1$. However, regulators may favor distressed sellers and require selective certification. Continuing with the same parameters, if regulators give zero consideration ($w = 0$) to the expected payoffs of non-distressed sellers, then the welfare maximizing certification cutoff $x_c = \overline{x} - \frac{(1+\sqrt{\pi})k}{(1-\sqrt{\pi})(1-w)} \approx 278.9$. This induces investment $I = \left[\frac{\beta(2f+\pi k)}{2f}\right]^{\frac{1}{1-\alpha}} \approx 101.4$, owner's expected payoff approximately equal to 98.8, and equilibrium uncertified asset price $P \approx 289.2$ (an uncertified price more favorable to distressed sellers). To get a sense of the impact of certification, we tabulate investment choices and expected payoffs for no and selective certification regulations and varied certification costs in table 9.14 and for full certification regulation and varied certification costs and stress likelihood in table 9.15.

Table 9.14: Investment choice and payoffs for no certification and selective certification

	$x_c = \underline{x}$, $k = 2$	$x_c = \underline{x}$, $k = 20$	$x_c = 200$, $k = 2$	$x_c = 200$, $k = 20$
π	0.7	0.7	0.7	0.7
I	100	100	101.4	114.5
P	191.1	191.1	246.2	251.9
$E[x-k]$	200	200	200.7	208
$E[V]$	100	100	99.3	93.5

Table 9.15: Investment choice and payoffs for full certification

	$x_c = \bar{x}$, $k = 2$	$x_c = \bar{x}$, $k = 20$	$x_c = \bar{x}$, $k = 2$	$x_c = \bar{x}$, $k = 20$
π	0.7	0.7	1.0	1.0
I	100	100	100	100
P	NA	NA	NA	NA
$E[x - k]$	198.6	186	198	180
$E[V]$	98.6	86	98	80

9.7.2 Full certification

The base case involves full certification $x_c = \bar{x}$ and all owners market their assets, $\pi = 1$. This setting ensures outcome data availability (excluding investment cost) which may be an issue when we relax these conditions. There are two firm types: one with low mean certification costs $\widehat{k}^L = 2$ and the other with high mean certification costs $\widehat{k}^H = 20$.

Full certification doesn't present an interesting experiment if owners anticipate full certification[10] but suppose owners choose their investment levels anticipating selective certification with $x_c = 200$ and forced sale is less than certain $\pi = 0.7$. Then, ex ante optimal investment levels for a selective certification environment are $I^L = 101.4$ (for low certification cost type) and $I^H = 114.5$ (for high certification cost type), and expected asset values including certification costs are $E\left[x^L - k^L\right] = 199.4$ and $E\left[x^H - k^H\right] = 194$. Treatment (investment level) is chosen based on ex ante beliefs of selective certification. As a result of two certification cost types, treatment is binary and the analyst observes low or high investment but not the investment level.[11] Treatment is denoted $D = 1$ when $I^L = 101.4$ while non-treatment is denoted $D = 0$ when $I^H = 114.5$. For this base case, outcome is ex post value in an always certify, always trade environment $Y_j = x^j - k^j$.

To summarize, the treatment effect of interest is the difference in outcome with treatment and outcome without treatment. For the base case, outcome with treatment is defined as realized value associated with the (ex ante) equilibrium investment choice when certification cost type is low (I^L). And, outcome with no treatment is defined as realized value associated with the (ex ante) equilibrium investment choice when certification cost type is high (I^H). Variations from the base case retain the definition for treatment (low versus high investment) but alter outcomes based on data availability given the setting (e.g., assets are not always traded so values may not be directly observed).

---

[10] As seen in the table, for full certification there is no variation in equilibrium investment level.
[11] If the analyst observes the investment level, then outcome includes investment cost and we work with a more complete measure of the owner's welfare.

Since the equilibrium investment choice for low certification cost type is treatment (I^L), the average treatment effect on the treated is

$$\begin{aligned} ATT &= E\left[Y_1 - Y_0 \mid D = 1\right] \\ &= E\left[x^L - k^L \mid D = 1\right] - E\left[x^H - k^L \mid D = 1\right] \\ &= E\left[x^L - x^H \mid D = 1\right] \\ &= 201.4 - 214 = -12.6 \end{aligned}$$

Similarly, the equilibrium investment choice for high certification cost type is no treatment (I^H). Therefore, the average treatment effect on the untreated is

$$\begin{aligned} ATUT &= E\left[Y_1 - Y_0 \mid D = 0\right] \\ &= E\left[x^L - k^H \mid D = 0\right] - E\left[x^H - k^H \mid D = 0\right] \\ &= E\left[x^L - x^H \mid D = 0\right] \\ &= 201.4 - 214 = -12.6 \end{aligned}$$

The above implies outcome is homogeneous,[12] $ATE = ATT = ATUT = -12.6$. With no covariates and outcome not mean independent of treatment, the OLS estimand is[13]

$$\begin{aligned} OLS &= E\left[Y_1 \mid D = 1\right] - E\left[Y_0 \mid D = 0\right] \\ &= E\left[x^L - k^L \mid D = 1\right] - E\left[x^H - k^H \mid D = 0\right] \\ &= 5.4 \end{aligned}$$

The regression is

$$E[Y \mid D] = \beta_0 + \beta_1 D$$

where $Y = D\left(x^L - k^L\right) + (1 - D)\left(x^H - k^H\right)$ (ex post payoff), β_1 is the estimand of interest, and

$$D = \begin{array}{ll} 1 & I^L = 101.4 \\ 0 & I^H = 114.5 \end{array}$$

A simple experiment supports the analysis above. We simulate 200 samples of 2,000 draws where traded market values are

$$x^j \sim uniform\left(\widehat{x}^j - 100, \widehat{x}^j + 100\right)$$

certification costs are

$$k^j \sim uniform\left(\widehat{k}^j - 1, \widehat{k}^j + 1\right)$$

---

[12] If k is unobservable, then outcome Y may be measured by x only (discussed later) and treatment effects represent gross rather than gains net of certification cost. In any case, we must exercise care in interpreting the treatment effects because of limitations in our outcome measure — more to come on the importance of outcome observability.

[13] Notice the difference in the treatment effects and what is estimated via OLS is $k^L - k^H = 2 - 20 = -18 = -12.6 - 5.4$.

9.7 Asset revaluation regulation example

and assignment of certification cost type is

$$L - type \sim Bernoulli\,(0.5)$$

Simulation results for the above *OLS* model including the estimated average treatment effect are reported in table 9.16. As simulation allows us to observe both the factual data and counterfactual data in the experiment, the sample statistics described in table 9.17 are "observed" average treatment effects.

Table 9.16: OLS results for full certification setting

statistics	β_0	$\beta_1\,(estATE)$
mean	193.8	5.797
median	193.7	5.805
stand.dev.	1.831	2.684
minimum	188.2	−1.778
maximum	198.9	13.32
$E\,[Y\mid D]=\beta_0+\beta_1 D$		

Table 9.17: Average treatment effect sample statistics for full certification setting

statistics	ATE	ATT	$ATUT$
mean	−12.54	−12.49	−12.59
median	−12.55	−12.44	−12.68
stand.dev.	1.947	2.579	2.794
minimum	−17.62	−19.53	−21.53
maximum	−7.718	−6.014	−6.083

OLS clearly produces biased estimates of the treatment effect in this simple base case. This can be explained as low or high certification cost type is a perfect predictor of treatment. That is, $\Pr\left(D=1\mid k^L\right)=1$ and $\Pr\left(D=1\mid k^H\right)=0$. Therefore, the common support condition for identifying counterfactuals fails and standard approaches (ignorable treatment or even instrumental variables) don't identify treatment effects.[14]

---

[14] An alternative analysis tests the common support condition. Suppose everything remains as above except $k^H \sim uniform\,(1,19)$ and sometimes the owners perceive certification cost to be low when it is high, hence $\Pr\,(D=1\mid type=H)=0.1$. This setup implies observed outcome is

$$Y = D\,[(Y_1\mid type=L)+(Y_1\mid type=H)] + (1-D)\,(Y_0\mid type=H)$$

such that

$$E\,[Y] = 0.5E\left[x^L-k^L\right] + 0.5\left\{0.1E\left[x^L-k^H\right] + 0.9E\left[x^H-k^H\right]\right\}$$

Suppose the analyst ex post observes the actual certification cost type and let $T=1$ if $type=L$. The common support condition is satisfied and the outcome mean is conditionally independent of treatment given T implies treatment is ignorable. *OLS* simulation results are tabulated below.

Adjusted outcomes

However, from the above we can manipulate the outcome variable to identify the treatment effects via *OLS*. Observed outcome is

$$Y = D\left(x^L - k^L\right) + (1-D)\left(x^H - k^H\right)$$
$$= \left(x^H - k^H\right) + D\left(x^L - x^H\right) - D\left(k^L - k^H\right)$$

Applying expectations, the first term is captured via the regression intercept and the second term is the average treatment effect. Therefore, if we add the last term $DE\left[k^L - k^H\right]$ to Y we can identify the treatment effect from the coefficient on D. If the analyst observes $k = Dk^L + (1-D)k^H$, then we can utilize a two-stage regression approach. The first stage is

$$E[k \mid D] = \alpha_0 + \alpha_1 D$$

where $\alpha_0 = E\left[k^H\right]$ and $\alpha_1 = E\left[k^L - k^H\right]$. Now, the second stage regression employs the sample statistic for α_1, $\widehat{\alpha}_1 = \overline{k}^L - \overline{k}^H$.

$$Y' = Y + D\widehat{\alpha}_1$$
$$= Y + D\left(\overline{k}^L - \overline{k}^H\right)$$

and estimate the treatment effect via the analogous regression to the above[15]

$$E[Y' \mid D] = \beta_0 + \beta_1 D$$

OLS parameter estimates with common support for full certification setting

statistics	β_0	β_1	β_2 (estATE)
mean	196.9	7.667	−5.141
median	196.9	7.896	−5.223
stand.dev.	1.812	6.516	6.630
minimum	191.5	−10.62	−23.54
maximum	201.6	25.56	14.25

$$E[Y \mid T, D] = \beta_0 + \beta_1 T + \beta_2 D$$

Average treatment effects sample statistics with common support for full certification setting

statistics	ATE	ATT	$ATUT$
mean	−5.637	−5.522	−5.782
median	−5.792	−5.469	−5.832
stand.dev.	1.947	2.361	2.770
minimum	−9.930	−12.05	−12.12
maximum	0.118	0.182	0.983

The estimated average treatment effect is slightly attenuated and has high variability that may compromise its finite sample utility. Nevertheless, the results are a dramatic departure and improvement from the results above where the common support condition fails.

[15]This is similar to a regression discontinuity design (for example, see Angrist and Lavy [1999] and Angrist and Pischke [2009]). However, the jump in cost of certification k^j violates the regression continuity in X condition (assuming $k = Dk^L + (1-D)k^H$ is observed and included in X). If the

9.7 Asset revaluation regulation example   181

Simulation results for the adjusted outcome *OLS* model are reported in table 9.18. With adjusted outcomes, *OLS* estimates correspond quite well with the treat-

Table 9.18: Adjusted outcomes OLS results for full certification setting

statistics	β_0	$\beta_1\,(estATE)$
mean	193.8	-12.21
median	193.7	-12.21
stand.dev.	1.831	2.687
minimum	188.2	-19.74
maximum	198.9	-64.691
$E\left[Y'\mid D\right]=\beta_0+\beta_1 D$		

ment effects. Next, we explore propensity score approaches.

Propensity score

Based on adjusted outcomes, the data are conditionally mean independent (i.e., satisfy ignorability of treatment). Therefore average treatment effects can be estimated via the propensity score as discussed earlier in chapter 9. Propensity score is the estimated probability of treatment conditional on the regressors $m_j = \Pr\left(D_j=1\mid Z_j\right)$. For simulation purposes, we employ an imperfect predictor in the probit regression

$$Z_j = z_{1j}D_j + z_{0j}\left(1-D_j\right) + \varepsilon_j$$

---

support of k^L and k^H is adjacent, then the regression discontinuity design

$$E\left[Y\mid X,D\right] = \beta_0 + \beta_1 k + \beta_2 D$$

effectively identifies the treatment effects but fails with the current *DGP*. Typical results for the current *DGP* (where *ATE* is the average treatment effect sample statistic for the simulation) are tabulated below.

OLS parameter estimates with jump in support for full certification setting

statistics	β_0	β_1	$\beta_2\,(estATE)$	ATE
mean	218.4	-1.210	-16.59	-12.54
median	213.2	-0.952	-12.88	-12.55
stand.dev.	42.35	2.119	38.26	1.947
minimum	122.9	-6.603	-115.4	-17.62
maximum	325.9	3.573	71.56	-7.718
$E\left[Y\mid k,D\right] = \beta_0 + \beta_1 k + \beta_2 D$				

The coefficient on D represents a biased and erratic estimate of the average treatment effect. Given the variability of the estimates, a regression discontinuity design has limited small sample utility for this *DGP*. However, we later return to regression discontinuity designs when modified *DGPs* are considered. For the current *DGP*, we employ the approach discussed above, which is essentially restricted least squares.

where

$$z_{1j} \sim Bernoulli\,(0.99)$$
$$z_{0j} \sim Bernoulli\,(0.01)$$
$$\varepsilon_j \sim N\,(0,1)$$

Some average treatment effects estimated via propensity score are

$$estATE = n^{-1} \sum_{j=1}^{n} \frac{(D_j - m_j) Y'_j}{m_j (1 - m_j)}$$

$$estATT = \frac{n^{-1} \sum_{j=1}^{n} \frac{(D_j - m_j) Y'_j}{(1 - m_j)}}{n^{-1} \sum_{j=1}^{n} D_j}$$

$$estATUT = \frac{n^{-1} \sum_{j=1}^{n} \frac{(D_j - m_j) Y'_j}{m_j}}{n^{-1} \sum_{j=1}^{n} (1 - D_j)}$$

Propensity score estimates of average treatment effects are reported in table 9.19. The estimates are somewhat more variable than we would like but they are

Table 9.19: Propensity score treatment effect estimates for full certification setting

statistics	estATE	estATT	estATUT
mean	−12.42	−13.96	−10.87
median	−12.50	−13.60	−11.40
stand.dev.	5.287	6.399	5.832
minimum	−31.83	−45.83	−25.61
maximum	−1.721	0.209	10.56

consistent with the sample statistics on average. Further, we cannot reject homogeneity even though the treatment effect means are not as similar as we might expect.

Propensity score matching

Propensity score matching is a simple and intuitively appealing approach where we match treated and untreated on propensity score then compute the average treatment effect based on the matched-pair outcome differences. We follow Sekhon [2008] by employing the "Matching" library for **R**.[16] We find optimal matches of

---

[16]We don't go into details regarding matching since we employ only one regressor in the propensity score model. Matching is a rich study in itself. For instance, Sekhon [2008] discusses a genetic matching algorithm. Heckman, Ichimura, and Todd [1998] discuss nonparametric kernel matching.

treated with untreated (within 0.01) using replacement sampling. Simulation results for propensity score matching average treatment effects are reported in table 9.20.[17] The matched propensity score results correspond well with the sample sta-

Table 9.20: Propensity score matching average treatment effect estimates for full certification setting

statistics	estATE	estATT	estATUT
mean	−12.46	−12.36	−12.56
median	−12.54	−12.34	−12.36
stand.dev.	3.530	4.256	4.138
minimum	−23.49	−24.18	−22.81
maximum	−3.409	−2.552	−0.659

tistics. In this setting, the matched propensity score estimates of causal effects are less variable than the previous propensity score results. Further, they are more uniform across treatment effects (consistent with homogeneity). Next, we turn to the more interesting, but potentially more challenging, selective certification setting.

9.7.3 Selective certification

Suppose the owners' ex ante perceptions of the certification threshold, $x_c = 200$, and likelihood of stress, $\pi = 0.7$, are consistent with ex post outcomes. Then, if outcomes x, k^j, and P^j for $j = L$ or H are fully observable to the analyst, expected outcome conditional on asset revaluation experience is[18]

$$\begin{aligned} E[Y \mid X] &= 251.93 \Pr\left(P^H\right) - 5.740 \Pr\left(P^L\right) D - 94.93 \Pr\left(\Im_c^H\right) \Im_c^H \\ &\quad -95.49 \Pr\left(\Im_c^L\right) \Im_c^L - 20 \Pr\left(\Im_{ck}^H\right) \Im_{ck}^H \\ &\quad -2 \Pr\left(\Im_{ck}^L\right) \Im_{ck}^L + 31.03 \Pr\left(\Im_u^H\right) \Im_u^H + 27.60 \Pr\left(\Im_u^L\right) \Im_u^L \\ E[Y \mid X] &= 251.93\,(0.477) - 5.740\,(0.424)\,D - 94.93\,(0.129)\,\Im_c^H \\ &\quad -95.49\,(0.148)\,\Im_c^L - 20\,(0.301)\,\Im_{ck}^H - 2\,(0.345)\,\Im_{ck}^L \\ &\quad +31.03\,(0.093)\,\Im_u^H + 27.60\,(0.083)\,\Im_u^L \end{aligned}$$

---

[17] ATE, ATT, and ATUT may be different because their regions of common support may differ. For example, ATT draws on common support only in the $D = 1$ region and ATUT draws on common support only in the $D = 0$ region.

[18] The probabilities reflect likelihood of the asset condition rather than incremental likelihood and hence sum to one for each investment level (treatment choice).

where the equilibrium price of traded, uncertified, high investment assets, P^H, is the reference outcome level, and X denotes the matrix of regressors

$$D = \begin{matrix} 1 & \text{low investment, } I^L \\ 0 & \text{high investment, } I^H \end{matrix}$$

$$\Im_c^j = \begin{matrix} 1 & \text{certified range, } x < x_c \\ 0 & \text{otherwise} \end{matrix}$$

$$\Im_{ck}^j = \begin{matrix} 1 & \text{certified traded} \\ 0 & \text{otherwise} \end{matrix}$$

$$\Im_u^j = \begin{matrix} 1 & \text{untraded asset, } x > P \quad j \in \{L, H\} \\ 0 & \text{otherwise} \end{matrix}$$

This implies the average treatment effect estimands are

$$ATT \equiv E\left[Y^L - Y^H \mid D = 1\right] = -12.7$$
$$ATUT \equiv E\left[Y^L - Y^H \mid D = 0\right] = -13.5$$

and

$$ATE \equiv E\left[Y^L - Y^H\right]$$
$$= \Pr(D=1) ATT + \Pr(D=0) ATUT = -13.1$$

Hence, in the selective certification setting we encounter modest heterogeneity. Why don't we observe self-selection through the treatment effects? Remember, we have a limited outcome measure. In particular, outcome excludes investment cost. If we include investment cost, then self-selection is supported by the average treatment effect estimands. That is, low investment outcome is greater than high investment outcome for low certification cost firms

$$ATT = -12.7 - (101.4 - 114.5) = 0.4 > 0$$

and high investment outcome is greater than low investment outcome for high certification cost firms

$$ATUT = -13.5 - (101.4 - 114.5) = -0.4 < 0$$

With this background for the selective certification setting, it's time to revisit identification. Average treatment effect identification is somewhat more challenging than the base case. For instance, the average treatment effect on the treated, *ATT*, is the difference between the mean of outcome with low investment and the

9.7 Asset revaluation regulation example

mean of outcome with high investment for low certification cost firms.

$$ATT = \pi \frac{1}{2f} \frac{1}{2} \left(x_c^2 - \left(\underline{x}^L \right)^2 \right) - k^L \left(x_c - \underline{x}^L \right) + P^L \left(\overline{x}^L - x_c \right)$$

$$+ (1 - \pi) \frac{1}{2f} \left[\begin{array}{c} \frac{1}{2} \left(x_c^2 - \left(\underline{x}^L \right)^2 \right) + P^L \left(P^L - x_c \right) \\ + \frac{1}{2} \left(\left(\overline{x}^L \right)^2 - \left(P^L \right)^2 \right) \end{array} \right]$$

$$- \pi \frac{1}{2f} \frac{1}{2} \left(x_c^2 - \left(\underline{x}^H \right)^2 \right) - k^L \left(x_c - \underline{x}^H \right) + P^H \left(\overline{x}^H - x_c \right)$$

$$- (1 - \pi) \frac{1}{2f} \left[\begin{array}{c} \frac{1}{2} \left(x_c^2 - \left(\underline{x}^H \right)^2 \right) + P^H \left(P^H - x_c \right) \\ + \frac{1}{2} \left(\left(\overline{x}^H \right)^2 - \left(P^H \right)^2 \right) \end{array} \right]$$

The average treatment effect on the untreated, *ATUT*, is the difference between the mean of outcome with low investment and the mean of outcome with high investment for high certification cost firms.

$$ATUT = \pi \frac{1}{2f} \frac{1}{2} \left(x_c^2 - \left(\underline{x}^L \right)^2 \right) - k^H \left(x_c - \underline{x}^L \right) + P^L \left(\overline{x}^L - x_c \right)$$

$$+ (1 - \pi) \frac{1}{2f} \left[\begin{array}{c} \frac{1}{2} \left(x_c^2 - \left(\underline{x}^L \right)^2 \right) + P^L \left(P^L - x_c \right) \\ + \frac{1}{2} \left(\left(\overline{x}^L \right)^2 - \left(P^L \right)^2 \right) \end{array} \right]$$

$$- \pi \frac{1}{2f} \frac{1}{2} \left(x_c^2 - \left(\underline{x}^H \right)^2 \right) - k^H \left(x_c - \underline{x}^H \right) + P^H \left(\overline{x}^H - x_c \right)$$

$$- (1 - \pi) \frac{1}{2f} \left[\begin{array}{c} \frac{1}{2} \left(x_c^2 - \left(\underline{x}^H \right)^2 \right) + P^H \left(P^H - x_c \right) \\ + \frac{1}{2} \left(\left(\overline{x}^H \right)^2 - \left(P^H \right)^2 \right) \end{array} \right]$$

But the *OLS* estimand is the difference between the mean of outcome with low investment for firms with low certification cost and the mean of outcome with high investment for firms with high certification cost.

$$OLS = \pi \frac{1}{2f} \frac{1}{2} \left(x_c^2 - \left(\underline{x}^L \right)^2 \right) - k^L \left(x_c - \underline{x}^L \right) + P^L \left(\overline{x}^L - x_c \right)$$

$$+ (1 - \pi) \frac{1}{2f} \left[\begin{array}{c} \frac{1}{2} \left(x_c^2 - \left(\underline{x}^L \right)^2 \right) + P^L \left(P^L - x_c \right) \\ + \frac{1}{2} \left(\left(\overline{x}^L \right)^2 - \left(P^L \right)^2 \right) \end{array} \right]$$

$$- \pi \frac{1}{2f} \frac{1}{2} \left(x_c^2 - \left(\underline{x}^H \right)^2 \right) - k^H \left(x_c - \underline{x}^H \right) + P^H \left(\overline{x}^H - x_c \right)$$

$$- (1 - \pi) \frac{1}{2f} \left[\begin{array}{c} \frac{1}{2} \left(x_c^2 - \left(\underline{x}^H \right)^2 \right) + P^H \left(P^H - x_c \right) \\ + \frac{1}{2} \left(\left(\overline{x}^H \right)^2 - \left(P^H \right)^2 \right) \end{array} \right]$$

As in the full certification setting, the key differences revolve around the costly certification terms. The costly certification term for the *ATT* estimand simplifies as

$$-\pi \frac{1}{2f} \left[k^L \left(x_c - \underline{x}^L \right) - k^L \left(x_c - \underline{x}^H \right) \right]$$
$$= -\pi \frac{\left(\underline{x}^H - \underline{x}^L \right)}{2f} k^L$$

and the costly certification term for the *ATUT* estimand simplifies as

$$-\pi \frac{1}{2f} \left[k^H \left(x_c - \underline{x}^L \right) - k^H \left(x_c - \underline{x}^H \right) \right]$$
$$= -\pi \frac{\left(\underline{x}^H - \underline{x}^L \right)}{2f} k^H$$

While the costly certification term in the estimand for *OLS* is

$$-\pi \frac{1}{2f} \left[k^L \left(x_c - \underline{x}^L \right) - k^H \left(x_c - \underline{x}^H \right) \right]$$

Adjusted outcomes

Similar to our approach in the full certification setting, we eliminate the costly certification term for *OLS* by adding this *OLS* bias to observed outcomes

$$Y' = Y + D\pi \frac{1}{2f} \left[k^L \left(x_c - \underline{x}^L \right) - k^H \left(x_c - \underline{x}^H \right) \right]$$

However, now we add back the terms to recover the average treatment effects

$$ATT = E\left[Y_1' - Y_0' \mid D = 1 \right] - \pi \frac{\left(\underline{x}^H - \underline{x}^L \right)}{2f} E\left[k^L \right]$$

$$ATUT = E\left[Y_1' - Y_0' \mid D = 0 \right] - \pi \frac{\left(\underline{x}^H - \underline{x}^L \right)}{2f} E\left[k^H \right]$$

$$\begin{aligned} ATE &= \Pr(D=1) \, ATT + \Pr(D=1) \, ATUT \\ &= E\left[Y_1' - Y_0' \right] - \Pr(D=1) \, \pi \frac{\left(\underline{x}^H - \underline{x}^L \right)}{2f} E\left[k^L \right] \\ &\quad - \Pr(D=0) \, \pi \frac{\left(\underline{x}^H - \underline{x}^L \right)}{2f} E\left[k^H \right] \end{aligned}$$

These terms account for heterogeneity in this asset revaluation setting but are likely to be much smaller than the *OLS* selection bias.[19]

---

[19] In our running numerical example, the certification cost term for *ATT* is -0.0882 and for *ATUT* is -0.882, while the *OLS* selection bias is 5.3298.

9.7 Asset revaluation regulation example

Conditional as well as unconditional average treatment effects can be identified from the following regression.

$$E[Y' \mid X] = \beta_0 + \beta_1 D + \beta_2 \Im_c^H + \beta_3 \Im_c^L$$
$$+ \beta_4 \Im_{ck}^H + \beta_5 \Im_{ck}^L + \beta_6 \Im_u^H + \beta_7 \Im_u^L$$

where

$$Y' = Y + D\pi \left[\Im_{ck}^L k^L - \overline{\Im}_{ck}^H \overline{k}^H \right]$$

$\overline{\Im}_{ck}^H$ and $\overline{k}^H$ are sample averages taken from the $D = 0$ regime.[20] The incremental impact on mean value of assets in the certification region is reflected in β_2 for high investment and β_3 for low investment firms, while the mean incremental impact of costly certification of assets, k^j, is conveyed via β_4 and β_5 for high and low investment firms, respectively. Finally, the mean incremental impact of untraded assets with values greater than the equilibrium price are conveyed via β_6 and β_7 for high and low investment firms, respectively.

Simulation results for the *OLS* model are reported in table 9.21 and sample treatment effect statistics are reported in table 9.22. *OLS* effectively estimates the average treatment effects (*ATE*, *ATT*, *ATUT*) in this (modestly heterogeneous) case. However, we're unlikely to be able to detect heterogeneity when the various treatment effect differences are this small. Note in this setting, while outcome is the ex post value net of certification cost, a random sample allows us to assess the owner's ex ante welfare excluding the cost of investment.[21]

Model-estimated treatment effects are derived in a non-standard manner as the regressors are treatment-type specific and we rely on sample evidence from each regime to estimate the probabilities associated with different ranges of support[22]

$$estATT = \beta_1 - \beta_2 \overline{\Im}_c^H + \beta_3 \overline{\Im}_c^L - \beta_4 \overline{\Im}_{ck}^H + \beta_5 \overline{\Im}_{ck}^L - \beta_6 \overline{\Im}_u^H + \beta_7 \overline{\Im}_u^L$$
$$- \pi \overline{k}^L \left(\overline{\Im}_{ck}^L - \overline{\Im}_{ck}^H \right)$$

$$estATUT = \beta_1 - \beta_2 \overline{\Im}_c^H + \beta_3 \overline{\Im}_c^L - \beta_4 \overline{\Im}_{ck}^H + \beta_5 \overline{\Im}_{ck}^L - \beta_6 \overline{\Im}_u^H + \beta_7 \overline{\Im}_u^L$$
$$- \pi \overline{k}^H \left(\overline{\Im}_{ck}^L - \overline{\Im}_{ck}^H \right)$$

and

$$estATE = \beta_1 - \beta_2 \overline{\Im}_c^H + \beta_3 \overline{\Im}_c^L - \beta_4 \overline{\Im}_{ck}^H + \beta_5 \overline{\Im}_{ck}^L - \beta_6 \overline{\Im}_u^H + \beta_7 \overline{\Im}_u^L$$
$$- \overline{D} \pi \overline{k}^L \left(\overline{\Im}_{ck}^L - \overline{\Im}_{ck}^H \right) - \left(1 - \overline{D} \right) \pi \overline{k}^H \left(\overline{\Im}_{ck}^L - \overline{\Im}_{ck}^H \right)$$

---

[20] Sample averages of certification cost, $\overline{k}^H$, and likelihood that an asset is certified and traded, $\overline{\Im}_{ck}^H$, for $D = 0$ (high investment) are employed as these are counterfactuals in the $D = 1$ (low investment) regime.

[21] Investment cost may also be observed or estimable by the analyst.

[22] Expected value of indicator variables equals the event probability and probabilities vary by treatment. Since there is no common support (across regimes) for the regressors, we effectively assume the analyst can extrapolate to identify counterfactuals (that is, from observed treated to unobserved treated and from observed untreated to unobserved untreated).

188   9. Treatment effects: ignorability

Table 9.21: OLS parameter estimates for selective certification setting

statistics	β_0	β_1	β_2	β_3
mean	251.9	−11.78	−94.78	−93.81
median	251.9	−11.78	−94.70	−93.85
stand.dev.	0.000	0.157	2.251	2.414
minimum	251.9	−12.15	−102.7	−100.9
maximum	251.9	−11.41	−88.98	−86.90
statistics	β_4	β_5	β_6	β_7
mean	−20.12	−2.087	31.20	27.66
median	−20.15	−2.160	31.23	27.72
stand.dev.	2.697	2.849	1.723	1.896
minimum	−28.67	−9.747	26.91	22.44
maximum	−12.69	8.217	37.14	32.81
statistics	$estATE$	$estATT$	$estATUT$	
mean	−12.67	−12.29	−13.06	
median	−12.73	−12.33	−13.10	
stand.dev.	2.825	2.686	2.965	
minimum	−21.25	−20.45	−22.03	
maximum	−3.972	−3.960	−3.984	

$$E\left[Y' \mid X\right] = \beta_0 + \beta_1 D + \beta_2 \Im_c^H + \beta_3 \Im_c^L \\ + \beta_4 \Im_{ck}^H + \beta_5 \Im_{ck}^L + \beta_6 \Im_u^H + \beta_7 \Im_u^L$$

where

$$\overline{\Im}_j^L = \frac{\sum D_i \Im_{ji}^L}{\sum D_i}$$

and

$$\overline{\Im}_j^H = \frac{\sum (1 - D_i) \Im_{ji}^H}{\sum (1 - D_i)}$$

for indicator j.

We can say a bit more about conditional average treatment effects from the above analysis. On average, owners who select high investment and trade the assets at their equilibrium price sell the assets for 11.78 more than owners who select low investment. Owners who select high investment and retain their assets earn $31.20 - 27.66 = 3.54$ higher proceeds, on average, than owners who select low investment. On the other hand, owners who select high investment and are forced to certify and sell their assets receive lower net proceeds by $20.12 - 2.09 = 18.03$, on average, than owners who select low investment. Recall all outcomes exclude investment cost which, of course, is an important component of owner's welfare.

As we can effectively randomize over the indicator variables, for simplicity, we focus on identification and estimation of unconditional average treatment effects and the remaining analyses are explored without covariates. Next, we demonstrate the above randomization claim via a reduced (no covariates except treatment) *OLS* model, then we explore propensity score approaches applied to selective certification.

9.7 Asset revaluation regulation example

Table 9.22: Average treatment effect sample statistics for selective certification setting

statistics	ATE	ATT	ATUT	OLS
mean	−13.01	−12.57	−13.45	−6.861
median	−13.08	−12.53	−13.46	−6.933
stand.dev.	1.962	2.444	2.947	2.744
minimum	−17.90	−19.52	−22.46	−15.15
maximum	−8.695	−5.786	−6.247	1.466

Reduced OLS model

We estimate unconditional average treatment effects via a reduced *OLS* model.

$$E[Y' \mid D] = \beta_0 + \beta_1 D$$

Results from the simulation, reported in table 9.23, indicate that reduced *OLS*, with the adjustments discussed above to recover the treatment effect, effectively recovers unconditional average treatment effects in the selective certification setting.

Table 9.23: Reduced OLS parameter estimates for selective certification setting

statistics	β_0	β_1	
mean	207.7	−12.21	
median	207.50	−12.24	
stand.dev.	1.991	2.655	
minimum	202.8	−20.28	
maximum	212.8	−3.957	
statistics	estATE	estATT	estATUT
mean	−12.67	−12.29	−13.06
median	−12.73	−12.33	−13.10
stand.dev.	2.825	2.686	2.965
minimum	−21.25	−20.45	−22.03
maximum	−3.972	−3.960	−3.984
$E[Y' \mid D] = \beta_0 + \beta_1 D$			

Propensity score

As in the full certification setting, propensity score, $\Pr(D = 1 \mid Z)$, is estimated via probit with predictor Z. Propensity score estimates, based on adjusted outcomes and treatment effect adjustments as discussed for *OLS*, of average treatment effects in the selective certification setting are reported in table 9.24. As in the full certification setting, the estimates are more variable than we prefer but, on average, correspond with the sample statistics. Again, homogeneity cannot be rejected but estimated differences in treatment effects do not correspond well with

Table 9.24: Propensity score average treatment effect estimates for selective certification setting

statistics	estATE	estATT	estATUT
mean	−12.84	−14.18	−11.47
median	−13.09	−13.71	−11.87
stand.dev.	5.680	6.862	6.262
minimum	−33.93	−49.88	−25.06
maximum	−0.213	1.378	13.80

the sample statistics (e.g., estimated *ATT* is the largest in absolute value but *ATT* is the smallest sample statistic as well as estimand).

Propensity score matching

Simulation results, based on outcome and treatment effect adjustments, for propensity score matching estimates of average treatment effects in the selective certification setting are reported in table 9.25. Again, propensity score matching results

Table 9.25: Propensity score matching average treatment effect estimates for selective certification setting

statistics	estATE	estATT	estATUT
mean	−12.90	−12.54	−13.27
median	−13.20	−12.89	−13.09
stand.dev.	3.702	4.478	4.335
minimum	−25.87	−25.54	−26.20
maximum	−4.622	−2.431	−2.532

correspond well with the sample statistics and are less variable than the propensity score approach above but cannot reject outcome homogeneity.

9.7.4 Outcomes measured by value x only

Now, we revisit selective certification when the analyst cannot observe the incremental cost of certification, k, but only asset value, x. Consequently, outcomes and therefore treatment effects reflect only $Y = x$. For instance, the *DGP* now

9.7 Asset revaluation regulation example

yields

$$\begin{aligned}
ATT &= E\left[Y^L - Y^H \mid D = 1\right] \\
&= E\left[x^L - x^H \mid D = 1\right] \\
&= 201.4 - 214 = -12.6 \\
ATUT &= E\left[Y^L - Y^H \mid D = 0\right] \\
&= E\left[x^L - x^H \mid D = 0\right] \\
&= 201.4 - 214 = -12.6 \\
ATE &= E\left[Y^L - Y^H\right] \\
&= E\left[x^L - x^H\right] \\
&= \Pr(D=1)\, ATT + \Pr(D=0)\, ATUT \\
&= 201.4 - 214 = -12.6 \\
OLS &= E\left[Y^L \mid D = 1\right] - E\left[Y^H \mid D = 0\right] \\
&= E\left[x^L \mid D = 1\right] - E\left[x^H \mid D = 0\right] \\
&= 201.4 - 214 = -12.6
\end{aligned}$$

The apparent advantage to high investment is even more distorted because not only are investment costs excluded but now also the incremental certification costs are excluded. In other words, we have a more limited outcome measure. We briefly summarize treatment effect analyses similar to those reported above but for the alternative, data limited, outcome measure $Y = x$. Notice, no outcome adjustment is applied.

OLS results

Simulation results for the *OLS* model are reported in table 9.26 and sample average treatment effect statistics are reported in table 9.27.

Table 9.26: OLS parameter estimates for Y=x in selective certification setting

statistics	β_0	$\beta_1\ (est ATE)$
mean	214.0	−12.70
median	214.1	−12.70
stand.dev.	1.594	2.355
minimum	209.3	−18.5
maximum	218.11	−5.430
$E[Y \mid D] = \beta_0 + \beta_1 D$		

OLS effectively estimates the treatment effects and outcome homogeneity is supported.

Propensity score

Propensity score estimates for average treatment effects are reported in table 9.28.

9. Treatment effects: ignorability

Table 9.27: Average treatment effect sample statistics for Y = x in selective certification setting

statistics	ATE	ATT	ATUT
mean	−12.73	−12.72	−12.75
median	−12.86	−12.78	−12.62
stand.dev.	1.735	2.418	2.384
minimum	−17.26	−19.02	−18.96
maximum	−7.924	−5.563	−6.636

Table 9.28: Propensity score average treatment effect for Y = x in selective certification setting

statistics	estATE	estATT	estATUT
mean	−13.02	−14.18	−11.86
median	−13.49	−13.96	−11.20
stand.dev.	5.058	5.764	5.680
minimum	−27.00	−34.39	−24.25
maximum	2.451	0.263	7.621

Similar to previous propensity score analyses, the limited outcome propensity score results are more variable than we'd like but generally correspond with average treatment effect sample statistics.

Propensity score matching

Propensity score matching simulation results are reported in table 9.29. Propen-

Table 9.29: Propensity score matching average treatment effect for Y = x in selective certification setting

statistics	estATE	estATT	estATUT
mean	−12.61	−12.43	−12.76
median	−12.83	−12.40	−13.10
stand.dev.	3.239	3.727	4.090
minimum	−20.57	−21.79	−24.24
maximum	−4.025	0.558	−1.800

sity score matching results are generally consistent with other results. For $Y = x$, matching effectively identifies average treatment effects, supports homogeneous outcome, and is less variable than the (immediately) above propensity score results.

Since outcome based on x only is more limited than $Y = x - k$, for the remaining discussion of this asset revaluation regulation example we refer to the broader outcome measure $Y = x - k$.

9.7.5 Selective certification with missing "factual" data

It is likely the analyst will not have access to ex post values when the assets are not traded. Then, the only outcome data observed is when assets are certified or when traded at the equilibrium price. In addition to not observing counterfactuals, we now face missing factual data. Missing outcome data produces a challenging treatment effect identification problem. The treatment effects are the same as the above observed data case but require some creative data augmentation to recover. We begin our exploration by examining model-based estimates if we ignore the missing data problem.

If we ignore missing data but adjust outcomes and treatment effects (as discussed earlier) and estimate the model via *OLS* we find the simulation results reported in table 9.30. The average model-estimated treatment effects are biased

Table 9.30: OLS parameter estimates ignoring missing data for selective certification setting

statistics	β_0	β_1	
mean	207.2	−9.992	
median	207.2	−9.811	
stand.dev.	2.459	3.255	
minimum	200.9	−18.30	
maximum	213.2	−2.627	
statistics	$estATE$	$estATT$	$estATUT$
mean	−10.45	−10.07	−10.81
median	−9.871	−5.270	−14.92
stand.dev.	3.423	3.285	3.561
minimum	−19.11	−18.44	−19.75
maximum	−2.700	−2.640	−2.762
$E\left[Y' \mid D\right] = \beta_0 + \beta_1 D$			

toward zero due to the missing outcome data.

Data augmentation

The above results suggest attending to the missing data. The observed data may not, in general, be representative of the missing factual data. We might attempt to model the missing data process and augment the observed data. Though, data augmentation might introduce more error than do the missing data and consequently generate poorer estimates of the average treatment effects. The observed data are

$$Y_1^o = \Im_{ck}^L \left(x^L - k^L\right) + \Im_p^L P^L$$

and

$$Y_0^o = \Im_{ck}^H \left(x^H - k^H\right) + \Im_p^H P^H$$

where

$$\Im_p^j = \begin{matrix} 1 & \text{asset traded at uncertified, equilibrium price for choice } j \\ 0 & \text{otherwise} \end{matrix}$$

and $\Im_{ck}^j$ refers to assets certified and traded for choice j, as before.

For the region $x < x_c$, we have outcome data for firms forced to sell, $x^j - k^j$, but we are missing untraded asset values, x^j. Based on the *DGP* for our continuing example, the contribution to treatment effects from this missing quantity is $22.289 - 20.253 = 2.036$. If we know k^j or can estimate it, we can model the missing data for this region. Since $I^H > I^L$, $E\left[x^H \mid x^H < x_c\right] > E\left[x^L \mid x^L < x_c\right]$ and $\Pr\left(x^L < x_c\right) > \Pr\left(x^H < x_c\right)$. That is, the adjustment to recover x_j is identified as

$$\frac{(1-\pi)}{\pi} \left(\frac{\sum \Im_{ck,i}^L \left(x_i^L - k_i^L\right)}{\sum D_i} - \frac{\sum \Im_{ck,i}^H \left(x_i^H - k_i^H\right)}{\sum (1 - D_i)} \right)$$
$$+ \frac{(1-\pi)}{\pi} \left(\frac{\sum \Im_{ck,i}^L k_i^L}{\sum D_i} - \frac{\sum \Im_{ck,i}^H k_i^H}{\sum (1 - D_i)} \right)$$

The other untraded assets region, $x^j > P^j$, is more delicate as we have no direct evidence, the conditional expectation over this region differs by investment choice, and $P^H > P^L$, it is likely $E\left[x^H \mid x^H > P^H\right] > E\left[x^L \mid x^L > P^L\right]$. Based on the *DGP* for our continuing example, the contribution to treatment effects from this missing quantity is $22.674 - 26.345 = -3.671$.

How do we model missing data in this region? This is not a typical censoring problem as we don't observe the sample size for either missing data region. Missing samples make estimating the probability of each mean level more problematic — recall this is important for estimating average treatment effects in the data observed, selective certification case.[23] Conditional expectations and probabilities of mean levels are almost surely related which implies any augmentation errors will be amplified in the treatment effect estimate.

We cannot infer the probability distribution for x by nonparametric methods since x is unobserved. To see this, recall the equilibrium pricing of uncertified assets satisfies

$$P = \frac{\pi \Pr\left(x_c < x < \bar{x}\right) E\left[x \mid x_c < x < \bar{x}\right]}{\pi \Pr\left(x_c < x < \bar{x}\right) + (1-\pi) \Pr\left(x_c < x < P\right)}$$
$$+ \frac{(1-\pi) \Pr\left(x_c < x < P\right) E\left[x \mid x_c < x < P\right]}{\pi \Pr\left(x_c < x < \bar{x}\right) + (1-\pi) \Pr\left(x_c < x < P\right)}$$

For instance, if all the probability mass in these intervals for x is associated with P, then the equilibrium condition is satisfied. But the equilibrium condition is

---

[23] As is typical, identification and estimation of average treatment effects is more delicate than identification and estimation of model parameters in this selective certification setting.

9.7 Asset revaluation regulation example 195

satisfied for other varieties of distributions for x as well. Hence, the distribution for x cannot be inferred when x is unobserved. If π is known we can estimate $\Pr(x_c < x < \overline{x})$ from certification frequency scaled by π. However, this still leaves much of the missing factual data process unidentified when x is unobserved or the distribution for x is unknown.

On the other hand, consistent probability assignment for x allows π to be inferred from observable data, P and x_c as well as the support for x: $\underline{x} < x < \overline{x}$. Further, consistent probability assignment for x enables us to model the *DGP* for the missing factual data. In particular, based on consistent probability assignment for x we can infer π and identify $\Pr(\underline{x} < x < x_c)$, $E[x \mid \underline{x} < x < x_c]$, $\Pr(P < x < \overline{x})$, and $E[x \mid P < x < \overline{x}]$.

To model missing factual data, suppose π is known and k^j is observed, consistent probability assignment suggests

$$\Pr(P < x < \overline{x}) = \Pr(x_c < x < P)$$

and

$$E[x \mid P < x < \overline{x}] = P + \frac{P - x_c}{2} = \frac{3P - x_c}{2}$$

are reasonable approximations. Then, our model for missing factual data suggests the following adjustments to estimate average treatment effects (*TE*).

$$\begin{aligned}
estTE = \ & TE \text{ estimated based on missing factual data} \\
& + \frac{(1-\pi)}{\pi} \left\{ \frac{\sum \Im^L_{ck,i} \left(x^L_i - k^L_i\right)}{\sum D_i} - \frac{\sum \Im^H_{ck,i} \left(x^H_i - k^H_i\right)}{\sum (1 - D_i)} \right\} \\
& + \frac{(1-\pi)}{\pi} \left[\frac{\sum \Im^L_{ck,i} k^L_i}{\sum D_i} - \frac{\sum \Im^H_{ck,i} k^H_i}{\sum (1 - D_i)} \right] \\
& + \frac{(1-\pi)}{1+\pi} \left[\frac{3P^L - x_c}{2} \frac{\sum \Im^L_{P,i}}{\sum D_i} - \frac{3P^H - x_c}{2} \frac{\sum \Im^H_{P,i}}{\sum (1 - D_i)} \right]
\end{aligned}$$

Results adjusted by the augmented factual missing data based on the previous *OLS* parameter estimates are reported in table 9.31. These augmented-*OLS* results

Table 9.31: Treatment effect OLS model estimates based on augmentation of missing data for selective certification setting

statistics	estATE	estATT	estATUT
mean	−11.80	−11.43	−12.18
median	−11.76	−11.36	−12.06
stand.dev.	3.165	3.041	3.290
minimum	−20.37	−19.58	−21.15
maximum	−2.375	−2.467	−2.280
$E[Y' \mid D] = \beta_0 + \beta_1 D$			

9.7.6 Sharp regression discontinuity design

Suppose the *DGP* is altered only in that

$$k^L \sim uniform\,(1, 3)$$

and

$$k^H \sim uniform\,(3, 37)$$

The means for k remain 2 and 20 but we have adjacent support. There is a crisp break at $k = 3$ but the regression function excluding the treatment effect (the regression as a function of k) is continuous. That is, the treatment effect fully accounts for the discontinuity in the regression function. This is a classic "sharp" regression discontinuity design (Trochim [1984] and Angrist and Pischke [2009]) where β_2 estimates the average treatment effect via *OLS*.

$$E\left[Y \mid k, D\right] = \beta_0 + \beta_1 k + \beta_2 D$$

With the previous *DGP*, there was discontinuity as a function of both the regressor k and treatment D. This creates a problem for the regression as least squares is unable to distinguish the treatment effect from the jump in the outcome regression and leads to poor estimation results. In this revised setting, we anticipate substantially improved (finite sample) results.

Full certification setting

Simulation results for the revised *DGP* in the full certification setting are reported in table 9.32 and average treatment effect sample statistics are reported in table

Table 9.32: Sharp RD OLS parameter estimates for full certification setting

$statistics$	β_0	β_1	$\beta_2\,(estATE)$
$mean$	214.2	-1.007	-12.93
$median$	214.5	-1.019	-13.04
$stand.dev.$	4.198	0.190	4.519
$minimum$	203.4	-1.503	-26.18
$maximum$	226.3	-0.539	-1.959
$E\left[Y \mid k, D\right] = \beta_0 + \beta_1 k + \beta_2 D$			

9.33.

Unlike the previous *DGP*, sharp regression discontinuity (*RD*) design effectively identifies the average treatment effect and *OLS* produces reliable estimates for the (simple) full certification setting. Next, we re-evaluate *RD* with the same adjacent support *DGP* but in the more challenging selective certification setting.

9.7 Asset revaluation regulation example

Table 9.33: Average treatment effect sample statistics for full certification setting

statistics	ATE	ATT	ATUT
mean	−12.54	−12.49	−12.59
median	−12.55	−12.44	−12.68
stand.dev.	1.947	2.579	2.794
minimum	−17.62	−19.53	−21.53
maximum	−7.718	−6.014	−6.083

Selective certification setting

To satisfy the continuity condition for the regression, suppose cost of certification $k = Dk^L + (1 − D) k^H$ is always observed whether assets are certified or not in the regression discontinuity analysis of selective certification. Simulation results for the revised *DGP* in the selective certification setting are reported in table 9.34.[24] In the selective certification setting, *RD* again identifies the average

Table 9.34: Sharp RD OLS parameter estimates for selective certification setting

statistics	β_0	β_1	$\beta_2\,(estATE)$
mean	214.2	−0.299	−13.00
median	214.5	−0.324	−12.89
stand.dev.	4.273	0.197	4.546
minimum	202.0	−0.788	−25.81
maximum	225.5	0.226	−1.886
$E\left[Y \mid k, D\right] = \beta_0 + \beta_1 k + \beta_2 D$			

treatment effect and *OLS* provides effective estimates. Next, we employ *RD* in the missing factual data setting.

Missing factual data

If some outcome data are unobserved by the analyst, it may be imprudent to ignore the issue. We employ the same missing data model as before and estimate the average treatment effect ignoring missing outcome data (β_2) and the average treatment effect adjusted for missing outcome data (β_2'). Simulation results for the revised *DGP* (with adjacent support) analyzed via a sharp RD design in the selective certification setting with missing outcome data are reported in table 9.35.

---

[24] We report results only for the reduced model. If the analyst knows where support changes (i.e., can identify the indicator variables) for the full model, the results are similar and the estimates have greater precision.

Table 9.35: Sharp RD OLS parameter estimates with missing data for selective certification setting

statistics	β_0	β_1	β_2	$\beta_2'\,(estATE)$
mean	214.4	−0.342	−11.35	−12.22
median	214.5	−0.336	−11.50	−12.47
stand.dev.	4.800	0.232	5.408	5.237
minimum	201.3	−0.928	−25.92	−26.50
maximum	227.9	0.325	2.542	1.383

$$E[Y \mid k, D] = \beta_0 + \beta_1 k + \beta_2 D$$

9.7.7 Fuzzy regression discontinuity design

Now, suppose the DGP is altered only in that support is overlapping as follows:

$$k^L \sim uniform\,(1, 3)$$

and

$$k^H \sim uniform\,(1, 39)$$

The means for k remain 2 and 20 but we have overlapping support. There is a crisp break in $E[D \mid k]$ at $k = 3$ but the regression function excluding the treatment effect (the regression as a function of k) is continuous. This leads to a fuzzy discontinuity regression design (van der Klaauw [2002]). Angrist and Lavy [1999] argue that *2SLS-IV* consistently estimates a local average treatment effect in such cases where

$$T = \begin{array}{ll} 1 & k \leq 3 \\ 0 & k > 3 \end{array}$$

serves as an instrument for treatment. In the first stage, we estimate the propensity score[25]

$$\widehat{D} \equiv E[D \mid k, T] = \gamma_0 + \gamma_1 k + \gamma_2 T$$

The second stage is then

$$E[Y \mid k, D] = \gamma_0 + \gamma_1 k + \gamma_2 \widehat{D}$$

Full certification setting

First, we estimate *RD* via *OLS* then we employ *2SLS-IV*. Simulation results for the overlapping support *DGP* in the full certification setting are reported in table 9.36.

Perhaps surprisingly, *OLS* effectively estimates the average treatment effect in this fuzzy *RD* setting. Recall the selection bias is entirely due to the expected difference in certification cost, $E\left[k^H - k^L\right]$. *RD* models outcome as a (regression)

---

[25] In this asset revaluation setting, the relations are linear. More generally, high order polynomial or nonparametric regressions are employed to accommodate nonlinearities (see Angrist and Pischke [2009]).

9.7 Asset revaluation regulation example   199

Table 9.36: Fuzzy RD OLS parameter estimates for full certification setting

statistics	β_0	β_1	$\beta_2\,(estATE)$
mean	214.3	−1.012	−12.79
median	214.2	−1.011	−12.56
stand.dev.	3.634	0.163	3.769
minimum	204.9	−1.415	−23.51
maximum	222.5	−0.625	−3.001
$E\left[Y\mid k,D\right]=\beta_0+\beta_1 k+\beta_2 D$			

function of k, $E\left[Y\mid k\right]$; hence, the selection bias is eliminated from the treatment effect. Next, we use *2SLS-IV* to estimate *LATE*.[26]

Binary instrument

Now, we utilize T as a binary instrument. Simulation results for the overlapping support *DGP* in the full certification setting are reported in table 9.37. As ex-

Table 9.37: Fuzzy RD 2SLS-IV parameter estimates for full certification setting

statistics	β_0	β_1	$\beta_2\,(estLATE)$
mean	214.5	−1.020	−13.07
median	214.6	−1.021	−13.27
stand.dev.	4.139	0.181	4.456
minimum	202.7	−1.461	−27.60
maximum	226.0	−0.630	−1.669
$E\left[Y\mid k,D\right]=\beta_0+\beta_1 k+\beta_2 \widehat{D}$			

pected, *2SLS-IV* effectively identifies *LATE* in this fuzzy *RD*, full certification setting. Next, we revisit selective certification with this overlapping support *DGP*.

9.7.8 Selective certification setting

First, we estimate *RD* via *OLS* then we employ *2SLS-IV*. Simulation results for the overlapping support *DGP* in the selective certification setting are reported in table 9.38. Since *RD* effectively controls the selection bias (as discussed above), *OLS* effectively estimates the average treatment effect.

Binary instrument

Using T as a binary instrument, *2SLS-IV* simulation results for the overlapping support *DGP* in the selective certification setting are reported in table 9.39. In the selective certification setting, *2SLS-IV* effectively estimates *LATE*, as anticipated.

---

[26]*LATE* is developed more fully in chapter 10.

Table 9.38: Fuzzy RD OLS parameter estimates for selective certification setting

statistics	β_0	β_1	$\beta_2\,(estATE)$
mean	214.3	-0.315	-12.93
median	214.1	-0.311	-12.73
stand.dev.	3.896	0.179	3.950
minimum	202.5	-0.758	-24.54
maximum	223.3	0.078	-3.201
$E\,[Y\mid k,D]=\beta_0+\beta_1 k+\beta_2 D$			

Table 9.39: Fuzzy RD 2SLS-IV parameter estimates for selective certification setting

statistics	β_0	β_1	$\beta_2\,(estLATE)$
mean	214.4	-0.321	-13.09
median	214.5	-0.317	-13.03
stand.dev.	4.438	0.200	4.631
minimum	201.1	-0.805	-27.23
maximum	225.6	-0.131	1.742
$E\,[Y\mid k,D]=\beta_0+\beta_1 k+\beta_2 \widehat{D}$			

Missing factual data

Continue with the overlapping support *DGP* and employ the same missing data model as before to address unobserved outcomes (by the analyst) when the assets are untraded. First, we report *OLS* simulation results in table 9.40 then we tabulate *2SLS-IV* simulation results where β_2 is the estimated for the local average treatment effect ignoring missing outcome data and β_2' is the local average treatment effect adjusted for missing outcome data. This *OLS RD* model for missing

Table 9.40: Fuzzy RD OLS parameter estimates with missing data for selective certification setting

statistics	β_0	β_1	β_2	$\beta_2'\,(estATE)$
mean	215.9	-0.426	-12.74	-13.60
median	216.2	-0.424	-12.63	-13.52
stand.dev.	4.765	0.223	4.792	4.612
minimum	201.9	-1.132	-24.20	-23.85
maximum	226.3	0.117	0.119	-0.817
$E\,[Y\mid k,D]=\beta_0+\beta_1 k+\beta_2 D$				

outcome data does not offer any clear advantages. Rather, the results seem to be slightly better without the missing data adjustments.

2SLS-IV with T as a binary instrument and missing outcome data adjustments are considered next. Simulation results for the overlapping support *DGP* in the

9.7 Asset revaluation regulation example

selective certification, missing outcome data setting are reported in table 9.41.

Table 9.41: Fuzzy RD 2SLS-IV parameter estimates with missing data for selective certification setting

statistics	β_0	β_1	β_2	β_2' (estLATE)
mean	217.7	−0.428	−12.80	−13.67
median	214.8	−0.425	−13.12	−14.30
stand.dev.	25.50	0.256	5.919	5.773
minimum	139.2	−1.147	−25.24	−25.97
maximum	293.9	0.212	6.808	6.010
$E[Y \mid k, D] = \beta_0 + \beta_1 k + \beta_2 \widehat{D}$				

Again, modeling the missing outcome data offers no apparent advantage in this fuzzy *RD*, *2SLS-IV* setting. In summary, when we have adjacent or overlapping support, sharp or fuzzy regression discontinuity designs appear to be very effective for controlling selection bias and identifying average treatment effects in this asset revaluation setting.

9.7.9 Common support

Standard identification conditions associated with ignorable treatment (and *IV* approaches as well) except for regression discontinuity designs include common support $0 < \Pr(D = 1 \mid X) < 1$. As indicated earlier, this condition fails in the asset revaluation setting as certification cost type is a perfect predictor of treatment $\Pr(D = 1 \mid T = 1) = 1$ and $\Pr(D = 1 \mid T = 0) = 0$ where $T = 1$ if type is L and zero otherwise. The foregoing discussion has addressed this issue in two ways. First, we employed an ad hoc adjustment of outcome to eliminate selection bias. This may be difficult or impractical to implement. Second, we employed a regression discontinuity design. The second approach may be unsatisfactory as the analyst needs full support access to adjacent or overlapping regressor k whether assets are certified or not.

However, if there is some noise in the relation between certification cost type and treatment (perhaps, due to nonpecuniary cost or benefit), then a third option may be available. We briefly illustrate this third possibility for the full certification setting.

Suppose everything remains as in the original full certification setting except $k^H \sim uniform(1, 19)$ and some owners select treatment (lower investment) when certification cost is high, hence $\Pr(D = 1 \mid \text{type} = H) = 0.1$. This setup implies observed outcome is

$$Y = D\left[(Y_1 \mid T = 1) + (Y_1 \mid T = 0)\right] + (1 - D)(Y_0 \mid T = 0)$$

such that

$$E[Y] = 0.5E\left[x^L - k^L\right] + 0.5\left\{0.1E\left[x^L - k^H\right] + 0.9E\left[x^H - k^H\right]\right\}$$

9. Treatment effects: ignorability

Suppose the analyst ex post observes the actual certification cost type. The common support condition is satisfied as $0 < \Pr(D = 1 \mid T = 0) < 1$ and if outcomes are conditionally mean independent of treatment given T then treatment is ignorable. The intuition is the type variable, T, controls the selection bias and allows D to capture the treatment effect. This involves a delicate balance as T and D must be closely but imperfectly related.

OLS common support results are reported in table 9.42 and simulation results for average treatment effect sample statistics are reported in table 9.43. The esti-

Table 9.42: Fuzzy RD OLS parameter estimates for full certification setting

statistics	β_0	β_1	$\beta_2\ (estATE)$
mean	196.9	7.667	-5.141
median	196.9	7.896	-5.223
stand.dev.	1.812	6.516	6.630
minimum	191.5	-10.62	-23.54
maximum	201.6	25.56	14.25
$E[Y \mid T, D] = \beta_0 + \beta_1 T + \beta_2 D$			

Table 9.43: Average treatment effect sample statistics for full certification setting

statistics	ATE	ATT	$ATUT$
mean	-5.637	-5.522	-5.782
median	-5.792	-5.469	-5.832
stand.dev.	1.947	2.361	2.770
minimum	-9.930	-12.05	-12.12
maximum	0.118	0.182	0.983

mated average treatment effect is slightly attenuated and has high variability that may compromise its finite sample utility. Nevertheless, the results are a dramatic departure and improvement from the results above where the common support condition fails and is ignored.

9.7.10 Summary

Outcomes at our disposal in this asset revaluation setting limit our ability to assess welfare implications for the owners. Nonetheless, the example effectively points to the importance of recognizing differences in data available to the analyst compared with information in the hands of the economic agents whose actions and welfare is the subject of study. To wit, treatment effects in this setting are uniformly negative. This is a product of comparing net gains associated with equilibrium investment levels, but net gains exclude investment cost. The benefits of higher investment when certification costs are low are not sufficient to overcome the cost of investment but this latter feature is not reflected in our outcome measure. Hence, if care is not exercised in interpreting the results we might draw erroneous conclusions from the data.

9.8 Control function approaches

Our final stop in the world of ignorable treatment involves the use of control functions. Control functions are functions that capture or control selection so effectively as to overcome the otherwise omitted, correlated variable concern created by endogenous selection. Various approaches can be employed. The simplest (strongest for the data) conditions employ conditional mean independence

$$E[Y_1 \mid X, D] = E[Y_1 \mid X]$$

and

$$E[Y_0 \mid X, D] = E[Y_0 \mid X]$$

and no expected individual-specific gain, $E[V_1 \mid X] = E[V_0 \mid X]$. Then,

$$E[Y \mid X, D] = \mu_0 + \alpha D + g_0(X)$$

where $g_0(X) = E[V_0 \mid X]$ is a control function and $\alpha = ATE = ATT = ATUT$.

9.8.1 Linear control functions

If we add the condition $E[V_0 \mid X] = g_0(X) = \eta_0 + h_0(X)\beta_0$ for some vector control function $h_0(X)$, then

$$E[Y \mid X, D] = \mu_0 + \eta_0 + \alpha D + h_0(X)\beta_0$$

That is, when the predicted individual-specific gain given X, $E[V_1 - V_0 \mid X]$, is zero and the control function is linear in its parameters, we can consistently estimate *ATE* via standard (linear) regression.

9.8.2 Control functions with expected individual-specific gain

Suppose we relax the restriction to allow expected individual specific-gain, that is allow $E[V_1 \mid X] \neq E[V_0 \mid X]$, then

$$E[Y \mid X, D] = \mu_0 + \alpha D + g_0(X) + D[g_1(X) - g_0(X)]$$

where $g_0(X) = E[V_0 \mid X]$ and $g_1(X) = E[V_1 \mid X]$ and $ATE = \alpha$ (but not necessarily equal to ATT).

9.8.3 Linear control functions with expected individual-specific gain

Continue with the idea that we allow expected individual specific-gain, $E[V_1 \mid X] \neq E[V_0 \mid X]$ and add the condition that the control functions are linear in parameters $E[V_0 \mid X] = g_0(X) = \eta_0 + h_0(X)\beta_0$ and $E[V_1 \mid X] = g_1(X) = \eta_1 + h_1(X)\beta_1$ for some vector control functions $h_0(X)$ and $h_1(X)$. Hence,

$$E[Y \mid X, D] = \phi + \alpha D + X\beta_0 + D(X - E[X])\delta$$

Now, conditional on X the average treatment effect, $ATE(X)$, is a function of X

$$\alpha + (X - E[X])\delta$$

When we average over all X, the second term is integrated out and $ATE = \alpha$. By similar reasoning, the average treatment effect on the treated can be estimated by integrating over the $D = 1$ subsample

$$ATT = \alpha + \left[\sum_{i=1}^{n} D_i\right]^{-1} \left[\sum_{i=1}^{n} D_i \left(X_i - \overline{X}\right) \delta\right]$$

and the average treatment effect on the untreated can be estimated by integrating over the $D = 0$ subsample

$$ATUT = \alpha - \left[\sum_{i=1}^{n} (1 - D_i)\right]^{-1} \left[\sum_{i=1}^{n} D_i \left(X_i - \overline{X}\right) \delta\right]$$

9.9 Summary

The key element for ignorable treatment identification of treatment effects is outcomes are conditionally mean independent of treatment given the regressors. How do we proceed when ignorable treatment (conditional mean independence) fails? A common response is to look for instruments and apply IV strategies to identify average treatment effects. Chapter 10 surveys some instrumental variable approaches and applies a subset of IV identification strategies in an accounting setting — report precision regulation.

9.10 Additional reading

Amemiya [1985] and Wooldridge [2002] provide extensive reviews of the econometrics of selection. Wooldridge [2002] discusses estimating average treatment effects in his chapter 18 (and sample selection earlier). Amemiya [1985] discusses qualitative response models in his chapter 9. Recent volumes of the *Handbook of*

Econometrics are filled with economic policy evaluation and treatment effects. Dawid [2000] offers an alternative view on causal inference.

Heckman, Ichimura, Smith, and Todd [1998] utilize experimental (as well as non-experimental) data to evaluate non-experimental methods (matching, differences - in - differences, and inverse-Mills selection models) for program evaluation. Their results indicate selection bias is mitigated, but not eliminated, by non-experimental methods that invoke common support and common weighting. In fact, they decompose conventional bias into (a) differences in the support of the regressors between treated and untreated, (b) differences in the shape of the distributions of regressors for the two groups in the region of common support, and (c) selection bias at common values of the regressors for both groups. Further, they find that matching cannot eliminate selection bias[27] but their data support the index sufficiency condition underlying standard control function models and a conditional version of differences-in-differences. Heckman and Navarro-Lozano [2004] succinctly review differences amongst matching, control function, and instrumental variable (the latter two are discussed in chapter 10 and the various strategies are compared in chapter 12) approaches to identification and estimation of treatment effects. In addition, they identify the bias produced by matching when the analyst's data fail to meet in the minimally sufficient information for ignorable treatment and when and how other approaches may be more robust to data omissions than matching. They also demonstrate that commonly-employed ad hoc "fixes" such as adding information to increase the goodness of fit of the propensity score model (when minimal information conditions are not satisfied) do not, in general, produce lower bias but rather may increase bias associated with matching.

---

[27]Heckman, Ichimura, and Todd [1997] find that matching sometimes increases selection bias, at least for some conditioning variables.

10
Treatment effects: *IV*

In this chapter we continue the discussion of treatment effects but replace ignorable treatment strategies in favor of instrumental variables and exclusion restrictions. Intuitively, instrumental variables are a standard econometric response to omitted, correlated variables so why not employ them to identify and estimate treatment effects. That is, we look for instruments that are highly related to the selection or treatment choice but unrelated to outcome. This is a bit more subtle than standard linear *IV* because of the counterfactual issue. The key is that exclusion restrictions allow identification of the counterfactuals as an individual's probability of receiving treatment can be manipulated without affecting potential outcomes.

We emphasize we're looking for good instruments. Recall that dropping variables from the outcome equations that should properly be included creates an omitted, correlated variable problem. There doesn't seem much advantage of swapping one malignant inference problem for another — the selection problem can also be thought of as an omitted, correlated variable problem.

10.1 Setup

The setup is the same as the previous chapter. We repeat it for convenience then relate it to common average treatment effects and the Roy model to facilitate interpretation. Suppose the *DGP* is

outcomes:[1]
$$Y_j = \mu_j(X) + V_j, j = 0, 1$$
selection mechanism:[2]
$$D^* = \mu_D(Z) - V_D$$
and observable response:
$$\begin{aligned} Y &= DY_1 + (1-D)Y_0 \\ &= \mu_0(X) + (\mu_1(X) - \mu_0(X))D + V_0 + (V_1 - V_0)D \end{aligned}$$
where
$$D = \begin{cases} 1 & D^* > 0 \\ 0 & otherwise \end{cases}$$

and Y_1 is (potential) outcome with treatment and Y_0 is (potential) outcome without treatment. The outcomes model is the Neyman-Fisher-Cox-Rubin model of potential outcomes (Neyman [1923], Fisher [1966], Cox]1958], and Rubin [1974]). It is also Quandt's [1972] switching regression model or Roy's income distribution model (Roy [1951] or Heckman and Honore [1990]).

10.2 Treatment effects

We address the same treatment effects but add a couple of additional effects to highlight issues related to unobservable heterogeneity. Heckman and Vytlacil [2005] describe the recent focus of the treatment effect literature as the heterogeneous response to treatment amongst otherwise observationally equivalent individuals. Unobservable heterogeneity is a serious concern whose analysis is challenging if not down right elusive.

In the binary case, the treatment effect is the effect on outcome of treatment compared with no treatment, $\Delta = Y_1 - Y_0$. Some typical treatment effects include: *ATE, ATT, ATUT, LATE*, and *MTE*. *ATE* refers to the average treatment effect, by iterated expectations, we can recover the unconditional average treatment effect from the conditional average treatment effect

$$\begin{aligned} ATE &= E_X[ATE(X)] \\ &= E_X[E[\Delta \mid X = x]] = E[Y_1 - Y_0] \end{aligned}$$

---

[1] Separating outcome into a constant and stochastic parts, yields
$$Y_j = \mu_j + U_j$$
Sometimes it will be instructive to write the stochastic part as a linear function of X
$$U_j = X\beta_j + V_j$$

[2] To facilitate discussion, we stick with binary choice for most of the discussion. We extend the discussion to multilevel discrete and continuous treatment later in chapter 11.

In other words, the average effect of treatment on outcome compared with no treatment for a random draw from the population.

ATT refers to the average treatment effect on the treated,

$$\begin{aligned} ATT &= E_X\left[ATT\left(X\right)\right] \\ &= E_X\left[E\left[\Delta \mid X=x, D=1\right]\right] = E\left[Y_1 - Y_0 \mid D=1\right] \end{aligned}$$

In other words, the average effect of treatment on outcome compared with no treatment for a random draw from the subpopulation selecting (or assigned) treatment.

$ATUT$ refers to the average treatment effect on the untreated,

$$\begin{aligned} ATUT &= E_X\left[ATUT\left(X\right)\right] \\ &= E_X\left[E\left[\Delta \mid X=x, D=0\right]\right] = E\left[Y_1 - Y_0 \mid D=0\right] \end{aligned}$$

In other words, the average effect of treatment on outcome compared with no treatment for a random draw from the subpopulation selecting (or assigned) no treatment.

For a binary instrument (to keep things simple), the local average treatment effect or *LATE* is

$$\begin{aligned} LATE &= E_X\left[LATE\left(X\right)\right] \\ &= E_X\left[E\left[\Delta \mid X=x, D_1 - D_0 = 1\right]\right] = E\left[Y_1 - Y_0 \mid D_1 - D_0 = 1\right] \end{aligned}$$

where D_j refers to the observed treatment conditional on the value j of the binary instrument. *LATE* refers to the local average or marginal effect of treatment on outcome compared with no treatment for a random draw from the subpopulation of "compliers" (Imbens and Angrist [1994]). That is, *LATE* is the (discrete) marginal effect on outcome for those individuals who would not choose treatment if the instrument takes a value of zero but would choose treatment if the instrument takes a value of one.

MTE (the marginal treatment effect) is a generalization of *LATE* as it represents the treatment effect for those individuals who are indifferent between treatment and no treatment.

$$MTE = E\left[Y_1 - Y_0 \mid X=x, V_D = v_D\right]$$

or following transformation $U_D = F_{V\mid X}\left(V\right)$, where $F_{V\mid X}\left(V\right)$ is the (cumulative) distribution function, we can work with $U_D \sim Uniform\left[0,1\right]$

$$MTE = E\left[Y_1 - Y_0 \mid X=x, U_D = u_D\right]$$

Treatment effect implications can be illustrated in terms of the generalized Roy model. The Roy model interpretation is discussed next.

10.3 Generalized Roy model

Roy [1951] introduced an equilibrium labor model where workers select between hunting and fishing. An individual's selection into hunting or fishing depends on his aptitude as well as supply of and demand for labor.[3] A modest generalization of the Roy model is a common framing of selection that frequently forms the basis for assessing treatment effects (Heckman and Robb [1986]).

Based on the *DGP* above, we identify the constituent pieces of the selection model.

Net benefit (or utility) from treatment is

$$\begin{aligned} D^* &= \mu_D(Z) - V_D \\ &= Y_1 - Y_0 - c(W) - V_c \\ &= \mu_1(X) - \mu_0(X) - c(W) + V_1 - V_0 - V_c \end{aligned}$$

Gross benefit of treatment is[4]

$$\mu_1(X) - \mu_0(X)$$

Cost associated with treatment is

$$c(W) + V_c$$

Observable cost associated with treatment is

$$c(W)$$

Observable net benefit of treatment is

$$\mu_1(X) - \mu_0(X) - c(W)$$

Unobservable net benefit of treatment is

$$-V_D = V_1 - V_0 - V_c$$

where the observables are $\begin{bmatrix} X & Z & W \end{bmatrix}$, typically Z contains variables not in X or W, and W is the subset of observables that speaks to cost of treatment.

Given a rich data generating process like above, the challenge is to develop identification strategies for the treatment effects of interest. The simplest *IV* approaches follow from the strongest conditions for the data and typically imply homogeneous response. Accommodating heterogeneous response holds economic appeal but also constitutes a considerable hurdle.

---

[3]Roy argues that self-selection leads to lesser earnings inequality than does random assignment. See Heckman and Honore [1990] for an extended discussion of the original Roy model including identification under various probability distribution assignments on worker skill (log skill).

[4]For linear outcomes, we have $\mu_1(X) - \mu_0(X) = (\mu_1 + X\beta_1) - (\mu_0 + X\beta_0)$.

10.4 Homogeneous response

Homogeneous response is attractive when pooling restrictions across individuals (or firms) are plausible. Homogeneous response implies the stochastic portion, U_j, is the same for individuals receiving treatment and not receiving treatment, $U_1 = U_0$. This negates the interaction term, $(U_1 - U_0)D$, in observed outcome and consequently rules out individual-specific gains. Accordingly, $ATE = ATT = ATUT = MTE$. Next, we review treatment effect identification conditions for a variety of homogeneous response models with endogenous treatment.

10.4.1 Endogenous dummy variable IV model

Endogenous dummy variable *IV* regression is a standard approach but not as robust in the treatment effect setting as we're accustomed in other settings. Let L be a linear projection of the leading argument into the column space of the conditioning variables where X includes the unity vector ι, that is,

$$\begin{aligned} L(Y \mid X) &= X\left(X^T X\right)^{-1} X^T Y \\ &= P_X Y \end{aligned}$$

and Z_i be a vector of instruments. Identification conditions are

Condition 10.1 $U_1 = U_0$ where $U_j = X\beta_j + V_j$, $j = 0, 1$,

Condition 10.2 $L(U_0 \mid X, Z) = L(U_0 \mid X)$, and

Condition 10.3 $L(D \mid X, Z) \neq L(D \mid X)$.

Condition 10.1 is homogeneous response while conditions 10.2 and 10.3 are exclusion restrictions. Conditions 10.1 and 10.2 imply observed outcome is

$$Y = \mu_0 + (\mu_1 - \mu_0)D + X\beta_0 + V_0$$

which can be written
$$Y = \delta + \alpha D + X\beta_0 + V_0$$

where $\alpha = ATE$ and $V_0 = U_0 - L(U_0 \mid X, Z)$. As D and V_0 are typically correlated (think of the Roy model interpretation), we effectively have an omitted, correlated variable problem and *OLS* is inconsistent.

However, condition 10.2 means that Z is properly excluded from the outcome equation. Unfortunately, this cannot be directly tested.[5] Under the above conditions, standard two stage least squares instrumental variable (*2SLS-IV*) estimation (see chapter 3) with $\{\iota, X, Z\}$ as instruments provides a consistent and asymptotically normal estimate for *ATE*. That is, the first stage discrete choice (say, logit

---

[5]Though we might be able to employ over-identifying tests of restrictions if we have multiple instruments. Of course, these tests assume that at least one is a legitimate instrument.

or probit) regression is

$$D = \gamma_0 + X\gamma_1 + Z\gamma_2 - V_D$$

and the second stage regression is

$$Y = \delta + \alpha \widehat{D} + X\beta_0 + V_0$$

where $\widehat{D} = \hat{\gamma}_0 + X\hat{\gamma}_1 + Z\hat{\gamma}_2$, predicted values from the first stage discrete choice regression.

10.4.2 Propensity score IV

Stronger conditions allow for a more efficient *IV* estimator. For instance, suppose the data satisfies the following conditions.

Condition 10.4 $U_1 = U_0$,

Condition 10.5 $E[U_0 \mid X, Z] = E(U_0 \mid X)$,

Condition 10.6 $\Pr(D = 1 \mid X, Z) \neq \Pr(D = 1 \mid X)$ *plus* $\Pr(D = 1 \mid X, Z) = G(X, Z, \gamma)$ *is a known parametric form (usually probit or logit), and*

Condition 10.7 $Var[U_0 \mid X, Z] = \sigma_0^2$.

The outcome equation is

$$Y = \delta + \alpha D + X\beta_0 + V_0$$

If we utilize $\{\iota, G(X, Z, \gamma), X\}$ as instruments, *2SLS-IV* is consistent asymptotically normal (*CAN*). Not only is this propensity score approach more efficient given the assumptions, but it is also more robust. Specifically, the link function doesn't have to be equal to G for *2SLS-IV* consistency but it does for *OLS* (see Wooldridge [2002], ch. 18).

10.5 Heterogeneous response and treatment effects

Frequently, homogeneity is implausible, $U_1 \neq U_0$. Idiosyncrasies emerge in both what is observed, say $X\beta_0 \neq X\beta_1$, (relatively straightforward to address) and what the analyst cannot observe, $V_0 \neq V_1$, (more challenging to address). Then observed outcome contains an individual-specific gain $(U_1 - U_0) D$ and, usually, $ATE \neq ATT \neq ATUT \neq MTE$. In general, the linear *IV* estimator (using Z or G as instruments) does not consistently estimate *ATE* (or *ATT*) when response is heterogeneous, $U_1 \neq U_0$. Next, we explore some *IV* estimators which may consistently estimate *ATE* even though response is heterogeneous.

10.5.1 Propensity score IV and heterogeneous response

First, we return to the propensity score and relax the conditions to accommodate heterogeneity. Let $U_j = X\beta_j + V_j$ where $E[V_j \mid X, Z] = 0$. Identification conditions are

Condition 10.8 *conditional mean redundancy,* $E[U_0 \mid X, Z] = E[U_0 \mid X]$ *and* $E[U_1 \mid X, Z] = E[U_1 \mid X]$,

Condition 10.9 $X\beta_1 - X\beta_0 = (X - E[X])\gamma$,

Condition 10.10 $V_1 = V_0$, *and*

Condition 10.11 $\Pr(D = 1 \mid X, Z) \neq \Pr(D = 1 \mid X)$ *and* $\Pr(D = 1 \mid X, Z) = G(X, Z, \gamma)$ *where again G is a known parametric form (usually probit or logit).*

If we utilize $\{\iota, G(X, Z, \gamma), X - \overline{X}\}$ as instruments in the regression

$$Y = \mu_0 + X\beta_0 + \alpha D + (X - \overline{X}) D\gamma + V_0$$

2SLS-IV is consistent asymptotically normal (*CAN*).

We can relax the above a bit if we replace condition 10.10, $V_1 = V_0$, by conditional mean independence

$$E[D(V_1 - V_0) \mid X, Z] = E[D(V_1 - V_0)]$$

While probably not efficient, α consistently identifies *ATE* for this two-stage propensity score *IV* strategy utilizing $\{\iota, G, X, G(X - E[X])\}$ as instruments.

10.5.2 Ordinate control function IV and heterogeneous response

Employing control functions to address the omitted, correlated variable problem created by endogenous selection is popular. We'll review two identification strategies: ordinate and inverse Mills *IV* control functions. The second one pioneered by Heckman [1979] is much more frequently employed. Although the first approach may be more robust.

Identification conditions are

Condition 10.12 *conditional mean redundancy,* $E[U_0 \mid X, Z] = E[U_0 \mid X]$ *and* $E[U_1 \mid X, Z] = E[U_1 \mid X]$,

Condition 10.13 $g_1(X) - g_0(X) = X\beta_1 - X\beta_0 = (X - E[X])\gamma$,

Condition 10.14 $V_1 - V_0$ *is independent of* $\{X, Z\}$ *and* $E[D \mid X, Z, V_1 - V_0] = h(X, Z) + k(V_1 - V_0)$ *for some functions h and k,*

Condition 10.15 $\Pr(D = 1 \mid X, Z, V_1 - V_0)$
$= \Phi(\theta_0 + X\theta_1 + Z\theta_2 + \varrho(V_1 - V_0))$, $\theta_2 \neq 0$, *and*

Condition 10.16 $V_1 - V_0 \sim N\left(0, \tau^2\right)$.

The model of observed outcome

$$Y = \mu_0 + \alpha D + X\beta_0 + D\left(X - E[X]\right)\gamma + \xi\phi + error$$

can be estimated by two-stage *IV* using instruments

$$\{\iota, \Phi, X, \Phi\left(X - E[X]\right), \phi\}$$

where Φ is the cumulative standard normal distribution function and ϕ is the ordinate from a standard normal each evaluated at $[X_i, Z_i]\widehat{\theta}$ from probit. With full common X support, *ATE* is consistently estimated by α since ϕ is a control function obtained via *IV* assumptions (hence the label ordinate control function).

10.5.3 Inverse Mills control function IV and heterogeneous response

Heckman's inverse Mills control function is closely related to the ordinate control function. Identification conditions are

Condition 10.17 *conditional mean redundancy,* $E[U_0 \mid X, Z] = E[U_0 \mid X]$ *and* $E[U_1 \mid X, Z] = E[U_1 \mid X]$,

Condition 10.18 $g_1(X) - g_0(X) = (X - E[X])\delta$,

Condition 10.19 (V_D, V_1, V_0) *is independent of* $\{X, Z\}$ *with joint normal distribution, especially* $V \sim N(0, 1)$, *and*

Condition 10.20 $D = I[\theta_0 + X\theta_1 + Z\theta_2 - V_D > 0]$ *where* I *is an indicator function equal to one when true and zero otherwise.*

While this can be estimated via *MLE*, Heckman's two-stage procedure is more common. First, estimate θ via a probit regression of D on $W = \{\iota, X, Z\}$ and identify observations with common support (that is, observations for which the regressors, X, for the treated overlap with regressors for the untreated). Second, regress Y onto

$$\left\{\iota, D, X, D\left(X - E[X]\right), D\left(\frac{\phi}{\Phi}\right), (1-D)\frac{-\phi}{1-\Phi}\right\}$$

for the overlapping subsample. With full support, the coefficient on D is a consistent estimator of *ATE*; with less than full common support, we have a local average treatment effect.[6]

---

[6]We should point out here that this second stage *OLS* does not provide valid estimates of standard errors. As Heckman [1979] points out there are two additional concerns: the errors are heteroskedastic (so an adjustment such as White suggested is needed) and θ has to be estimated (so we must account for this added variation). Heckman [1979] identifies a valid variance estimator for this two-stage procedure.

10.5 Heterogeneous response and treatment effects

The key ideas behind treatment effect identification via control functions can be illustrated by reference to this case.

$$E[Y_j \mid X, D = j] = \mu_j + X\beta_j + E[V_j \mid D = j]$$

Given the conditions, $E[V_j \mid D = j] \neq 0$ unless $Corr(V_j, V_D) = \rho_{jV_D} = 0$. For $\rho_{jV_D} \neq 0$,

$$E[V_1 \mid D = 1] = \rho_{1V_D}\sigma_1 E[V_D \mid V_D > -W\theta]$$
$$E[V_0 \mid D = 1] = \rho_{0V_D}\sigma_0 E[V_D \mid V_D > -W\theta]$$
$$E[V_1 \mid D = 0] = \rho_{1V_D}\sigma_1 E[V_D \mid V_D \leq -W\theta]$$

and

$$E[V_0 \mid D = 0] = \rho_{0V_D}\sigma_0 E[V_D \mid V_D \leq -W\theta]$$

The final term in each expression is the expected value of a truncated standard normal random variate where

$$E[V_D \mid V_D > -W\theta] = \frac{\phi(-W\theta)}{1 - \Phi(-W\theta)} = \frac{\phi(W\theta)}{\Phi(W\theta)}$$

and

$$E[V_D \mid V_D \leq -W\theta] = -\frac{\phi(-W\theta)}{\Phi(-W\theta)} = -\frac{\phi(W\theta)}{1 - \Phi(W\theta)}$$

Putting this together, we have

$$E[Y_1 \mid X, D = 1] = \mu_1 + X\beta_1 + \rho_{1V_D}\sigma_1 \frac{\phi(W\theta)}{\Phi(W\theta)}$$

$$E[Y_0 \mid X, D = 0] = \mu_0 + X\beta_0 - \rho_{0V_D}\sigma_0 \frac{\phi(W\theta)}{1 - \Phi(W\theta)}$$

and counterfactuals

$$E[Y_0 \mid X, D = 1] = \mu_0 + X\beta_0 + \rho_{0V_D}\sigma_0 \frac{\phi(W\theta)}{\Phi(W\theta)}$$

and

$$E[Y_1 \mid X, D = 0] = \mu_1 + X\beta_1 - \rho_{1V_D}\sigma_1 \frac{\phi(W\theta)}{1 - \Phi(W\theta)}$$

The affinity for Heckman's inverse Mills ratio approach can be seen in its estimation simplicity and the ease with which treatment effects are then identified. Of course, this doesn't justify the identification conditions — only our understanding of the data can do that.

$$ATT(X, Z) = \mu_1 - \mu_0 + X(\beta_1 - \beta_0) + \left(\rho_{1V_D}\sigma_1 - \rho_{0V_D}\sigma_0\right) \frac{\phi(W\theta)}{\Phi(W\theta)}$$

216   10. Treatment effects: *IV*

by iterated expectations (with full support), we have

$$ATT = \mu_1 - \mu_0 + E[X](\beta_1 - \beta_0) + \left(\rho_{1V_D}\sigma_1 - \rho_{0V_D}\sigma_0\right) E\left[\frac{\phi(W\theta)}{\Phi(W\theta)}\right]$$

Also,

$$ATUT(X,Z) = \mu_1 - \mu_0 + X(\beta_1 - \beta_0) - \left(\rho_{1V_D}\sigma_1 - \rho_{0V_D}\sigma_0\right) \frac{\phi(W\theta)}{1 - \Phi(W\theta)}$$

by iterated expectations, we have

$$ATUT = \mu_1 - \mu_0 + E[X](\beta_1 - \beta_0) - \left(\rho_{1V_D}\sigma_1 - \rho_{0V_D}\sigma_0\right) E\left[\frac{\phi(W\theta)}{1 - \Phi(W\theta)}\right]$$

Since

$$\begin{aligned}ATE(X,Z) &= \Pr(D=1 \mid X,Z) ATT(X,Z) \\ &\quad + \Pr(D=0 \mid X,Z) ATUT(X,Z) \\ &= \Phi(W\theta) ATT(X,Z) + (1 - \Phi(W\theta)) ATUT(X,Z)\end{aligned}$$

we have

$$\begin{aligned}ATE(X,Z) &= \mu_1 - \mu_0 + X(\beta_1 - \beta_0) \\ &\quad + \left(\rho_{1V}\sigma_1 - \rho_{0V_D}\sigma_0\right)\phi(W\theta) - \left(\rho_{1V}\sigma_1 - \rho_{0V_D}\sigma_0\right)\phi(W\theta) \\ &= \mu_1 - \mu_0 + X(\beta_1 - \beta_0)\end{aligned}$$

by iterated expectations (with full common support), we have

$$ATE = \mu_1 - \mu_0 + E[X](\beta_1 - \beta_0)$$

Wooldridge [2002, p. 631] suggests identification of

$$ATE = \mu_1 - \mu_0 + E[X](\beta_1 - \beta_0)$$

via α in the following regression

$$\begin{aligned}E[Y \mid X, Z] &= \mu_0 + \alpha D + X\beta_0 + D(X - E[X])(\beta_1 - \beta_0) \\ &\quad + D\rho_{1V_D}\sigma_1 \frac{\phi(W\theta)}{\Phi(W\theta)} - (1-D)\rho_{0V_D}\sigma_0 \frac{\phi(W\theta)}{1 - \Phi(W\theta)}\end{aligned}$$

This follows from the observable response

$$\begin{aligned}Y &= D(Y_1 \mid D=1) + (1-D)(Y_0 \mid D=0) \\ &= (Y_0 \mid D=0) + D[(Y_1 \mid D=1) - (Y_0 \mid D=0)]\end{aligned}$$

and applying conditional expectations

$$E[Y_1 \mid X, D=1] = \mu_1 + X\beta_1 + \rho_{1V_D}\sigma_1 \frac{\phi(W\theta)}{\Phi(W\theta)}$$

$$E[Y_0 \mid X, D=0] = \mu_0 + X\beta_0 - \rho_{0V_D}\sigma_0 \frac{\phi(W\theta)}{1 - \Phi(W\theta)}$$

Simplification produces Wooldridge's result.

10.5.4 Heterogeneity and estimating ATT by IV

Now we discuss a general approach for estimating *ATT* by *IV* in the face of unobservable heterogeneity.

$$\begin{aligned} ATT(X) &= E\left[Y_1 - Y_0 \mid X, D = 1\right] \\ &= \mu_1 - \mu_0 + E\left[U_1 - U_0 \mid X, D = 1\right] \end{aligned}$$

Identification (data) conditions are

Condition 10.21 $E\left[U_0 \mid X, Z\right] = E\left[U_0 \mid X\right]$,

Condition 10.22 $E\left[U_1 - U_0 \mid X, Z, D = 1\right] = E\left[U_1 - U_0 \mid X, D = 1\right]$, and

Condition 10.23 $\Pr(D = 1 \mid X, Z) \neq \Pr(D = 1 \mid X)$ and $\Pr(D = 1 \mid X, Z) = G(X, Z; \gamma)$ *is a known parametric form (usually probit or logit).*

Let

$$\begin{aligned} Y_j &= \mu_j + U_j \\ &= \mu_j + g_j(X) + V_j \end{aligned}$$

and write

$$\begin{aligned} Y &= \mu_0 + g_0(X) + D\left\{(\mu_1 - \mu_0) + E\left[U_1 - U_0 \mid X, D = 1\right]\right\} \\ &\quad + D\left\{(U_1 - U_0) - E\left[U_1 - U_0 \mid X, D = 1\right]\right\} + V_0 \\ &= \mu_0 + g_0(X) + ATT(X)D + a + V_0 \end{aligned}$$

where $a = D\left\{(U_1 - U_0) - E\left[U_1 - U_0 \mid X, D = 1\right]\right\}$. Let $r = a + V_0$, the data conditions imply $E\left[r \mid X, Z\right] = 0$. Now, suppose $\mu_0(X) = \eta_0 + h(X)\beta_0$ and $ATT(X) = \tau + f(X)\delta$ for some functions $h(X)$ and $f(X)$. Then, we can write

$$Y = \gamma_0 + h(X)\beta_0 + \tau D + Df(X)\delta + r$$

where $\gamma_0 = \mu_0 + \eta_0$. The above equation can be estimated by *IV* using any functions of $\{X, Z\}$ as instruments. Averaging $\tau + f(X)\delta$ over observations with $D = 1$ yields a consistent estimate for *ATT*, $\frac{\sum D_i(\tau_i + f(X_i)\delta)}{\sum D_i}$. By similar reasoning, *ATUT* can be estimated by averaging over the $D = 0$ observations, $\frac{\sum D_i(\tau_i + f(X_i)\delta)}{\sum (1 - D_i)}$.

10.5.5 LATE and linear IV

Concerns regarding lack of robustness (logical inconsistency) of ignorable treatment, or, for instance, the sometimes logical inconsistency of normal probability assignment to unobservable expected utility (say, with Heckman's inverse Mills *IV* control function strategy) have generated interest in alternative *IV* approaches.

One that has received considerable attention is linear *IV* estimation of local average treatment effects (*LATE*; Imbens and Angrist [1994]). We will focus on the binary instrument case to highlight identification issues and aid intuition. First, we provide a brief description then follow with a more extensive treatment. As this is a discrete version of the marginal treatment effect, it helps provide intuition for how instruments, more generally, can help identify treatment effects.

For binary instrument Z,

$$LATE = E\left[Y_1 - Y_0 \mid D_1 - D_0 = 1\right]$$

where $D_1 = (D \mid Z = 1)$ and $D_0 = (D \mid Z = 0)$. That is, *LATE* is the expected gain from treatment of those individuals who switch from no treatment to treatment when the instrument Z changes from 0 to 1. Angrist, Imbens, and Rubin [1996] refer to this subpopulation as the "compliers". This treatment effect is only identified for this subpopulation and because it involves counterfactuals the subpopulation cannot be identified from the data. Nonetheless, the approach has considerable appeal as it is reasonably robust even in the face of unobservable heterogeneity.

Setup

The usual exclusion restriction (existence of instrument) applies. Identification conditions are

Condition 10.24 $\{Y_1, Y_0\}$ *independent of* Z,

Condition 10.25 $D_1 \geq D_0$ *for each individual, and*

Condition 10.26 $\Pr(D = 1 \mid Z = 1) \neq \Pr(D = 1 \mid Z = 0)$.

Conditions 10.24 and 10.26 are usual instrumental variables conditions. Conditional 10.25 is a uniformity condition. For the subpopulation of "compliers" the instrument induces a change to treatment when Z takes a value of 1 but not when $Z = 0$.

Identification

LATE provides a straightforward opportunity to explore *IV* identification of treatment effects. Identification is a thought experiment regarding whether an estimand, the population parameter associated with an estimator, can be uniquely identified from the data. *IV* approaches rely on exclusion restrictions to identify population characteristics of counterfactuals. Because of the counterfactual problem, it is crucial to our *IV* identification thought experiment that we be able to manipulate treatment choice without impacting outcomes. Hence, the exclusion restriction or existence of an instrument (or instruments) is fundamental. Once identification is secured we can focus on matters of estimation (such as consistency and efficiency). Next, we discuss *IV* identification of *LATE*. This is followed

10.5 Heterogeneous response and treatment effects

by discussion of the implication of exclusion restriction failure for treatment effect identification.

For simplicity there are no covariates and two points of support $Z_i = 1$ and $Z_i = 0$ where

$$\Pr(D_i = 1 \mid Z_i = 1) > \Pr(D_i = 1 \mid Z_i = 0)$$

Compare the outcome expectations

$$\begin{aligned}&E[Y_i \mid Z_i = 1] - E[Y_i \mid Z_i = 0] \\ =\ &E[D_i Y_{1i} + (1 - D_i) Y_{0i} \mid Z_i = 1] \\ &- E[D_i Y_{1i} + (1 - D_i) Y_{0i} \mid Z_i = 0]\end{aligned}$$

$\{Y_1, Y_0\}$ independent of Z implies

$$\begin{aligned}&E[Y_i \mid Z_i = 1] - E[Y_i \mid Z_i = 0] \\ =\ &E[D_{1i} Y_{1i} + (1 - D_{1i}) Y_{0i}] - E[D_{0i} Y_{1i} + (1 - D_{0i}) Y_{0i}]\end{aligned}$$

rearranging yields

$$E[(D_{1i} - D_{0i}) Y_{1i} - (D_{1i} - D_{0i}) Y_{0i}]$$

combining terms produces

$$E[(D_{1i} - D_{0i})(Y_{1i} - Y_{0i})]$$

utilizing the sum and product rules of Bayes' theorem gives

$$\begin{aligned}&\Pr(D_{1i} - D_{0i} = 1) E[Y_{1i} - Y_{0i} \mid D_{1i} - D_{0i} = 1] \\ &- \Pr(D_{1i} - D_{0i} = -1) E[Y_{1i} - Y_{0i} \mid D_{1i} - D_{0i} = -1]\end{aligned}$$

How do we interpret this last expression? Even for a strictly positive causal effect of D on Y for all individuals, the average treatment effect is ambiguous as it can be positive, zero, or negative. That is, the treatment effect of those who switch from nonparticipation to participation when Z changes from 0 to 1 can be offset by those who switch from participation to nonparticipation. Therefore, identification of average treatment effects requires additional data conditions. *LATE* invokes uniformity in response to the instrument for all individuals. Uniformity eliminates the second term above as $\Pr(D_{1i} - D_{0i} = -1) = 0$. Then, we can replace $\Pr(D_{1i} - D_{0i} = 1)$ with $E[D_i \mid Z_i = 1] - E[D_i \mid Z_i = 0]$ and

$$\begin{aligned}&\Pr(D_{1i} - D_{0i} = 1) E[Y_{1i} - Y_{0i} \mid D_{1i} - D_{0i} = 1] \\ =\ &(E[D_i \mid Z_i = 1] - E[D_i \mid Z_i = 0]) E[Y_{1i} - Y_{0i} \mid D_{1i} - D_{0i} = 1] \\ =\ &(E[D_i \mid Z_i = 1] - E[D_i \mid Z_i = 0])(E[Y_i \mid Z_i = 1] - E[Y_i \mid Z_i = 0])\end{aligned}$$

From the above we can write

$$\frac{E[Y_i \mid Z_i = 1] - E[Y_i \mid Z_i = 0]}{E[D_i \mid Z_i = 1] - E[D_i \mid Z_i = 0]}$$
$$= \frac{\Pr(D_{1i} - D_{0i} = 1) E[Y_{1i} - Y_{0i} \mid D_{1i} - D_{0i} = 1]}{E[D_i \mid Z_i = 1] - E[D_i \mid Z_i = 0]}$$
$$= \frac{(E[D_i \mid Z_i = 1] - E[D_i \mid Z_i = 0]) E[Y_{1i} - Y_{0i} \mid D_{1i} - D_{0i} = 1]}{E[D_i \mid Z_i = 1] - E[D_i \mid Z_i = 0]}$$
$$= E[Y_{1i} - Y_{0i} \mid D_{1i} - D_{0i} = 1]$$

and since

$$LATE = E[Y_{1i} - Y_{0i} \mid D_{1i} - D_{0i} = 1]$$

we can identify *LATE* by extracting

$$\frac{E[Y_i \mid Z_i = 1] - E[Y_i \mid Z_i = 0]}{E[D_i \mid Z_i = 1] - E[D_i \mid Z_i = 0]}$$

from observables. This is precisely what standard *2SLS-IV* estimates with a binary instrument (developed more fully below).

As *IV* identification of treatment effects differs from standard applications of linear *IV*,[7] this seems an appropriate juncture to explore *IV* identification. The foregoing discussion of *LATE* identification provides an attractive vehicle to illustrate the nuance of identification with an exclusion restriction. Return to the above approach, now suppose condition 10.24 fails, $\{Y_1, Y_0\}$ not independent of Z. Then,

$$E[Y_i \mid Z_i = 1] - E[Y_i \mid Z_i = 0]$$
$$= E[D_{1i}Y_{1i} + (1 - D_{1i})Y_{0i} \mid Z_i = 1]$$
$$\quad - E[D_{0i}Y_{1i} + (1 - D_{0i})Y_{0i} \mid Z_i = 0]$$

but $\{Y_1, Y_0\}$ not independent of Z implies

$$E[Y_i \mid Z_i = 1] - E[Y_i \mid Z_i = 0]$$
$$= E[D_{1i}Y_{1i} + (1 - D_{1i})Y_{0i} \mid Z_i = 1]$$
$$\quad - E[D_{0i}Y_{1i} + (1 - D_{0i})Y_{0i} \mid Z_i = 0]$$
$$= \{E[D_{1i}Y_{1i} \mid Z_i = 1] - E[D_{0i}Y_{1i} \mid Z_i = 0]\}$$
$$\quad - \{E[D_{1i}Y_{0i} \mid Z_i = 1] - E[D_{0i}Y_{0i} \mid Z_i = 0]\}$$
$$\quad + \{E[Y_{0i} \mid Z_i = 1] - E[Y_{0i} \mid Z_i = 0]\}$$

Apparently, the first two terms cannot be rearranged and simplified to identify any treatment effect and the last term does not vanish (recall from above when $\{Y_1, Y_0\}$ independent of Z, this term equals zero). Hence, when the exclusion

---

[7] Heckman and Vytlacil [2005, 2007a, 2007b] emphasize this point.

restriction fails we apparently cannot identify any treatment effects without appealing to other strong conditions.

Sometimes *LATE* can be directly connected to other treatment effects. For example, if $\Pr(D_0 = 1) = 0$, then $LATE = ATT$. Intuitively, the only variation in participation and therefore the only source of overlaps from which to extrapolate from factuals to counterfactuals occurs when $Z_i = 1$. When treatment is accepted, we're dealing with compliers and the group of compliers participate when $Z_i = 1$. Hence, $LATE = ATT$.

Also, if $\Pr(D_1 = 1) = 1$, then $LATE = ATUT$. Similarly, the only variation in participation and therefore the only source of overlaps from which to extrapolate from factuals to counterfactuals occurs when $Z_i = 0$. When treatment is declined, we're dealing with compliers and the group of compliers don't participate when $Z_i = 0$. Hence, $LATE = ATUT$.

Linear *IV* estimation

As indicated above, *LATE* can be estimated via standard *2SLS-IV*. Here, we develop the idea more completely. For Z binary, the estimand for the regression of Y on Z is

$$\frac{E[Y \mid Z = 1] - E[Y \mid Z = 0]}{1 - 0} = E[Y \mid Z = 1] - E[Y \mid Z = 0]$$

and the estimand for the regression of D on Z is

$$\frac{E[D \mid Z = 1] - E[D \mid Z = 0]}{1 - 0} = E[D \mid Z = 1] - E[D \mid Z = 0]$$

Since Z is a scalar the estimand for *IV* estimation is their ratio

$$\frac{E[Y \mid Z = 1] - E[Y \mid Z = 0]}{E[D \mid Z = 1] - E[D \mid Z = 0]}$$

which is the result utilized above to identify *LATE*, the marginal treatment effect for the subpopulation of compliers. Next, we explore some examples illustrating *IV* estimation of *LATE* with a binary instrument.

Tuebingen-style examples

We return to the Tuebingen-style examples introduced in chapter 8 by supplementing them with a binary instrument Z. Likelihood assignment to treatment choice maintains the state-by-state probability structure. Uniformity dictates that we assign zero likelihood that an individual is a defier,[8]

$$p_D \equiv \Pr(s, D_0 = 1, D_1 = 0) = 0.0$$

---

[8]This assumption preserves the identification link between *LATE* and *IV* estimation. Uniformity is a natural consequence of an index-structured propensity score, say $\Pr(D_i \mid W_i) = G\left(W_i^T \gamma\right)$. Case 1b below illustrates how the presence of defiers in the sample confounds *IV* identification of *LATE*.

Then, we assign the likelihoods that an individual is a complier,
$$p_C \equiv \Pr(s, D_0 = 0, D_1 = 1)$$
an individual never selects treatment,
$$p_N \equiv \Pr(s, D_0 = 0, D_1 = 0)$$
and an individual always selects treatment,
$$p_A \equiv \Pr(s, D_0 = 1, D_1 = 1)$$
such that state-by-state
$$p_1 \equiv \Pr(s, D_1 = 1) = p_C + p_A$$
$$p_0 \equiv \Pr(s, D_0 = 1) = p_D + p_A$$
$$q_1 \equiv \Pr(s, D_1 = 0) = p_D + p_N$$
$$q_0 \equiv \Pr(s, D_0 = 0) = p_C + p_N$$

Since $(Y_j \mid D = 1, s) = (Y_j \mid D = 0, s)$ for $j = 0$ or 1, the exclusion restriction is satisfied if
$$\Pr(s \mid Z = 1) = \Pr(s \mid Z = 0)$$
and
$$\Pr(s \mid Z = 1) = p_1 + q_1$$
$$= p_C + p_A + p_D + p_N$$
equals
$$\Pr(s \mid Z = 0) = p_0 + q_0$$
$$= p_D + p_A + p_C + p_N$$

probability assignment for compliance determines the remaining likelihood structure given $\Pr(s, D)$, $\Pr(Z)$, and $p_D = 0$. For instance,
$$\Pr(s, D = 0, Z = 0) = (p_C + p_N) \Pr(Z = 0)$$
and
$$\Pr(s, D = 0, Z = 1) = (p_D + p_N) \Pr(Z = 1)$$
since
$$\Pr(s, D = 0) = (p_C + p_N) \Pr(Z = 0) + (p_D + p_N) \Pr(Z = 1)$$
implies
$$p_N = \Pr(s, D = 0) - p_C \Pr(Z = 0) - p_D \Pr(Z = 1)$$
By similar reasoning,
$$p_A = \Pr(s, D = 1) - p_C \Pr(Z = 1) - p_D \Pr(Z = 0)$$
Now we're prepared to explore some specific examples.

10.5 Heterogeneous response and treatment effects   223

Table 10.1: Tuebingen IV example treatment likelihoods for case 1: ignorable treatment

state (s)	one	two	three
$\Pr(s)$	0.04	0.32	0.64
$\Pr(D=1 \mid s)$	0.32	0.0	0.08
compliers: $\Pr(s, D_0 = 0, D_1 = 1)$	0.0128	0.0	0.0512
never treated: $\Pr(s, D_0 = 0, D_1 = 0)$	0.01824	0.32	0.55296
always treated: $\Pr(s, D_0 = 1, D_1 = 1)$	0.00896	0.0	0.03584
defiers: $\Pr(s, D_0 = 1, D_1 = 0)$	0.0	0.0	0.0
$\Pr(Z=1) = 0.3$			

Table 10.2: Tuebingen IV example outcome likelihoods for case 1: ignorable treatment

state (s)	one		two		three	
$\Pr\begin{pmatrix} Y,D,s, \\ Z=0 \end{pmatrix}$	0.021728	0.006272	0.224	0.0	0.422912	0.025088
$\Pr\begin{pmatrix} Y,D,s, \\ Z=1 \end{pmatrix}$	0.005472	0.006528	0.096	0.0	0.165888	0.026112
D	0	1	0	1	0	1
Y	0	1	1	1	2	1
Y_0	0	0	1	1	2	2
Y_1	1	1	1	1	1	1

Case 1

Given $\Pr(Z=1) = 0.3$, treatment likelihood assignments for case 1 are described in table 10.1. Then, from

$$\begin{aligned}\Pr(s, D=1) &= (p_C + p_A)\Pr(Z=1) + (p_D + p_A)\Pr(Z=0) \\ &= \Pr(D=1, Z=1) + \Pr(D=1, Z=0)\end{aligned}$$

and

$$\begin{aligned}\Pr(s, D=0) &= (p_D + p_N)\Pr(Z=1) + (p_C + p_N)\Pr(Z=0) \\ &= \Pr(D=0, Z=1) + \Pr(D=0, Z=0)\end{aligned}$$

the *DGP* for case 1, ignorable treatment, is identified in table 10.2. Various treatment effects including *LATE* and the *IV*-estimand for case 1 are reported in table 10.3. Case 1 illustrates homogeneous response — all treatment effects, including *LATE*, are the same. Further, endogeneity of treatment is ignorable as Y_1 and Y_0 are conditionally mean independent of D; hence, *OLS* identifies the treatment effects.

Case 1b

Suppose everything remains the same as above except treatment likelihood includes a nonzero defier likelihood as defined in table 10.4. This case highlights

Table 10.3: Tuebingen IV example results for case 1: ignorable treatment

Results	Key components
$LATE = E[Y_1 - Y_0 \mid D_1 - D_0 = 1]$ $= -0.6$	$p = \Pr(D = 1) = 0.064$
$IV - estimand = \frac{E[Y\mid Z=1] - E[Y\mid Z=0]}{E[D\mid Z=1] - E[D\mid Z=0]}$ $= -0.6$	$\Pr(D = 1 \mid Z = 1) = 0.1088$
	$\Pr(D = 1 \mid Z = 0) = 0.0448$
	$E[Y_1 \mid D = 1] = 1.0$
	$E[Y_1 \mid D = 0] = 1.0$
$OLS = \begin{matrix} E[Y_1 \mid D = 1] \\ -E[Y_0 \mid D = 0] \end{matrix} = -0.6$	$E[Y_1] = 1.0$
$ATT = E[Y_1 - Y_0 \mid D = 1] = -0.6$	$E[Y_0 \mid D = 1] = 1.6$
$ATUT = E[Y_1 - Y_0 \mid D = 0] = -0.6$	$E[Y_0 \mid D = 0] = 1.6$
$ATE = E[Y_1 - Y_0] = -0.6$	$E[Y_0] = 1.6$

Table 10.4: Tuebingen IV example treatment likelihoods for case 1b: uniformity fails

state (s)	one	two	three
$\Pr(s)$	0.04	0.32	0.64
$\Pr(D = 1 \mid s)$	0.32	0.0	0.08
compliers: $\Pr(s, D_0 = 0, D_1 = 1)$	0.0064	0.0	0.0256
never treated: $\Pr(s, D_0 = 0, D_1 = 0)$	0.02083	0.32	0.56323
always treated: $\Pr(s, D_0 = 1, D_1 = 1)$	0.00647	0.0	0.02567
defiers: $\Pr(s, D_0 = 1, D_1 = 0)$	0.0063	0.0	0.0255
$\Pr(Z = 1) = 0.3$			

Table 10.5: Tuebingen IV example treatment likelihoods for case 2: heterogeneous response

state (s)	one	two	three
$\Pr(s)$	0.04	0.32	0.64
$\Pr(D = 1 \mid s)$	0.32	0.3	0.08
compliers: $\Pr(D_0 = 0, D_1 = 1)$	0.01	0.096	0.0512
never treated: $\Pr(D_0 = 0, D_1 = 0)$	0.0202	0.1568	0.55296
always treated: $\Pr(D_0 = 1, D_1 = 1)$	0.0098	0.0672	0.03584
defiers: $\Pr(D_0 = 1, D_1 = 0)$	0.0	0.0	0.0
$\Pr(Z = 1) = 0.3$			

the difficulty of identifying treatment effects when uniformity of selection with respect to the instrument fails even though in this ignorable treatment setting all treatment effects are equal. Uniformity failure means some individuals who were untreated when $Z = 0$ opt for treatment when $Z = 1$ but other individuals who were treated when $Z = 0$ opt for no treatment when $Z = 1$.

From the identification discussion, the difference in expected observed outcome when the instrument changes is

$$E[Y_i \mid Z_i = 1] - E[Y_i \mid Z_i = 0]$$
$$= \Pr(D_{1i} - D_{0i} = 1) E[Y_{1i} - Y_{0i} \mid D_{1i} - D_{0i} = 1]$$
$$+ \Pr(D_{1i} - D_{0i} = -1) E[-(Y_{1i} - Y_{0i}) \mid D_{1i} - D_{0i} = -1]$$
$$= 0.032(-0.6) + 0.0318(0.6038) = 0.0$$

The effects
$$E[Y_{1i} - Y_{0i} \mid D_{1i} - D_{0i} = 1] = -0.6$$
and
$$E[-(Y_{1i} - Y_{0i}) \mid D_{1i} - D_{0i} = -1] = 0.6038$$

are offsetting and seemingly hopelessly confounded. *2SLS-IV* estimates

$$\frac{E[Y_i \mid Z_i = 1] - E[Y_i \mid Z_i = 0]}{E[D_i \mid Z_i = 1] - E[D_i \mid Z_i = 0]} = \frac{0.0}{0.0002} = 0.0$$

which differs from $LATE = E[Y_{1i} - Y_{0i} \mid D_i(1) - D_i(0) = 1] = -0.6$. Therefore, we may be unable to identify *LATE*, the marginal treatment effect for compliers, via *2SLS-IV* when defiers are present in the sample.

Case 2

Case 2 perturbs the probabilities resulting in non-ignorable, inherently endogenous treatment and heterogeneous treatment effects. Treatment adoption likelihoods, assuming the likelihood an individual is a defier equals zero and $\Pr(Z = 1) = 0.3$, are assigned in table 10.5. These treatment likelihoods imply the data structure in table 10.6. Various treatment effects including *LATE* and the *IV*-estimand

Table 10.6: Tuebingen IV example outcome likelihoods for case 2: heterogeneous response

state (s)	one		two		three	
$\Pr\begin{pmatrix} Y,D,s, \\ Z=0 \end{pmatrix}$	0.021728	0.006272	0.224	0.0	0.422912	0.025088
$\Pr\begin{pmatrix} Y,D,s, \\ Z=1 \end{pmatrix}$	0.005472	0.006528	0.096	0.0	0.165888	0.026112
D	0	1	0	1	0	1
Y	0	1	1	1	2	1
Y_0	0	0	1	1	2	2
Y_1	1	1	1	1	1	1

Table 10.7: Tuebingen IV example results for case 2: heterogeneous response

Results	Key components
$LATE = E[Y_1 - Y_0 \mid D_1 - D_0 = 1]$ $= -0.2621$	$p = \Pr(D=1) = 0.16$
$IV - estimand = \frac{E[Y\mid Z=1] - E[Y\mid Z=0]}{E[D\mid Z=1] - E[D\mid Z=0]}$ $= -0.2621$	$\Pr(D=1 \mid Z=1) = 0.270$
	$\Pr(D=1 \mid Z=0) = 0.113$
	$E[Y_1 \mid D=1] = 1.0$
	$E[Y_1 \mid D=0] = 1.0$
$OLS = \begin{matrix} E[Y_1 \mid D=1] \\ -E[Y_0 \mid D=0] \end{matrix} = -0.669$	$E[Y_1] = 1.0$
$ATT = E[Y_1 - Y_0 \mid D=1] = -0.24$	$E[Y_0 \mid D=1] = 1.24$
$ATUT = E[Y_1 - Y_0 \mid D=0] = -0.669$	$E[Y_0 \mid D=0] = 1.669$
$ATE = E[Y_1 - Y_0] = -0.6$	$E[Y_0] = 1.6$

10.5 Heterogeneous response and treatment effects

Table 10.8: Tuebingen IV example treatment likelihoods for case 2b: LATE = ATT

state (s)	one	two	three
$\Pr(s)$	0.04	0.32	0.64
$\Pr(D=1 \mid s)$	0.3	0.3	0.08
compliers: $\Pr(s, D_0 = 0, D_1 = 1)$	0.04	0.32	0.17067
never treated: $\Pr(s, D_0 = 0, D_1 = 0)$	0.0	0.0	0.46933
always treated: $\Pr(s, D_0 = 1, D_1 = 1)$	0.0	0.0	0.0
defiers: $\Pr(s, D_0 = 1, D_1 = 0)$	0.0	0.0	0.0
$\Pr(Z=1) = 0.3$			

Table 10.9: Tuebingen IV example outcome likelihoods for case 2b: LATE = ATT

state (s)	one		two		three	
$\Pr(Y, D, s, Z = 0)$	0.028	0.0	0.224	0.0	0.448	0.0
$\Pr(Y, D, s, Z = 1)$	0.0	0.012	0.0	0.096	0.1408	0.0512
D	0	1	0	1	0	1
Y	0	1	1	1	2	1
Y_0	0	0	1	1	2	2
Y_1	1	1	1	1	1	1

for case 2 are reported in table 10.7. In contrast to case 1, for case 2 all treatment effects (*ATE, ATT, ATUT,* and *LATE*) differ which, of course, means *OLS* cannot identify all treatment effects (though it does identify *ATUT* in this setting). Importantly, the *IV*-estimand identifies *LATE* for the subpopulation of compliers.

Case 2b

If we perturb the probability structure such that

$$\Pr(D = 1 \mid Z = 0) = 0$$

then $LATE = ATT$.[9] For $\Pr(Z = 1) = 0.3$, treatment adoption likelihoods are assigned in table 10.8. Then, the data structure is as indicated in table 10.9. Various treatment effects including *LATE* and the *IV*-estimand for case 2b are reported in table 10.10. With this perturbation of likelihoods but maintenance of independence between Z and (Y_1, Y_0), *LATE=ATT* and *LATE* is identified via the *IV*-estimand but is not identified via *OLS*. Notice the evidence on counterfactuals draws from $Z = 1$ as no one adopts treatment when $Z = 0$.

Case 3

Case 3 maintains the probability structure of case 2 but adds some variation to outcomes with treatment Y_1. For $\Pr(Z = 1) = 0.3$, treatment adoption likelihoods

---

[9] We also perturbed $\Pr(D = 1 \mid s = one) = 0.3$ rather than 0.32 to maintain the exclusion restriction and a proper (non-negative) probability distribution.

228   10. Treatment effects: *IV*

Table 10.10: Tuebingen IV example results for case 2b: LATE = ATT

Results	Key components
$LATE = E[Y_1 - Y_0 \mid D_1 - D_0 = 1]$ $= -0.246$	$p = \Pr(D = 1) = 0.1592$
$IV-estimand = \frac{E[Y\mid Z=1]-E[Y\mid Z=0]}{E[D\mid Z=1]-E[D\mid Z=0]}$ $= -0.246$	$\Pr(D = 1 \mid Z = 1) = 0.5307$
	$\Pr(D = 1 \mid Z = 0) = 0.0$ $E[Y_1 \mid D = 1] = 1.0$ $E[Y_1 \mid D = 0] = 1.0$
$OLS = \begin{matrix} E[Y_1 \mid D = 1] \\ -E[Y_0 \mid D = 0] \end{matrix} = -0.667$	$E[Y_1] = 1.0$
$ATT = E[Y_1 - Y_0 \mid D = 1] = -0.246$	$E[Y_0 \mid D = 1] = 1.246$
$ATUT = E[Y_1 - Y_0 \mid D = 0] = -0.667$	$E[Y_0 \mid D = 0] = 1.667$
$ATE = E[Y_1 - Y_0] = -0.6$	$E[Y_0] = 1.6$

Table 10.11: Tuebingen IV example treatment likelihoods for case 3: more heterogeneity

state (s)	one	two	three
$\Pr(s)$	0.04	0.32	0.64
$\Pr(D = 1 \mid s)$	0.32	0.3	0.08
compliers: $\Pr(s, D_0 = 0, D_1 = 1)$	0.01	0.096	0.0512
never treated: $\Pr(s, D_0 = 0, D_1 = 0)$	0.0202	0.1568	0.55296
always treated: $\Pr(s, D_0 = 1, D_1 = 1)$	0.0098	0.0672	0.03584
defiers: $\Pr(s, D_0 = 1, D_1 = 0)$	0.0	0.0	0.0
$\Pr(Z = 1) = 0.3$			

are assigned in table 10.11. Then, the data structure is defined in table 10.12 where Z_0 refers to $Z = 0$ and Z_1 refers to $Z = 1$. Various treatment effects including *LATE* and the *IV*-estimand for case 3 are reported in table 10.13. *OLS* doesn't identify any treatment effect but the *IV*-estimand identifies the discrete marginal treatment effect, *LATE*, for case 3.

Case 3b

Suppose the probability structure of case 3 is perturbed such that

$$\Pr(D = 1 \mid Z = 1) = 1$$

then *LATE=ATUT*.[10] For $\Pr(Z = 1) = 0.3$, treatment adoption likelihoods are assigned in table 10.14. Then, the data structure is as defined in table 10.15. Various treatment effects including *LATE* and the *IV*-estimand for case 3b are reported in table 10.16. The *IV*-estimand identifies *LATE* and $LATE = ATUT$ since treat-

---

[10]We assign $\Pr(D = 1 \mid s = three) = 0.6$ rather than 0.08 to preserve the exclusion restriction.

10.5 Heterogeneous response and treatment effects

Table 10.12: Tuebingen IV example outcome likelihoods for case 3: more heterogeneity

state (s)	one		two		three	
$\Pr \begin{pmatrix} Y \\ D \\ s \\ Z_0 \end{pmatrix}$	0.02114	0.00686	0.17696	0.04704	0.422912	0.025088
$\Pr \begin{pmatrix} Y \\ D \\ s \\ Z_1 \end{pmatrix}$	0.00606	0.00594	0.04704	0.04896	0.165888	0.026112
D	0	1	0	1	0	1
Y	0	1	1	1	2	0
Y_0	0	0	1	1	2	2
Y_1	1	1	1	1	0	0

Table 10.13: Tuebingen IV example results for case 3: more heterogeneity

Results	Key components
$LATE = E[Y_1 - Y_0 \mid D_1 - D_0 = 1]$ $= -0.588$	$p = \Pr(D = 1) = 0.16$
$IV-estimand = \frac{E[Y\mid Z=1] - E[Y\mid Z=0]}{E[D\mid Z=1] - E[D\mid Z=0]}$ $= -0.588$	$\Pr(D = 1 \mid Z = 1) = 0.270$
	$\Pr(D = 1 \mid Z = 0) = 0.113$
	$E[Y_1 \mid D = 1] = 0.68$
	$E[Y_1 \mid D = 0] = 0.299$
$OLS = \frac{E[Y_1 \mid D = 1]}{-E[Y_0 \mid D = 0]} = -0.989$	$E[Y_1] = 0.36$
$ATT = E[Y_1 - Y_0 \mid D = 1] = -0.56$	$E[Y_0 \mid D = 1] = 1.24$
$ATUT = E[Y_1 - Y_0 \mid D = 0] = -1.369$	$E[Y_0 \mid D = 0] = 1.669$
$ATE = E[Y_1 - Y_0] = -1.24$	$E[Y_0] = 1.6$

ment is always selected when $Z = 1$. Also, notice *OLS* is close to *ATE* even though this is a case of inherent endogeneity. This suggests comparing *ATE* with *OLS* provide an inadequate test for the existence of endogeneity.

Case 4

Case 4 employs a richer set of outcomes but the probability structure for (D, Y, s) employed in case 1 and yields the Simpson's paradox result noted in chapter 8. For $\Pr(Z = 1) = 0.3$, assignment of treatment adoption likelihoods are described in table 10.17. Then, the data structure is identified in table 10.18. Various treatment effects including *LATE* and the *IV*-estimand for case 4 are reported in table 10.19. *OLS* estimates a negative effect while all the standard average treatment effects are positive. Identification conditions are satisfied and the *IV*-estimand identifies *LATE*.

Table 10.14: Tuebingen IV example treatment likelihoods for case 3b: LATE = ATUT

state (s)	one	two	three
$\Pr(s)$	0.04	0.32	0.64
$\Pr(D = 1 \mid s)$	0.32	0.3	0.6
compliers: $\Pr(s, D_0 = 0, D_1 = 1)$	0.038857	0.32	0.365714
never treated: $\Pr(s, D_0 = 0, D_1 = 0)$	0.0	0.0	0.0
always treated: $\Pr(s, D_0 = 1, D_1 = 1)$	0.001143	0.0	0.274286
defiers: $\Pr(s, D_0 = 1, D_1 = 0)$	0.0	0.0	0.0
$\Pr(Z = 1) = 0.3$			

Table 10.15: Tuebingen IV example outcome likelihoods for case 3b: LATE = ATUT

state (s)	one		two		three	
$\Pr(Y, D, s, Z = 0)$	0.0272	0.0008	0.224	0.0	0.256	0.192
$\Pr(Y, D, s, Z = 1)$	0.0	0.0012	0.0	0.096	0.0	0.192
D	0	1	0	1	0	1
Y	0	1	1	1	2	0
Y_0	0	0	1	1	2	2
Y_1	1	1	1	1	0	0

Case 4b

For $Z = D$ and $\Pr(Z = 1) = \Pr(D = 1) = 0.16$, case 4b explores violation of the exclusion restriction. Assignment of treatment adoption likelihoods are described in table 10.20. However, as indicated earlier the exclusion restriction apparently can only be violated in this binary instrument setting if treatment alters the outcome distributions. To explore the implications of this variation, we perturb outcomes with treatment slightly as defined in table 10.21. Various treatment effects including *LATE* and the *IV*-estimand for case 4b are reported in table 10.22. Since the exclusion restriction is not satisfied the *IV*-estimand fails to identify *LATE*. In fact, *OLS* and *2SLS-IV* estimates are both negative while *ATE* and *LATE* are positive. As $Z = D$, the entire population consists of compliers, and it is difficult to assess the counterfactuals as there is no variation in treatment when either $Z = 0$ or $Z = 1$. Hence, it is critical to treatment effect identification that treatment not induce a shift in the outcome distributions but rather variation in the instruments produces a change in treatment status only.

Case 5

Case 5 involves $\Pr(z = 1) = 0.3$, and non-overlapping support:

$$\Pr(s = one, D = 0) = 0.04$$

$$\Pr(s = two, D = 1) = 0.32$$

10.5 Heterogeneous response and treatment effects   231

Table 10.16: Tuebingen IV example results for case 3b: LATE = ATUT

Results	Key components
$LATE = E\left[Y_1 - Y_0 \mid D_1 - D_0\right] = 1]$ $= -0.9558$	$p = \Pr(D = 1) = 0.4928$
$IV - estimand = \frac{E[Y\mid Z=1] - E[Y\mid Z=0]}{E[D\mid Z=1] - E[D\mid Z=0]}$ $= -0.9558$	$\Pr(D = 1 \mid Z = 1) = 1.0$
	$\Pr(D = 1 \mid Z = 0) = 0.2754$ $E\left[Y_1 \mid D = 1\right] = 0.2208$ $E\left[Y_1 \mid D = 0\right] = 0.4953$
$OLS = \begin{array}{c} E\left[Y_1 \mid D = 1\right] \\ -E\left[Y_0 \mid D = 0\right] \end{array} = -1.230$	$E\left[Y_1\right] = 0.36$
$ATT = E\left[Y_1 - Y_0 \mid D = 1\right] = -1.5325$	$E\left[Y_0 \mid D = 1\right] = 1.7532$
$ATUT = E\left[Y_1 - Y_0 \mid D = 0\right] = -0.9558$	$E\left[Y_0 \mid D = 0\right] = 1.4511$
$ATE = E\left[Y_1 - Y_0\right] = -1.24$	$E\left[Y_0\right] = 1.6$

Table 10.17: Tuebingen IV example treatment likelihoods for case 4: Simpson's paradox

state (s)	one	two	three
$\Pr(s)$	0.04	0.32	0.64
$\Pr(D = 1 \mid s)$	0.32	0.3	0.08
compliers: $\Pr(s, D_0 = 0, D_1 = 1)$	0.01	0.096	0.0512
never treated: $\Pr(s, D_0 = 0, D_1 = 0)$	0.0202	0.1568	0.55296
always treated: $\Pr(s, D_0 = 1, D_1 = 1)$	0.0098	0.0672	0.03584
defiers: $\Pr(s, D_0 = 1, D_1 = 0)$	0.0	0.0	0.0
$\Pr(Z = 1) = 0.3$			

and

$$\Pr(s = three, D = 0) = 0.64$$

as assigned in table 10.23.

There is no positive complier likelihood for this setting. The intuition for this is as follows. Compliers elect no treatment when the instrument takes a value of zero but select treatment when the instrument is unity. With the above likelihood structure there is no possibility for compliance as each state is singularly treatment or no treatment irrespective of the instrument as described in table 10.24.

Various treatment effects including *LATE* and the *IV*-estimand for case 5 are reported in table 10.25. Case 5 illustrates the danger of lack of common support. Common support concerns extend to other standard ignorable treatment and *IV* identification approaches beyond *LATE*. Case 5b perturbs the likelihoods slightly to recover *IV* identification of *LATE*.

Table 10.18: Tuebingen IV example outcome likelihoods for case 4: Simpson's paradox

state (s)	one		two		three	
$\Pr \begin{pmatrix} Y \\ D \\ s \\ Z_0 \end{pmatrix}$	0.02114	0.00686	0.17696	0.04704	0.422912	0.025088
$\Pr \begin{pmatrix} Y \\ D \\ s \\ Z_1 \end{pmatrix}$	0.00606	0.00594	0.04704	0.04896	0.165888	0.026112
D	0	1	0	1	0	1
Y	0.0	1.0	1.0	1.0	2.0	2.3
Y_0	0.0	0.0	1.0	1.0	2.0	2.0
Y_1	1.0	1.0	1.0	1.0	2.3	2.3

Table 10.19: Tuebingen IV example results for case 4: Simpson's paradox

Results	Key components
$LATE = E[Y_1 - Y_0 \mid D_1 - D_0 = 1]$ $= 0.161$	$p = \Pr(D = 1) = 0.16$
$IV - estimand = \frac{E[Y\mid Z=1] - E[Y\mid Z=0]}{E[D\mid Z=1] - E[D\mid Z=0]}$ $= 0.161$	$\Pr(D = 1 \mid Z = 1) = 0.27004$
	$\Pr(D = 1 \mid Z = 0) = 0.11284$
	$E[Y_1 \mid D = 1] = 1.416$
	$E[Y_1 \mid D = 0] = 1.911$
$OLS = \begin{array}{l} E[Y_1 \mid D = 1] \\ -E[Y_0 \mid D = 0] \end{array} = -0.253$	$E[Y_1] = 1.832$
$ATT = E[Y_1 - Y_0 \mid D = 1] = 0.176$	$E[Y_0 \mid D = 1] = 1.24$
$ATUT = E[Y_1 - Y_0 \mid D = 0] = 0.243$	$E[Y_0 \mid D = 0] = 1.669$
$ATE = E[Y_1 - Y_0] = 0.232$	$E[Y_0] = 1.6$

Case 5b

Case 5b perturbs the probabilities slightly such that

$$\Pr(s = two, D = 1) = 0.3104$$

and

$$\Pr(s = two, D = 0) = 0.0096$$

as depicted in table 10.26; everything else remains as in case 5. This slight perturbation accommodates treatment adoption likelihood assignments as defined in table 10.27. Various treatment effects including *LATE* and the *IV*-estimand for case 5b are reported in table 10.28. Even though there is a very small subpopulation of compliers, *IV* identifies *LATE*. The common support issue was discussed in

10.5 Heterogeneous response and treatment effects

Table 10.20: Tuebingen IV example treatment likelihoods for case 4b: exclusion restriction violated

state (s)	one	two	three
$\Pr(s)$	0.04	0.32	0.64
$\Pr(D=1 \mid s)$	0.32	0.3	0.08
compliers: $\Pr(s, D_0=0, D_1=1)$	0.04	0.32	0.64
never treated: $\Pr(s, D_0=0, D_1=0)$	0.0	0.0	0.0
always treated: $\Pr(s, D_0=1, D_1=1)$	0.0	0.0	0.0
defiers: $\Pr(s, D_0=1, D_1=0)$	0.0	0.0	0.0
$\Pr(Z=1) = 0.16$			

Table 10.21: Tuebingen IV example outcome likelihoods for case 4b: exclusion restriction violated

state (s)	one		two		three	
$\Pr\begin{pmatrix} Y,D,s, \\ Z=0 \end{pmatrix}$	0.0336	0.0	0.2688	0.0	0.5376	0.0
$\Pr\begin{pmatrix} Y,D,s, \\ Z=1 \end{pmatrix}$	0.0	0.0064	0.0	0.0512	0.0	0.1024
D	0	1	0	1	0	1
Y	0.0	3.0	1.0	1.0	2.0	1.6
Y_0	0.0	0.0	1.0	1.0	2.0	2.0
Y_1	1.0	1.0	1.0	1.0	2.3	1.6

the context of the asset revaluation regulation example in chapter 9 and comes up again in the discussion of regulated report precision example later in this chapter.

Discussion of *LATE*

Linear *IV* estimation of *LATE* has considerable appeal. Given the existence of instruments, it is simple to implement (*2SLS-IV*) and robust; it doesn't rely on strong distributional conditions and can accommodate unobservable heterogeneity. However, it also has drawbacks. We cannot identify the subpopulation of compliers due to unobservable counterfactuals. If the instruments change, it's likely that the treatment effect (*LATE*) and the subpopulation of compliers will change. This implies that different analysts are likely to identify different treatment effects — an issue of concern to Heckman and Vytlacil [2005]. Continuous or multi-level discrete instruments and/or regressors produce a complicated weighted average of marginal treatment effects that are again dependent on the particular instrument chosen as discussed in the next chapter. Finally, the treatment effect literature is asymmetric. Outcome heterogeneity can be accommodated but uniformity (or homogeneity) of treatment is fundamental. This latter limitation applies to all *IV* approaches including local *IV* (*LIV*) estimation of *MTE* which is discussed in chapter 11.

Table 10.22: Tuebingen IV example results for case 4b: exclusion restriction violated

Results	Key components
$LATE = E\left[Y_1 - Y_0 \mid D_1 - D_0 = 1\right]$ $= 0.160$	$p = \Pr(D = 1) = 0.16$
$IV-estimand = \frac{E[Y\mid Z=1] - E[Y\mid Z=0]}{E[D\mid Z=1] - E[D\mid Z=0]}$ $= -0.216$	$\Pr(D = 1 \mid Z = 1) = 1.0$
	$\Pr(D = 1 \mid Z = 0) = 0.0$
	$E\left[Y_1 \mid D = 1\right] = 1.192$
	$E\left[Y_1 \mid D = 0\right] = 1.911$
$OLS = \begin{array}{c} E\left[Y_1 \mid D = 1\right] \\ -E\left[Y_0 \mid D = 0\right] \end{array} = -0.477$	$E\left[Y_1\right] = 1.796$
$ATT = E\left[Y_1 - Y_0 \mid D = 1\right] = -0.048$	$E\left[Y_0 \mid D = 1\right] = 1.24$
$ATUT = E\left[Y_1 - Y_0 \mid D = 0\right] = 0.243$	$E\left[Y_0 \mid D = 0\right] = 1.669$
$ATE = E\left[Y_1 - Y_0\right] = 0.196$	$E\left[Y_0\right] = 1.6$

Table 10.23: Tuebingen IV example outcome likelihoods for case 5: lack of common support

state (s)	one		two		three	
$\Pr(Y, D, s, Z = 0)$	0.028	0.0	0.0	0.224	0.448	0.0
$\Pr(Y, D, s, Z = 1)$	0.012	0.0	0.0	0.096	0.192	0.0
D	0	1	0	1	0	1
Y	0	1	1	2	2	0
Y_0	0	0	1	1	2	2
Y_1	1	1	2	2	0	0

Censored regression and *LATE*

Angrist [2001] discusses identification of *LATE* in the context of censored regression.[11] He proposes a non-negative transformation $\exp(X\beta)$ combined with linear *IV* to identify a treatment effect. Like the discussion of *LATE* above, the approach is simplest and most easily interpreted when the instrument is binary and there are no covariates. Angrist extends the discussion to cover quantile treatment effects based on censored quantile regression combined with Abadie's [2000] causal *IV*.

---

[11] This is not to be confused with sample selection. Here, we refer to cases in which the observed outcome follows a switching regression that permits identification of counterfactuals.

10.5 Heterogeneous response and treatment effects

Table 10.24: Tuebingen IV example treatment likelihoods for case 5: lack of common support

state	one	two	three
compliers: $\Pr(D_0 = 0, D_1 = 1)$	0.0	0.0	0.0
never treated: $\Pr(D_0 = 0, D_1 = 0)$	0.04	0.0	0.64
always treated: $\Pr(D_0 = 1, D_1 = 1)$	0.0	0.32	0.0
defiers: $\Pr(D_0 = 1, D_1 = 0)$	0.0	0.0	0.0
$\Pr(Z = 1) = 0.3$			

Table 10.25: Tuebingen IV example results for case 5: lack of common support

Results	Key components
$LATE = E[Y_1 - Y_0 \mid D_1 - D_0 = 1]$ $= NA$	$p = \Pr(D = 1) = 0.32$
$IV-estimand = \frac{E[Y\mid Z=1] - E[Y\mid Z=0]}{E[D\mid Z=1] - E[D\mid Z=0]}$ $= \frac{0}{0}$	$\Pr(D = 1 \mid Z = 1) = 0.32$
	$\Pr(D = 1 \mid Z = 0) = 0.32$
	$E[Y_1 \mid D = 1] = 2.0$
	$E[Y_1 \mid D = 0] = 0.0588$
$OLS = \begin{array}{c} E[Y_1 \mid D = 1] \\ -E[Y_0 \mid D = 0] \end{array} = 0.118$	$E[Y_1] = 0.68$
$ATT = E[Y_1 - Y_0 \mid D = 1] = 1.0$	$E[Y_0 \mid D = 1] = 1.0$
$ATUT = E[Y_1 - Y_0 \mid D = 0] = -1.824$	$E[Y_0 \mid D = 0] = 1.882$
$ATE = E[Y_1 - Y_0] = -0.92$	$E[Y_0] = 1.6$

For (Y_{0i}, Y_{1i}) independent of $(D_i \mid X_i, D_{1i} > D_{0i})$ Abadie defines the causal IV effect, *LATE*.

$$\begin{aligned} LATE &= E[Y_i \mid X_i, D_i = 1, D_{1i} > D_{0i}] \\ &\quad - E[Y_i \mid X_i, D_i = 0, D_{1i} > D_{0i}] \\ &= E[Y_{1i} - Y_{0i} \mid X_i, D_{1i} > D_{0i}] \end{aligned}$$

Then, for binary instrument Z, Abadie shows

$$E\left[\left(E[Y_i \mid X_i, D_i, D_{1i} > D_{0i}] - X_i^T b - aD_i\right)^2 \mid D_{1i} > D_{0i}\right]$$
$$= \frac{E\left[\kappa_i \left(E[Y_i \mid X_i, D_i, D_{1i} > D_{0i}] - X_i^T b - aD_i\right)^2\right]}{\Pr(D_{1i} > D_{0i})}$$

where

$$\kappa_i = 1 - \frac{D_i(1 - Z_i)}{\Pr(Z_i = 0 \mid X_i)} - \frac{(1 - D_i)Z_i}{\Pr(Z_i = 1 \mid X_i)}$$

Table 10.26: Tuebingen IV example outcome likelihoods for case 5b: minimal common support

state (s)	one		two		three	
$\Pr(Y, D, s, Z=0)$	0.028	0.0	0.0082	0.21518	0.448	0.0
$\Pr(Y, D, s, Z=1)$	0.012	0.0	0.00078	0.09522	0.192	0.0
D	0	1	0	1	0	1
Y	0	1	1	2	2	0
Y_0	0	0	1	1	2	2
Y_1	1	1	2	2	0	0

Table 10.27: Tuebingen IV example outcome likelihoods for case 5b: minimal common support

state	one	two	three
compliers: $\Pr(D_0 = 0, D_1 = 1)$	0.0	0.01	0.0
never treated: $\Pr(D_0 = 0, D_1 = 0)$	0.04	0.0026	0.64
always treated: $\Pr(D_0 = 1, D_1 = 1)$	0.0	0.3074	0.0
defiers: $\Pr(D_0 = 1, D_1 = 0)$	0.0	0.0	0.0
$\Pr(Z=1) = 0.30$			

Since κ_i can be estimated from the observable data, one can employ minimum "weighted" least squares to estimate a and b. That is,

$$\min_{a,b} E\left[\kappa_i \left(Y_i - X_i^T b - aD_i\right)^2\right]$$

Notice for compliers $Z_i = D_i$ (for noncompliers, $Z_i \neq D_i$) and κ_i always equals one for compliers and is unequal to one (in fact, negative) for noncompliers. Intuitively, Abadie's causal IV estimator weights the data such that the residuals are small for compliers but large (in absolute value) for noncompliers. The coefficient on D, a, is the treatment effect. We leave remaining details for the interested reader to explore. In chapter 11, we discuss a unified strategy, proposed by Heckman and Vytlacil [2005, 2007a, 2007b] and Heckman and Abbring [2007], built around marginal treatment effects for addressing means as well as distributions of treatment effects.

10.6 Continuous treatment

Suppressing covariates, the average treatment effect for continuous treatment can be defined as

$$ATE = E\left[\frac{\partial}{\partial d} Y\right]$$

Table 10.28: Tuebingen IV example results for case 5b: minimal common support

Results	Key components
$LATE = E\left[Y_1 - Y_0 \mid D_1 - D_0 = 1\right]$ $= 1.0$	$p = \Pr\left(D = 1\right) = 0.3104$
$IV - estimand = \frac{E[Y\mid Z=1] - E[Y\mid Z=0]}{E[D\mid Z=1] - E[D\mid Z=0]}$ $= 1.0$	$\Pr\left(D = 1 \mid Z = 1\right) = 0.3174$
	$\Pr\left(D = 1 \mid Z = 0\right) = 0.3074$
	$E\left[Y_1 \mid D = 1\right] = 2.0$
	$E\left[Y_1 \mid D = 0\right] = 0.086$
$OLS = \frac{E\left[Y_1 \mid D = 1\right]}{-E\left[Y_0 \mid D = 0\right]} = 0.13$	$E\left[Y_1\right] = 0.68$
$ATT = E\left[Y_1 - Y_0 \mid D = 1\right] = 1.0$	$E\left[Y_0 \mid D = 1\right] = 1.0$
$ATUT = E\left[Y_1 - Y_0 \mid D = 0\right] = -1.784$	$E\left[Y_0 \mid D = 0\right] = 1.870$
$ATE = E\left[Y_1 - Y_0\right] = -0.92$	$E\left[Y_0\right] = 1.6$

Often, the more economically-meaningful effect, the average treatment effect on treated for continuous treatment is

$$ATT = E\left[\frac{\partial}{\partial d}Y \mid D = d\right]$$

Wooldridge [1997, 2003] provides conditions for identifying continuous treatment effects via *2SLS-IV*. This is a classic correlated random coefficients setting (see chapter 3) also pursued by Heckman [1997] and Heckman and Vytlacil [1998] (denoted HV in this subsection). As the parameters or coefficients are random, the model accommodates individual heterogeneity. Further, correlation between the treatment variable and the treatment effect parameter accommodates unobservable heterogeneity.

Let y be the outcome variable and $\mathbf{D}$ be a vector of G treatment variables.[12] The structural model[13] written in expectation form is

$$E\left[y \mid a, b, D\right] = a + \mathbf{b}\mathbf{D}$$

or in error form, the model is

$$y = a + \mathbf{b}\mathbf{D} + e$$

where $E\left[e \mid a, b, \mathbf{D}\right] = 0$. It's instructive to rewrite the model in error form for random draw i

$$y_i = a_i + \mathbf{D}_i \mathbf{b}_i + e_i$$

The model suggests that the intercept, a_i, and slopes, b_{ij}, $j = 1, \ldots, G$, can be individual-specific and depend on observed covariates or unobserved heterogeneity. Typically, we focus on the average treatment effect, $\boldsymbol{\beta} \equiv E\left[\mathbf{b}\right] = E\left[\mathbf{b}_i\right]$, as $\mathbf{b}$

---

[12] For simplicity as well as clarity, we'll stick with Wooldridge's [2003] setting and notation.
[13] The model is structural in the sense that the partial effects of D_j on the mean response are identified after controlling for the factor determining the intercept and slope parameters.

is likely a function of unobserved heterogeneity and we cannot identify the vector of slopes, $\mathbf{b}_i$, for any individual i.

Suppose we have K covariates $\mathbf{x}$ and L instrumental variables $\mathbf{z}$. As is common with *IV* strategies, identification utilizes an exclusion restriction. Specifically, the identification conditions are

Condition 10.27 *The covariates $\mathbf{x}$ and instruments $\mathbf{z}$ are redundant for the outcome y.*
$$E[y \mid a, \mathbf{b}, \mathbf{D}, \mathbf{x}, \mathbf{z}] = E[y \mid a, \mathbf{b}, \mathbf{D}]$$

Condition 10.28 *The instruments $\mathbf{z}$ are redundant for a and $\mathbf{b}$ conditional on $\mathbf{x}$.*
$$\begin{aligned} E[a \mid \mathbf{x}, \mathbf{z}] &= E[a \mid \mathbf{x}] = \gamma_0 + \mathbf{x}\boldsymbol{\gamma} \\ E[b_j \mid \mathbf{x}, \mathbf{z}] &= E[b_j \mid \mathbf{x}] = \beta_{0j} + (\mathbf{x} - E[\mathbf{x}])\boldsymbol{\delta}_j, j = 1, \ldots, G \end{aligned}$$

Let the error form of a and b be
$$\begin{aligned} a &= \gamma_0 + \mathbf{x}\boldsymbol{\gamma} + c, & E[c \mid \mathbf{x}, \mathbf{z}] &= 0 \\ b_j &= \beta_{0j} + (\mathbf{x} - E[\mathbf{x}])\boldsymbol{\delta}_j + v_j, & E[v_j \mid \mathbf{x}, \mathbf{z}] &= 0, \quad j = 1, \ldots, G \end{aligned}$$

When plugged into the outcome equation this yields
$$y = \gamma_0 + \mathbf{x}\boldsymbol{\gamma} + \mathbf{D}\boldsymbol{\beta}_0 + D_1(\mathbf{x} - E[\mathbf{x}])\boldsymbol{\delta}_1 + \ldots + D_G(\mathbf{x} - E[\mathbf{x}])\boldsymbol{\delta}_G + c + \mathbf{D}\mathbf{v} + e$$

where $\mathbf{v} = (v_1, \ldots, v_G)^T$. The composite error $\mathbf{D}\mathbf{v}$ is problematic as, generally, $E[\mathbf{D}\mathbf{v} \mid x, z] \neq 0$ but as discussed by Wooldridge [1997] and HV [1998], it is possible that the conditional covariances do not depend on $(\mathbf{x}, \mathbf{z})$. This is the third identification condition.

Condition 10.29 *The conditional covariances between $\mathbf{D}$ and $\mathbf{v}$ do not depend on $(\mathbf{x}, \mathbf{z})$.*
$$E[D_j v_j \mid \mathbf{x}, \mathbf{z}] = \alpha_j \equiv Cov(D_j, v_j) = E[D_j v_j], \quad j = 1, \ldots, G$$

Let $\alpha_0 = \alpha_1 + \cdots + \alpha_G$ and $r = \mathbf{D}\mathbf{v} - E[\mathbf{D}\mathbf{v} \mid x, z]$ and write the outcome equation as
$$y = (\gamma_0 + \alpha_0) + x\boldsymbol{\gamma} + \mathbf{D}\boldsymbol{\beta}_0 + D_1(x - E[x])\boldsymbol{\delta}_1 + \ldots + D_G(x - E[x])\boldsymbol{\delta}_G + c + r + e$$

Since the composite error $u \equiv c + r + e$ has zero mean conditional on (x, z), we can use any function of (x, z) as instruments in the outcome equation
$$y = \theta_0 + \mathbf{x}\boldsymbol{\gamma} + \mathbf{D}\boldsymbol{\beta}_0 + D_1(\mathbf{x} - E[\mathbf{x}])\boldsymbol{\delta}_1 + \ldots + D_G(\mathbf{x} - E[\mathbf{x}])\boldsymbol{\delta}_G + u$$

Wooldridge [2003, p. 189] argues *2SLS-IV* is more robust than HV's plug-in estimator and the standard errors are simpler to obtain. Next, we revisit the third accounting setting from chapter 2, regulated report precision, and explore various treatment effect strategies within this richer accounting context.

10.7 Regulated report precision

Now, consider the report precision example introduced in chapter 2. Recall regulators set a target report precision as regulation increases report precision and improves the owner's welfare relative to private precision choice. However, regulation also invites transaction design (commonly referred to as earnings management) which produces deviations from regulatory targets. The owner's expected utility including the cost of transaction design, $\alpha_d \left(\hat{b} - \sigma_2^2\right)^2$, is

$$EU(\sigma_2) = \mu - \beta \frac{\sigma_1^2 \bar{\sigma}_2^2}{\sigma_1^2 + \bar{\sigma}_2^2} - \gamma \frac{\sigma_1^4 \left(\sigma_1^2 + \sigma_2^2\right)}{\left(\sigma_1^2 + \bar{\sigma}_2^2\right)^2} - \alpha \left(b - \sigma_2^2\right)^2 - \alpha_d \left(\hat{b} - \sigma_2^2\right)^2$$

Outcomes Y are reflected in exchange values or prices and accordingly reflect only a portion of the owner's expected utility.

$$Y = P(\bar{\sigma}_2) = \mu + \frac{\sigma_1^2}{\sigma_1^2 + \bar{\sigma}_2^2}(s - \mu) - \beta \frac{\sigma_1^2 \bar{\sigma}_2^2}{\sigma_1^2 + \bar{\sigma}_2^2}$$

In particular, cost may be hidden from the analysts' view; cost includes the explicit cost of report precision, $\alpha \left(b - \sigma_2^2\right)^2$, cost of any transaction design, $\alpha_d \left(\hat{b} - \sigma_2^2\right)^2$, and the owner's risk premia, $\gamma \frac{\sigma_1^4 \left(\sigma_1^2 + \sigma_2^2\right)}{\left(\sigma_1^2 + \bar{\sigma}_2^2\right)^2}$. Further, outcomes (prices) reflect realized draws from the accounting system, s, whereas the owner's expected utility is based on anticipated reports and her knowledge of the distribution for (s, EU). The causal effect of treatment (report precision choice) on outcomes is the subject under study and is almost surely endogenous. Our analysis entertains variations of treatment data including binary choice that is observed (by the analyst) binary, a continuum of choices that is observed binary, and continuous treatment that is observed from a continuum of choices.

10.7.1 Binary report precision choice

Suppose there are two types of owners, those with low report precision cost parameter α_d^L, and those with high report precision cost parameter α_d^H. An owner chooses report precision based on maximizing her expected utility, a portion of which is unobservable (to the analyst). For simplicity, we initially assume report precision is binary and observable to the analyst.

Base case

Focus attention on the treatment effect of report precision. To facilitate this exercise, we simulate data by drawing 200 samples of 2,000 observations for normally distributed reports with mean μ and variance $\sigma_1^2 + \sigma_2^2$. Parameter values are tabulated below

Base case parameter values
$\mu = 1,000$
$\sigma_1^2 = 100$
$\beta^L = \beta^H = \beta = 7$
$b = 150$
$\widehat{b} = 128.4$
$\gamma = 2.5$
$\alpha = 0.02$
$\alpha_d^L \sim N\left(0.02, 0.005^2\right)$
$\alpha_d^H \sim N\left(0.04, 0.01^2\right)$

The random α_d^j draws are not observed by firm owners until after their report precision choices are made.[14] On the other hand, the analyst observes α_d^j draws ex post but their mean is unknown.[15] The owner chooses inverse report precision (report variance) $\left\{\left(\sigma_2^L\right)^2 = 133.5, \left(\sigma_2^H\right)^2 = 131.7\right\}$ to maximize her expected utility given her type, $E\left[\alpha_d^L\right]$, or $E\left[\alpha_d^H\right]$.

The report variance choices described above are the Nash equilibrium strategies for the owner and investors. That is, for α_d^L, investors' conjecture $\left(\bar{\sigma}_2^L\right)^2 = 133.5$ and the owner's best response is $\left(\sigma_2^L\right)^2 = 133.5$. While for α_d^H, investors' conjecture $\left(\bar{\sigma}_2^H\right)^2 = 131.7$ and the owner's best response is $\left(\sigma_2^H\right)^2 = 131.7$. Hence, the owner's expected utility associated with low variance reports given α_d^L is $(EU_1 \mid D = 1) = 486.8$ while the owner's expected utility associated with high variance reports given α_d^L is lower, $(EU_0 \mid D = 1) = 486.6$. Also, the owner's expected utility associated with high variance reports given α_d^H is $(EU_0 \mid D = 0) = 487.1$ while the owner's expected utility associated with low variance reports given α_d^H is lower, $(EU_1 \mid D = 0) = 486.9$.

Even though treatment choice is driven by cost of transaction design, α_d, observable outcomes are traded values, P, and don't reflect cost of transaction design. To wit, the observed treatment effect on the treated is

$$\begin{aligned} TT &= \left(P^L \mid D = 1\right) - \left(P^H \mid D = 1\right) = (Y_1 \mid D = 1) - (Y_0 \mid D = 1) \\ &= \left(\mu + \frac{\sigma_1^2}{\sigma_1^2 + \left(\bar{\sigma}_2^L\right)^2}\left(s^L - \mu\right) - \beta \frac{\sigma_1^2 \left(\bar{\sigma}_2^L\right)^2}{\sigma_1^2 + \left(\bar{\sigma}_2^L\right)^2}\right) \\ &\quad - \left(\mu + \frac{\sigma_1^2}{\sigma_1^2 + \left(\bar{\sigma}_2^L\right)^2}\left(s^H - \mu\right) - \beta \frac{\sigma_1^2 \left(\bar{\sigma}_2^L\right)^2}{\sigma_1^2 + \left(\bar{\sigma}_2^L\right)^2}\right) \end{aligned}$$

Since $E\left[s^L - \mu\right] = E\left[s^H - \mu\right] = 0$,

$$E\left[TT\right] = ATT = 0$$

---

[14] For the simulation, type is drawn from a Bernoulli distribution with probability 0.5.
[15] Consequently, even if other parameters are observed by the analyst, there is uncertainty associated with selection due to α_d^j.

10.7 Regulated report precision

Also, the observed treatment effect on the untreated is

$$TUT = \left(P^L \mid D=0\right) - \left(P^H \mid D=0\right) = (Y_1 \mid D=0) - (Y_0 \mid D=0)$$

$$= \left[\mu + \frac{\sigma_1^2}{\sigma_1^2 + \left(\bar{\sigma}_2^H\right)^2}\left(s^L - \mu\right) - \beta\frac{\sigma_1^2\left(\bar{\sigma}_2^H\right)^2}{\sigma_1^2 + \left(\bar{\sigma}_2^H\right)^2}\right]$$

$$- \left[\mu + \frac{\sigma_1^2}{\sigma_1^2 + \left(\bar{\sigma}_2^H\right)^2}\left(s^H - \mu\right) - \beta\frac{\sigma_1^2\left(\bar{\sigma}_2^H\right)^2}{\sigma_1^2 + \left(\bar{\sigma}_2^H\right)^2}\right]$$

and

$$E\left[TUT\right] = ATUT = 0$$

Therefore, the average treatment effect is

$$ATE = 0$$

However, the OLS estimand is

$$\begin{aligned}OLS &= E\left[\left(P^L \mid D=1\right) - \left(P^H \mid D=0\right)\right]\\ &= E\left[(Y_1 \mid D=1) - (Y_0 \mid D=0)\right]\\ &= \left[\mu + \frac{\sigma_1^2}{\sigma_1^2 + \left(\bar{\sigma}_2^L\right)^2}E\left[s^L - \mu\right] - \beta\frac{\sigma_1^2\left(\bar{\sigma}_2^L\right)^2}{\sigma_1^2 + \left(\bar{\sigma}_2^L\right)^2}\right]\\ &\quad - \left[\mu + \frac{\sigma_1^2}{\sigma_1^2 + \left(\bar{\sigma}_2^H\right)^2}E\left[s^H - \mu\right] - \beta\frac{\sigma_1^2\left(\bar{\sigma}_2^H\right)^2}{\sigma_1^2 + \left(\bar{\sigma}_2^H\right)^2}\right]\\ &= \beta\frac{\sigma_1^2\left(\bar{\sigma}_2^H\right)^2}{\sigma_1^2 + \left(\bar{\sigma}_2^H\right)^2} - \beta\frac{\sigma_1^2\left(\bar{\sigma}_2^L\right)^2}{\sigma_1^2 + \left(\bar{\sigma}_2^L\right)^2}\end{aligned}$$

For the present example, the OLS bias is nonstochastic

$$\beta\left[\frac{\sigma_1^2\left(\bar{\sigma}_2^H\right)^2}{\sigma_1^2 + \left(\bar{\sigma}_2^H\right)^2} - \frac{\sigma_1^2\left(\bar{\sigma}_2^L\right)^2}{\sigma_1^2 + \left(\bar{\sigma}_2^L\right)^2}\right] = -2.33$$

Suppose we employ a naive (unsaturated) regression model, ignoring the OLS bias,

$$E\left[Y \mid s, D\right] = \beta_0 + \beta_1 s + \beta_2 D$$

or even a saturated regression model that ignores the OLS bias

$$E\left[Y \mid s, D\right] = \beta_0 + \beta_1 s + \beta_2 Ds + \beta_3 D$$

where

$$D = \begin{array}{ll} 1 & \text{if } EU^L > EU^H \\ 0 & \text{if } EU^L < EU^H \end{array}$$

Table 10.29: Report precision OLS parameter estimates for binary base case

statistic	β_0	β_1	$\beta_2\ (estATE)$
mean	172.2	0.430	−2.260
median	172.2	0.430	−2.260
std.dev.	0.069	0.0001	0.001
minimum	172.0	0.430	−2.264
maximum	172.4	0.430	−2.257
$E\left[Y \mid D, s\right] = \beta_0 + \beta_1 s + \beta_2 D$			

Table 10.30: Report precision average treatment effect sample statistics for binary base case

statistic	ATT	ATUT	ATE
mean	0.024	−0.011	0.006
median	0.036	0.002	0.008
std.dev.	0.267	0.283	0.191
minimum	−0.610	−0.685	−0.402
maximum	0.634	0.649	0.516

$$EU^j = \mu - \beta^j \frac{\sigma_1^2 \left(\bar{\sigma}_2^j\right)^2}{\sigma_1^2 + \left(\bar{\sigma}_2^j\right)^2} - \gamma \frac{\sigma_1^4 \left(\sigma_1^2 + \left(\sigma_2^j\right)^2\right)}{\left(\sigma_1^2 + \left(\bar{\sigma}_2^j\right)^2\right)^2}$$

$$-\alpha \left(b - \left(\sigma_2^j\right)^2\right)^2 - E\left[\alpha_d^j\right] \left(\hat{b} - \left(\sigma_2^j\right)^2\right)^2$$

$$Y = DY^L + (1-D)Y^H$$

$$Y^j = \mu + \frac{\sigma_1^2}{\sigma_1^2 + \left(\bar{\sigma}_2^j\right)^2}(s^j - \mu) - \beta^j \frac{\sigma_1^2 \left(\bar{\sigma}_2^j\right)^2}{\sigma_1^2 + \left(\bar{\sigma}_2^j\right)^2}$$

and

$$s = Ds^L + (1-D)s^H$$

$$s^j \sim N\left(\mu, \sigma_1^2 + \left(\sigma_2^j\right)^2\right)$$

for $j \in \{L, H\}$. Estimation results for the above naive regression are reported in table 10.29. Since this is simulation, we have access to the "missing" data and can provide sample statistics for average treatment effects. Sample statistics for standard average treatment effects, *ATE*, *ATT*, and *ATUT*, are reported in table 10.30. Estimation results for the above saturated regression are reported in table 10.31. As expected, the results indicate substantial *OLS* selection bias in both regressions. Clearly, to effectively estimate any treatment effect, we need to eliminate this *OLS* selection bias from outcome.

10.7 Regulated report precision

Table 10.31: Report precision saturated OLS parameter estimates for binary base case

statistic	β_0	β_1	β_2
mean	602.1	0.432	−0.003
median	602.1	0.432	0.003
std.dev.	0.148	0.000	0.000
minimum	601.7	0.432	−0.003
maximum	602.6	0.432	−0.003
statistic	$estATT$	$estATUT$	$\beta_3\,(estATE)$
mean	−2.260	−2.260	−2.260
median	−2.260	−2.260	−2.260
std.dev.	0.001	0.001	0.001
minimum	−2.264	−2.265	−2.264
maximum	−2.255	−2.256	−2.257
$E\left[Y \mid D, s\right] = \beta_0 + \beta_1 s + \beta_2 Ds + \beta_3 D$			

Adjusted outcomes

It's unusual to encounter *nonstochastic* selection bias.[16] Normally, nonstochastic bias is easily eliminated as it's captured in the intercept but here the selection bias is perfectly aligned with the treatment effect of interest. Consequently, we must decompose the two effects — we separate the selection bias from the treatment effect. Since the components of selection bias are proportional to the coefficients on the reports and these coefficients are consistently estimated when selection bias is nonstochastic, we can utilize the estimates from the coefficients on s^L and s^H. For example, the coefficient on s^L is $\omega_{s^L} = \frac{\sigma_1^2}{\sigma_1^2 + \left(\bar{\sigma}_2^L\right)^2}$. Then, $\left(\bar{\sigma}_2^L\right)^2 = \frac{\sigma_1^2(1-\omega_{s^L})}{\omega_{s^L}}$ and $\frac{\sigma_1^2\left(\bar{\sigma}_2^L\right)^2}{\sigma_1^2+\left(\bar{\sigma}_2^L\right)^2} = \omega_{s^L}\frac{\sigma_1^2(1-\omega_{s^L})}{\omega_{s^L}} = \sigma_1^2(1-\omega_{s^L})$. Hence, the OLS selection bias

$$bias = \beta \left(\frac{\sigma_1^2\left(\bar{\sigma}_2^H\right)^2}{\sigma_1^2 + \left(\bar{\sigma}_2^H\right)^2} - \frac{\sigma_1^2\left(\bar{\sigma}_2^L\right)^2}{\sigma_1^2 + \left(\bar{\sigma}_2^L\right)^2} \right)$$

can be written

$$bias = \beta\sigma_1^2\left(\omega_{s^L} - \omega_{s^H}\right)$$

This decomposition suggests we work with adjusted outcome

$$Y' = Y - \beta\sigma_1^2\left(D\omega_{s^L} - (1-D)\omega_{s^H}\right)$$

---

[16] Like the asset revaluation setting (chapter 9), the explanation lies in the lack of common support for identifying counterfactuals. In this base case, cost of transaction design type (L or H) is a perfect predictor of treatment. That is, $\Pr\left(D = 1 \mid \text{type} = L\right) = 1$ and $\Pr\left(D = 1 \mid \text{type} = H\right) = 0$. In subsequent settings, parameter variation leads to common support and selection bias is resolved via more standard *IV* approaches.

The adjustment can be estimated as follows. Estimate ω_{sL} and ω_{sH} from the regression

$$E\left[Y \mid D, s^L, s^H\right] = \omega_0 + \omega_1 D + \omega_{sL} D s^L + \omega_{sH}(1-D) s^H$$

Then, since

$$\begin{aligned} Y^j &= \mu + \frac{\sigma_1^2}{\sigma_1^2 + \left(\bar{\sigma}_2^j\right)^2}\left(s^j - \mu\right) - \beta^j \frac{\sigma_1^2 \left(\bar{\sigma}_2^j\right)^2}{\sigma_1^2 + \left(\bar{\sigma}_2^j\right)^2} \\ &= \mu + \omega_{s^j}\left(s^j - \mu\right) - \beta^j \sigma_1^2 \left(1 - \omega_{s^j}\right) \end{aligned}$$

we can recover the weight, $\omega = -\beta \sigma_1^2$, on $(1 - \omega_{s^j})$ utilizing the "restricted" regression

$$E\left[\begin{array}{c} Y - \omega_0 - \omega_{sL} D\left(s^L - \mu\right) \\ -\omega_{sH}(1-D)\left(s^H - \mu\right) \end{array} \mid D, s^L, s^H, \omega_{sL}, \omega_{sH}\right]$$
$$= \omega\left[D\left(1 - \omega_{sL}\right) + (1-D)\left(1 - \omega_{sH}\right)\right]$$

Finally, adjusted outcome is determined by plugging the estimates for ω, ω_{sL}, and ω_{sH} into

$$Y' = Y + \omega\left(D \omega_{sL} - (1-D) \omega_{sH}\right)$$

Now, we revisit the saturated regression employing the adjusted outcome Y'.

$$E\left[Y' \mid D, s\right] = \beta_0 + \beta_1 (s - \mu) + \beta_2 D (s - \mu) + \beta_3 D$$

The coefficient on D, β_2, estimates the average treatment effect. Estimation results for the saturated regression with adjusted outcome are reported in table 10.32.

As there is no residual uncertainty, response is homogeneous and the sample statistics for standard treatment effects, *ATE*, *ATT*, and *ATUT*, are of very similar magnitude — certainly within sampling variation. No residual uncertainty (in adjusted outcome) implies treatment is ignorable.

Heterogeneous response

Now, we explore a more interesting setting. Everything remains as in the base case except there is unobserved (by the analyst) variation in β the parameter controlling the discount associated with uncertainty in the buyer's ability to manage the assets. In particular, β^L, β^H are independent normally distributed with mean 7 and unit variance.[17] These β^L, β^H draws are observed by the owner in conjunction with the known mean for α_d^L, α_d^L when selecting report precision. In this setting, it is as if the owners choose equilibrium inverse-report precision, σ_2^L or σ_2^H, based on the combination of β^L and α_d^L or β^H and α_d^H with greatest expected utility.[18]

---

[17] Independent identically distributed draws of β for *L*-type and *H*-type firms ensure the variance-covariance matrix for the unobservables/errors is nonsingular.

[18] Notice the value of β does not impact the value of the welfare maximizing report variance. Therefore, the optimal inverse report precision choices correspond to $\left(\alpha, \gamma, E\left[\alpha_d^j\right]\right)$ as in the base case but the binary choice σ_2^L or σ_2^H does depend on β^j.

10.7 Regulated report precision

Table 10.32: Report precision adjusted outcome OLS parameter estimates for binary base case

statistic	β_0	β_1	β_2
mean	1000	0.432	−0.000
median	1000	0.432	0.000
std.dev.	0.148	0.000	0.001
minimum	999.6	0.432	−0.004
maximum	1001	0.432	0.003
statistic	$estATT$	$estATUT$	$\beta_3\,(estATE)$
mean	−0.000	−0.000	−0.000
median	0.000	−0.000	0.000
std.dev.	0.001	0.002	0.001
minimum	−0.004	−0.005	−0.004
maximum	0.005	0.004	0.003

$E\left[Y' \mid D, s\right] = \beta_0 + \beta_1\left(s - \bar{s}\right) + \beta_2 D\left(s - \bar{s}\right) + \beta_3 D$

Therefore, unlike the base case, common support is satisfied, i.e., there are no perfect predictors of treatment, $0 < \Pr\left(D = 1 \mid \beta^j, \alpha_d^j\right) < 1$. Plus, the choice equation and price regressions have correlated, stochastic unobservables.[19] In fact, this correlation in the errors[20] creates a classic endogeneity concern addressed by Heckman [1974, 1975, 1978, 1979].

First, we define average treatment effect estimands for this heterogeneity setting, then we simulate results for various treatment effect identification strategies. The average treatment effect on the treated is

$$\begin{aligned}
ATT &= E\left[Y_1 - Y_0 \mid D = 1, \beta^H, \beta^L\right] \\
&= E\left[\begin{array}{l}\left\{\mu + \dfrac{\sigma_1^2}{\sigma_1^2 + \left(\bar{\sigma}_2^L\right)^2}\left(s^L - \mu\right) - \beta^L \dfrac{\sigma_1^2\left(\bar{\sigma}_2^L\right)^2}{\sigma_1^2 + \left(\bar{\sigma}_2^L\right)^2}\right\} \\ -\left(\mu + \dfrac{\sigma_1^2}{\sigma_1^2 + \left(\bar{\sigma}_2^L\right)^2}\left(s^H - \mu\right) - \beta^H \dfrac{\sigma_1^2\left(\bar{\sigma}_2^L\right)^2}{\sigma_1^2 + \left(\bar{\sigma}_2^L\right)^2}\right)\end{array}\right] \\
&= \left(\beta^H - \beta^L\right) \dfrac{\sigma_1^2\left(\bar{\sigma}_2^L\right)^2}{\sigma_1^2 + \left(\bar{\sigma}_2^L\right)^2}
\end{aligned}$$

---

[19] The binary nature of treatment may seem a bit forced with response heterogeneity. This could be remedied by recognizing that owners' treatment choice is continuous but observed by the analyst to be binary. In later discussions, we explore such a setting with a richer *DGP*.

[20] The two regression equations and the choice equation have trivariate normal error structure.

The average treatment effect on the untreated is

$$\begin{aligned}ATUT &= E\left[Y_1 - Y_0 \mid D = 0, \beta^H, \beta^L\right] \\ &= E\left[\begin{array}{l}\left(\mu + \frac{\sigma_1^2}{\sigma_1^2 + (\bar{\sigma}_2^H)^2}\left(s^L - \mu\right) - \beta^L \frac{\sigma_1^2(\bar{\sigma}_2^H)^2}{\sigma_1^2 + (\bar{\sigma}_2^H)^2}\right) \\ -\left(\mu + \frac{\sigma_1^2}{\sigma_1^2 + (\bar{\sigma}_2^H)^2}\left(s^H - \mu\right) - \beta^H \frac{\sigma_1^2(\bar{\sigma}_2^H)^2}{\sigma_1^2 + (\bar{\sigma}_2^H)^2}\right)\end{array}\right] \\ &= \left(\beta^H - \beta^L\right)\frac{\sigma_1^2\left(\bar{\sigma}_2^H\right)^2}{\sigma_1^2 + \left(\bar{\sigma}_2^H\right)^2}\end{aligned}$$

OLS

Our first simulation for this heterogeneous setting attempts to estimate average treatment effects via *OLS*

$$E[Y \mid s, D] = \beta_0 + \beta_1 (s - \bar{s}) + \beta_2 (s - \bar{s}) D + \beta_3 D$$

Following Wooldridge, the coefficient on D, β_3, is the model-based average treatment effect (under strong identification conditions). Throughout the remaining discussion $(s - \bar{s})$ is the regressor of interest (based on our structural model). The model-based average treatment effect on the treated is

$$estATT = \beta_3 + \frac{\sum_i D_i (s_i - \bar{s}) \beta_2}{\sum_i D_i}$$

and the model-based average treatment effect on the untreated is

$$estATUT = \beta_3 - \frac{\sum_i D_i (s_i - \bar{s}) \beta_2}{\sum_i (1 - D_i)}$$

Simulation results, including model-based estimates and sample statistics for standard treatment effects, are reported in table 10.33. Average treatment effect sample statistics from the simulation for this binary heterogenous case are reported in table 10.34. Not surprisingly, *OLS* performs poorly. The key *OLS* identification condition is ignorable treatment but this is not sustained by the *DGP*. *OLS* model-based estimates of *ATE* are not within sampling variation of the average treatment effect. Further, the data are clearly heterogeneous and *OLS* (ignorable treatment) implies homogeneity.

IV approaches

Poor instruments

Now, we consider various *IV* approaches for addressing endogeneity. First, we explore various linear *IV* approaches. The analyst observes D and α_d^L if $D = 1$

Table 10.33: Report precision adjusted outcome OLS parameter estimates for binary heterogeneous case

statistic	β_0	β_1	β_2
mean	634.2	0.430	−0.003
median	634.2	0.429	−0.007
std.dev.	1.534	0.098	0.137
minimum	629.3	0.197	−0.458
maximum	637.7	0.744	0.377
statistic	$\beta_3\,(estATE)$	$estATT$	$estATUT$
mean	−2.227	−2.228	−2.225
median	−2.236	−2.257	−2.207
std.dev.	2.208	2.210	2.207
minimum	−6.672	−6.613	−6.729
maximum	3.968	3.971	3.966
$E\left[Y \mid s, D\right] = \beta_0 + \beta_1\left(s - \bar{s}\right) + \beta_2\left(s - \bar{s}\right)D + \beta_3 D$			

Table 10.34: Report precision average treatment effect sample statistics for binary heterogeneous case

statistic	ATE	ATT	ATUT
mean	0.189	64.30	−64.11
median	0.298	64.19	−64.10
std.dev.	1.810	1.548	1.462
minimum	−4.589	60.47	−67.80
maximum	4.847	68.38	−60.90

248   10. Treatment effects: *IV*

Table 10.35: Report precision poor 2SLS-IV estimates for binary heterogeneous case

statistic	β_0	β_1	β_2
mean	634.2	0.433	−0.010
median	634.4	0.439	−0.003
std.dev.	1.694	0.114	0.180
minimum	629.3	0.145	−0.455
maximum	638.2	0.773	0.507
statistic	$\beta_3\,(estATE)$	$estATT$	$estATUT$
mean	−2.123	−2.125	−2.121
median	−2.212	−2.217	−2.206
std.dev.	2.653	2.650	2.657
minimum	−7.938	−7.935	−7.941
maximum	6.425	6.428	6.423
$E\left[Y\mid s,D\right]=\beta_0+\beta_1\left(s-\bar{s}\right)+\beta_2\left(s-\bar{s}\right)D+\beta_3D$			

or α_d^H if $D=0$. Suppose the analyst employs $\alpha_d=D\alpha_d^L+(1-D)\alpha_d^H$ as an "instrument." As desired, α_d is related to report precision selection, unfortunately α_d is not conditionally mean independent, $E\left[y^j\mid s,\alpha_d\right]\neq E\left[y^j\mid s\right]$. To see this, recognize the outcome errors are a function of β^j and while α_d^j and β^j are independent, only α_d and not α_d^j is observed. Since α_d and β^j are related through selection D, α_d is a poor instrument. Two stage least squares instrumental variable estimation (*2SLS-IV*) produces the results reported in table 10.35 where β_3 is the model estimate for *ATE*. These results differ little from the *OLS* results except the *IV* model-based interval estimates of the treatment effects are wider as is expected even of a well-specified *IV* model. The results serve as a reminder of how little consolation comes from deriving similar results from two or more poorly-specified models.

Weak instruments

Suppose we have a "proper" instrument z_α in the sense that z_α is conditional mean independent. For purposes of the simulation, we construct the instrument z_α as the residuals from a regression of α_d onto

$$U^L=-\left(\beta^L-E\left[\beta\right]\right)\left[D\frac{\sigma_1^2\left(\sigma_2^L\right)^2}{\sigma_1^2+\left(\sigma_2^L\right)^2}+(1-D)\frac{\sigma_1^2\left(\sigma_2^H\right)^2}{\sigma_1^2+\left(\sigma_2^H\right)^2}\right]$$

and

$$U^H=-\left(\beta^H-E\left[\beta\right]\right)\left[D\frac{\sigma_1^2\left(\sigma_2^L\right)^2}{\sigma_1^2+\left(\sigma_2^L\right)^2}+(1-D)\frac{\sigma_1^2\left(\sigma_2^H\right)^2}{\sigma_1^2+\left(\sigma_2^H\right)^2}\right]$$

But, we wish to explore the implications for treatment effect estimation if the instrument is only weakly related to treatment. Therefore, we create a noisy instrument by adding an independent normal random variable ε with mean zero and

10.7 Regulated report precision

Table 10.36: Report precision weak 2SLS-IV estimates for binary heterogeneous case

statistic	β_0	β_1	β_2
mean	628.5	-0.605	2.060
median	637.3	0.329	0.259
std.dev.	141.7	7.678	15.52
minimum	-856.9	-73.00	-49.60
maximum	915.5	24.37	153.0
statistic	$\beta_3\,(estATE)$	$estATT$	$estATUT$
mean	8.770	8.139	9.420
median	-6.237	-6.532	-6.673
std.dev.	276.8	273.2	280.7
minimum	-573.3	-589.4	-557.7
maximum	2769	2727	2818
$E[Y \mid s, D] = \beta_0 + \beta_1(s - \bar{s}) + \beta_2(s - \bar{s})D + \beta_3 D$			

standard deviation 0.1. This latter perturbation ensures the instrument is weak. This instrument $z_\alpha + \varepsilon$ is employed to generate model-based estimates of some standard treatment effects via *2SLS-IV*. Results are provided in table 10.36 where β_3 is the model estimate for *ATE*. The weak *IV* model-estimates are extremely noisy. Weak instruments frequently are suspected to plague empirical work. In a treatment effects setting, this can be a serious nuisance as evidenced here.

A stronger instrument

Suppose z_α is available and employed as an instrument. Model-based treatment effect estimates are reported in table 10.37 where β_3 is the model estimate for *ATE*. These results are far less noisy but nonetheless appear rather unsatisfactory. The results, on average, diverge from sample statistics for standard treatment effects and provide little or no evidence of heterogeneity. Why? As Heckman and Vytlacil [2005, 2007] discuss, it is very difficult to identify what treatment effect linear *IV* estimates and different instruments produce different treatment effects. Perhaps then, it is not surprising that we are unable to connect the *IV* treatment effect to *ATE*, *ATT*, or *ATUT*.

Propensity score as an instrument

A popular ignorable treatment approach implies homogeneous response[21] and uses the propensity score as an instrument. We estimate the propensity score via a probit regression of D onto instruments z_α and z_σ, where z_α is (as defined above) the residuals of $\alpha_d = D\alpha_d^L + (1-D)\alpha_d^H$ onto U^L and U^H and z_σ is the residuals from a regression of $\sigma_2 = D\sigma_2^L + (1-D)\sigma_2^H$ onto U^L and U^H. Now, use

---

[21] An exception, propensity score with heterogeneous response, is discussed in section 10.5.1. However, this *IV*-identification strategy doesn't accommodate the kind of unobservable heterogeneity present in this report precision setting.

250   10. Treatment effects: *IV*

Table 10.37: Report precision stronger 2SLS-IV estimates for binary heterogeneous case

statistic	β_0	β_1	β_2
mean	634.3	0.427	0.005
median	634.2	0.428	0.001
std.dev.	2.065	0.204	0.376
minimum	629.2	−0.087	−0.925
maximum	639.8	1.001	1.005
statistic	$\beta_3\,(estATE)$	$estATT$	$estATUT$
mean	−2.377	−2.402	−2.351
median	−2.203	−2.118	−2.096
std.dev.	3.261	3.281	3.248
minimum	−10.15	−10.15	−10.15
maximum	6.878	6.951	6.809
$E\left[Y\mid s,D\right]=\beta_0+\beta_1\left(s-\bar{s}\right)+\beta_2\left(s-\bar{s}\right)D+\beta_3 D$			

the estimated probabilities $m=\Pr\left(D=1\mid z_\alpha,z_\sigma\right)$ in place of D to estimate the treatment effects.

$$E\left[Y\mid s,D\right]=\beta_0+\beta_1\left(s-\bar{s}\right)+\beta_2\left(s-\bar{s}\right)m+\beta_3 m$$

Model-based estimates of the treatment effects are reported in table 10.38 with β_3 corresponding to *ATE*. These results also are very unsatisfactory and highly erratic. Poor performance of the propensity score *IV* for estimating average treatment effects is not surprising as the data are inherently heterogeneous and the key propensity score *IV* identification condition is ignorability of treatment.[22] Next, we explore propensity score matching followed by two *IV* control function approaches.

Propensity score matching

Propensity score matching estimates of average treatment effects are reported in table 10.39.[23] While not as erratic as the previous results, these results are also unsatisfactory. Estimated *ATT* and *ATUT* are the opposite sign of one another as expected but reversed of the underlying sample statistics (based on simulated counterfactuals). This is not surprising as ignorability of treatment is the key identifying condition for propensity score matching.

Ordinate IV control function

Next, we consider an ordinate control function *IV* approach. The regression is

$$E\left[Y\mid s,D,\phi\right]=\beta_0+\beta_1\left(s-\bar{s}\right)+\beta_2 D\left(s-\bar{s}\right)+\beta_3\phi\left(Z\theta\right)+\beta_4 D$$

---

[22] Ignorable treatment implies homogeneous response, $ATE=ATT=ATUT$, except for common support variations.

[23] Propensity scores within 0.02 are matched using Sekhon's [2008] matching **R** package.

Table 10.38: Report precision propensity score estimates for binary heterogeneous case

statistic	β_0	β_1	β_2	β_3
mean	634.4	0.417	0.024	−2.610
median	634.3	0.401	0.039	−2.526
std.dev.	1.599	0.151	0.256	2.075
minimum	630.9	−0.002	−0.617	−7.711
maximum	638.9	0.853	0.671	2.721
statistic	estATE	estATT	estATUT	
mean	−74.64	−949.4	−799.8	
median	7.743	−386.1	412.8	
std.dev.	1422	2400	1503	
minimum	−9827	−20650	57.75	
maximum	7879	−9.815	17090	

$E\left[Y \mid s, m\right] = \beta_0 + \beta_1\left(s - \overline{s}\right) + \beta_2\left(s - \overline{s}\right)m + \beta_3 m$

Table 10.39: Report precision propensity score matching estimates for binary heterogeneous case

statistic	estATE	estATT	estATUT
mean	−2.227	−39.88	35.55
median	−2.243	−39.68	35.40
std.dev.	4.247	5.368	4.869
minimum	−14.00	−52.00	23.87
maximum	12.43	−25.01	46.79

252   10. Treatment effects: *IV*

Table 10.40: Report precision ordinate control IV estimates for binary heterogeneous case

statistic	β_0	β_1	β_2	β_3
mean	598.6	0.410	0.030	127.6
median	598.5	0.394	0.049	127.1
std.dev.	3.503	0.139	0.237	12.08
minimum	590.0	0.032	−0.595	91.36
maximum	609.5	0.794	0.637	164.7
statistic	β_4 (estATE)	estATT	estATUT	
mean	−2.184	33.41	−37.91	
median	−2.130	33.21	−37.83	
std.dev.	1.790	3.831	3.644	
minimum	−6.590	22.27	−48.56	
maximum	2.851	43.63	−26.01	
$E[Y \mid s, D, \phi] = \beta_0 + \beta_1 (s - \bar{s}) + \beta_2 D (s - \bar{s}) + \beta_3 \phi(Z\theta) + \beta_4 D$				

and is estimated via *IV* where instruments $\{\iota, (s - \bar{s}), m(s - \bar{s}), \phi(Z\theta), m\}$ are employed and $m = \Pr(D = 1 \mid Z = \begin{bmatrix} \iota & z_\alpha & z_\sigma \end{bmatrix})$ is estimated via probit. *ATE* is estimated via β_4, the coefficient on D. Following the general *IV* identification of *ATT*, *ATT* is estimated as

$$estATT = \beta_4 + \frac{\sum D_i \beta_3 \phi(Z_i \theta)}{\sum D_i}$$

and *ATUT* is estimated as

$$estATUT = \beta_4 - \frac{\sum D_i \beta_3 \phi(Z_i \theta)}{\sum (1 - D_i)}$$

Simulation results are reported in table 10.40. The ordinate control function results are clearly the most promising so far but still underestimate the extent of heterogeneity. Further, an important insight is emerging. If we only compare *OLS* and *ATE* estimates, we might conclude endogeneity is a minor concern. However, estimates of *ATT* and *ATUT* and their support of self-selection clearly demonstrate the false nature of such a conclusion.

Inverse-Mills IV

Heckman's control function approach, utilizing inverse-Mills ratios as the control function for conditional expectations, employs the regression

$$E[Y \mid s, D, \lambda] = \beta_0 + \beta_1 (1 - D)(s - \bar{s}) + \beta_2 D (s - \bar{s})$$
$$+ \beta_3 (1 - D) \lambda^H + \beta_4 D \lambda^L + \beta_5 D$$

where $\bar{s}$ is the sample average of s, $\lambda^H = -\frac{\phi(Z\theta)}{1 - \Phi(Z\theta)}$, $\lambda^L = \frac{\phi(Z\theta)}{\Phi(Z\theta)}$, and θ is the estimated parameter vector from a probit regression of report precision choice D

Table 10.41: Report precision inverse Mills IV estimates for binary heterogeneous case

statistic	β_0	β_1	β_2	β_3	β_4
mean	603.2	0.423	0.433	−56.42	56.46
median	603.1	0.416	0.435	−56.72	56.63
std.dev.	1.694	0.085	0.089	2.895	2.939
minimum	598.7	0.241	0.188	−65.40	48.42
maximum	607.8	0.698	0.652	−47.53	65.59
statistic		$\beta_5\,(estATE)$	$estATT$	$estATUT$	
mean		−2.155	59.65	−64.14	
median		−2.037	59.59	−64.09	
std.dev.		1.451	2.950	3.039	
minimum		−6.861	51.36	−71.19	
maximum		1.380	67.19	−56.10	

$$E[Y \mid s, D, \lambda] = \beta_0 + \beta_1 (1-D)(s-\bar{s}) + \beta_2 D(s-\bar{s}) \\ + \beta_3 (1-D)\lambda^H + \beta_4 D\lambda^L + \beta_5 D$$

on $Z = \begin{bmatrix} \iota & z_\alpha & z_\sigma \end{bmatrix}$ (ι is a vector of ones). The coefficient on D, β_5, is the model-based estimate of the average treatment effect, *ATE*. The average treatment effect on the treated is estimated as

$$ATT = \beta_5 + (\beta_2 - \beta_1) E[s - \bar{s}] + (\beta_4 - \beta_3) E\left[\lambda^L\right]$$

While the average treatment effect on the untreated is estimated as

$$ATUT = \beta_5 + (\beta_2 - \beta_1) E[s - \bar{s}] + (\beta_4 - \beta_3) E\left[\lambda^H\right]$$

Simulation results including model-estimated average treatment effects on treated (*estATT*) and untreated (*estATUT*) are reported in table 10.41. The inverse-Mills treatment effect estimates correspond nicely with their sample statistics. Next, we explore a variation on treatment.

10.7.2 Continuous report precision but observed binary

Heterogeneous response

Now, suppose the analyst only observes high or low report precision but there is considerable variation across firms. In other words, wide variation in parameters across firms is reflected in a continuum of report precision choices.[24] Specifically, variation in the cost of report precision parameter α, the discount parameter associated with the buyer's uncertainty in his ability to manage the asset, β, and the

---

[24]It is not uncommon for analysts to observe discrete choices even though there is a richer underlying choice set. Any discrete choice serves our purpose here, for simplicity we work with the binary case.

owner's risk premium parameter γ produces variation in owners' optimal report precision $\frac{1}{\sigma_2}$.

Variation in α_d is again not observed by the owners prior to selecting report precision. However, α_d is observed ex post by the analyst where α_d^L is normally distributed with mean 0.02 and standard deviation 0.005, while α_d^H is normally distributed with mean 0.04 and standard deviation 0.01. There is unobserved (by the analyst) variation in β the parameter controlling the discount associated with uncertainty in the buyer's ability to manage the assets such that β is independent normally distributed with mean 7 and variance 0.2. Independent identically distributed draws of β are taken for L-type and H-type firms so that the variance-covariance matrix for the unobservables/errors is nonsingular. On the contrary, draws for "instruments" α (normally distributed with mean 0.03 and standard deviation 0.005) and γ (normally distributed with mean 5 and standard deviation 1) are not distinguished by type to satisfy *IV* assumptions. Otherwise, conditional mean independence of the outcome errors and instruments is violated.[25] For greater unobservable variation (that is, variation through the β term), the weaker are the instruments, and the more variable is estimation of the treatment effects. Again, endogeneity is a first-order consideration as the choice equation and price (outcome) regression have correlated, stochastic unobservables.

OLS

First, we explore treatment effect estimation via the following *OLS* regression

$$E[Y \mid s, D] = \beta_0 + \beta_1 (s - \bar{s}) + \beta_2 D (s - \bar{s}) + \beta_3 D$$

Simulation results are reported in table 10.42. Average treatment effect sample statistics from the simulation are reported in table 10.43. In this setting, *OLS* effectively estimates the average treatment effect, *ATE*, for a firm/owner drawn at random. This is readily explained by noting the sample statistic estimated by *OLS* is within sampling variation of the sample statistic for *ATE* but *ATE* is indistinguishable from zero. However, if we're interested in response heterogeneity and other treatment effects, *OLS*, not surprisingly, is sorely lacking. *OLS* provides inconsistent estimates of treatment effects on the treated and untreated and has almost no diagnostic power for detecting response heterogeneity — notice there is little variation in *OLS*-estimated *ATE*, *ATT*, and *ATUT*.

Propensity score as an instrument

Now, we estimate the propensity score via a probit regression of D onto instruments α and γ, and use the estimated probabilities

$$m = \Pr(D = 1 \mid z_\alpha, z_\sigma)$$

---

[25] As we discuss later, these conditions are sufficient to establish α and γ as instruments — though weak instruments.

10.7 Regulated report precision

Table 10.42: Continuous report precision but observed binary OLS parameter estimates

statistic	β_0	β_1	β_2
mean	634.3	0.423	0.004
median	634.3	0.425	0.009
std.dev.	1.486	0.096	0.144
minimum	630.7	0.151	−0.313
maximum	638.4	0.658	0.520
statistic	$\beta_3\,(estATE)$	$estATT$	$estATUT$
mean	−1.546	−1.544	−1.547
median	−1.453	−1.467	−1.365
std.dev.	2.083	2.090	2.078
minimum	−8.108	−8.127	−8.088
maximum	5.170	5.122	5.216
$E\left[Y \mid s, D\right] = \beta_0 + \beta_1\left(s - \bar{s}\right) + \beta_2 D\left(s - \bar{s}\right) + \beta_3 D$			

Table 10.43: Continuous report precision but observed binary average treatment effect sample statistics

statistic	ATE	ATT	$ATUT$
mean	0.194	64.60	−64.20
median	0.215	64.55	−64.18
std.dev.	1.699	1.634	1.524
minimum	−4.648	60.68	−68.01
maximum	4.465	68.70	−60.18

Table 10.44: Continuous report precision but observed binary propensity score parameter estimates

statistic	β_0	β_1	β_2	β_3
mean	612.2	0.095	0.649	42.80
median	619.9	0.309	0.320	24.43
std.dev.	248.2	4.744	9.561	499.2
minimum	−1693	−29.80	−46.64	−1644
maximum	1441	23.35	60.58	4661
statistic	estATE	estATT	estATUT	
mean	−1.558	−1.551	−1.565	
median	−1.517	−1.515	−1.495	
std.dev.	2.086	2.090	2.085	
minimum	−8.351	−8.269	−8.437	
maximum	5.336	5.300	5.370	
$E\left[Y \mid s, m\right] = \beta_0 + \beta_1 \left(s - \bar{s}\right) + \beta_2 \left(s - \bar{s}\right) m + \beta_3 m$				

Table 10.45: Continuous report precision but observed binary propensity score matching parameter estimates

statistic	estATE	estATT	estATUT
mean	−1.522	−1.612	−1.430
median	−1.414	−1.552	−1.446
std.dev.	2.345	2.765	2.409
minimum	−7.850	−8.042	−8.638
maximum	6.924	9.013	4.906

in place of D to estimate the treatment effects.

$$E\left[Y \mid s, m\right] = \beta_0 + \beta_1 \left(s - \bar{s}\right) + \beta_2 \left(s - \bar{s}\right) m + \beta_3 m$$

Model-based estimates of the treatment effects are reported in 10.44. These results again are very unsatisfactory and highly variable. As before, poor performance of the propensity score *IV* for estimating average treatment effects is not surprising as the data are inherently heterogeneous and the key propensity score *IV* identification condition is ignorability of treatment (conditional mean redundancy).

Propensity score matching

Propensity score matching estimates of average treatment effects are reported in table 10.45.[26] While not as erratic as the previous results, these results are also unsatisfactory. Estimated *ATT* and *ATUT* are nearly identical even though the data are quite heterogeneous. The poor performance is not surprising as ignorability

---

[26] Propensity scores within 0.02 are matched using Sekhon's [2008] **R** matching package. Other bin sizes (say, 0.01) produce similar results though fewer matches..

Table 10.46: Continuous report precision but observed binary ordinate control IV parameter estimates

statistic	β_0	β_1	β_2	β_3
mean	-11633	5.798	-10.68	30971
median	772.7	0.680	-0.497	-390.8
std.dev.	176027	36.08	71.36	441268
minimum	-2435283	-58.78	-663.3	-1006523
maximum	404984	325.7	118.6	6106127
statistic	$\beta_4\ (estATE)$	$estATT$	$estATUT$	
mean	-173.7	12181	-12505	
median	-11.21	-168.6	176.3	
std.dev.	1176	176015	175648	
minimum	-11237	-407049	-2431259	
maximum	2598	2435846	390220	
$E\left[Y \mid s, D, \phi\right] = \beta_0 + \beta_1 (s - \bar{s}) + \beta_2 D (s - \bar{s}) + \beta_3 \phi(Z\theta) + \beta_4 D$				

of treatment (conditional stochastic independence, or at least, conditional mean independence) is the key identifying condition for propensity score matching.

Ordinate IV control

Now, we consider two *IV* approaches for addressing endogeneity. The ordinate control function regression is

$$E\left[Y \mid s, D, \phi\right] = \beta_0 + \beta_1 (s - \bar{s}) + \beta_2 D (s - \bar{s}) + \beta_3 \phi(Z\theta) + \beta_4 D$$

and is estimated via *IV* where instruments

$$\{\iota, (s - \bar{s}), m(s - \bar{s}), \phi(Z\theta), m\}$$

are employed and

$$m = \Pr\left(D = 1 \mid Z = \begin{bmatrix} \iota & \alpha & \gamma \end{bmatrix}\right)$$

is estimated via probit. *ATE* is estimated via β_4, the coefficient on D. Simulation results are reported in table 10.46. The ordinate control function results are inconsistent and extremely noisy. Apparently, the instruments, α and γ, are sufficiently weak that the propensity score is a poor instrument. If this conjecture holds, we should see similar poor results in the second *IV* control function approach as well.

Inverse-Mills IV

The inverse-Mills *IV* control function regression is

$$\begin{aligned} E\left[Y \mid s, D, \lambda\right] &= \beta_0 + \beta_1 (1 - D)(s - \bar{s}) + \beta_2 D (s - \bar{s}) \\ &\quad + \beta_3 D \lambda^H + \beta_4 (1 - D) \lambda^L + \beta_5 D \end{aligned}$$

Table 10.47: Continuous report precision but observed binary inverse Mills IV parameter estimates

statistic	β_0	β_1	β_2	β_3	β_4
mean	633.7	0.423	0.427	−0.926	−55.41
median	642.2	0.424	0.418	9.178	−11.44
std.dev.	198.6	0.096	0.106	249.9	407.9
minimum	−1141	0.152	0.164	−2228	−3676
maximum	1433	0.651	0.725	1020	1042
statistic		$\beta_5\,(estATE)$	$estATT$	$estATUT$	
mean		43.38	−0.061	86.87	
median		23.46	−16.03	17.39	
std.dev.		504.2	399.1	651.0	
minimum		−1646	−1629	−1663	
maximum		12.50	3556	5867	

$$E[Y \mid s, D, \lambda] = \beta_0 + \beta_1(1-D)(s-\bar{s}) + \beta_2 D(s-\bar{s})$$
$$+ \beta_3 D \lambda^H + \beta_4(1-D)\lambda^L + \beta_5 D$$

where $\bar{s}$ is the sample average of s, $\lambda^H = -\frac{\phi(Z\theta)}{1-\Phi(Z\theta)}$, $\lambda^L = \frac{\phi(Z\theta)}{\Phi(Z\theta)}$, and θ is the estimated parameters from a probit regression of precision choice D on $Z = \begin{bmatrix} \iota & \alpha & \gamma \end{bmatrix}$ (ι is a vector of ones). The coefficient on D, β_5, is the estimate of the average treatment effect, *ATE*. The average treatment effect on the treated is estimated as

$$ATT = \beta_5 + (\beta_2 - \beta_1)E[s-\bar{s}] + (\beta_4 - \beta_3)E\left[\lambda^L\right]$$

While the average treatment effect on the untreated is estimated as

$$ATUT = \beta_5 + (\beta_2 - \beta_1)E[s-\bar{s}] + (\beta_4 - \beta_3)E\left[\lambda^H\right]$$

Simulation results including estimated average treatment effects on treated (*estATT*) and untreated (*estATUT*) are reported in table 10.47. While not as variable as ordinate control function model estimates, the inverse-Mills *IV* estimates are inconsistent and highly variable. It's likely, we are unable to detect endogeneity or diagnose heterogeneity based on this strategy as well.

The explanation for the problem lies with our supposed instruments, α and γ. Conditional mean independence may be violated due to variation in report precision or the instruments may be weak. That is, optimal report precision is influenced by variation in α and γ and variation in report precision is reflected in outcome error variation

$$U^L = -\left(\beta^L - E[\beta]\right)\left[D\frac{\sigma_1^2\left(\bar{\sigma}_2^L\right)^2}{\sigma_1^2 + \left(\bar{\sigma}_2^L\right)^2} + (1-D)\frac{\sigma_1^2\left(\bar{\sigma}_2^H\right)^2}{\sigma_1^2 + \left(\bar{\sigma}_2^H\right)^2}\right]$$

and

$$U^H = -\left(\beta^H - E[\beta]\right)\left[D\frac{\sigma_1^2\left(\bar{\sigma}_2^L\right)^2}{\sigma_1^2 + \left(\bar{\sigma}_2^L\right)^2} + (1-D)\frac{\sigma_1^2\left(\bar{\sigma}_2^H\right)^2}{\sigma_1^2 + \left(\bar{\sigma}_2^H\right)^2}\right]$$

10.7 Regulated report precision

Table 10.48: Continuous report precision but observed binary sample correlations

statistic	$r(\alpha, U^L)$	$r(\alpha, U^H)$	$r(\gamma, U^L)$	$r(\gamma, U^H)$
mean	−0.001	−0.002	0.003	−0.000
median	−0.001	−0.004	0.003	0.001
std.dev.	0.020	0.024	0.023	0.024
minimum	−0.052	−0.068	−0.079	−0.074
maximum	0.049	0.053	0.078	0.060
statistic	$r(\alpha, D)$	$r(\gamma, D)$	$r(w_1, D)$	$r(w_2, D)$
mean	−0.000	0.001	−0.365	0.090
median	−0.001	0.003	−0.365	0.091
std.dev.	0.021	0.025	0.011	0.013
minimum	−0.046	−0.062	−0.404	0.049
maximum	0.050	0.075	−0.337	0.122

To investigate the poor instrument problem we report in table 10.48 sample correlation statistics $r(\cdot, \cdot)$ for α and γ determinants of optimal report precision with *unobservable* outcome errors U^L and U^H. We also report sample correlations between potential instruments, α, γ, w_1, w_2, and treatment D to check for weak instruments. The problem with the supposed instruments, α and γ, is apparently that they're weak and not that they're correlated with U^L and U^H. On the other hand, w_1 and w_2 (defined below) hold some promise. We experiment with these instruments next.

Stronger instruments

To further investigate this explanation, we employ stronger instruments, w_1 (the component of α_d independent of U^L and U^H) and w_2 (the component of $\sigma_2^D \equiv D\sigma_2^L + (1-D)\sigma_2^H$ independent of U^L and U^H),[27] and reevaluate propensity score as an instrument.[28]

Propensity score as an instrument. Now, we use the estimated probabilities

$$m = \Pr(D = 1 \mid w_1, w_2)$$

from the above propensity score in place of D to estimate the treatment effects.

$$E[Y \mid s, m] = \beta_0 + \beta_1(s - \bar{s}) + \beta_2(s - \bar{s})m + \beta_3 m$$

Model-based estimates of the treatment effects are reported in table 10.49. These results again are very unsatisfactory and highly variable. As before, poor performance of the propensity score *IV* for estimating average treatment effects is not surprising as the data are inherently heterogeneous and the key propensity score

---

[27] For purposes of the simulation, these are constructed from the residuals of regressions of α_d and σ_2^D on unobservables U^H and U^L.

[28] A complementary possibility is to search for measures of nonpecuniary satisfaction as instruments. That is, measures which impact report precision choice but are unrelated to outcomes.

Table 10.49: Continuous report precision but observed binary stronger propensity score parameter estimates

statistic	β_0	β_1	β_2	β_3
mean	637.1	0.419	0.012	−7.275
median	637.1	0.419	−0.007	−7.215
std.dev.	2.077	0.203	0.394	3.455
minimum	631.8	−0.183	−0.820	−16.61
maximum	1441	23.35	60.58	4661
statistic	estATE	estATT	estATUT	
mean	−70.35	−99.53	−41.10	
median	−69.73	−97.19	−41.52	
std.dev.	12.92	21.04	7.367	
minimum	−124.0	−188.0	−58.59	
maximum	5.336	5.300	5.370	
$E\left[Y \mid s, m\right] = \beta_0 + \beta_1 \left(s - \bar{s}\right) + \beta_2 \left(s - \bar{s}\right) m + \beta_3 m$				

Table 10.50: Continuous report precision but observed binary stronger propensity score matching parameter estimates

statistic	estATE	estATT	estATUT
mean	2.291	−7.833	13.80
median	2.306	−8.152	13.74
std.dev.	2.936	3.312	3.532
minimum	−6.547	−17.00	5.189
maximum	12.38	4.617	24.94

IV identification condition is ignorability of treatment (conditional mean independence).

Propensity score matching

Propensity score matching estimates of average treatment effects are reported in table 10.50.[29] While not as erratic as the previous results, these results are also unsatisfactory. Estimated *ATT* and *ATUT* are opposite their sample statistics. The poor performance is not surprising as ignorability of treatment is the key identifying condition for propensity score matching.

Ordinate IV control function. The ordinate control function regression is

$$E\left[Y \mid s, D, \phi\right] = \beta_0 + \beta_1 \left(s - \bar{s}\right) + \beta_2 D \left(s - \bar{s}\right) + \beta_3 \phi\left(Z\theta\right) + \beta_4 D$$

and is estimated via *IV* where instruments

$$\{\iota, \left(s - \bar{s}\right), m \left(s - \bar{s}\right), \phi\left(Z\theta\right), m\}$$

---

[29] Propensity scores within 0.02 are matched.

10.7 Regulated report precision   261

Table 10.51: Continuous report precision but observed binary stronger ordinate control IV parameter estimates

statistic	β_0	β_1	β_2	β_3
mean	616.0	0.419	0.010	66.21
median	616.5	0.418	−0.006	65.24
std.dev.	7.572	0.202	0.381	24.54
minimum	594.0	−0.168	−0.759	1.528
maximum	635.5	0.885	1.236	147.3
statistic	$\beta_4\,(estATE)$	$estATT$	$estATUT$	
mean	−11.91	12.52	−36.35	
median	−11.51	12.31	−36.53	
std.dev.	4.149	7.076	12, 14	
minimum	−24.68	−5.425	−77.47	
maximum	−2.564	32.37	−4.535	
$E\left[Y \mid s, D, \phi\right] = \beta_0 + \beta_1\left(s - \bar{s}\right) + \beta_2 D\left(s - \bar{s}\right) + \beta_3 \phi\left(Z\theta\right) + \beta_4 D$				

are employed and

$$m = \Pr\left(D = 1 \mid Z = \begin{bmatrix} \iota & w_1 & w_2 \end{bmatrix}\right)$$

is estimated via probit. *ATE* is estimated via β_4, the coefficient on D. Simulation results are reported in table 10.51. The ordinate control function results are markedly improved relative to those obtained with poor instruments, α and γ. Model-estimated average treatment effects are biased somewhat toward zero. Nonetheless, the ordinate control *IV* approach might enable us to detect endogeneity via heterogeneity even though *OLS* and *ATE* are within sampling variation of one another. The important point illustrated here is that the effectiveness of *IV* control function approaches depend heavily on strong instruments. It's important to remember proper instruments in large part have to be evaluated ex ante — sample evidence is of limited help due to unobservability of counterfactuals.

Inverse-Mills IV

The inverse-Mills *IV* regression is

$$E\left[Y \mid s, D, \lambda\right] = \beta_0 + \beta_1\left(1 - D\right)\left(s - \bar{s}\right) + \beta_2 D\left(s - \bar{s}\right)$$
$$+ \beta_3 D \lambda^H + \beta_4\left(1 - D\right)\lambda^L + \beta_5 D$$

where $\bar{s}$ is the sample average of s, $\lambda^H = -\frac{\phi(Z\theta)}{1-\Phi(Z\theta)}$, $\lambda^L = \frac{\phi(Z\theta)}{\Phi(Z\theta)}$, and θ is the estimated parameters from a probit regression of precision choice D on $Z = \begin{bmatrix} \iota & w_1 & w_2 \end{bmatrix}$ (ι is a vector of ones). The coefficient on D, β_5, is the estimate of the average treatment effect, *ATE*. Simulation results including estimated average treatment effects on treated (*estATT*) and untreated (*estATUT*) are reported in table 10.52. While the inverse-Mills *IV* average treatment effect estimates come closest of any strategies (so far considered) to maintaining the

Table 10.52: Continuous report precision but observed binary stronger inverse Mills IV parameter estimates

statistic	β_0	β_1	β_2	β_3	β_4
mean	611.6	0.423	0.428	-32.03	80.04
median	611.5	0.431	0.422	-32.12	79.84
std.dev.	2.219	0.093	0.099	3.135	6.197
minimum	606.6	0.185	0.204	-41.47	62.39
maximum	617.5	0.635	0.721	-20.70	98.32
statistic	$\beta_5\,(estATE)$		$estATT$	$estATUT$	
mean	-35.55		43.77	-114.8	
median	-35.11		43.80	-114.7	
std.dev.	3.868		4.205	8.636	
minimum	-47.33		30.02	-142.0	
maximum	-26.00		57.97	-90.55	

$$E\left[Y \mid s, D, \lambda\right] = \beta_0 + \beta_1 \left(1 - D\right)\left(s - \overline{s}\right) + \beta_2 D \left(s - \overline{s}\right)$$
$$+ \beta_3 D \lambda^H + \beta_4 \left(1 - D\right) \lambda^L + \beta_5 D$$

spread between and direction of *ATT* and *ATUT*, all average treatment effect estimates are biased downward and the spread is somewhat exaggerated. Nevertheless, we are able to detect endogeneity and diagnose heterogeneity by examining estimated *ATT* and *ATUT*. Importantly, this derives from employing strong instruments, w_1 (the component of α_d independent of U^L and U^H) and w_2 (the component of $\sigma_2^D = D\sigma_2^L + (1 - D)\sigma_2^H$ independent of U^L and U^H). The next example reexamines treatment effect estimation in a setting where *OLS* and *ATE* differ markedly and estimates of *ATE* may help detect endogeneity.

Simpson's paradox

Suppose a firm's owner receives nonpecuniary and unobservable (to the analyst) satisfaction associated with report precision choice. This setting highlights a deep concern when analyzing data — perversely omitted, correlated variables which produce a Simpson's paradox result.

Consider α_d^L is normally distributed with mean 1.0 and standard deviation 0.25, while α_d^H is normally distributed with mean 0.04 and standard deviation 0.01.[30] As with β^j, these differences between *L* and *H*-type cost parameters are perceived or observed by the owner; importantly, β^L has standard deviation 2 while β^H has standard deviation 0.2 and each has mean 7. The unpaid cost of transaction design is passed on to the firm and its investors by *L*-type owners. Investors are aware of this (and price the firm accordingly) but the analyst is not (hence it's unobserved). *L*-type owners get nonpecuniary satisfaction from transaction design such that their personal cost is only 2% of $\alpha_d^L \left(\widehat{b} - \sigma_2^2\right)^2$, while *H*-type owners receive

---

[30]The labels seem reversed, but bear with us.

no nonpecuniary satisfaction — hence the labels.[31] Other features remain as in the previous setting. Accordingly, expected utility for *L*-type owners who choose treatment is

$$EU^L\left(\sigma_2^L\right) = \mu - \beta^L \frac{\sigma_1^2\left(\bar{\sigma}_2^L\right)^2}{\sigma_1^2 + \left(\bar{\sigma}_2^L\right)^2} - \gamma \frac{\sigma_1^4\left(\sigma_1^2 + \left(\bar{\sigma}_2^L\right)^2\right)}{\left(\sigma_1^2 + \left(\bar{\sigma}_2^L\right)^2\right)^2}$$

$$-\alpha\left(b - \left(\sigma_2^L\right)^2\right)^2 - 0.02\alpha_d^L\left(\hat{b} - \left(\sigma_2^L\right)^2\right)^2$$

while expected utility for *H*-type owners who choose no treatment is

$$EU^H\left(\sigma_2^H\right) = \mu - \beta^H \frac{\sigma_1^2\left(\bar{\sigma}_2^H\right)^2}{\sigma_1^2 + \left(\bar{\sigma}_2^H\right)^2} - \gamma \frac{\sigma_1^4\left(\sigma_1^2 + \left(\bar{\sigma}_2^H\right)^2\right)}{\left(\sigma_1^2 + \left(\bar{\sigma}_2^H\right)^2\right)^2}$$

$$-\alpha\left(b - \left(\sigma_2^H\right)^2\right)^2 - \alpha_d^H\left(\hat{b} - \left(\sigma_2^H\right)^2\right)^2$$

Also, outcomes or prices for owners who choose treatment include the cost of transaction design and accordingly are

$$Y^L = P\left(\bar{\sigma}_2^L\right) = \mu + \frac{\sigma_1^2}{\sigma_1^2 + \left(\bar{\sigma}_2^L\right)^2}\left(s^L - \mu\right) - \beta^L \frac{\sigma_1^2\left(\bar{\sigma}_2^L\right)^2}{\sigma_1^2 + \left(\bar{\sigma}_2^L\right)^2} - \alpha_d^L\left(\hat{b} - \left(\sigma_2^L\right)^2\right)^2$$

OLS

An *OLS* regression is

$$E[Y \mid s, D] = \beta_0 + \beta_1(s - \bar{s}) + \beta_2 D(s - \bar{s}) + \beta_3 D$$

Simulation results are reported in table 10.53. The average treatment effect sample statistics from the simulation are reported in table 10.54. Clearly, *OLS* produces poor estimates of the average treatment effects. As other ignorable treatment strategies fair poorly in settings of rich heterogeneity, we skip propensity score strategies and move ahead to control function strategies.

Ordinate IV control

We consider two *IV* control function approaches for addressing endogeneity. An ordinate control function regression is

$$E[Y \mid s, D, \phi] = \beta_0 + \beta_1(s - \bar{s}) + \beta_2 D(s - \bar{s}) + \beta_3 \phi(Z\theta) + \beta_4 D$$

---

[31] The difference in variability between β^L and β^H creates the spread between *ATE* and the effect estimated via *OLS* while nonpecuniary reward creates a shift in their mean outcomes such that *OLS* is positive and *ATE* is negative.

Table 10.53: Continuous report precision but observed binary OLS parameter estimates for Simpson's paradox DGP

statistic	β_0	β_1	β_2
mean	603.2	0.434	−0.014
median	603.2	0.434	−0.007
std.dev.	0.409	0.023	0.154
minimum	602.2	0.375	−0.446
maximum	604.4	0.497	0.443
statistic	$\beta_3\,(estATE)$	$estATT$	$estATUT$
mean	54.03	54.03	54.04
median	53.89	53.89	53.91
std.dev.	2.477	2.474	2.482
minimum	46.17	46.26	46.08
maximum	62.31	62.25	62.37
$E\left[Y\mid s,D\right]=\beta_0+\beta_1\left(s-\overline{s}\right)+\beta_2 D\left(s-\overline{s}\right)+\beta_3 D$			

Table 10.54: Continuous report precision but observed binary average treatment effect sample statistics for Simpson's paradox DGP

statistic	ATE	ATT	$ATUT$
mean	−33.95	57.76	−125.4
median	−34.06	57.78	−125.4
std.dev.	2.482	2.386	2.363
minimum	−42.38	51.15	−131.3
maximum	−26.57	66.49	−118.5

10.7 Regulated report precision   265

Table 10.55: Continuous report precision but observed binary ordinate control IV parameter estimates for Simpson's paradox DGP

statistic	β_0	β_1	β_2	β_3
mean	561.0	0.441	−0.032	266.3
median	561.5	0.479	−0.041	263.7
std.dev.	9.703	0.293	0.497	31.41
minimum	533.5	−0.442	−1.477	182.6
maximum	585.7	1.305	1.615	361.5
statistic	$\beta_4\,(estATE)$	$estATT$	$estATUT$	
mean	−48.72	48.45	−145.6	
median	−49.02	47.97	−143.0	
std.dev.	8.190	10.43	16.58	
minimum	−71.88	21.53	−198.0	
maximum	−25.12	84.89	−99.13	

$$E\left[Y\mid s, D, \phi\right] = \beta_0 + \beta_1\left(s - \bar{s}\right) + \beta_2 D\left(s - \bar{s}\right) + \beta_3 \phi\left(Z\theta\right) + \beta_4 D$$

and is estimated via *IV* where instruments

$$\{\iota, s, m\left(s - \bar{s}\right) m, \phi\left(Z\theta\right), m\}$$

are employed and

$$m = \Pr\left(D = 1 \mid Z = \begin{bmatrix} \iota & w_1 & w_2 \end{bmatrix}\right)$$

is estimated via probit. *ATE* is estimated via β_4, the coefficient on D. Simulation results are reported in table 10.55. As expected, the ordinate control function fairs much better than *OLS*. Estimates of *ATUT* are biased somewhat away from zero and, as expected, more variable than the sample statistic, but estimates are within sampling variation. Nevertheless, the ordinate control *IV* model performs better than in previous settings. Next, we compare results with the inverse-Mills *IV* strategy.

Inverse-Mills IV

The inverse-Mills *IV* control function regression is

$$\begin{aligned}E\left[Y\mid s, D, \lambda\right] &= \beta_0 + \beta_1\left(1 - D\right)\left(s - \bar{s}\right) + \beta_2 D\left(s - \bar{s}\right) \\ &\quad + \beta_3\left(1 - D\right)\lambda^H + \beta_4 D \lambda^L + \beta_5 D\end{aligned}$$

where $\bar{s}$ is the sample average of s, $\lambda^H = -\frac{\phi(Z\theta)}{1-\Phi(Z\theta)}$, $\lambda^L = \frac{\phi(Z\theta)}{\Phi(Z\theta)}$, and θ is the estimated parameters from a probit regression of precision choice D on $Z = \begin{bmatrix} \iota & w_1 & w_2 \end{bmatrix}$ (ι is a vector of ones). The coefficient on D, β_5, is the estimate of the average treatment effect, *ATE*. Simulation results including estimated average treatment effects on treated (*estATT*) and untreated (*estATUT*) are reported in table 10.56. As with the ordinate control function approach, inverse-Mills estimates of the treatment effects (especially *ATUT*) are somewhat biased

266   10. Treatment effects: IV

Table 10.56: Continuous report precision but observed binary inverse Mills IV parameter estimates for Simpson's paradox DGP

statistic	β_0	β_1	β_2	β_3	β_4
mean	603.3	0.434	0.422	0.057	182.8
median	603.2	0.434	0.425	0.016	183.0
std.dev.	0.629	0.023	0.128	0.787	11.75
minimum	601.1	0.375	0.068	−2.359	151.8
maximum	604.9	0.497	0.760	1.854	221.7
statistic		$\beta_5\,(estATE)$	$estATT$	$estATUT$	
mean		−74.17	53.95	−201.9	
median		−74.46	53.88	−201.3	
std.dev.		8.387	2.551	16.58	
minimum		−99.78	45.64	−256.7	
maximum		−52.65	61.85	−159.1	

$$E\left[Y \mid s, D, \lambda\right] = \beta_0 + \beta_1\left(1 - D\right)\left(s - \bar{s}\right) + \beta_2 D\left(s - \bar{s}\right)$$
$$+ \beta_3\left(1 - D\right)\lambda^H + \beta_4 D \lambda^L + \beta_5 D$$

away from zero and, as expected, more variable than the sample statistics. However, the model supplies strong evidence of endogeneity (*ATE* along with *ATT* and *ATUT* differ markedly from *OLS* estimates) and heterogeneous response ($ATE \neq ATT \neq ATUT$). Importantly, mean and median estimates reveal a Simpson's paradox result—*OLS* estimates a positive average treatment effect while endogeneity of selection produces a negative average treatment effect.[32]

10.7.3 Observable continuous report precision choice

Now we consider the setting where the analyst observes a continuum of choices based on the investors' (equilibrium) conjecture of the owner's report precision $\bar{\tau} = \frac{1}{\sigma_1^2 + \bar{\sigma}_2^2}$. This plays out as follows. The equilibrium strategy is the fixed point where the owner's expected utility maximizing report precision, $\frac{1}{\sigma^2} = \frac{1}{\sigma_1^2 + \sigma_2^2}$, equals investors' conjectured best response report precision, $\bar{\tau} = \frac{1}{\sigma_1^2 + \bar{\sigma}_2^2}$. Let conjectured report variance be denoted $\bar{\sigma}^2 \equiv \sigma_1^2 + \bar{\sigma}_2^2$. The owner's expected utility is

$$EU\left(\sigma_2\right) = \mu - \beta \frac{\sigma_1^2 \bar{\sigma}_2^2}{\sigma_1^2 + \bar{\sigma}_2^2} - \gamma \frac{\sigma_1^4\left(\sigma_1^2 + \sigma_2^2\right)}{\left(\sigma_1^2 + \bar{\sigma}_2^2\right)^2} - \alpha\left(b - \sigma_2^2\right)^2 - \alpha_d\left(\hat{b} - \sigma_2^2\right)^2$$

---

[32] As noted previously, untabulated results using weak instruments (α and γ) reveal extremely erratic estimates of the treatment effects.

substitution of $\bar{\sigma}^2$ for $\sigma_1^2 + \bar{\sigma}_2^2$ yields

$$EU(\sigma_2) = \mu - \beta \frac{\sigma_1^2 (\bar{\sigma}^2 - \sigma_1^2)}{\bar{\sigma}^2} - \gamma \frac{\sigma_1^4 \sigma^2}{\bar{\sigma}^4}$$
$$- \alpha \left(b - \sigma^2 + \sigma_1^2\right)^2 - \alpha_d \left(\hat{b} - \sigma^2 + \sigma_1^2\right)^2$$

The first order condition combined with the equilibrium condition is

$$\sigma^2 = \frac{\alpha b + \alpha_d \hat{b} - \gamma \frac{\sigma_1^4}{2\bar{\sigma}^4}}{\alpha + \alpha_d}$$
$$s.t. \ \sigma^2 = \bar{\sigma}^2$$

As the outcome equation

$$Y = P(\bar{\sigma}_2^2) = \mu + \frac{\sigma_1^2}{\sigma_1^2 + \bar{\sigma}_2^2} (s - \mu) - \beta \frac{\sigma_1^2 \bar{\sigma}_2^2}{\sigma_1^2 + \bar{\sigma}_2^2}$$
$$= P(\bar{\tau}) = \mu + \sigma_1^2 (s - \mu) \bar{\tau} - \beta \sigma_1^2 (1 - \sigma_1^2 \bar{\tau})$$

is not directly affected by the owner's report precision choice (but rather by the conjectured report precision), we exploit the equilibrium condition to define an average treatment effect

$$ATE(\bar{\tau}) = E \left[\frac{\partial Y}{\partial \bar{\tau}} \right] = \beta \sigma_1^4$$

and an average treatment effect on the treated[33]

$$ATT(\bar{\tau}) = E \left[\frac{\partial Y}{\partial \bar{\tau}} \mid \bar{\tau} = \tau_j \right] = \beta_j \sigma_1^4$$

If β differs across firms, as is likely, the outcome equation

$$Y_j = \left[\mu - \beta_j \sigma_1^2 \right] + \left[\sigma_1^2 \right] (s_j - \mu) \bar{\tau}_j + \left[\beta_j \sigma_1^4 \right] \bar{\tau}_j$$

is a random coefficients model. And, if $\beta_j \sigma_1^4$ and $\bar{\tau}_j = \frac{1}{\sigma_1^2 + (\bar{\sigma}_{2j})^2}$ are related, then we're dealing with a correlated random coefficients model.

For our experiment, a simulation based on 200 samples of (balanced) panel data with $n = 200$ individuals and $T = 10$ periods (sample size, $nT = 2,000$) is employed. Three data variations are explored.

---

[33] As Heckman [1997] suggests the average treatment effect based on a random draw from the population of firms often doesn't address a well-posed economic question whether treatment is continuous or discrete.

Table 10.57: Continuous treatment OLS parameter estimates and average treatment effect estimates and sample statistics with only between individual variation

statistic	ω_0	ω_1	$\omega_2\,(estATE)$	ATE	$corr\,(\omega_{2i}, \bar{\tau}_i)$
mean	300.4	100.3	69916.	70002.	−0.001
median	300.4	100.3	69938.	70007.	0.002
std.dev.	7.004	1.990	1616	73.91	0.067
minimum	263.1	93.44	61945.	69779.	−0.194
maximum	334.9	106.2	78686.	70203.	0.140
$E\left[Y \mid s, \bar{\tau}\right] = \omega_0 + \omega_1\left(s - \bar{s}\right)\bar{\tau} + \omega_2\bar{\tau}$					

Between individual variation

First, we explore a setting involving only variation in report precision between individuals. The following independent stochastic parameters characterize the data

Stochastic components	
parameters	number of draws
$\alpha \sim N\,(0.02, 0.005)$	n
$\alpha_d \sim N\,(0.02, 0.005)$	n
$\gamma \sim N\,(2.5, 1)$	n
$\beta \sim N\,(7, 0.1)$	n
$s \sim N\,(1000, \sigma)$	nT

where σ is the equilibrium report standard deviation; σ varies across firms but is constant through time for each firm.

First, we suppose treatment is ignorable and estimate the average treatment effect via *OLS*.

$$E\left[Y \mid s, \bar{\tau}\right] = \omega_0 + \omega_1\left(s - \mu\right)\bar{\tau} + \omega_2\bar{\tau}$$

Then, we accommodate unobservable heterogeneity (allow treatment and treatment effect to be correlated) and estimate the average treatment effect via *2SLS-IV*.

Hence, the *DGP* is

$$Y = 300 + 100\left(s - \mu\right)\bar{\tau} + \left(70,000 + \varepsilon_\beta\right)\bar{\tau}$$

where $\varepsilon_\beta = \beta_j - E\left[\beta_j\right] \sim N\left(0, 1\right)$, $j = 1, \ldots, n$.

OLS

Results for *OLS* along with sample statistics for *ATE* and the correlation between treatment and treatment effect are reported in table 10.57 where ω_2 is the estimate of *ATE*. The *OLS* results correspond quite well with the *DGP* and the average treatment effect sample statistics. This is not surprising given the lack of correlation between treatment and treatment effect.

10.7 Regulated report precision

Table 10.58: Continuous treatment 2SLS-IV parameter and average treatment effect estimates with only between individual variation

statistic	ω_0	ω_1	$\omega_2\,(estATE)$
mean	300.4	100.3	69916.
median	300.4	100.2	69915.
std.dev.	7.065	1.994	1631
minimum	262.7	93.44	61308.
maximum	337.6	106.2	78781.
$E\left[Y \mid s, \bar{\tau}\right] = \omega_0 + \omega_1\left(s - \bar{s}\right)\bar{\tau} + \omega_2\bar{\tau}$			

2SLS-IV

On the other hand, as suggested by Wooldridge [1997, 2003], *2SLS-IV* consistently estimates *ATE* in this random coefficients setting. We employ the residuals from regressions of $(s - \mu)\bar{\tau}$ and $\bar{\tau}$ on U as instruments, z_1 and z_2; these are strong instruments. Results for *2SLS-IV* are reported in table 10.58. The *IV* results correspond well with the *DGP* and the sample statistics for *ATE*. Given the lack of correlation between treatment and treatment effect, it's not surprising that *IV* (with strong instruments) and *OLS* results are very similar.

Modest within individual variation

Second, we explore a setting involving within individual as well as between individuals report variation. Within individual variation arises through modest variation through time in the cost parameter associated with transaction design. The following independent stochastic parameters describe the data

Stochastic components	
parameters	number of draws
$\alpha \sim N\left(0.02, 0.005\right)$	n
$\alpha_d \sim N\left(0.02, 0.0005\right)$	nT
$\gamma \sim N\left(2.5, 1\right)$	n
$\beta \sim N\left(7, 0.1\right)$	n
$\beta_i = \beta + N\left(0, 0.0001\right)$	nT
$s \sim N\left(1000, \sigma\right)$	nT

where σ is the equilibrium report standard deviation; σ varies across firms and through time for each firm and unobserved β_i produces residual uncertainty.

OLS

This setting allows identification of *ATE* and *ATT* where $ATT\left(\bar{\tau} = median\left[\bar{\tau}\right]\right)$. First, we estimate the average treatment effects via *OLS* where individual specific intercepts and slopes are accommodated.

$$E\left[Y \mid s_i, \bar{\tau}_i\right] = \sum_{i=1}^{n} \omega_{0i} + \omega_{1i}\left(s_i - \mu\right)\bar{\tau}_i + \omega_{2i}\bar{\tau}_i$$

270   10. Treatment effects: *IV*

Table 10.59: Continuous treatment OLS parameter and average treatment effect estimates for modest within individual report precision variation setting

statistic	estATE	estATT ($\bar{\tau} = median\ [\bar{\tau}]$)
mean	70306.	70152.
median	70193.	70368.
std.dev.	4625.	2211.
minimum	20419.	64722.
maximum	84891.	75192.
$E\left[Y \mid s_i, \bar{\tau}_i\right] = \sum_{i=1}^{n} \omega_{0i} + \omega_{1i}\left(s_i - \mu\right)\bar{\tau}_i + \omega_{2i}\bar{\tau}_i$		

Table 10.60: Continuous treatment ATE and ATT sample statistics and correlation between treatment and treatment effect for modest within individual report precision variation setting

statistic	ATE	ATT ($\bar{\tau} = median\ [\bar{\tau}]$)	$corr\left(\omega_{2it}, \bar{\tau}_{it}\right)$
mean	70014.	70026	−0.0057
median	70014.	69993	−0.0063
std.dev.	65.1	974.	0.072
minimum	69850.	67404	−0.238
maximum	70169.	72795	0.173

We report the simple average of ω_2 for $estATE$, and ω_{2i} for the median (of average $\bar{\tau}_i$ by individuals) as $estATT$ in table 10.59. That is, we average $\bar{\tau}_i$ for each individual, then select the median value of the individual averages as the focus of treatment on treated. Panel data allow us to focus on the average treatment effect for an individual but the median reported almost surely involves different individuals across simulated samples.

Sample statistics for *ATE* and *ATT* ($\bar{\tau} = median\ [\bar{\tau}]$) along with the correlation between treatment and the treatment effect are reported in table 10.60. There is good correspondence between the average treatment effect estimates and sample statistics. The interval estimates for *ATT* are much tighter than those for *ATE*. Correlations between treatment and treatment effect suggest there is little to be gained from *IV* estimation. We explore this next.

2SLS-IV

Here, we follow Wooldridge [1997, 2003], and estimate average treatment effects via *2SLS-IV* in this random coefficients setting. We employ the residuals from regressions of $(s - \mu)\bar{\tau}$ and $\bar{\tau}$ on U as strong instruments, z_1 and z_2. Results for *2SLS-IV* are reported in table 10.61. The *IV* results correspond well with the *DGP* and the sample statistics for the average treatment effects. Also, as expected given the low correlation between treatment and treatment effect, *IV* produces similar results to those for *OLS*.

10.7 Regulated report precision

Table 10.61: Continuous treatment 2SLS-IV parameter and average treatment effect estimates for modest within individual report precision variation setting

statistic	estATE	estATT ($\bar{\tau} = median\,[\bar{\tau}]$)
mean	69849.	70150.
median	70096.	70312.
std.dev.	5017	2210
minimum	35410.	64461.
maximum	87738.	75467.

$$E\left[Y \mid s_i, \bar{\tau}_i\right] = \sum_{i=1}^{n} \omega_{0i} + \omega_{1i}\left(s_i - \mu\right)\bar{\tau}_i + \omega_{2i}\bar{\tau}_i$$

More variation

Finally, we explore a setting with greater between individuals report variation as well as continued within individual variation. The independent stochastic parameters below describe the data

Stochastic components	
parameters	number of draws
$\alpha \sim N\left(0.02, 0.005\right)$	n
$\alpha_d \sim N\left(0.02, 0.0005\right)$	nT
$\gamma \sim N\left(2.5, 1\right)$	n
$\beta \sim N\left(7, 1\right)$	n
$\beta_i = \beta + N\left(0, 0.001\right)$	nT
$s \sim N\left(1000, \sigma\right)$	nT

where σ is the equilibrium report standard deviation; σ varies across firms and through time for each firm and greater unobserved β_i variation produces increased residual uncertainty.

OLS

This setting allows identification of *ATE* and *ATT* where $ATT\left(\bar{\tau} = median\,[\bar{\tau}]\right)$. First, we estimate the average treatment effects via *OLS* where individual specific intercepts and slopes are accommodated.

$$E\left[Y \mid s_i, \bar{\tau}_i\right] = \sum_{i=1}^{n} \omega_{0i} + \omega_{1i}\left(s_i - \mu\right)\bar{\tau}_i + \omega_{2i}\bar{\tau}_i$$

We report the simple average of ω_{2i} for *estATE* and ω_{2i} for the median of average $\bar{\tau}_i$ by individuals as *estATT* in table 10.62.

Sample statistics for *ATE* and $ATT\left(\bar{\tau} = median\,[\bar{\tau}]\right)$ along with the correlation between treatment and the treatment effect are reported in table 10.63. As expected with greater residual variation, there is weaker correspondence between the average treatment effect estimates and sample statistics. Correlations between treatment and treatment effect again suggest there is little to be gained from *IV* estimation. We explore *IV* estimation next.

10. Treatment effects: *IV*

Table 10.62: Continuous treatment OLS parameter and average treatment effect estimates for the more between and within individual report precision variation setting

statistic	estATE	estATT ($\bar{\tau} = median\,[\bar{\tau}]$)
mean	71623.	67870.
median	70011.	68129.
std.dev.	34288.	22360.
minimum	−20220.	−8934.
maximum	223726.	141028.
$E\left[Y \mid s_i, \bar{\tau}_i\right] = \sum_{i=1}^{n} \omega_{0i} + \omega_{1i}\left(s_i - \mu\right)\bar{\tau}_i + \omega_{2i}\bar{\tau}_i$		

Table 10.63: Continuous treatment ATE and ATT sample statistics and correlation between treatment and treatment effect for the more between and within individual report precision variation setting

statistic	ATE	ATT ($\bar{\tau} = median\,[\bar{\tau}]$)	$corr\left(\omega_{2it}, \bar{\tau}_{it}\right)$
mean	69951.	69720.	−0.0062
median	69970.	70230.	−0.0129
std.dev.	709.	10454.	0.073
minimum	67639.	34734	−0.194
maximum	71896.	103509	0.217

2SLS-IV

Again, we follow Wooldridge's [1997, 2003] random coefficients analysis, and estimate average treatment effects via *2SLS-IV*. We employ the residuals from regressions of $(s - \mu)\bar{\tau}$ and $\bar{\tau}$ on U as strong instruments, z_1 and z_2. Results for *2SLS-IV* are reported in table 10.64. The *IV* results are similar to those for *OLS* as expected given the near zero correlation between treatment and treatment effect.

Table 10.64: Continuous treatment 2SLS-IV parameter and average treatment effect estimates for the more between and within individual report precision variation setting

statistic	estATE	estATT ($\bar{\tau} = median\,[\bar{\tau}]$)
mean	66247.	67644.
median	68998.	68004.
std.dev.	36587	22309.
minimum	−192442.	−9387.
maximum	192722.	141180.
$E\left[Y \mid s_i, \bar{\tau}_i\right] = \sum_{i=1}^{n} \omega_{0i} + \omega_{1i}\left(s_i - \mu\right)\bar{\tau}_i + \omega_{2i}\bar{\tau}_i$		

10.8 Summary

This chapter has surveyed some *IV* approaches for identifying and estimating average treatment effects and illustrated them in a couple of ways. The Tuebingen-style examples illustrate critical features for *IV* identification then we added accounting context. The endogenous selection of report precision examples highlight several key features in the econometric analysis of accounting choice. First, reliable results follow from carefully linking theory and data. For instance, who observes which data is fundamental. When the analysis demands instruments (ignorable treatment conditions are typically not satisfied by the data in this context), their identification and collection is critical. Poor instruments (exclusion restriction fails) or weak instruments (weakly associated with selection) can lead to situations where the "cure" is worse than the symptom. *IV* results can be less reliable (more prone to generate logical inconsistencies) than *OLS* when faced with endogeneity if we employ faulty instruments. Once again, we see there is no substitute for task-appropriate data. Finally, two (or more) poor analyses don't combine to produce one satisfactory analysis.

10.9 Additional reading

Wooldridge [2002] (chapter 18 is heavily drawn upon in these pages), Amemiya [1985, chapter 9], and numerous other econometric texts synthesize *IV* treatment effect identification strategies. Recent volumes of *Handbook of Econometrics* (especially volumes 5 and 6b) report extensive reviews as well as recent results.

11
Marginal treatment effects

In this chapter, we review policy evaluation and Heckman and Vytlacil's [2005, 2007a] (*HV*) strategy for linking marginal treatment effects to other average treatment effects including policy-relevant treatment effects. Recent innovations in the treatment effects literature including dynamic and general equilibrium considerations are mentioned briefly but in-depth study of these matters is not pursued. *HV*'s marginal treatment effects strategy is applied to the regulated report precision setting introduced in chapter 2, discussed in chapter 10, and continued in the next chapter. This analysis highlights the relative importance of probability distribution assignment to unobservables and quality of instruments.

11.1 Policy evaluation and policy invariance conditions

Heckman and Vytlacil [2007a] discuss causal effects and policy evaluation. Following the lead of Bjorklund and Moffitt [1987], *HV* base their analysis on marginal treatment effects. *HV*'s marginal treatment effects strategy combines the strengths of the treatment effect approach (simplicity and lesser demands on the data) and the Cowles Commission's structural approach (utilize theory to help extrapolate results to a broader range of settings). *HV* identify three broad classes of policy evaluation questions.

(P-1) Evaluate the impact of historically experienced and documented policies on outcomes via counterfactuals. Outcome or welfare evaluations may be objective (inherently ex post) or subjective (may be ex ante or ex post). P-1 is an *inter-*

nal validity problem (Campbell and Stanley [1963]) — the problem of identifying treatment parameter(s) in a given environment.

(P-2) Forecasting the impact of policies implemented in one environment by extrapolating to other environments via counterfactuals. This is the *external validity* problem (Campbell and Stanley [1963]).

(P-3) Forecasting the impact of policies never historically experienced to various environments via counterfactuals. This is the most ambitious policy evaluation problem.

The study of policy evaluation frequently draws on some form of policy invariance. Policy invariance allows us to characterize outcomes without fully specifying the structural model including incentives, assignment mechanisms, and choice rules. The following policy invariance conditions support this relaxation.[1]

(PI-1) For a given choice of treatment, outcomes are invariant to variations in incentive schedules or assignment mechanisms. PI-1 is a strong condition. It says that randomized assignment or threatening with a gun to gain cooperation has no impact on outcomes for a given treatment choice (see Heckman and Vytlacil [2007b] for evidence counter to the condition).

(PI-2) The actual mechanism used to assign treatment does not impact outcomes. This rules out general equilibrium effects (see Abbring and Heckman [2007]).

(PI-3) Utilities are unaffected by variations in incentive schedules or assignment mechanisms. This is the analog to (PI-1) but for utilities or subjective evaluations in place of outcomes. Again, this is a strong condition (see Heckman and Vytlacil [2007b] for evidence counter to the condition).

(PI-4) The actual mechanism used to assign treatment does not impact utilities. This is the analog to (PI-2) but for utilities or subjective evaluations in place of outcomes. Again, this rules out general equilibrium effects.

It's possible to satisfy (PI-1) and (PI-2) but not (PI-3) and (PI-4) (see Heckman and Vytlacil [2007b]). Next, we discuss marginal treatment effects and begin the exploration of how they unify policy evaluation.

Briefly, Heckman and Vytlacil's [2005] local instrumental variable (*LIV*) estimator is a more ambitious endeavor than the methods discussed in previous chapters. *LIV* estimates the marginal treatment effect (*MTE*) under standard *IV* conditions. *MTE* is the treatment effect associated with individuals who are indifferent between treatment and no treatment. Heckman and Vytlacil identify weighted distributions (Rao [1986] and Yitzhaki [1996]) that connect *MTE* to a variety of other treatment effects including *ATE, ATT, ATUT, LATE*, and policy-relevant treatment effects (*PRTE*).

MTE is a generalization of *LATE* as it represents the treatment effect for those individuals who are indifferent between treatment and no treatment.

$$MTE = E\left[Y_1 - Y_0 \mid X = x, V_D = v_D\right]$$

---

[1]Formal statements regarding policy invariance are provided in Heckman and Vytlacil [2007a].

Or, the marginal treatment effect can alternatively be defined by a transformation of unobservable V by $U_D = F_{V|X}(V)$ so that we can work with $U_D \sim Unif[0,1]$

$$MTE = E[Y_1 - Y_0 \mid X = x, U_D = u_D]$$

11.2 Setup

The setup is the same as the previous chapters. We repeat it for convenience. Suppose the *DGP* is
outcome equations:
$$Y_j = \mu_j(X) + V_j, j = 0, 1$$
selection equation:
$$D^* = \mu_D(Z) - V_D$$
observable response:
$$\begin{aligned} Y &= DY_1 + (1-D)Y_0 \\ &= \mu_0(X) + (\mu_1(X) - \mu_0(X))D + V_0 + (V_1 - V_0)D \end{aligned}$$
where
$$D = \begin{array}{cc} 1 & D^* > 0 \\ 0 & otherwise \end{array}$$

and Y_1 is (potential) outcome with treatment while Y_0 is the outcome without treatment. The outcomes model is the Neyman-Fisher-Cox-Rubin model of potential outcomes (Neyman [1923], Fisher [1966], Cox]1958], and Rubin [1974]). It is also Quandt's [1972] switching regression model or Roy's income distribution model (Roy [1951] or Heckman and Honore [1990]).

The usual exclusion restriction and uniformity applies. That is, if instrument changes from z to z' then everyone either moves toward or away from treatment. Again, the treatment effects literature is asymmetric; heterogeneous outcomes are permitted but homogeneous treatment is required for identification of estimators. Next, we repeat the generalized Roy model — a useful frame for interpreting causal effects.

11.3 Generalized Roy model

Roy [1951] introduced an equilibrium model for work selection (hunting or fishing).[2] An individual's selection into hunting or fishing depends on his/her aptitude

---

[2] The *basic* Roy model involves no cost of treatment. The *extended* Roy model includes only observed cost of treatment. While the *generalized* Roy model includes both observed and unobserved cost of treatment (see Heckman and Vytlacil [2007a, 2007b]).

as well as supply of and demand for product of labor. A modest generalization of the Roy model is a common framing of self-selection that forms the basis for assessing treatment effects (Heckman and Robb [1986]).

Based on the *DGP* above, we identify the constituent pieces of the selection model.

Net benefit (or utility) from treatment is

$$\begin{aligned} D^* &= \mu_D(Z) - V_D \\ &= Y_1 - Y_0 - c(W) - V_c \\ &= \mu_1(X) - \mu_0(X) - c(W) + V_1 - V_0 - V_C \end{aligned}$$

Gross benefit of treatment is

$$\mu_1(X) - \mu_0(X)$$

Cost associated with treatment is[3]

$$c(W) + V_C$$

Observable cost associated with treatment is

$$c(W)$$

Observable net benefit of treatment is

$$\mu_1(X) - \mu_0(X) - c(W)$$

Unobservable net benefit of treatment is

$$-V_D = V_1 - V_0 - V_C$$

where the observables are $\begin{bmatrix} X & Z & W \end{bmatrix}$, typically Z contains variables not in X or W and W is the subset of observables that speak to cost of treatment.

11.4 Identification

Marginal treatment effects are defined conditional on the regressors X and unobserved utility V_D

$$MTE = E[Y_1 - Y_0 \mid X = x, V_D = v_D]$$

or transformed unobserved utility U_D.

$$MTE = E[Y_1 - Y_0 \mid X = x, U_D = u_D]$$

HV describe the following identifying conditions.

---

[3]The model is called the *original* or *basic* Roy model if the cost term is omitted. If the cost is constant ($V_C = 0$ so that cost is the same for everyone) it is called the *extended* Roy model.

11.4 Identification

Condition 11.1 $\{U_0, U_1, V_D\}$ are independent of Z conditional on X (conditional independence),

Condition 11.2 $\mu_D(Z)$ is a nondegenerate random variable conditional on X (rank condition),

Condition 11.3 the distribution of V_D is continuous,

Condition 11.4 the values of $E[||Y_0||]$ and $E[||Y_1||]$ are finite (finite means),

Condition 11.5 $0 < Pr(D = 1 \mid X) < 1$ (common support).

These are the base conditions for *MTE*. They are augmented below to facilitate interpretation.[4] Condition 11.7 applies specifically to policy-relevant treatment effects where p and p' refer to alternative policies.

Condition 11.6 Let X_0 denote the counterfactual value of X that would be observed if D is set to 0. X_1 is defined analogously. Assume $X_d = X$ for $d = 0, 1$. (The X_D are invariant to counterfactual manipulations.)

Condition 11.7 The distribution of $(Y_{0,p}, Y_{1,p}, V_{D,p})$ conditional on $X_p = x$ is the same as the distribution of $(Y_{0,p'}, Y_{1,p'}, V_{D,p'})$ conditional on $X_{p'} = x$ (policy invariance of the distribution).

Under the above conditions, *MTE* can be estimated by local *IV* (*LIV*)

$$LIV = \left.\frac{\partial E[Y|X=x,P(Z)=p]}{\partial p}\right|_{p=u_D}$$

where $P(Z) \equiv Pr(D \mid Z)$. To see the connection between *MTE* and *LIV* rewrite the numerator of *LIV*

$$E[Y \mid X = x, P(Z) = p] = E[Y_0 + (Y_1 - Y_0)D \mid X = x, P(Z) = p]$$

by conditional independence and Bayes' theorem we have

$$E[Y_0 \mid X = x] + E[Y_1 - Y_0 \mid X = x, D = 1] Pr(D = 1 \mid Z = z)$$

transforming V_D such that U_D is distributed uniform$[0, 1]$ produces

$$E[Y_0 \mid X = x] + \int_0^p E[Y_1 - Y_0 \mid X = x, U_D = u_D] \, du_D$$

Now, the partial derivative of this expression with respect to p evaluated at $p = u_D$ is

$$\left.\frac{\partial E[Y|X=x,P(Z)=p]}{\partial p}\right|_{p=u_D} = E[Y_1 - Y_0 \mid X = x, U_D = u_D]$$

---

[4] The conditions remain largely the same for *MTE* analysis of alternative settings including multi-level discrete treatment, continuous treatment, and discrete outcomes. Modifications are noted in the discussions of each.

Hence, LIV identifies MTE.

With homogeneous response, MTE is constant and equal to ATE, ATT, and $ATUT$. With unobservable heterogeneity, MTE is typically a nonlinear function of u_D (where u_D continues to be distributed uniform$[0, 1]$). The intuition for this is individuals who are less likely to accept treatment require a larger potential gain from treatment to induce treatment selection than individuals who are more likely to participate.

11.5 *MTE* connections to other treatment effects

Heckman and Vytlacil show that MTE can be connected to other treatment effects (*TE*) by weighted distributions $h_{TE}(\cdot)$ (Rao [1986] and Yitzhaki [1996]).[5] Broadly speaking and with full support

$$TE(x) = \int_0^1 MTE(x, u_D) h_{TE}(x, u_D) \, du_D$$

and integrating out x yields the population moment

$$Average(TE) = \int_0^1 TE(x) \, dF(x)$$

If full support exists, then the weight distribution for the average treatment effect is

$$h_{ATE}(x, u_D) = 1$$

Let f be the density function of observed utility $\tilde{W} = \mu_D(Z)$, then the weighted distribution to recover the treatment effect on the treated from MTE is

$$\begin{aligned} h_{TT}(x, u_D) &= \left[\int_{u_D}^1 f(p \mid X = x) \, dp \right] \frac{1}{E[p \mid X = x]} \\ &= \frac{\Pr\left(P\left(\tilde{W}\right) > u_D \mid X = x\right)}{\int_0^1 \Pr\left(P\left(\tilde{W}\right) > u_D \mid X = x\right) du_d} \end{aligned}$$

where $P\left(\tilde{W}\right) \equiv \Pr\left(D = 1 \mid \tilde{W} = w\right)$. Similarly, the weighted distribution to recover the treatment effect on the untreated from MTE is

$$\begin{aligned} h_{TUT}(x, u_D) &= \left[\int_0^{u_D} f(p \mid X = x) \, dp \right] \frac{1}{E[1 - p \mid X = x]} \\ &= \frac{\Pr\left(P\left(\tilde{W}\right) \leq u_D \mid X = x\right)}{\int_0^1 \Pr\left(P\left(\tilde{W}\right) \leq u_D \mid X = x\right) du_d} \end{aligned}$$

---

[5]Weight functions are nonnegative and integrate to one (like density functions).

11.5 MTE connections to other treatment effects

Figure 11.1 depicts MTE ($\Delta_{MTE}(u_D)$) and weighted distributions for treatment on treated $h_{TT}(u_D)$ and treatment on the untreated $h_{TUT}(u_D)$ with regressors suppressed.

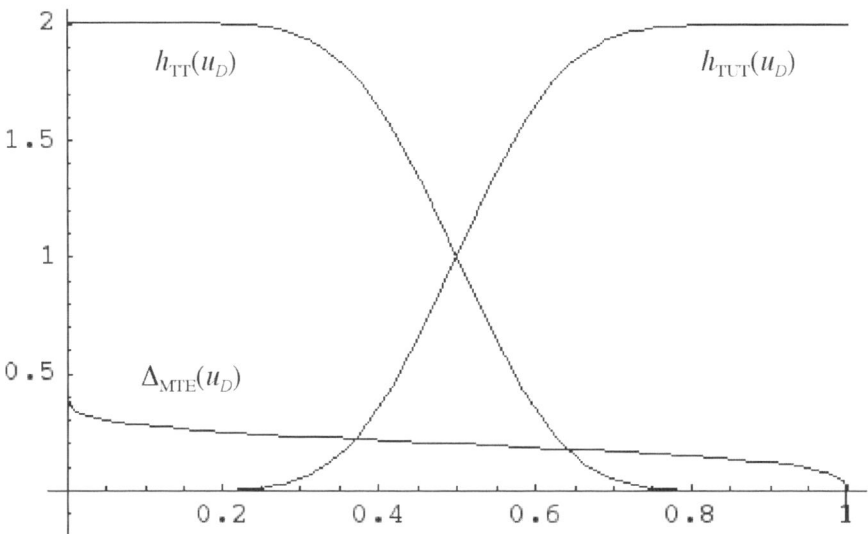

Figure 11.1: MTE and weight functions for other treatment effects

Applied work determines the weights by estimating

$$\Pr\left(P\left(\tilde{W}\right) > u_D \mid X = x\right)$$

Since $\Pr\left(P\left(\tilde{W}\right) > u_D \mid X = x\right) = \Pr\left(I\left[P\left(\tilde{W}\right) > u_D\right] = 1 \mid X = x\right)$ where $I\left[\cdot\right]$ is an indicator function, we can use our selection or choice model (say, probit) to estimate

$$\Pr\left(I\left[P\left(\tilde{W}\right) > u_D\right] = 1 \mid X = x\right)$$

for each value of u_D. As the weighted distributions integrate to one, we use their sum to determine the normalizing constant (the denominator). The analogous idea applies to $h_{TUT}(x, u_D)$.

However, it is rare that full support is satisfied as this implies both treated and untreated samples would be evidenced at all probability levels for some model of treatment (e.g., probit). Often, limited support means the best we can do is estimate a local average treatment effect.

$$LATE(x) = \frac{1}{u' - u} \int_u^{u'} MTE(x, u_D) \, du_D$$

In the limit as the interval becomes arbitrarily small $LATE$ converges to MTE.

11.5.1 Policy-relevant treatment effects vs. policy effects

What is the average gross gain from treatment following policy intervention? This is a common question posed in the study of accounting. Given uniformity (one way flows into or away from participation in response to a change in instrument) and policy invariance, *IV* can identify the average treatment effect for policy a compared with policy a', that is, a policy-relevant treatment effect (*PRTE*). Policy invariance means the policy impacts the likelihood of treatment but not the potential outcomes (that is, the distributions of $\{y_{1a}, y_{0a}, V_{Da} \mid X_a = x\}$ and $\{y_{1a'}, y_{0a'}, V_{Da'} \mid X_{a'} = x\}$ are equal).

The policy-relevant treatment effect is

$$PRTE = E[Y \mid X = x, a] - E[Y \mid X = x, a']$$
$$= \int_0^1 MTE(x, u_D) \left[F_{P(a')\mid X}(u_D \mid x) - F_{P(a)\mid X}(u_D \mid x) \right] du_D$$

where $F_{P(a)\mid X}(u_D \mid x)$ is the distribution of P, the probability of treatment conditional on $X = x$, and the weight function is $h_{PRTE}(x, u_D)$.[6]

$$h_{PRTE}(x, u_D) = \left[F_{P(a')\mid X}(u_D \mid x) - F_{P(a)\mid X}(u_D \mid x) \right]$$

Intuition for the above connection can be seen as follows, where conditioning on X is implicit.

$$E[Y \mid a] = \int_0^1 E[Y \mid P(Z) = p] dF_{P(a)}(p)$$
$$= \int_0^1 \left(\begin{array}{c} \int_0^1 \mathfrak{I}_{[0,p]}(u_D) E(Y_1 \mid U = u_D) \\ + \mathfrak{I}_{(p,1]}(u_D) E(Y_0 \mid U = u) du_D \end{array} \right) dF_{P(a)}(p)$$
$$= \int_0^1 \left(\begin{array}{c} [1 - F_{P(a)}(u_D)] E[Y_1 \mid U = u_D] \\ + F_{P(a)}(u_D) E[Y_0 \mid U = u_D] \end{array} \right) du_D$$

---

[6]Heckman and Vytlacil [2005] also identify the per capita weight for policy-relevant treatment as

$$\frac{\Pr\left(P\left(\tilde{W}\right) \leq u_D \mid X = x, a'\right) - \Pr\left(P\left(\tilde{W}\right) \leq u_D \mid X = x, a\right)}{\int_0^1 \Pr\left(P\left(\tilde{W}\right) \leq u_D \mid X = x, a'\right) du_d - \int_0^1 \Pr\left(P\left(\tilde{W}\right) \leq u_D \mid X = x, a\right) du_d}$$

where $\Im_A(u_D)$ is an indicator function for the event $u_D \in A$. Hence, comparing policy a to a', we have

$$\begin{aligned}
&E[Y \mid X = x, a] - E[Y \mid X = x, a'] \\
&= \int_0^1 \left(\begin{array}{l} [1 - F_{P(a)}(u_D)] E[Y_1 \mid U = u_D] \\ + F_{P(a)}(u_D) E[Y_0 \mid U = u_D] \end{array} \right) du_D \\
&\quad - \int_0^1 \left(\begin{array}{l} [1 - F_{P(a')}(u_D)] E[Y_1 \mid U = u_D] \\ + F_{P(a')}(u_D) E[Y_0 \mid U = u_D] \end{array} \right) du_D \\
&= \int_0^1 [F_{P(a')}(u_D) - F_{P(a)}(u_D)] E[Y_1 - Y_0 \mid U = u_D] du_D \\
&= \int_0^1 [F_{P(a')}(u_D) - F_{P(a)}(u_D)] MTE(U = u_D) du_D
\end{aligned}$$

On the other hand, we might be interested in the policy effect or net effect of a policy change rather than the treatment effect. In which case it is perfectly sensible to estimate the net impact with some individuals leaving and some entering, this is a policy effect not a treatment effect. The policy effect parameter is $E[Y \mid Z_{a'} = z'] - E[Y \mid Z_a = z]$

$$\begin{aligned}
&= E[Y_1 - Y_0 \mid D(z') > D(z)] \Pr(D(z') > D(z)) \\
&\quad - E[Y_1 - Y_0 \mid D(z') \leq D(z)] \Pr(D(z') \leq D(z))
\end{aligned}$$

Notice the net impact may be positive, negative, or zero as two way flows are allowed (see Heckman and Vytlacil [2006]).

11.5.2 Linear IV weights

As mentioned earlier, *HV* argue that linear *IV* produces a complex weighting of effects that can be difficult to interpret and depends on the instruments chosen. This argument is summarized by their linear *IV* weight distribution. Let $J(Z)$ be any function of Z such that $Cov[J(Z), D] \neq 0$. The population analog of the *IV* estimator is $\frac{Cov[J(Z), Y]}{Cov[J(Z), D]}$. Consider the numerator.

$$\begin{aligned}
Cov[J(Z), Y] &= E[(J(Z) - E[J(Z)]) Y] \\
&= E[(J(Z) - E[J(Z)])(Y_0 + D(Y_1 - Y_0))] \\
&= E[(J(Z) - E[J(Z)]) D(Y_1 - Y_0)]
\end{aligned}$$

Define $\tilde{J}(Z) = J(Z) - E[J(Z)]$. Then, $Cov[J(Z), Y]$

$$\begin{aligned}
&= E\left[\tilde{J}(Z) D (Y_1 - Y_0)\right] \\
&= E\left[\tilde{J}(Z) I [U_D \leq P(Z)] (Y_1 - Y_0)\right] \\
&= E\left[\tilde{J}(Z) I [U_D \leq P(Z)] E [(Y_1 - Y_0) \mid Z, V_D]\right] \\
&= E\left[\tilde{J}(Z) I [U_D \leq P(Z)] E [(Y_1 - Y_0) \mid V_D]\right] \\
&= E_{V_D}\left[E_Z\left[\tilde{J}(Z) I [U_D \leq P(Z)] \mid U_D\right] E [(Y_1 - Y_0) \mid U_D]\right] \\
&= \int_0^1 E\left[\tilde{J}(Z) \mid P(Z) \geq u_D\right] \Pr(P(Z) \geq u_D) \\
&\quad \times E [(Y_1 - Y_0) \mid U_D = u_D] \, du_D \\
&= \int_0^1 \Delta_{MTE}(x, u_D) E\left[\tilde{J}(Z) \mid P(Z) \geq u_D\right] \Pr(P(Z) \geq u_D) \, du_D
\end{aligned}$$

where $P(Z)$ is propensity score utilized as an instrument.

For the denominator we have, by iterated expectations,

$$Cov[J(Z), D] = Cov[J(Z), P(Z)]$$

Hence,

$$h_{IV}(x, u_D) = \frac{E\left[\tilde{J}(Z) \mid P(Z) \geq u_D\right] \Pr(P(Z) \geq u_D)}{Cov[J(Z), P(Z)]}$$

where $Cov[J(Z), P(Z)] \neq 0$. Heckman, Urzua, and Vytlacil [2006] illustrate the sensitivity of treatment effects identified via linear *IV* to choice of instruments.

11.5.3 OLS weights

It's instructive to identify the effect exogenous dummy variable *OLS* estimates as a function of *MTE*. While not a true weighted distribution (as the weights can be negative and don't necessarily sum to one), for consistency we'll write $h_{OLS}(x, u_D) =$

$$\begin{cases} 1 + \frac{E[Y_1 \mid x, u_D] h_{ATT}(x, u_D) - E[v_0 \mid x, u_D] h_{ATUT}(x, u_D)}{MTE(x, u_D)} & MTE(x, u_D) \neq 0 \\ 0 & otherwise \end{cases}$$

Table 11.1: Comparison of identification conditions for common econometric strategies (adapted from Heckman and Navarro-Lozano's [2004] table 3)

Method	Exclusion required?	Separability of observables and unobservables in outcome equations?
Matching	no	no
Control function	yes, for nonparametric identification	conventional, but not required
IV (linear)	yes	yes
LIV	yes	no
Method	Functional form required?	Marginal = Average (given X, Z)?
Matching	no	yes
Control function	conventional, but not required	no
IV (linear)	no	no (yes, in standard case)
LIV	no	no
Method	Key identification conditions for means (assuming separability)	
Matching	$E[U_1 \mid X, D=1, Z] = E[U_1 \mid X, Z]$ $E[U_0 \mid X, D=1, Z] = E[U_0 \mid X, Z]$	
Control function	$E[U_0 \mid X, D=1, Z]$ and $E[U_1 \mid X, D=1, Z]$ can be varied independently of $\mu_0(X)$ and $\mu_1(X)$, respectively, and intercepts can be identified through limit arguments (identification at infinity), or symmetry assumptions	
IV (linear)	$E[U_0 + D(U_1 - U_0) \mid X, Z] = E[U_0 + D(U_1 - U_0) \mid X]$ (ATE) $E[U_0 + D(U_1 - U_0) - E[U_0 + D(U_1 - U_0) \mid X] \mid P(W), X]$ $= E[U_0 + D(U_1 - U_0) - E[U_0 + D(U_1 - U_0) \mid X] \mid X]$ (ATT)	
LIV	(U_0, U_1, U_D) independent of $Z \mid X$	
Method	Key identification conditions for propensity score	
Matching	$0 < \Pr(D=1 \mid Z, X) < 1$	
Control function	$0 \leq \Pr(D=1 \mid Z, X) \leq 1$ is a nontrivial function of Z for each X	
IV (linear)	not needed	
LIV	$0 < \Pr(D=1 \mid X) < 1$ $0 \leq \Pr(D=1 \mid Z, X) \leq 1$ is a nontrivial function of Z for each X	

11.6 Comparison of identification strategies

Following Heckman and Navarro-Lozano [2004], we compare and report in table 11.1 treatment effect identification strategies for four common econometric approaches: matching (especially, propensity score matching), control functions (selection models), conventional (linear) instrumental variables (*IV*), and local instrumental variables (*LIV*).

All methods define treatment parameters on common support — the intersection of the supports of X given $D = 1$ and X given $D = 0$, that is,

$$Support\,(X \mid D = 1) \cap Support\,(X \mid D = 0)$$

LIV employs common support of the propensity score — overlaps in $P(X, Z)$ for $D = 0$ and $D = 1$. Matching breaks down if there exists an explanatory variable that serves as a perfect classifier. On the other hand, control functions exploit limit arguments for identification,[7] hence, avoiding the perfect classifier problem. That is, identification is secured when $P(X, Z) = 1$ for some $Z = z$ but there exists $P(X, Z) < 1$ for some $Z = z'$. Similarly, when $P(W) = 0$, where $W = (X, Z)$, for some $Z = z$ there exists $P(X, Z) > 0$ for some $Z = z''$.

11.7 *LIV* estimation

We've laid the groundwork for the potential of marginal treatment effects to address various treatment effects in the face of unobserved heterogeneity, it's time to discuss estimation. Earlier, we claimed *LIV* can estimate *MTE*

$$\left. \frac{\partial E[Y|X=x, P(Z)=p]}{\partial p} \right|_{p=u_D} = E\left[Y_1 - Y_0 \mid X = x, U_D = u_D\right]$$

For the linear separable model we have

$$Y_1 = \delta + \alpha + X\beta_1 + V_1$$

and

$$Y_0 = \delta + X\beta_0 + V_0$$

Then,

$$E\left[Y \mid X = x, P(Z) = p\right] = X\beta_0 + X(\beta_1 - \beta_0)\Pr(Z) + \kappa(p)$$

where

$$\kappa(p) = \alpha \Pr(Z) + E\left[v_0 \mid \Pr(Z) = p\right] + E\left[v_1 - v_0 \mid D = 1, \Pr(Z) = p\right]\Pr(Z)$$

Now, *LIV* simplifies to

$$LIV = X(\beta_1 - \beta_0) + \left. \frac{\partial \kappa(p)}{\partial p} \right|_{p=u_D}$$

---

[7] This is often called "identification at infinity."

Since *MTE* is based on the partial derivative of expected outcome with respect to p

$$\frac{\partial}{\partial p} E[Y \mid X = x, P(Z) = p] = X(\beta_1 - \beta_0) + \frac{\partial \kappa(p)}{\partial p},$$

the objective is to estimate $(\beta_1 - \beta_0)$ and the derivative of $\kappa(p)$. Heckman, Urzua, and Vytlacil's [2006] local *IV* estimation strategy employs a relaxed distributional assignment based on the data and accommodates unobservable heterogeneity. *LIV* employs nonparametric (local linear kernel density; see chapter 6) regression methods.

LIV Estimation proceeds as follows:

Step 1: Estimate the propensity score, $P(Z)$, via probit, nonparametric discrete choice, etc.

Step 2: Estimate β_0 and $(\beta_1 - \beta_0)$ by employing a nonparametric version of *FWL* (double residual regression). This involves a local linear regression (*LLR*) of each regressor in X and $X * P(Z)$ onto $P(Z)$. *LLR* for X_k (the kth regressor) is $\{\tau_{0k}(p), \tau_{1k}(p)\} =$

$$\arg\min_{\{\tau_0(p), \tau_1(p)\}} \left\{ \sum_{j=1}^n (X_k(j) - \tau_0 - \tau_1(P(Z_j) - p))^2 K\left(\left|\frac{P(Z_j) - p}{h}\right|\right) \right\}$$

where $K(W)$ is a (Gaussian, biweight, or Epanechnikov) kernel evaluated at W. The bandwidth h is estimated by leave-one out generalized cross-validation based on the nonparametric regression of $X_k(j)$ onto $(\tau_{0k} + \tau_{1k}p)$.

For each regressor in X and $X * P(Z)$ and for the response variable y estimate the residuals from *LLR*. Denote the matrix of residuals from the regressors (ordered with X followed by $X * P(Z)$) as e_X and the residuals from Y, e_Y.

Step 3: Estimate $[\beta_0, \beta_1 - \beta_0]$ from a no-intercept linear regression of e_Y onto e_X. That is, $\left[\widehat{\beta_0}, \widehat{\beta_1 - \beta_0}\right]^T = \left[e_X^T e_X\right]^{-1} e_X^T e_Y$.

Step 4: For $E[Y \mid X = x, P(Z) = p]$, we've effectively estimated $\beta_0 X_i + (\beta_1 - \beta_0) X_i * P(Z_i)$. What remains is to estimate the derivative of $\kappa(p)$. We complete nonparametric *FWL* by defining the restricted response as follows.

$$\tilde{Y}_i = Y_i - \widehat{\beta_0} X_i - \left(\widehat{\beta_1 - \beta_0}\right) X_i * P(Z_i)$$

The intuition for utilizing the restricted response is as follows. In the textbook linear model case

$$Y = X\beta + Z\gamma + \varepsilon$$

FWL produces

$$E[Y \mid X, Z] = P_Z Y + (I - P_Z) Xb$$

where b is the *OLS* estimator for β and P_Z is the projection matrix $Z(Z^T Z)^{-1} Z^T$. Rewriting we can identify the estimator for γ, g, from

$$E[Y \mid X, Z] = Xb + P_Z(Y - Xb) = Xb + Zg$$

Hence, $g = \left(Z^T Z\right)^{-1} Z^T (Y - Xb)$. That is, g is estimated from a regression of the restricted response $(Y - Xb)$ onto the regressor Z. *LIV* employs the nonparametric analog.

Step 5: Estimate $\tau_1(p) = \frac{\partial \kappa(p)}{\partial p}$ by *LLR* of $Y_i - \widehat{\beta_0} X_i - \left(\widehat{\beta_1 - \beta_0}\right) X_i * P(Z_i)$ onto $P(Z_i)$ for each observation i in the set of overlaps. The set of overlaps is the region for which *MTE* is identified — the subset of common support of $P(Z)$ for $D = 1$ and $D = 0$.

Step 6: The *LIV* estimator of $MTE(x, u_D)$ is $\left(\widehat{\beta_1 - \beta_0}\right) X + \widehat{\tau_1(p)}$.

MTE depends on the propensity score p as well as X. In the homogeneous response setting, MTE is constant and $MTE = ATE = ATT = ATUT$. While in the heterogeneous response setting, *MTE* is nonlinear in p.

11.8 Discrete outcomes

Aakvik, Heckman, and Vytlacil [2005] (*AHV*) describe an analogous *MTE* approach for the discrete outcomes case. The setup is analogous to the continuous case discussed above except the following modifications are made to the potential outcomes model.

$$Y_1 = \mu_1(X, U_1)$$
$$Y_0 = \mu_0(X, U_0)$$

A linear latent index is assumed to generate discrete outcomes

$$\mu_j(X, U_j) = I\left[X\beta_j \geq U_j\right]$$

AHV describe the following identifying conditions.

Condition 11.8 (U_0, V_D) and (U_1, V_D) are independent of (Z, X) (conditional independence),

Condition 11.9 $\mu_D(Z)$ is a nondegenerate random variable conditional on X (rank condition),

Condition 11.10 (V_0, V_D) and (V_1, V_D) are continuous,

Condition 11.11 the values of $E[|Y_0|]$ and $E[|Y_1|]$ are finite (finite means is trivially satisfied for discrete outcomes),

Condition 11.12 $0 < Pr(D = 1 \mid X) < 1$.

Mean treatment parameters for dichotomous outcomes are

$$\begin{align}
MTE\,(x,u) &= \Pr(Y_1 = 1 \mid X = x, U_D = u) \\
&\quad - \Pr(Y_0 = 1 \mid X = x, U_D = u) \\
ATE\,(x) &= \Pr(Y_1 = 1 \mid X = x) - \Pr(Y_0 = 1 \mid X = x) \\
ATT\,(x, D = 1) &= \Pr(Y_1 = 1 \mid X = x, D = 1) \\
&\quad - \Pr(Y_0 = 1 \mid X = x, D = 1) \\
ATUT\,(x, D = 0) &= \Pr(Y_1 = 1 \mid X = x, D = 0) \\
&\quad - \Pr(Y_0 = 1 \mid X = x, D = 0)
\end{align}$$

AHV also discuss and empirically estimate treatment effect distributions utilizing a (single) factor-structure strategy for model unobservables.[8]

11.8.1 Multilevel discrete and continuous endogenous treatment

To this point, our treatment effects discussion has been limited to binary treatment. In this section, we'll briefly discuss extensions to the multilevel discrete (ordered and unordered) case (Heckman and Vytlacil [2007b]) and continuous treatment case (Florens, Heckman, Meghir, and Vytlacil [2003] and Heckman and Vytlacil [2007b]). Identification conditions are similar for all cases of multinomial treatment.

FHMV and *HV* discuss conditions under which control function, *IV*, and *LIV* equivalently identify *ATE* via the partial derivative of the outcome equation with respect to (continuous) treatment. This is essentially the homogeneous response case. In the heterogenous response case, *ATE* can be identified by a control function or *LIV* but under different conditions. *LIV* allows relaxation of the standard single index (uniformity) assumption. Refer to *FHMV* for details. Next, we return to *HV*'s *MTE* framework and briefly discuss how it applies to ordered choice, unordered choice, and continuous treatment.

Ordered choice

Consider an ordered choice model where there are S choices. Potential outcomes are
$$Y_s = \mu_s\,(X, U_s) \quad \text{for } s = 1, \ldots, S$$
Observed choices are
$$D_s = \mathbf{1}\left[C_{s-1}\,(W_{s-1}) < \mu_D\,(Z) - V_D < C_s\,(W_s)\right]$$
for latent index $U = \mu_D\,(Z) - V_D$ and cutoffs $C_s\,(W_s)$ where Z shift the index generally and W_s affect s-specific transitions. Intuitively, one needs an instrument

---

[8]Carneiro, Hansen, and Heckman [2003] extend this by analyzing panel data, allowing for multiple factors, and more general choice processes.

(or source of variation) for each transition. Identifying conditions are similar to those above.

Condition 11.13 (U_s, V_D) *are independent of* (Z, W) *conditional on* X *for* $s = 1, \ldots, S$ *(conditional independence),*

Condition 11.14 $\mu_D(Z)$ *is a nondegenerate random variable conditional on* (X, W) *(rank condition),*

Condition 11.15 *the distribution of* V_D *is continuous,*

Condition 11.16 *the values of* $E[|Y_s|]$ *are finite for* $s = 1, \ldots, S$ *(finite means),*

Condition 11.17 $0 < Pr(D_s = 1 \mid X) < 1$ *for* $s = 1, \ldots, S$ *(in large samples, there are some individuals in each treatment state).*

Condition 11.18 *For* $s = 1, \ldots, S - 1$, *the distribution of* $C_s(W_s)$ *conditional on* (X, Z) *and the other* $C_j(W_j)$, $j = 1, \ldots S$, $j \neq s$, *is nondegenerate and continuous.*

The transition-specific *MTE* for the transition from s to $s + 1$ is

$$\Delta_{s,s+1}^{MTE}(x, v) = E[Y_{s+1} - Y_s \mid X = x, V_D = v] \quad \text{for } s = 1, \ldots, S - 1$$

Unordered choice

The parallel conditions for evaluating causal effects in multilevel unordered discrete treatment models are:

Condition 11.19 (U_s, V_D) *are independent of* Z *conditional on* X *for* $s = 1, \ldots, S$ *(conditional independence),*

Condition 11.20 *for each* Z_j *there exists at least one element* $Z^{[j]}$ *that is not an element of* Z_k, $j \neq k$, *and such that the distribution of* $\mu_D(Z)$ *conditional on* $(X, Z^{[-j]})$ *is not degenerate,*

or

Condition 11.21 *for each* Z_j *there exists at least one element* $Z^{[j]}$ *that is not an element of* Z_k, $j \neq k$, *and such that the distribution of* $\mu_D(Z)$ *conditional on* $(X, Z^{[-j]})$ *is continuous.*

Condition 11.22 *the distribution of* V_D *is continuous,*

Condition 11.23 *the values of* $E[|Y_s|]$ *are finite for* $s = 1, \ldots, S$ *(finite means),*

Condition 11.24 $0 < Pr(D_s = 1 \mid X) < 1$ *for* $s = 1, \ldots, S$ *(in large samples, there are some individuals in each treatment state).*

The treatment effect is $Y_j - Y_k$ where $j \neq k$. And regime j can be compared with the best alternative, say k, or other variations.

Continuous treatment

Continue with our common setup except assume outcome Y_d is continuous in d. This implies that for d and d' close so are Y_d and $Y_{d'}$. The average treatment effect can be defined as

$$ATE_d(x) = E\left[\frac{\partial}{\partial d}Y_d \mid X = x\right]$$

The average treatment effect on treated is

$$ATT_d(x) = E\left[\frac{\partial}{\partial d_1}Y_{d_1} \mid D = d_2, X = x\right]\bigg|_{d=d_1=d_2}$$

And the marginal treatment effect is

$$MTE_d(x, u) = E\left[\frac{\partial}{\partial d}Y_d \mid X = x, U_D = u\right]$$

See Florens, Heckman, Meghir, and Vytlacil [2003] and Heckman and Vytlacil [2007b, pp.5021-5026] for additional details regarding semiparametric identification of treatment effects.

11.9 Distributions of treatment effects

A limitation of the discussion to this juncture is we have focused on population means of treatment effects. This prohibits discussion of potentially important properties such as the proportion of individuals who benefit or who suffer from treatment.

Abbring and Heckman [2007] discuss utilization of factor models to identify the joint distribution of outcomes (including counterfactual distributions) and accordingly the distribution of treatment effects $Y_1 - Y_0$. Factor models are a type of replacement function (Heckman and Robb [1986]) where conditional on the factors, outcomes and choice equations are independent. That is, we rely on a type of conditional independence for identification. A simple one-factor model illustrates. Let θ be a scalar factor that produces dependence amongst the unobservables (unobservables are assumed to be independent of (X, Z)). Let M be a proxy measure for θ where $M = \mu_M(X) + \alpha_M \theta + \varepsilon_M$

$$\begin{aligned} V_0 &= \alpha_0 \theta + \varepsilon_0 \\ V_1 &= \alpha_1 \theta + \varepsilon_1 \\ V_D &= \alpha_D \theta + \varepsilon_D \end{aligned}$$

$\varepsilon_0, \varepsilon_1, \varepsilon_D, \varepsilon_M$ are mutually independent and independent of θ, all with mean zero. To fix the scale of the unobserved factor, normalize one coefficient (loading) to, say, $\alpha_M = 1$. The key is to exploit the notion that all of the dependence arises from θ.

$$Cov[Y_0, M \mid X, Z] = \alpha_0 \alpha_M \sigma_\theta^2$$
$$Cov[Y_1, M \mid X, Z] = \alpha_1 \alpha_M \sigma_\theta^2$$
$$Cov[Y_0, D^* \mid X, Z] = \alpha_0 \frac{\alpha_D}{\sigma_{U_D}} \sigma_\theta^2$$
$$Cov[Y_1, D^* \mid X, Z] = \alpha_1 \frac{\alpha_D}{\sigma_{U_D}} \sigma_\theta^2$$
$$Cov[D^*, M \mid X, Z] = \frac{\alpha_D}{\sigma_{U_D}} \alpha_M \sigma_\theta^2$$

From the ratio of $Cov[Y_1, D^* \mid X, Z]$ to $Cov[D^*, M \mid X, Z]$, we find α_1 ($\alpha_M = 1$ by normalization). From $\frac{Cov[Y_1, D^* \mid X, Z]}{Cov[Y_0, D^* \mid X, Z]} = \frac{\alpha_1}{\alpha_0}$, we determine α_0. Finally, from either $Cov[Y_0, M \mid X, Z]$ or $Cov[Y_1, M \mid X, Z]$ we determine scale σ_θ^2. Since $Cov[Y_0, Y_1 \mid X, Z] = \alpha_0 \alpha_1 \sigma_\theta^2$, the joint distribution of objective outcomes is identified.

See Abbring and Heckman [2007] for additional details, including use of proxies, panel data and multiple factors for identification of joint distributions of subjective outcomes, and references.

11.10 Dynamic timing of treatment

The foregoing discussion highlights one time (now or never) static analysis of the choice of treatment. In some settings it's important to consider the impact of acquisition of information on the option value of treatment. It is important to distinguish what information is available to decision makers and when and what information is available to the analyst. Distinctions between ex ante and ex post impact and subjective versus objective gains to treatment are brought to the fore.

Policy invariance (P-1 through P-4) as well as the distinction between the evaluation problem and the selection problem lay the foundation for identification. The evaluation problem is one where we observe the individual in one treatment state but wish to determine the individual's outcome in another state. The selection problem is one where the distribution of outcomes for an individual we observe in a given state is not the same as the marginal outcome distribution we would observe if the individual is randomly assigned to the state. Policy invariance simplifies the dynamic evaluation problem to (a) identifying the dynamic assignment of treatments under the policy, and (b) identifying dynamic treatment effects on individual outcomes.

Dynamic treatment effect analysis typically takes the form of a duration model (or time to treatment model; see Heckman and Singer [1986] for an early and extensive review of the problem). A variety of conditional independence, matching, or dynamic panel data analyses supply identification conditions. Discrete-time and continuous-time as well as reduced form and structural approaches have been proposed. Abbring and Heckman [2007] summarize this work, and provide additional details and references.

11.11 General equilibrium effects

Policy invariance pervades the previous discussion. Sometimes policies or programs to be evaluated are so far reaching to invalidate policy invariance. Interactions among individuals mediated by markets can be an important behavioral consideration that invalidates the partial equilibrium restrictions discussed above and mandates general equilibrium considerations (for example, changing prices and/or supply of inputs as a result of policy intervention). As an example, Heckman, Lochner, and Tabor [1998a, 1998b, 1998c] report that static treatment effects overstate the impact of college tuition subsidy on future wages by ten times compared to their general equilibrium analysis. See Abbring and Heckman [2007] for a review of the analysis of general equilibrium effects.

In any social setting, policy invariance conditions PI-2 and PI-4 are very strong. They effectively claim that untreated individuals are unaffected by who does receive treatment. Relaxation of invariance conditions or entertainment of general equilibrium effects is troublesome for standard approaches like difference - in - difference estimators as the "control group" is affected by policy interventions but a difference-in-difference estimator fails to identify the impact. Further, in stark contrast to conventional uniformity conditions of microeconometric treatment effect analysis, general equilibrium analysis must accommodate two way flows.

11.12 Regulated report precision example

LIV estimation of marginal treatment effects is illustrated for the regulated report precision example from chapter 10. We don't repeat the setup here but rather refer the reader to chapters 2 and 10. Bayesian data augmentation and analysis of marginal treatment effects are discussed and illustrated for regulated report precision in chapter 12.

11.12.1 Apparent nonnormality and MTE

We explore the impact of apparent nonnormality on the analysis of report precision treatment effects. In our simulation, α_d is observed by the owner prior to selecting report precision, α_d^L is drawn from an exponential distribution with rate $\frac{1}{0.02}$ (reciprocal of the mean), α_d^H is drawn from an exponential distribution with rate $\frac{1}{0.04}$, α is drawn from an exponential distribution with rate $\frac{1}{0.03}$ and γ is drawn from an exponential distribution with rate $\frac{1}{5}$.[9] This means the unobservable (by the analyst) portion of the choice equation is apparently nonnormal. Setting parameters are summarized below.

Stochastic parameters
$\alpha_d^L \sim \exp\left(\frac{1}{0.02}\right)$
$\alpha_d^H \sim \exp\left(\frac{1}{0.04}\right)$
$\alpha \sim \exp\left(\frac{1}{0.03}\right)$
$\gamma \sim \exp\left(\frac{1}{5}\right)$
$\beta^L \sim N(7,1)$
$\beta^H \sim N(7,1)$

First, we report benchmark *OLS* results and results from *IV* strategies developed in chapter 10. Then, we apply *LIV* to identify *MTE*-estimated average treatment effects.

OLS results

Benchmark *OLS* simulation results are reported in table 11.2 and sample statistics for average treatment effects in table 11.3. Although there is little difference between *ATE* and *OLS*, *OLS* estimates of other average treatment effects are poor, as expected. Further, *OLS* cannot detect outcome heterogeneity. *IV* strategies may be more effective.

Ordinate *IV* control model

The ordinate control function regression is

$$E[Y \mid s, D, \phi] = \beta_0 + \beta_1(s - \bar{s}) + \beta_2 D(s - \bar{s}) + \beta_3 \phi(Z\theta) + \beta_4 D$$

and is estimated via two stage *IV* where instruments

$$\{\iota, (s - \bar{s}), m(s - \bar{s}), \phi(Z\theta), m\}$$

are employed and

$$m = \Pr(D = 1 \mid Z = [\begin{array}{ccc} \iota & w_1 & w_2 \end{array}])$$

is estimated via probit. The coefficient on D, β_4, estimates *ATE*. Simulation results are reported in table 11.4. Although, on average, the rank ordering of *ATT*

---

[9]Probability as logic implies that if we only know the mean and support is nonnegative, then we conclude α_d has an exponential distribution. Similar reasoning implies knowledge of the variance leads to a Gaussian distribution (see Jaynes [2003] and chapter 13).

11.12 Regulated report precision example

Table 11.2: Continuous report precision but observed binary OLS parameter estimates for apparently nonnormal DGP

statistic	β_0	β_1	β_2
mean	635.0	0.523	$-.006$
median	635.0	0.526	-0.066
std.dev.	1.672	0.105	0.148
minimum	630.1	0.226	-0.469
maximum	639.6	0.744	0.406
statistic	$\beta_3\,(estATE)$	$estATT$	$estATUT$
mean	4.217	4.244	4.192
median	4.009	4.020	4.034
std.dev.	2.184	2.183	2.187
minimum	-1.905	-1.887	-1.952
maximum	10.25	10.37	10.13
$E\left[Y\mid s,D\right]=\beta_0+\beta_1\left(s-\bar{s}\right)+\beta_2 D\left(s-\bar{s}\right)+\beta_3 D$			

Table 11.3: Continuous report precision but observed binary average treatment effect sample statistics for apparently nonnormal DGP

statistic	ATE	ATT	$ATUT$
mean	-1.053	62.04	-60.43
median	-1.012	62.12	-60.44
std.dev.	1.800	1.678	1.519
minimum	-6.007	58.16	-64.54
maximum	3.787	65.53	-56.94

Table 11.4: Continuous report precision but observed binary ordinate control IV parameter estimates for apparently nonnormal DGP

statistic	β_0	β_1	β_2	β_3
mean	805.7	-2.879	5.845	54.71
median	765.9	-2.889	5.780	153.3
std.dev.	469.8	1.100	1.918	1373
minimum	-482.7	-5.282	0.104	-3864
maximum	2135	0.537	10.25	3772
statistic	$\beta_4\,(estATE)$	$estATT$	$estATUT$	
mean	-391.4	-369.6	-411.7	
median	-397.9	-336.5	-430.7	
std.dev.	164.5	390.4	671.2	
minimum	-787.4	-1456	-2190	
maximum	130.9	716.0	1554	
$E\left[Y\mid s,D,\phi\right]=\beta_0+\beta_1\left(s-\bar{s}\right)+\beta_2 D\left(s-\bar{s}\right)+\beta_3\phi\left(Z\theta\right)+\beta_4 D$				

Table 11.5: Continuous report precision but observed binary inverse Mills IV parameter estimates for apparently nonnormal DGP

statistic	β_0	β_1	β_2	β_3	β_4
mean	636.7	0.525	0.468	2.074	0.273
median	636.1	0.533	0.467	0.610	−4.938
std.dev.	30.61	0.114	0.114	39.74	41.53
minimum	549.2	0.182	0.108	−113.5	−118.4
maximum	724.4	0.809	0.761	116.0	121.4
statistic		$\beta_5\ (estATE)$	$estATT$	$estATUT$	
mean		2.168	0.687	3.555	
median		5.056	0.439	12.26	
std.dev.		48.44	63.22	66.16	
minimum		−173.4	−181.4	−192.9	
maximum		117.8	182.6	190.5	

$$E[Y \mid s, D, \lambda] = \beta_0 + \beta_1(1-D)(s-\bar{s}) + \beta_2 D(s-\bar{s})$$
$$+ \beta_3(1-D)\lambda^H + \beta_4 D\lambda^L + \beta_5 D$$

and *ATUT* is consistent with the sample statistics,.the ordinate control function treatment effect estimates are inconsistent (biased downward) and extremely variable, In other words, the evidence suggests nonnormality renders the utility of a normality-based ordinate control function approach suspect.

Inverse-Mills *IV* model

Heckman's inverse-Mills ratio regression is

$$E[Y \mid s, D, \lambda] = \beta_0 + \beta_1(1-D)(s-\bar{s}) + \beta_2 D(s-\bar{s})$$
$$+ \beta_3(1-D)\lambda^H + \beta_4 D\lambda^L + \beta_5 D$$

where $\bar{s}$ is the sample average of s, $\lambda^H = -\frac{\phi(Z\theta)}{1-\Phi(Z\theta)}$, $\lambda^L = \frac{\phi(Z\theta)}{\Phi(Z\theta)}$, and θ is the estimated parameters from a probit regression of precision choice D on $Z = \begin{bmatrix} \iota & w_1 & w_2 \end{bmatrix}$ (ι is a vector of ones). The coefficient on D, β_5, is the estimate of the average treatment effect, *ATE*. Simulation results including estimated average treatment effects on treated (*estATT*) and untreated (*estATUT*) are reported in table 11.5. The inverse-Mills estimates of the treatment effects are inconsistent and sufficiently variable that we may not detect nonzero treatment effects — though estimated treated effects are not as variable as those estimated by the ordinate control *IV* model. Further, the inverse-Mills results suggest greater homogeneity (all treatment effects are negative, on average) which suggests we likely would be unable to identify outcome heterogeneity based on this control function strategy.

MTE estimates via *LIV*

Next, we employ Heckman's *MTE* approach for estimating the treatment effects via a semi-parametric local instrumental variable estimator (*LIV*). Our *LIV* semi-

11.12 Regulated report precision example

Table 11.6: Continuous report precision but observed binary LIV parameter estimates for apparently nonnormal DGP

$statistic$	β_1	β_2	$estATE$	$estATT$	$estATUT$
$mean$	1.178	-1.390	17.98	14.73	25.79
$std.dev.$	0.496	1.009	23.54	26.11	38.08
$minimum$	0.271	-3.517	-27.63	-32.86	-55.07
$maximum$	2.213	0.439	64.67	69.51	94.19
$E\left[Y \mid s, D, \tau_1\left(p\right)\right] = \beta_1\left(s - \bar{s}\right) + \beta_2 D\left(s - \bar{s}\right) + \tau_1\left(p\right)$					

parametric approach only allows us to recover estimates from the outcome equations for β_1 and β_2 where the reference regression is

$$E\left[Y \mid s, D, \tau_1\left(p\right)\right] = \beta_1\left(s - \bar{s}\right) + \beta_2 D\left(s - \bar{s}\right) + \tau_1\left(p\right)$$

We employ semi-parametric methods to estimate the outcome equation. Estimated parameters and treatment effects based on bootstrapped semi-parametric weighted *MTE* are in table 11.6.[10] While the *MTE* results may more closely approximate the sample statistics than their parametric counterpart *IV* estimators, their high variance and apparent bias compromises their utility. Could we reliably detect endogeneity or heterogeneity? Perhaps — however the ordering of the estimated treatment effects doesn't correspond well with sample statistics for the average treatment effects.

Are these results due to nonnormality of the unobservable features of the selection equation? Perhaps, but a closer look suggests that our original thinking applied to this *DGP* is misguided. While expected utility associated with low (or high) inverse report precision equilibrium strategies are distinctly nonnormal, selection involves their relative ranking or, in other words, the unobservable of interest comes from the difference in unobservables. Remarkably, their difference (V_D) is not distinguishable from Gaussian draws (based on descriptive statistics, plots, etc.).

Then, what is the explanation? It is partially explained by the analyst observing binary choice when there is a multiplicity of inverse report precision choices. However, we observed this in an earlier case (see chapter 10) with a lesser impact than demonstrated here. Rather, the feature that stands out is the quality of the instruments. The same instruments are employed in this "nonnormal" case as previously employed but, apparently, are much weaker instruments in this allegedly nonnormal setting. In table 11.7 we report the analogous sample correlations to those reported in chapter 10 for Gaussian draws. Correlations between the instruments, w_1 and w_2, and treatment, D, are decidedly smaller than the examples reported in chapter 10. Further, α and γ offer little help.

---

[10]Unlike other simulations which are developed within R, these results are produced using Heckman, Urzua, and Vytlacil's *MTE* program. Reported results employ a probit selection equation. Similar results obtain when either a linear probability or nonparametric regression selection equation is employed.

11. Marginal treatment effects

Table 11.7: Continuous report precision but observed binary sample correlations for apparently nonnormal DGP

statistic	$r(\alpha, U^L)$	$r(\alpha, U^H)$	$r(\gamma, U^L)$	$r(\gamma, U^H)$
mean	−0.004	0.000	0.005	−0.007
median	−0.005	−0.001	0.007	−0.006
std.dev.	0.022	0.024	0.023	0.022
minimum	−0.081	−0.056	−0.048	−0.085
maximum	0.054	0.064	0.066	0.039
statistic	$r(\alpha, D)$	$r(\gamma, D)$	$r(w_1, D)$	$r(w_2, D)$
mean	0.013	−0.046	−0.114	0.025
median	0.013	−0.046	−0.113	0.024
std.dev.	0.022	0.021	0.012	0.014
minimum	−0.042	−0.106	−0.155	−0.011
maximum	0.082	0.017	−0.080	0.063

Stronger instruments

To further explore this explanation, we create a third and stronger instrument, w_3, and utilize it along with w_1 in the selection equation where $W = \begin{bmatrix} w_1 & w_3 \end{bmatrix}$. This third instrument is the residuals of a binary variable

$$\Im\left(EU\left(\sigma_2^L, \overline{\sigma}_2^L\right) > EU\left(\sigma_2^H, \overline{\sigma}_2^L\right)\right)$$

regressed onto U^L and U^H where $\Im(\cdot)$ is an indicator function. Below we report in table 11.8 ordinate control function results. Average treatment effect sample statistics for this simulation including the *OLS* effect are reported in table 11.9. Although the average treatment effects are attenuated a bit toward zero, these results are a marked improvement of the previous, wildly erratic results. Inverse-Mills results are reported in table 11.10. These results correspond quite well with treatment effect sample statistics. Hence, we're reminded (once again) the value of strong instruments for logically consistent analysis cannot be over-estimated.

Finally, we report in table 11.11 *LIV*-estimated average treatment effects derived from *MTE* with this stronger instrument, w_3. Again, the results are improved relative to those with the weaker instruments but as before the average treatment effects are attenuated.[11] Average treatment on the untreated along with the average treatment effect correspond best with their sample statistics. Not surprisingly, the results are noisier than the parametric results. For this setting, we conclude that strong instruments are more important than relaxed distributional assignment (based on the data) for identifying and estimating various average treatment effects.

---

[11] Reported results employ a probit regression for the selection equations (as is the case for the foregoing parametric analyses). Results based on a nonparametric regression for the treatment equation are qualitatively unchanged.

11.12 Regulated report precision example   299

Table 11.8: Continuous report precision but observed binary stronger ordinate control IV parameter estimates for apparently nonnormal DGP

statistic	β_0	β_1	β_2	β_3
mean	596.8	0.423	0.024	137.9
median	597.0	0.414	0.025	138.2
std.dev.	4.168	0.140	0.238	14.87
minimum	586.8	−0.012	−0.717	90.56
maximum	609.8	0.829	0.728	179.2
statistic	$\beta_4\ (estATE)$	$estATT$	$estATUT$	
mean	−2.494	40.35	−43.77	
median	−2.449	40.07	−43.58	
std.dev.	2.343	−4.371	5.598	
minimum	−8.850	28.50	−58.91	
maximum	4.162	52.40	−26.60	

$$E\left[Y \mid s, D, \phi\right] = \beta_0 + \beta_1\left(s - \bar{s}\right) + \beta_2 D\left(s - \bar{s}\right) + \beta_3 \phi\left(W\theta\right) + \beta_4 D$$

Table 11.9: Continuous report precision but observed binary average treatment effect sample statistics for apparently nonnormal DGP

statistic	ATE	ATT	ATUT	OLS
mean	−0.266	64.08	−62.26	0.578
median	−0.203	64.16	−62.30	0.764
std.dev.	1.596	1.448	1.584	2.100
minimum	−5.015	60.32	−66.64	−4.980
maximum	3.746	67.48	−57.38	6.077

Table 11.10: Continuous report precision but observed binary stronger inverse Mills IV parameter estimates for apparently nonnormal DGP

statistic	β_0	β_1	β_2	β_3	β_4
mean	608.9	0.432	0.435	−48.27	61.66
median	608.9	0.435	0.438	−48.55	61.60
std.dev.	1.730	0.099	0.086	2.743	3.949
minimum	603.8	0.159	0.238	−54.85	51.27
maximum	613.3	0.716	0.652	−40.70	72.70
statistic	$\beta_5\ (estATE)$		$estATT$	$estATUT$	
mean	−8.565		57.61	−72.28	
median	−8.353		57.44	−72.28	
std.dev.	2.282		3.294	4.628	
minimum	−15.51		48.44	−85.37	
maximum	−2.814		67.11	−60.39	

$$E\left[Y \mid s, D, \lambda\right] = \beta_0 + \beta_1\left(1 - D\right)\left(s - \bar{s}\right) + \beta_2 D\left(s - \bar{s}\right)$$
$$+ \beta_3\left(1 - D\right)\lambda^H + \beta_4 D\lambda^L + \beta_5 D$$

Table 11.11: Continuous report precision but observed binary stronger LIV parameter estimates for apparently nonnormal DGP

statistic	β_1	β_2	estATE	estATT	estATUT
mean	0.389	0.220	−7.798	9.385	−24.68
std.dev.	0.159	0.268	9.805	14.17	16.38
minimum	0.107	−0.330	−26.85	−17.69	−57.14
maximum	0.729	0.718	11.58	37.87	−26.85
statistic		OLS	ATE	ATT	ATUT
mean		3.609	1.593	63.76	−61.75
median		3.592	1.642	63.91	−61.70
std.dev.		2.484	1.894	1.546	1.668
minimum		−3.057	−4.313	59.58	−66.87
maximum		11.28	5.821	67.12	−58.11
$E\left[Y \mid s, D, \tau_1\right] = \beta_1 (s - \bar{s}) + \beta_2 D (s - \bar{s}) + \tau_1 (p)$					

11.13 Additional reading

There are numerous contributions to this literature. We suggest beginning with Heckman's [2001] Nobel lecture, Heckman and Vytlacil [2005, 2007a, 2007b], and Abbring and Heckman [2007]. These papers provide extensive discussions and voluminous references. This chapter has provided at most a thumbnail sketch of this extensive and important work. A FORTRAN program and documentation for estimating Heckman, Urzua, and Vytlacil's [2006] marginal treatment effect can be found at URL: http://jenni.uchicago.edu/underiv/.

12
Bayesian treatment effects

We continue with the selection setting discussed in the previous three chapters and apply Bayesian analysis. Bayesian augmentation of the kind proposed by Albert and Chib [1993] in the probit setting (see chapter 5) can be extended to selection analysis of treatment effects (Li, Poirier, and Tobias [2004]). An advantage of the approach is treatment effect distributions can be identified by bounding the unidentified parameter. As counterfactuals are not observed, the correlation between outcome errors is unidentified. However, Poirier and Tobias [2003] and Li, Poirier, and Tobias [2004] suggest using the positive definiteness of the variance-covariance matrix (for the selection equation error and the outcome equations' errors) to bound the unidentified parameter. This is a computationally-intensive complementary strategy to Heckman's factor analytic approach (see chapter 11 and Abbring and Heckman [2007]) which may be accessible even when factors cannot be identified.[1] Marginal treatment effects identified by Bayesian analysis are employed in a prototypical selection setting as well as the regulated report precision setting introduced in chapter 2 and continued in chapters 10 and 11. Also, policy-relevant treatment effects discussed in chapter 11 are revisited in this chapter including Bayesian applications to regulated versus unregulated report precision.

---

[1] We prefer to think of classic and Bayesian approaches as complementary strategies. Together, they may help us to better understand the *DGP*.

12.1 Setup

The setup is the same as the previous chapters. We repeat it for convenience. Suppose the *DGP* is

outcome equations:
$$Y_j = \mu_j(X) + V_j, j = 0, 1$$

selection equation:
$$D^* = \mu_D(Z) - V_D$$

observable response:
$$\begin{aligned} Y &= DY_1 + (1-D)Y_0 \\ &= \mu_0(X) + (\mu_1(X) - \mu_0(X))D + V_0 + (V_1 - V_0)D \end{aligned}$$

where
$$D = \begin{cases} 1 & D^* > 0 \\ 0 & otherwise \end{cases}$$

and Y_1 is the (potential) outcome with treatment while Y_0 is the outcome without treatment. The usual *IV* restrictions apply as Z contains some variable(s) not included in X.

There are effectively three sources of missing data: the latent utility index D^*, and the two counterfactuals $(Y_1 \mid D = 0)$ and $(Y_0 \mid D = 1)$. If these data were observable identification of treatment effects $\Delta \equiv Y_1 - Y_0$ (including distributions) would be straightforward.

12.2 Bounds and learning

Even if we know Δ is normally distributed, unobservability of the counterfactuals creates a problem for identifying the distribution of Δ as

$$Var[\Delta \mid X] = Var[V_1] + Var[V_0] - 2Cov[V_1, V_0]$$

and $\rho_{10} \equiv Corr[V_1, V_0]$ is unidentified.[2] Let $\eta \equiv [V_D, V_1, V_0]^T$ then

$$\Sigma \equiv Var[\eta] = \begin{bmatrix} 1 & \rho_{D1}\sigma_1 & \rho_{D0}\sigma_0 \\ \rho_{D1}\sigma_1 & \sigma_1^2 & \rho_{10}\sigma_1\sigma_0 \\ \rho_{D0}\sigma_0 & \rho_{10}\sigma_1\sigma_0 & \sigma_0^2 \end{bmatrix}$$

From the positivity of the determinant (or eigenvalues) of Σ we can bound the unidentified correlation

$$\rho_{D1}\rho_{D0} - \left[(1-\rho_{D1}^2)(1-\rho_{D0}^2)\right]^{\frac{1}{2}} \leq \rho_{10} \leq \rho_{D1}\rho_{D0} + \left[(1-\rho_{D1}^2)(1-\rho_{D0}^2)\right]^{\frac{1}{2}}$$

This allows learning about ρ_{10} and, in turn, identification of the distribution of treatment effects. Notice the more pressing is the endogeneity problem (ρ_{D1}, ρ_{D0} large in absolute value) the tighter are the bounds.

---

[2] The variables are never simultaneously observed as needed to identify correlation.

12.3 Gibbs sampler

As in the case of Albert and Chib's augmented probit, we work with conditional posterior distributions for the augmented data. Define the complete or augmented data as

$$r_i^* = \begin{bmatrix} D_i^* & D_i Y_i + (1 - D_i) Y_i^{miss} & D_i Y_i^{miss} + (1 - D_i) Y_i \end{bmatrix}^T$$

Also, let

$$H_i = \begin{bmatrix} Z_i & 0 & 0 \\ 0 & X_i & 0 \\ 0 & 0 & X_i \end{bmatrix}$$

and

$$\beta = \begin{bmatrix} \theta \\ \beta_1 \\ \beta_0 \end{bmatrix}$$

12.3.1 Full conditional posterior distributions

Let Γ_{-x} denote all parameters other than x. The full conditional posteriors for the augmented outcome data are

$$Y_i^{miss} \mid \Gamma_{-Y_i^{miss}}, Data \sim N\left((1 - D_i)\mu_{1i} + D_i \mu_{0i}, (1 - D_i)\omega_{1i} + D_i \omega_{0i}\right)$$

where standard multivariate normal theory is applied to derive means and variances conditional on the draw for latent utility and the other outcome

$$\mu_{1i} = X_i \beta_1 + \frac{\sigma_0^2 \sigma_{D1} - \sigma_{10} \sigma_{D0}}{\sigma_0^2 - \sigma_{D0}^2}(D_i^* - Z_i \theta) + \frac{\sigma_{10} - \sigma_{D1} \sigma_{D0}}{\sigma_0^2 - \sigma_{D0}^2}(Y_i - X_i \beta_0)$$

$$\mu_{0i} = X_i \beta_0 + \frac{\sigma_1^2 \sigma_{D0} - \sigma_{10} \sigma_{D1}}{\sigma_1^2 - \sigma_{D1}^2}(D_i^* - Z_i \theta) + \frac{\sigma_{10} - \sigma_{D1} \sigma_{D0}}{\sigma_1^2 - \sigma_{D1}^2}(Y_i - X_i \beta_1)$$

$$\omega_{1i} = \sigma_1^2 - \frac{\sigma_{D1}^2 \sigma_0^2 - 2\sigma_{10} \sigma_{D1} \sigma_{D0} + \sigma_{10}^2}{\sigma_0^2 - \sigma_{D0}^2}$$

$$\omega_{0i} = \sigma_0^2 - \frac{\sigma_{D0}^2 \sigma_1^2 - 2\sigma_{10} \sigma_{D1} \sigma_{D0} + \sigma_{10}^2}{\sigma_1^2 - \sigma_{D1}^2}$$

Similarly, the conditional posterior for the latent utility is

$$D_i^* \mid \Gamma_{-D_i^*}, Data \sim \begin{matrix} TN_{(0,\infty)}\left(\mu_{D_i}, \omega_D\right) & if\ D_i = 1 \\ TN_{(-\infty,0)}\left(\mu_{D_i}, \omega_D\right) & if\ D_i = 0 \end{matrix}$$

where $TN(\cdot)$ refers to the truncated normal distribution with support indicated via the subscript and the arguments are parameters of the untruncated distribution.

Applying multivariate normal theory for $(D_i^* \mid Y_i)$ we have

$$\mu_{D_i} = Z_i\theta + \left(D_iY_i + (1-D_i)Y_i^{miss} - X_i\beta_1\right)\frac{\sigma_0^2\sigma_{D1} - \sigma_{10}\sigma_{D0}}{\sigma_1^2\sigma_0^2 - \sigma_{10}^2}$$

$$+ \left(D_iY_i^{miss} + (1-D_i)Y_i - X_i\beta_0\right)\frac{\sigma_1^2\sigma_{D0} - \sigma_{10}\sigma_{D1}}{\sigma_1^2\sigma_0^2 - \sigma_{10}^2}$$

$$\omega_D = 1 - \frac{\sigma_{D1}^2\sigma_0^2 - 2\sigma_{10}\sigma_{D1}\sigma_{D0} + \sigma_{D0}^2\sigma_1^2}{\sigma_1^2\sigma_0^2 - \sigma_{10}^2}$$

The conditional posterior distribution for the parameters is

$$\beta \mid \Gamma_{-\beta}, Data \sim N\left(\mu_\beta, \omega_\beta\right)$$

where by the *SUR* (seemingly-unrelated regression) generalization of Bayesian regression (see chapter 7)

$$\mu_\beta = \left[H^T\left(\Sigma^{-1}\otimes I_n\right)H + V_\beta^{-1}\right]^{-1}\left[H^T\left(\Sigma^{-1}\otimes I_n\right)r^* + V_\beta^{-1}\beta_0\right]$$

$$\omega_\beta = \left[H^T\left(\Sigma^{-1}\otimes I_n\right)H + V_\beta^{-1}\right]^{-1}$$

and the prior distribution is $p(\beta) \sim N(\beta_0, V_\beta)$. The conditional distribution for the trivariate variance-covariance matrix is

$$\Sigma \mid \Gamma_{-\Sigma}, Data \sim G^{-1}$$

where

$$G \sim Wishart\left(n + \rho, S + \rho R\right)$$

with prior $p(G) \sim Wishart(\rho, \rho R)$, and $S = \sum_{i=1}^{n}(r_i^* - H_i\beta)(r_i^* - H_i\beta)^T$.

As usual, starting values for the Gibbs sampler are varied to test convergence of parameter posterior distributions.

Nobile's algorithm

Recall σ_D^2 is normalized to one. This creates a slight complication as the conditional posterior is no longer inverse-Wishart. Nobile [2000] provides a convenient algorithm for random Wishart (multivariate χ^2) draws with a restricted element. The algorithm applied to the current setting results in the following steps:

1. Exchange rows and columns one and three in $S + \rho R$, call this matrix V.

2. Find L such that $V = \left(L^{-1}\right)^T L^{-1}$.

3. Construct a lower triangular matrix A with
   a. a_{ii} equal to the square root of χ^2 random variates, $i = 1, 2$.
   b. $a_{33} = \frac{1}{l_{33}}$ where l_{33} is the third row-column element of L.
   c. a_{ij} equal to $N(0,1)$ random variates, $i > j$.

4. Set $V' = \left(L^{-1}\right)^T \left(A^{-1}\right)^T A^{-1} L^{-1}$.

5. Exchange rows and columns one and three in V' and denote this draw Σ.

Prior distributions

Li, Poirier, and Tobias choose relatively diffuse priors such that the data dominates the posterior distribution. Their prior distribution for β is $p(\beta) \sim N(\beta_0, V_\beta)$ where $\beta_0 = 0, V_\beta = 4I$ and their prior for Σ^{-1} is $p(G) \sim Wishart(\rho, \rho R)$ where $\rho = 12$ and R is a diagonal matrix with elements $\left\{\frac{1}{12}, \frac{1}{4}, \frac{1}{4}\right\}$.

12.4 Predictive distributions

The above Gibbs sampler for the selection problem can be utilized to generate treatment effect predictive distributions $\left(Y_1^f - Y_0^f\right)$ conditional on X^f and Z^f using the post-convergence parameter draws. That is, the predictive distribution for the treatment effect is

$$p\left(Y_1^f - Y_0^f \mid X^f\right) \sim N\left(X^f [\beta_1 - \beta_0], \gamma_2\right)$$

where $\gamma_2 \equiv Var\left[Y_1^f - Y_0^f \mid X^f\right] = \sigma_1^2 + \sigma_0^2 - 2\sigma_{10}$.

Using Bayes' theorem, we can define the predictive distribution for the treatment effect on the treated as

$$p\left(Y_1^f - Y_0^f \mid X^f, D\left(Z^f\right) = 1\right)$$
$$= \frac{p\left(Y_1^f - Y_0^f \mid X^f\right) p\left(D\left(Z^f\right) = 1 \mid Y_1^f - Y_0^f, X^f\right)}{p\left(D\left(Z^f\right) = 1\right)}$$

where

$$p\left(D\left(Z^f\right) = 1 \mid Y_1^f - Y_0^f, X^f\right)$$
$$= \Phi\left(\frac{Z^f \theta + \frac{\gamma_1}{\gamma_2}\left[\left(Y_1^f - Y_0^f\right) - X^f(\beta_1 - \beta_0)\right]}{\sqrt{1 - \frac{\gamma_1^2}{\gamma_2}}}\right)$$

and

$$\gamma_1 = Cov\left[Y_1^f - Y_0^f, D^* \mid X, Z\right] = \sigma_{D1} - \sigma_{D0}$$

Analogously, the predictive distribution for the treatment effect on the untreated is

$$p\left(Y_1^f - Y_0^f \mid X^f, D\left(Z^f\right) = 0\right)$$
$$= \frac{p\left(Y_1^f - Y_0^f \mid X^f\right)\left[1 - p\left(D\left(Z^f\right) = 1 \mid Y_1^f - Y_0^f, X^f\right)\right]}{1 - p\left(D\left(Z^f\right) = 1\right)}$$

Also, the predictive distribution for the local treatment effect is

$$p\left(Y_1^f - Y_0^f \mid X^f, D\left(Z'^f\right) = 1, D\left(Z^f\right) = 0\right)$$

$$= \frac{p\left(Y_1^f - Y_0^f \mid X^f\right)}{p\left(D\left(Z'^f\right) = 1\right) - p\left(D\left(Z^f\right) = 0\right)}$$

$$\times \left[\begin{array}{c} p\left(D\left(Z'^f\right) = 1 \mid Y_1^f - Y_0^f, X^f\right) \\ -p\left(D\left(Z^f\right) = 0 \mid Y_1^f - Y_0^f, X^f\right) \end{array} \right]$$

12.4.1 Rao-Blackwellization

The foregoing discussion focuses on identifying predictive distributions conditional on the parameters Γ. "Rao-Blackwellization" efficiently utilizes the evidence to identify unconditional predictive distributions (see Rao-Blackwell theorem in the appendix). That is, density ordinates are averaged over parameter draws

$$p\left(Y_1^f - Y_0^f \mid X^f\right) = \frac{1}{m} \sum_{i=1}^m p\left(Y_1^f - Y_0^f \mid X^f, \Gamma = \Gamma_i\right)$$

where Γ_i is the ith post-convergence parameter draw out of m such draws.

12.5 Hierarchical multivariate Student t variation

Invoking a common over-dispersion practice, for example Albert and Chib [1993], Li, Poirier, and Tobias [2003] add a mixing variable or hyperparameter, λ, to extend their Gaussian analysis to a multivariate Student t distribution on marginalizing λ. λ is assigned an inverted gamma prior density, $\lambda \sim IG(a, b)$ where

$$p(\lambda) \propto \lambda^{-(a+1)} \exp\left[-\frac{1}{b\lambda}\right]$$

For computational purposes, Li, Poirier, and Tobias [2003] scale all variables by λ in the non-lambda conditionals to convert back to Gaussians and proceed as with the Gaussian *McMC* selection analysis except for the addition of sampling from the conditional posterior for the mixing parameter, λ, where the unscaled data are employed.

12.6 Mixture of normals variation

We might be concerned about robustness to departure from normality in this selection analysis. Li, Poirier, and Tobias suggest exploring a mixture of normals. For a two component mixture the likelihood function is

$$p\left(r_i^* \mid \Gamma\right) = \pi_1 \phi\left(r_i^*; H_i \beta^1, \Sigma^1\right) + \pi_2 \phi\left(r_i^*; H_i \beta^2, \Sigma^2\right)$$

where each component has its own parameter vector β^j and variance-covariance matrix Σ^j, and $\pi_1 + \pi_2 = 1$.[3]

Conditional posterior distributions for the components are

$$c_i \mid \Gamma_{-c}, Data \sim Multinomial\left(1, p^1, p^2\right)\text{[4]}$$

where

$$p^j = \frac{\pi_j |\Sigma|^{-\frac{1}{2}} \exp\left[-\frac{1}{2}\left(r_i^* - H_i\beta^j\right)^T \left(\Sigma^j\right)^{-1} \left(r_i^* - H_i\beta^j\right)\right]}{\sum_{j=1}^{2} \pi_j |\Sigma|^{-\frac{1}{2}} \exp\left[-\frac{1}{2}\left(r_i^* - H_i\beta^j\right)^T \left(\Sigma^j\right)^{-1} \left(r_i^* - H_i\beta^j\right)\right]}$$

The conditional posterior distribution for component probabilities follows a Dirichlet distribution (see chapter 7 to review properties of the Dirichlet distribution).

$$\pi_i \mid \Gamma_{-\pi}, Data \sim Dirichlet\left(n_1 + \alpha_1, n_2 + \alpha_2\right)$$

with prior hyperparameter α_j and $n_j = \sum_{i=1}^{n} c_{ji}$.

Conditional predictive distributions for the mixture of normals selection analysis are the same as above except that we condition on the component and utilize parameters associated with each component. The predictive distribution is then based on a probability weighted average of the components.

12.7 A prototypical Bayesian selection example

An example may help fix ideas regarding *McMC* Bayesian data augmentation procedures in the context of selection and missing data on the counterfactuals. Here we consider a prototypical selection problem with the following *DGP*. A decision maker faces a binary choice where the latent choice equation (based on expected utility, *EU*, maximization) is

$$\begin{aligned} EU &= \gamma_0 + \gamma_1 x + \gamma_2 z + V \\ &= -1 + x + z + V \end{aligned}$$

x is an observed covariate, z is an observed instrument (both x and z have mean 0.5), and V is unobservable (to the analyst) contributions to expected utility. The outcome equations are

$$\begin{aligned} Y_1 &= \beta_0^1 + \beta_1^1 x + U_1 \\ &= 2 + 10x + U_1 \\ Y_0 &= \beta_0^0 + \beta_1^0 x + U_0 \\ &= 1 + 2x + U_0 \end{aligned}$$

---

[3] Li, Poirier, and Tobias [2004] specify identical priors for all Σ^j.
[4] A binomial distribution suffices for the two component mixture.

Unobservables $\begin{bmatrix} V & U_1 & U_0 \end{bmatrix}^T$ are jointly normally distributed with expected value $\begin{bmatrix} 0 & 0 & 0 \end{bmatrix}^T$ and variance $\Sigma = \begin{bmatrix} 1 & 0.7 & -0.7 \\ 0.7 & 1 & -0.1 \\ -0.7 & -0.1 & 1 \end{bmatrix}$. Clearly, the average treatment effect is

$$ATE = (2 + 10 * 0.5) - (1 + 2 * 0.5) = 5.$$

Even though *OLS* estimates the same quantity as *ATE*,

$$OLS = E[Y_1 \mid D = 1] - E[Y_0 \mid D = 0] = 7.56 - 2.56 = 5$$

selection is inherently endogenous. Further, outcomes are heterogeneous as[5]

$$ATT = E[Y_1 \mid D = 1] - E[Y_0 \mid D = 1] = 7.56 - 1.44 = 6.12$$

and

$$ATUT = E[Y_1 \mid D = 0] - E[Y_0 \mid D = 0] = 6.44 - 2.56 = 3.88$$

12.7.1 Simulation

To illustrate we generate 20 samples of 5,000 observations each. For the simulation, x and z are independent and uniformly distributed over the interval $(0, 1)$, and $\begin{bmatrix} V & U_1 & U_0 \end{bmatrix}$ are drawn from a joint normal distribution with zero mean and variance Σ. If $EU_j > 0$, then $D_j = 1$, otherwise $D_j = 0$. Relatively diffuse priors are employed with mean zero and variance $100I$ for the parameters $\begin{bmatrix} \beta^1 & \beta^0 & \gamma \end{bmatrix}$ and trivariate error $\begin{bmatrix} V & U_1 & U_0 \end{bmatrix}$ distribution degrees of freedom parameter $\rho = 12$ and sums of squares variation ρI.[6] Data augmentation produces missing data for the latent choice variable EU plus counterfactuals $(Y_1 \mid D = 0)$ and $(Y_0 \mid D = 1)$.[7] Data augmentation permits collection of statistical evidence directly on the treatment effects. The following treatment effect

---

[5]We can connect the dots by noting the average of the inverse Mills ratio is approximately 0.8 and recalling

$$\begin{aligned} ATE &= \Pr(D=1) ATT + \Pr(D=0) ATUT \\ &= 0.5(6.12) + 0.5(3.88) = 5 \end{aligned}$$

[6]Initialization of the trivariate variance matrix for the Gibbs sampler is set equal to $100I$. Burn-in takes care of initialization error.

[7]Informativeness of the priors for the trivariate error variance is controlled by ρ. If ρ is small compared to the number of observations in the sample, the likelihood dominates the data augmentation.

statistics are collected:

$$estATE = \frac{1}{n}\sum_{j=1}^{n}\left(Y_{1j}^{*} - Y_{0j}^{*}\right)$$

$$estATT = \frac{\sum_{j=1}^{n} D_j \left(Y_{1j}^{*} - Y_{0j}^{*}\right)}{\sum_{j=1}^{n} D_j}$$

$$estATUT = \frac{\sum_{j=1}^{n}(1 - D_j)\left(Y_{1j}^{*} - Y_{0j}^{*}\right)}{\sum_{j=1}^{n}(1 - D_j)}$$

where Y_j^* is the augmented response. That is,

$$Y_{1j}^{*} = D_j Y_1 + (1 - D_j)(Y_1 \mid D = 0)$$

and

$$Y_{0j}^{*} = D_j (Y_0 \mid D = 1) + (1 - D_j) Y_0$$

12.7.2 Bayesian data augmentation and MTE

With a strong instrument in hand, this is an attractive setting to discuss a version of Bayesian data augmentation-based estimation of marginal treatment effects (*MTE*). As data augmentation generates repeated draws for unobservables V_j, $(Y_{1j} \mid D_j = 0)$, and $(Y_{0j} \mid D_j = 1)$, we exploit repeated samples to describe the distribution for $MTE(u_D)$ where V is transformed to uniform $(0, 1)$, $u_D = p_v$. For each draw, $V = v$, we determine $u_D = \Phi(v)$ and calculate $MTE(u_D) = E[Y_1 - Y_0 \mid u_D]$.

MTE is connected to standard population-level treatment effects, *ATE*, *ATT*, and *ATUT*, via non-negative weights whose sum is one

$$w_{ATE}(u_D) = \frac{\sum_{j=1}^{n} I(u_{Dj})}{n}$$

$$w_{ATT}(u_D) = \frac{\sum_{j=1}^{n} I(u_{Dj}) D_j}{\sum_{j=1}^{n} D_j}$$

$$w_{ATUT}(u_D) = \frac{\sum_{j=1}^{n} I(u_{Dj})(1 - D_j)}{\sum_{j=1}^{n}(1 - D_j)}$$

where probabilities p_k refer to bins from 0 to 1 by increments of 0.01 for indicator variable

$$\begin{array}{ll} I(u_{Dj}) = 1 & u_{Dj} = p_k \\ I(u_{Dj}) = 0 & u_{Dj} \neq p_k \end{array}$$

12. Bayesian treatment effects

Table 12.1: McMC parameter estimates for prototypical selection

statistic	β_0^1	β_1^1	β_0^0	β_1^0
mean	2.118	9.915	1.061	2.064
median	2.126	9.908	1.059	2.061
std.dev.	0.100	0.112	0.063	0.102
minimum	1.709	9.577	0.804	1.712
maximum	2.617	10.283	1.257	2.432
statistic	γ_0	γ_1	γ_2	
mean	-1.027	1.001	1.061	
median	-1.025	0.998	1.061	
std.dev.	0.066	0.091	0.079	
minimum	-1.273	0.681	0.729	
maximum	-0.783	1.364	1.362	
statistic	$cor(V, U_1)$	$cor(V, U_0)$	$cor(U_1, U_0)$	
mean	0.621	-0.604	-0.479	
median	0.626	-0.609	-0.481	
std.dev.	0.056	0.069	0.104	
minimum	0.365	-0.773	-0.747	
maximum	0.770	-0.319	0.082	

$$Y_1 = \beta_0^1 + \beta_1^1 x + U_1$$
$$Y_0 = \beta_0^0 + \beta_1^0 x + U_0$$
$$EU = \gamma_0 + \gamma_1 x + \gamma_2 z + V$$

Simulation results

Since the Gibbs sampler requires a burn-in period for convergence, for each sample we take 4,000 conditional posterior draws, treat the first 3,000 as the burn-in period, and retain the final 1,000 draws for each sample, in other words, a total of 20,000 draws are retained. Parameter estimates for the simulation are reported in table 12.1. McMC estimated average treatment effects are reported in table 12.2 and sample statistics are reported in table 12.3. The treatment effect estimates are consistent with their sample statistics despite the fact that bounding the unidentified correlation between U_1 and U_0 produces a rather poor estimate of this parameter.

Table 12.2: McMC estimates of average treatment effects for prototypical selection

statistic	estATE	estATT	estATUT
mean	4.992	6.335	3.635
median	4.996	6.329	3.635
std.dev.	0.087	0.139	0.117
minimum	4.703	5.891	3.209
maximum	5.255	6.797	4.067

Table 12.3: McMC average treatment effect sample statistics for prototypical selection

statistic	ATE	ATT	ATUT	OLS
mean	5.011	6.527	3.481	5.740
median	5.015	6.517	3.489	5.726
std.dev.	0.032	0.049	0.042	0.066
minimum	4.947	6.462	3.368	5.607
maximum	5.088	6.637	3.546	5.850

Table 12.4: McMC MTE-weighted average treatment effects for prototypical selection

statistic	estATE	estATT	estATUT
mean	4.992	5.861	4.114
median	4.980	5.841	4.115
std.dev.	0.063	0.088	0.070
minimum	4.871	5.693	3.974
maximum	5.089	6.003	4.242

In addition, we report results on marginal treatment effects. First, $MTE(u_D)$ versus $u_D = p_v$ is plotted. The conditional mean $MTE(u_D)$ over the 20,000 draws is plotted below versus $u_D = p_v$. Figure 12.1 depicts the mean at each p_v. Nonconstancy, indeed nonlinearity, of MTE is quite apparent from the plot. Table 12.4 reports simulation statistics from weighted averages of MTE employed to recover standard population-level treatment effects, ATE, ATT, and $ATUT$. Nonconstancy of $MTE(u_D)$ along with marked differences in $estATE$, $estATT$, and $estATUT$ provide support for heterogeneous response. The MTE-weighted average treatment effect estimates are very comparable (perhaps slightly dampened) to the previous estimates and average treatment effect sample statistics.

12.8 Regulated report precision example

Consider the report precision example initiated in chapter 2. The owner's expected utility is

$$EU(\sigma_2) = \mu - \beta \frac{\sigma_1^2 \bar{\sigma}_2^2}{\sigma_1^2 + \bar{\sigma}_2^2} - \gamma \frac{\sigma_1^4 (\sigma_1^2 + \sigma_2^2)}{(\sigma_1^2 + \bar{\sigma}_2^2)^2} - \alpha (b - \sigma_2^2)^2 - \alpha_d (\hat{b} - \sigma_2^2)^2$$

selection is binary (to the analyst)

$$D = \begin{matrix} 1 & (\sigma_2^L) \\ 0 & (\sigma_2^H) \end{matrix} \quad \begin{matrix} \text{if } EU(\sigma_2^L) - EU(\sigma_2^H) > 0 \\ \text{otherwise} \end{matrix}$$

312   12. Bayesian treatment effects

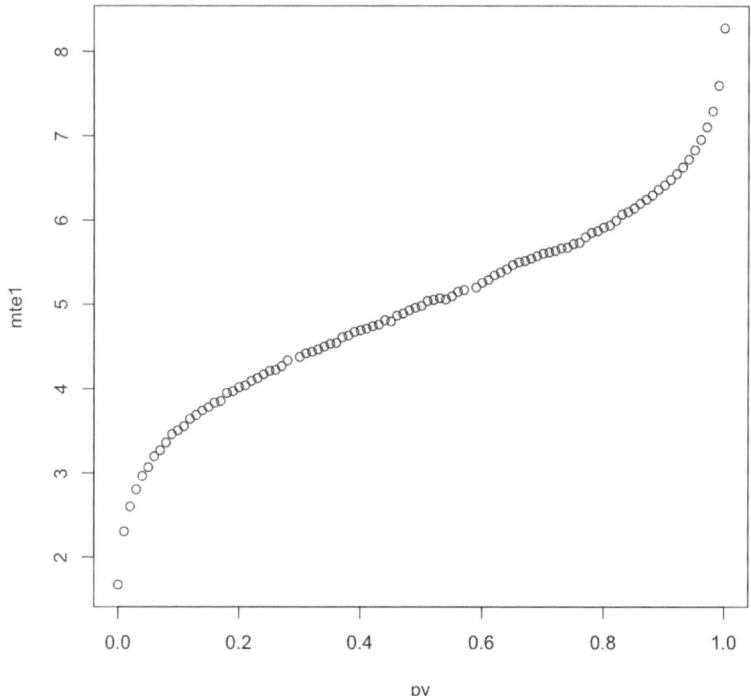

Figure 12.1: $MTE\,(u_D)$ versus $u_D = p_\nu$ for prototypical selection

outcomes are

$$\begin{aligned}
Y &\equiv P\left(\bar{\sigma}_2\right) \\
&= D\left(Y_1 \mid D=1\right) + (1-D)\left(Y_0 \mid D=0\right) \\
&= \mu + D\left\{\frac{\sigma_1^2}{\sigma_1^2 + \left(\bar{\sigma}_2^L\right)^2}\left(s^L - \mu\right) - \beta^L \frac{\sigma_1^2 \left(\bar{\sigma}_2^L\right)^2}{\sigma_1^2 + \left(\bar{\sigma}_2^L\right)^2}\right\} \\
&\quad + (1-D)\left\{\frac{\sigma_1^2}{\sigma_1^2 + \left(\bar{\sigma}_2^H\right)^2}\left(s^H - \mu\right) - \beta^H \frac{\sigma_1^2 \left(\bar{\sigma}_2^H\right)^2}{\sigma_1^2 + \left(\bar{\sigma}_2^H\right)^2}\right\}
\end{aligned}$$

observed outcomes are

$$Y^j = \begin{cases}(Y_1 \mid D=1) & j=L \\ (Y_0 \mid D=0) & j=H\end{cases}$$

$$Y^j = \beta_0^j + \beta_1^j\left(s^j - \mu\right) + U^j$$

and factual and counterfactual outcomes with treatment are

$$Y_1 = D\left(Y_1 \mid D=1\right) + (1-D)\left(Y_1 \mid D=0\right)$$

12.8 Regulated report precision example

and without treatment are

$$Y_0 = D(Y_0 \mid D = 1) + (1 - D)(Y_0 \mid D = 0)$$

Notice, factual observations, $(Y_1 \mid D = 1)$ and $(Y_0 \mid D = 0)$, are outcomes associated with equilibrium strategies while the counterfactuals, $(Y_1 \mid D = 0)$ and $(Y_0 \mid D = 1)$, are outcomes associated with off-equilibrium strategies.

We now investigate Bayesian *McMC* estimation of treatment effects associated with report precision selection. We begin with the binary choice (to the owner) and normal unobservables setting.

12.8.1 Binary choice

The following parameters characterize the binary choice setting:

Binary choice parameters
$\mu = 1,000$
$\sigma_1 = 10$
$\gamma = 2.5$
$\alpha = 0.02$
$b = 150$
$\widehat{b} = 128.4$
$\alpha_d^L \sim N(0.02, 0.005)$
$\alpha_d^H \sim N(0.04, 0.01)$
$\beta^L, \beta^H \stackrel{iid}{\sim} N(7, 1)$
$s^j \sim N\left(\mu, \sigma_1^2 + \left(\sigma_2^j\right)^2\right) \quad j = L \text{ or } H$

The owners know the expected value of α_d^j when report precision is selected but not the draw.

While the draws for β^L and β^H impact the owner's choice of high or low precision, the numerical value of inverse report precision

$$\sigma_2^L = f\left(\alpha, \gamma, E\left[\alpha_d^L\right] \mid \overline{\sigma}_2^L\right)$$

or

$$\sigma_2^H = f\left(\alpha, \gamma, E\left[\alpha_d^H\right] \mid \overline{\sigma}_2^H\right)$$

that maximizes her expected utility is independent of the β^j draws

$$EU\left(\sigma_2^L, \beta^L \mid \overline{\sigma}_2^L\right) - EU\left(\sigma_2^H, \beta^H \mid \overline{\sigma}_2^H\right) > 0 \quad D = 1$$
$$\text{otherwise} \quad D = 0$$

$$\sigma_2^L = \arg\max_{\sigma_2} EU\left(\sigma_1, \alpha, \gamma, E\left[\alpha_d^L\right] \mid \overline{\sigma}_2^L\right)$$
$$\sigma_2^H = \arg\max_{\sigma_2} EU\left(\sigma_1, \alpha, \gamma, E\left[\alpha_d^H\right] \mid \overline{\sigma}_2^H\right)$$

The analyst only observes report precision selection $D = 1$ (for σ_2^L) or $D = 0$ (for σ_2^H), outcomes $Y = DY^L + (1 - D)Y^H$, covariate s, and instruments w_1 and w_2. The instruments, w_1 and w_2, are the components of $\alpha_d = D\alpha_d^L + (1 - D)\alpha_d^H$, and $\sigma_2 = D\sigma_2^L + (1 - D)\sigma_2^H$ orthogonal to

$$U^L = -\left(\beta^L - E[\beta]\right)\left[D\frac{\sigma_1^2\left(\sigma_2^L\right)^2}{\sigma_1^2 + \left(\sigma_2^L\right)^2} + (1 - D)\frac{\sigma_1^2\left(\sigma_2^H\right)^2}{\sigma_1^2 + \left(\sigma_2^H\right)^2}\right]$$

and

$$U^H = -\left(\beta^H - E[\beta]\right)\left[D\frac{\sigma_1^2\left(\sigma_2^L\right)^2}{\sigma_1^2 + \left(\sigma_2^L\right)^2} + (1 - D)\frac{\sigma_1^2\left(\sigma_2^H\right)^2}{\sigma_1^2 + \left(\sigma_2^H\right)^2}\right]$$

The keys to selection are the relations between V and U^L, and V and U^H where

$$\begin{aligned}
V = &-\left(\beta^L - \beta^H\right)\left[D\frac{\sigma_1^2\left(\sigma_2^L\right)^2}{\sigma_1^2 + \left(\sigma_2^L\right)^2} + (1 - D)\frac{\sigma_1^2\left(\sigma_2^H\right)^2}{\sigma_1^2 + \left(\sigma_2^H\right)^2}\right]\\
&-\gamma\left(\frac{\sigma_1^4\left(\sigma_1^2 + \left(\sigma_2^L\right)^2\right)}{\left(\sigma_1^2 + \left(\bar{\sigma}_2^L\right)^2\right)^2} - \frac{\sigma_1^4\left(\sigma_1^2 + \left(\sigma_2^H\right)^2\right)}{\left(\sigma_1^2 + \left(\bar{\sigma}_2^H\right)^2\right)^2}\right)\\
&-\alpha\left(\left(b - \left(\sigma_2^L\right)^2\right)^2 - \left(b - \left(\sigma_2^H\right)^2\right)^2\right)\\
&-\left(E\left[\alpha_d^L\right]\left(\hat{b} - \left(\sigma_2^L\right)^2\right)^2 - E\left[\alpha_d^H\right]\left(\hat{b} - \left(\sigma_2^H\right)^2\right)^2\right)\\
&-\left(\gamma_0 + \gamma_1 w_1 + \gamma_2 w_2\right)
\end{aligned}$$

Simulation results

We take $4,000$ conditional posterior draws, treat the first $3,000$ as the burn-in period, and retain the final $1,000$ draws for each sample. Twenty samples for a total of $20,000$ draws are retained. Parameter estimates are reported in table 12.5, average treatment effect estimates are reported in table 12.6, and average treatment effect sample statistics for the simulation are reported in table 12.7. The correspondence of estimated and sample statistics for average treatment effects is quite good. Model estimates of average treatment effects closely mirror their counterpart sample statistics (based on simulated counterfactuals). Further, the model provides evidence supporting heterogeneity. As seems to be typical, the unidentified correlation parameter is near zero but seriously misestimated by our bounding approach.

A plot depicting the nonconstant nature of the simulation average marginal treatment effect is depicted in figure 12.2. Consistent with outcome heterogeneity, the plot is distinctly nonconstant (and nonlinear). Weighted-*MTE* estimates of population-level average treatment effects are reported in table 12.8.

12.8 Regulated report precision example

Table 12.5: Binary report precision McMC parameter estimates for heterogeneous outcome

statistic	β_0^1	β_1^1	β_0^0	β_1^0
mean	603.4	0.451	605.1	0.459
median	603.5	0.462	605.2	0.475
std.dev.	1.322	0.086	1.629	0.100
minimum	597.7	0.172	599.3	0.118
maximum	607.2	0.705	610.0	0.738

statistic	γ_0	γ_1	γ_2
mean	0.002	-0.899	38.61
median	-0.003	0.906	38.64
std.dev.	0.038	2.276	1.133
minimum	-0.123	-10.13	33.96
maximum	0.152	7.428	42.73

statistic	$cor(V, U_L)$	$cor(V, U_H)$	$cor(U_L, U_H)$
mean	0.858	-0.859	-1.000
median	0.859	-0.859	-1.000
std.dev.	0.010	0.010	0.000
minimum	0.820	-0.888	-1.000
maximum	0.889	-0.821	-0.998

$$Y^L = \beta_0^L + \beta_1^L \left(s^L - \overline{s}\right) + U_L$$
$$Y^H = \beta_0^H + \beta_1^H \left(s^H - \overline{s}\right) + U_H$$
$$EU = \gamma_0 + \gamma_1 w_1 + \gamma_2 w_2 + V$$

Table 12.6: Binary report precision McMC average treatment effect estimates for heterogeneous outcome

statistic	estATE	estATT	estATUT
mean	-1.799	55.51	-58.92
median	-1.801	54.99	-58.57
std.dev.	1.679	1.942	1.983
minimum	-5.848	51.83	-63.21
maximum	2.148	59.42	-54.88

Table 12.7: Binary report precision McMC average treatment effect sample statistics for heterogeneous outcome

statistic	ATE	ATT	ATUT	OLS
mean	-0.008	64.39	-64.16	-2.180
median	0.147	64.28	-64.08	-2.218
std.dev.	1.642	0.951	0.841	1.195
minimum	-2.415	62.43	-65.71	-4.753
maximum	4.653	66.07	-62.53	-0.267

316   12. Bayesian treatment effects

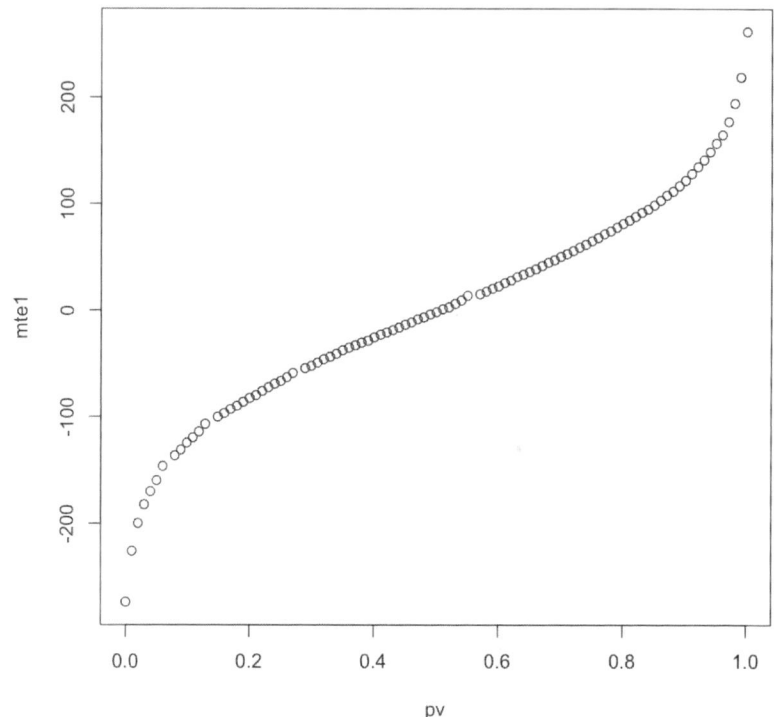

Figure 12.2: $MTE\left(u_D\right)$ versus $u_D = p_\nu$ for binary report precision

These *MTE*-weighted estimation results are similar to the estimates above, though a bit dampened. Next, we consider a multitude of report precision choices to the owners but observed as binary (high or low inverse report precision) by the analyst.

12.8.2 Continuous report precision but observed binary selection

The owners' report precision selection is highly varied as it depends on the realized draws for α, γ, α_d^L, and α_d^H but the analyst observes binary (high or low) report precision.[8] That is, the owner chooses between inverse report precision

$$\bar{\sigma}_2^L \equiv \arg\max_{\sigma_2} EU\left(\sigma_1, \alpha, \gamma, E\left[\alpha_d^L\right] \mid \bar{\sigma}_2^L\right)$$

---

[8] We emphasize the distinction from the binary selection case. Here, α and γ refer to realized draws from a normal distribution whereas in the binary case they are constants (equal to the means of their distributions in this setting).

12.8 Regulated report precision example

Table 12.8: Binary report precision McMC MTE-weighted average treatment effect estimates for heterogeneous outcome

statistic	estATE	estATT	estATUT
mean	−1.799	52.13	−55.54
median	−1.757	51.35	−55.45
std.dev.	1.710	1.941	1.818
minimum	−5.440	48.86	−58.87
maximum	1.449	55.21	−52.92

and

$$\sigma_2^H \equiv \arg\max_{\sigma_2} EU\left(\sigma_1, \alpha, \gamma, E\left[\alpha_d^H\right] \mid \overline{\sigma}_2^H\right)$$

to maximize her expected utility

$$D = \begin{matrix} 1 \\ 0 \end{matrix} \quad \begin{matrix} EU\left(\sigma_2^L, \beta^L \mid \overline{\sigma}_2^L\right) - EU\left(\sigma_2^H, \beta^H \mid \overline{\sigma}_2^H\right) > 0 \\ \text{otherwise} \end{matrix}$$

The analyst only observes report precision selection $D = 1$ (σ_2^L) or $D = 0$ (σ_2^H), outcomes Y, covariate s, instruments w_1 and w_2, and $\alpha_d = D\alpha_d^L + (1-D)\alpha_d^H$ where draws are $\alpha_d^L \sim N(0.02, 0.005)$ and $\alpha_d^H \sim N(0.04, 0.01)$. As discussed earlier, the instruments are the components of α_d and $\sigma_2 = D\sigma_2^L + (1-D)\sigma_2^H$ orthogonal to

$$U^L = -\left(\beta^L - E[\beta]\right)\left[D\frac{\sigma_1^2\left(\overline{\sigma}_2^L\right)^2}{\sigma_1^2 + \left(\overline{\sigma}_2^L\right)^2} + (1-D)\frac{\sigma_1^2\left(\overline{\sigma}_2^H\right)^2}{\sigma_1^2 + \left(\overline{\sigma}_2^H\right)^2}\right]$$

and

$$U^H = -\left(\beta^H - E[\beta]\right)\left[D\frac{\sigma_1^2\left(\overline{\sigma}_2^L\right)^2}{\sigma_1^2 + \left(\overline{\sigma}_2^L\right)^2} + (1-D)\frac{\sigma_1^2\left(\overline{\sigma}_2^H\right)^2}{\sigma_1^2 + \left(\overline{\sigma}_2^H\right)^2}\right]$$

The following parameters characterize the setting:

Continuous choice but observed binary parameters
$\mu = 1,000$
$\sigma_1 = 10$
$b = 150$
$\widehat{b} = 128.4$
$\gamma \sim N(2.5, 1)$
$\alpha \sim N(0.02, 0.005^2)$
$\alpha_d^L \sim N(0.02, 0.005)$
$\alpha_d^H \sim N(0.04, 0.01)$
$\beta^L, \beta^H \overset{iid}{\sim} N(7, 1)$
$s^j \sim N\left(\mu, \sigma_1^2 + \left(\sigma_2^j\right)^2\right) \quad j = L \text{ or } H$

318   12. Bayesian treatment effects

Table 12.9: Continuous report precision but observed binary selection McMC parameter estimates

statistic	β_0^1	β_1^1	β_0^0	β_1^0
mean	586.0	0.515	589.7	0.370
median	586.5	0.520	589.8	0.371
std.dev.	2.829	0.086	2.125	0.079
minimum	575.6	0.209	581.1	0.078
maximum	592.1	0.801	596.7	0.660
statistic	γ_0	γ_1	γ_2	
mean	-0.055	-39.49	0.418	
median	-0.055	-39.51	0.418	
std.dev.	0.028	2.157	0.161	
minimum	-0.149	-47.72	-0.222	
maximum	0.041	-31.51	1.035	
statistic	$cor(V, U_L)$	$cor(V, U_H)$	$cor(U_L, U_H)$	
mean	0.871	-0.864	-0.997	
median	0.870	-0.864	-1.000	
std.dev.	0.014	0.015	0.008	
minimum	0.819	-0.903	-1.000	
maximum	0.917	-0.807	-0.952	

$$Y^L = \beta_0^L + \beta_1^L \left(s^L - \bar{s}\right) + U_L$$
$$Y^H = \beta_0^H + \beta_1^H \left(s^H - \bar{s}\right) + U_H$$
$$EU = \gamma_0 + \gamma_1 w_1 + \gamma_2 w_2 + V$$

Again, we take 4,000 conditional posterior draws, treat the first 3,000 as the burn-in period, and retain the final 1,000 draws for each sample, a total of 20,000 draws are retained. *McMC* parameter estimates are reported in table 12.9, average sample statistics are reported in table 12.10, and average treatment effect sample statistics are reported in table 12.11. The correspondence of estimated and sample statistics for average treatment effects is not quite as strong as for the binary case. While model estimated *ATE* mirrors its counterpart sample statistic, estimates of both *ATT* and *ATUT* are over-estimated relative to their sample statistics. However, the model provides evidence properly supporting heterogeneity. As seems to

Table 12.10: Continuous report precision but observed binary selection McMC average treatment effect estimates

statistic	estATE	estATT	estATUT
mean	-3.399	87.49	-93.86
median	-2.172	87.66	-92.59
std.dev.	3.135	3.458	4.551
minimum	-13.49	76.86	-106.2
maximum	1.124	96.22	-87.21

12.8 Regulated report precision example 319

Table 12.11: Continuous report precision but observed binary selection McMC average treatment effect sample statistics

statistic	ATE	ATT	$ATUT$	OLS
mean	0.492	64.68	−63.40	−1.253
median	0.378	64.67	−63.32	−1.335
std.dev.	1.049	0.718	0.622	1.175
minimum	−1.362	63.42	−64.70	−4.065
maximum	2.325	65.77	−62.44	0.970

Table 12.12: Continuous report precision but observed binary selection McMC MTE-weighted average treatment effect estimates

statistic	$estATE$	$estATT$	$estATUT$
mean	−3.399	84.08	−90.46
median	−2.194	84.00	−89.08
std.dev.	3.176	3.430	4.602
minimum	−11.35	79.45	−101.2
maximum	0.529	92.03	−84.96

be typical, the unidentified correlation parameter is near zero but poorly estimated by our bounding approach.

Figure 12.3 depicts the nonconstant nature of the simulation average marginal treatment effect. Consistent with outcome heterogeneity, the plot is distinctly nonconstant (and nonlinear). Table 12.12 reports weighted-*MTE* estimates of population-level average treatment effects. These *MTE*-weighted estimation results are very similar to the estimates above.

12.8.3 Apparent nonnormality of unobservable choice

The following parameters characterize the setting where the analyst observes binary selection but the owners' selection is highly varied and choice is apparently

320   12. Bayesian treatment effects

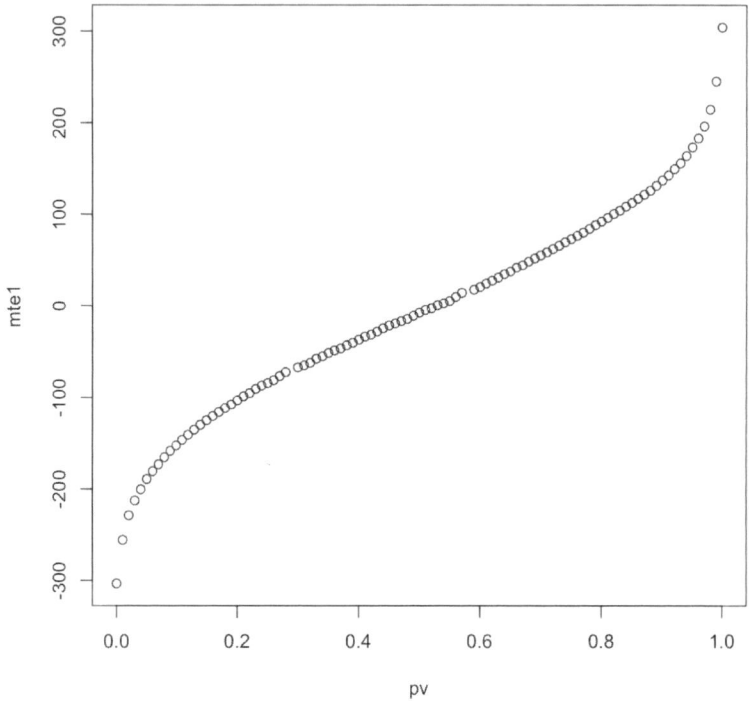

Figure 12.3: $MTE\left(u_D\right)$ versus $u_D = p_\nu$ for continuous report precision but binary selection

nonnormal as unobservables $\alpha, \gamma,$ and α_d^j are exponential random variables:

Apparent nonnormality parameters
$\mu = 1,000$
$\sigma_1 = 10$
$b = 150$
$\widehat{b} = 128.4$
$\gamma \sim \exp\left(\frac{1}{5}\right)$
$\alpha \sim \exp\left(\frac{1}{0.03}\right)$
$\alpha_d^L \sim \exp\left(\frac{1}{0.02}\right)$
$\alpha_d^H \sim \exp\left(\frac{1}{0.04}\right)$
$\beta^L, \beta^H \stackrel{iid}{\sim} N\left(7,1\right)$
$s^j \sim N\left(\mu, \sigma_1^2 + \left(\sigma_2^j\right)^2\right) \quad j = L \text{ or } H$

The owner chooses between inverse report precision

$$\overline{\sigma}_2^L \equiv \arg\max_{\sigma_2} EU\left(\sigma_1, \alpha, \gamma, E\left[\alpha_d^L\right] \mid \overline{\sigma}_2^L\right)$$

and

$$\overline{\sigma}_2^H \equiv \arg\max_{\sigma_2} EU\left(\sigma_1, \alpha, \gamma, E\left[\alpha_d^H\right] \mid \overline{\sigma}_2^H\right)$$

to maximize her expected utility

$$D = \begin{cases} 1 & EU\left(\overline{\sigma}_2^L, \beta^L \mid \overline{\sigma}_2^L\right) - EU\left(\overline{\sigma}_2^H, \beta^H \mid \overline{\sigma}_2^H\right) > 0 \\ 0 & \text{otherwise} \end{cases}$$

The analyst does not observe $\alpha_d = D\alpha_d^L + (1-D)\alpha_d^H$. Rather, the analyst only observes report precision selection $D = 1$ (for $\overline{\sigma}_2^L$) or $D = 0$ (for $\overline{\sigma}_2^H$), outcomes Y, covariate s, and instruments w_1 and w_2. As discussed earlier, the instruments are the components of α_d and $\sigma_e = D\overline{\sigma}_2^L + (1-D)\overline{\sigma}_2^H$ orthogonal to

$$U^L = -\left(\beta^L - E[\beta]\right)\left[D\frac{\sigma_1^2\left(\overline{\sigma}_2^L\right)^2}{\sigma_1^2 + \left(\overline{\sigma}_2^L\right)^2} + (1-D)\frac{\sigma_1^2\left(\overline{\sigma}_2^H\right)^2}{\sigma_1^2 + \left(\overline{\sigma}_2^H\right)^2}\right]$$

and

$$U^H = -\left(\beta^H - E[\beta]\right)\left[D\frac{\sigma_1^2\left(\overline{\sigma}_2^L\right)^2}{\sigma_1^2 + \left(\overline{\sigma}_2^L\right)^2} + (1-D)\frac{\sigma_1^2\left(\overline{\sigma}_2^H\right)^2}{\sigma_1^2 + \left(\overline{\sigma}_2^H\right)^2}\right]$$

Again, we take 4,000 conditional posterior draws, treat the first 3,000 as the burn-in period, and retain the final 1,000 draws for each sample, a total of 20,000 draws are retained. *McMC* parameter estimates are reported in table 12.13, average sample statistics are reported in table 12.14, and average treatment effect sample statistics are reported in table 12.15. These results evidence some bias which is not surprising as we assume a normal likelihood function even though the *DGP* employed for selection is apparently nonnormal. The unidentified correlation parameter is near zero but again poorly estimated by our bounding approach. Not surprisingly, the model provides support for heterogeneity (but we might be concerned that it would erroneously support heterogeneity when the *DGP* is homogeneous).

The simulation average marginal treatment effect is plotted in figure 12.4. The plot is distinctly nonconstant and, in fact, nonlinear — a strong indication of outcome heterogeneity. Weighted-*MTE* population-level average treatment effects are reported in table 12.16. These results are very similar to those reported above and again over-state the magnitude of self-selection (*ATT* is upward biased and *ATUT* is downward biased).

Stronger instrument

As discussed in the classical selection analysis of report precision in chapter 10, the poor results obtained for this case is not likely a result of nonnormality of

322   12. Bayesian treatment effects

Table 12.13: Continuous report precision but observed binary selection McMC parameter estimates for nonnormal DGP

statistic	β_0^1	β_1^1	β_0^0	β_1^0
mean	568.9	0.478	575.9	0.426
median	569.1	0.474	576.1	0.430
std.dev.	2.721	0.101	2.709	0.093
minimum	561.2	0.184	565.8	0.093
maximum	577.0	0.872	583.8	0.738
statistic	γ_0	γ_1	γ_2	
mean	−0.066	−3.055	0.052	
median	−0.064	−3.043	0.053	
std.dev.	0.028	0.567	0.052	
minimum	−0.162	−5.126	−0.139	
maximum	0.031	−1.062	0.279	
statistic	$cor(V, U_1)$	$cor(V, U_0)$	$cor(U_1, U_0)$	
mean	0.917	−0.917	−1.000	
median	0.918	−0.918	−1.000	
std.dev.	0.007	0.007	0.000	
minimum	0.894	−0.935	−1.000	
maximum	0.935	−0.894	−0.999	

$$Y^L = \beta_0^L + \beta_1^L \left(s^L - \bar{s}\right) + U_L$$
$$Y^H = \beta_0^H + \beta_1^H \left(s^H - \bar{s}\right) + U_H$$
$$EU = \gamma_0 + \gamma_1 w_1 + \gamma_2 w_2 + V$$

Table 12.14: Continuous report precision but observed binary selection McMC average treatment effect estimates for nonnormal DGP

statistic	estATE	estATT	estATUT
mean	−6.575	124.6	−127.5
median	−5.722	124.4	−127.4
std.dev.	2.987	4.585	3.962
minimum	−12.51	117.3	−135.8
maximum	−1.426	136.3	−117.9

Table 12.15: Continuous report precision but observed binary selection McMC average treatment effect sample statistics for nonnormal DGP

statistic	ATE	ATT	ATUT	OLS
mean	−0.183	61.79	−57.32	1.214
median	−0.268	61.70	−57.49	0.908
std.dev.	0.816	0.997	0.962	1.312
minimum	−2.134	59.79	−58.60	−0.250
maximum	1.409	63.64	−55.19	4.962

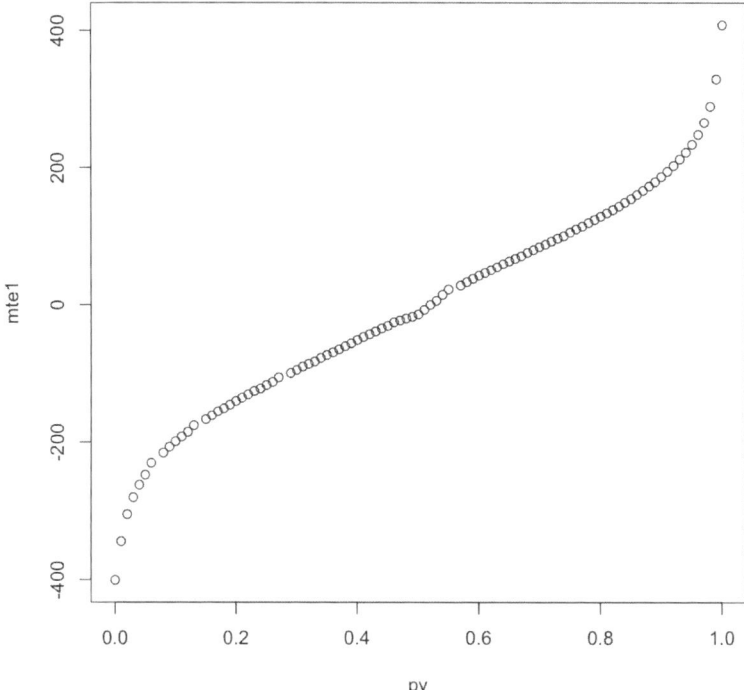

Figure 12.4: $MTE(u_D)$ versus $u_D = p_\nu$ for nonnormal DGP

unobservable utility but rather weak instruments. We proceed by replacing instrument w_2 with a stronger instrument, w_3, and repeat the Bayesian selection analysis.[9] *McMC* parameter estimates are reported in table 12.17, average sample statistics are reported in table 12.18, and average treatment effect sample statistics are reported in table 12.19. The average treatment effect estimates correspond nicely with their sample statistics even though the unidentified correlation parameter is near zero but again poorly estimated by our bounding approach. The model strongly (and appropriately) supports outcome heterogeneity.

The simulation average marginal treatment effect is plotted in figure 12.5. The plot is distinctly nonconstant and nonlinear — a strong indication of outcome heterogeneity. Weighted-*MTE* population-level average treatment effects are reported in table 12.20. These results are very similar to those reported above.

---

[9]Instrument w_3 is constructed, for simulation purposes, from the residuals of a regression of the indicator variable $\Im\left(EU\left(\sigma_2^L, \bar{\sigma}_2^L\right) > EU\left(\sigma_2^H, \bar{\sigma}_2^L\right)\right)$ onto U^L and U^H.

Table 12.16: Continuous report precision but observed binary selection McMC MTE-weighted average treatment effect estimates for nonnormal DGP

statistic	estATE	estATT	estATUT
mean	−5.844	123.7	−125.7
median	−6.184	123.7	−127.0
std.dev.	4.663	3.714	5.404
minimum	−17.30	117.9	−137.6
maximum	3.211	131.9	−116.2

Table 12.17: Continuous report precision but observed binary selection stronger McMC parameter estimates

statistic	β_0^1	β_1^1	β_0^0	β_1^0
mean	603.8	0.553	605.5	0.536
median	603.7	0.553	605.4	0.533
std.dev.	1.927	0.095	1.804	0.080
minimum	598.1	0.192	600.4	0.270
maximum	610.3	0.880	611.6	0.800
statistic	γ_0	γ_1	γ_2	
mean	−0.065	−0.826	2.775	
median	−0.063	−0.827	2.768	
std.dev.	0.030	0.659	0.088	
minimum	−0.185	−3.647	2.502	
maximum	0.037	1.628	3.142	
statistic	$cor(V, U_1)$	$cor(V, U_0)$	$cor(U_1, U_0)$	
mean	0.832	−0.832	−1.000	
median	0.833	−0.833	−1.000	
std.dev.	0.014	0.014	0.000	
minimum	0.775	−0.869	−1.000	
maximum	0.869	−0.775	−0.998	

$$Y^L = \beta_0^L + \beta_1^L (s^L - \bar{s}) + U_L$$
$$Y^H = \beta_0^H + \beta_1^H (s^H - \bar{s}) + U_H$$
$$EU = \gamma_0 + \gamma_1 w_1 + \gamma_2 w_2 + V$$

Table 12.18: Continuous report precision but observed binary selection stronger McMC average treatment effect estimates

statistic	estATE	estATT	estATUT
mean	−1.678	62.29	−61.03
median	−1.936	62.49	−61.27
std.dev.	1.783	2.370	3.316
minimum	−6.330	56.86	−66.97
maximum	2.593	67.19	−52.46

Table 12.19: Continuous report precision but observed binary selection stronger McMC average treatment effect sample statistics

statistic	ATE	ATT	$ATUT$	OLS
mean	0.151	62.53	−57.72	0.995
median	−0.042	62.41	−57.36	1.132
std.dev.	1.064	1.086	1.141	1.513
minimum	−1.918	60.92	−56.96	−1.808
maximum	1.904	64.63	−55.60	3.527

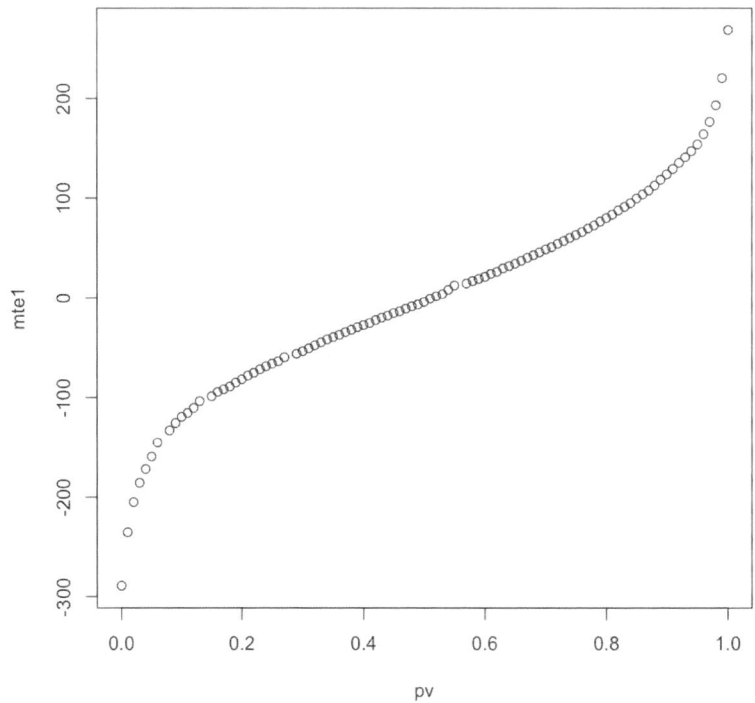

Figure 12.5: $MTE(u_D)$ versus $u_D = p_\nu$ with stronger instruments

Table 12.20: Continuous report precision but observed binary selection stronger McMC MTE-weighted average treatment effect estimates

statistic	estATE	estATT	estATUT
mean	−1.678	57.14	−56.26
median	−1.965	57.35	−56.36
std.dev.	1.814	2.301	3.147
minimum	−5.764	52.86	−60.82
maximum	2.319	60.24	−49.43

12.8.4 Policy-relevant report precision treatment effect

An important question involves the treatment effect induced by policy intervention. Regulated report precision seems a natural setting to explore policy intervention. What is the impact of treatment on outcomes when the reporting environment changes from unregulated (private report precision selection) to regulated (with the omnipresent possibility of costly transaction design). This issue was first raised in the marginal treatment effect discussion of chapter 11. The policy-relevant treatment effect and its connection to MTE when policy intervention affects the likelihood of treatment but not the distribution of outcomes is

$$PRTE = E[Y \mid X = x, a] - E[Y \mid X = x, a']$$
$$= \int_0^1 MTE(x, u_D) \left[F_{P(a')|X}(u_D \mid x) - F_{P(a)|X}(u_D \mid x) \right] du_D$$

where $F_{P(a)|X}(u_D \mid x)$ is the distribution function for the probability of treatment, P, and policy a refers to regulated report precision while policy a' denotes unregulated reporting precision.

Treatment continues to be defined by high or low inverse-report precision. However, owner type is now defined by high or low report precision cost $\alpha^H = 0.04$ or $\alpha^L = 0.02$ rather than transaction design cost $\alpha_d = 0.02$ to facilitate comparison between regulated and unregulated environments. Since transaction design cost of deviating from the standard does not impact the owner's welfare in an unregulated report environment, high and low inverse report precision involves different values for $\sigma_2^L = \sqrt{139.1}$ or $\sigma_2^H = \sqrt{144.8}$ for unregulated privately selected report precision than for regulated report precision, $\sigma_2^L = \sqrt{133.5}$ or $\sigma_2^H = \sqrt{139.2}$. Consequently, marginal treatment effects can be estimated from either the unregulated privately selected report precision data or the regulated report precision data. As the analyst likely doesn't observe this discrepancy or any related divergence in outcome distributions, the treatment effect analysis is potentially confounded. We treat the data source for estimating MTE as an experimental manipulation in the simulation reported below.

We explore the implications of this potentially confounded policy intervention induced treatment effect via simulation. In particular, we compare the sample sta-

Table 12.21: Policy-relevant average treatment effects with original precision cost parameters

statistic	$estPRTE\,(a)$	$estPRTE\,(a')$	$PRTE$ sample statistic
mean	0.150	0.199	6.348
median	0.076	0.115	6.342
std.dev.	0.542	0.526	0.144
minimum	−0.643	−0.526	6.066
maximum	1.712	1.764	6.725

tistic for $E\,[Y \mid X = x, a] - E\,[Y \mid X = x, a']$ ($PRTE$ sample statistic)[10] with the treatment effects estimated via the regulated $MTE\,(a)$ data

$$estPRTE\,(a) = \int_0^1 MTE\,(x, u_D, a)$$
$$\times \left[F_{P(a')|X}\,(u_D \mid x) - F_{P(a)|X}\,(u_D \mid x)\right] du_D$$

and via the unregulated $MTE\,(a')$ data

$$estPRTE\,(a') = \int_0^1 MTE\,(x, u_D, a')$$
$$\times \left[F_{P(a')|X}\,(u_D \mid x) - F_{P(a)|X}\,(u_D \mid x)\right] du_D$$

The simulation employs 20 samples of 5,000 draws. 4,000 conditional posterior draws are draws with 3,000 discarded as burn-in. As before, $\gamma = 2.5$, w_3 is employed as an instrument,[11] and stochastic variation is generated via independent draws for β^L and β^H normal with mean seven and unit variance. Marginal treatment effects and their estimated policy-relevant treatment effects for the unregulated and regulated regimes are quite similar. However, (interval) estimates of the policy-relevant treatment effect based on MTE from the regulated regime ($estPRTE\,(a)$) or from the unregulated regime ($estPRTE\,(a')$) diverge substantially from the sample statistic as reported in table 12.21. These results suggest it is difficult to satisfy the MTE ceteris paribus (policy invariance) conditions associated with policy-relevant treatment parameters in this report precision setting.

---

[10]In this report precision setting, this is what a simple difference-in-difference regression estimates. For instance,

$$E\,[Y \mid s, D, \Im(p_a)] = \beta_0 + \beta_1 D + \beta_2 \Im(p_a) + \beta_3 D\Im(p_a)$$
$$+\beta_4 (s - \bar{s}) + \beta_5 (s - \bar{s}) D + \beta_6 (s - \bar{s}) \Im(p_a)$$
$$+\beta_7 (s - \bar{s}) D\Im(p_a)$$

where $\Im(p_a)$ is an indicator for change in policy from $a\prime$ to a and β_3 is the parameter of interest. Of course, in general it may be difficult to identify and control for factors that cause changes in outcome in the absence of policy intervention.

[11]The instrument, w_3, is constructed from the residuals of $D\sigma_2^L + (1 - D) \sigma_2^H$ regressed onto U^L and U^H.

Table 12.22: Policy-relevant average treatment effects with revised precision cost parameters

statistic	$estPRTE(a)$	$estPRTE(a')$	$PRTE$ sample statistic
mean	1.094	1.127	0.109
median	0.975	0.989	0.142
std.dev.	0.538	0.545	0.134
minimum	−0.069	−0.085	−0.136
maximum	2.377	2.409	0.406

Now, suppose σ_2^L and σ_2^H are the same under both policy a' and a, say as the result of some sort of fortuitous general equilibrium effect. Let $\alpha^L = 0.013871$ rather than 0.02 and $\alpha^H = 0.0201564$ rather than 0.04 leading to $\sigma_2^L = \sqrt{133.5}$ or $\sigma_2^H = \sqrt{139.2}$ under both policies. Of course, this raises questions about the utility of policy a. But what does the analyst estimate from data? Table 12.22 reports results from repeating the above simulation but with this input for α. As before, *MTE* is very similar under both policies. Even with these "well-behaved" data, the *MTE* estimates of the policy-relevant treatment effect are biased upward somewhat, on average, relative to their sample statistic. Nonetheless, in this case the intervals overlap with the sample statistic and include zero, as expected. Unfortunately, this might suggest a positive, albeit modest, treatment effect induced by the policy change when there likely is none. A plot of the average marginal treatment effect, confirming once again heterogeneity of outcomes, is depicted in figure 12.6.

12.8.5 Summary

A few observations seem clear. First, choice of instruments is vital to any *IV* selection analysis. Short changing identification of strong instruments risks seriously compromising the analysis. Second, estimation of variance via Wishart draws requires substantial data and even then the quality of variance estimates is poorer than estimates of other parameters.[12] In spite of these concerns, Bayesian data augmentation with a multivariate Gaussian likelihood compares well with other classical analysis of the selection problem. Perhaps, the explanation involves building the analysis around what we know (see the discussion in the next section). Finally, policy invariance is difficult to satisfy in many accounting contexts; hence, there is more work to be done to address policy-relevant treatment effects of accounting regulations. Perhaps, an appropriately modified (to match what is known about the setting) difference-in-difference approach such as employed by Heckman, Ichimura, Smith, and Todd [1998] provides consistent evaluation of the evidence and background knowledge.

---

[12]Our experiments with mixtures of normals, not reported, are seriously confounded by variance estimation difficulties.

12.8 Regulated report precision example 329

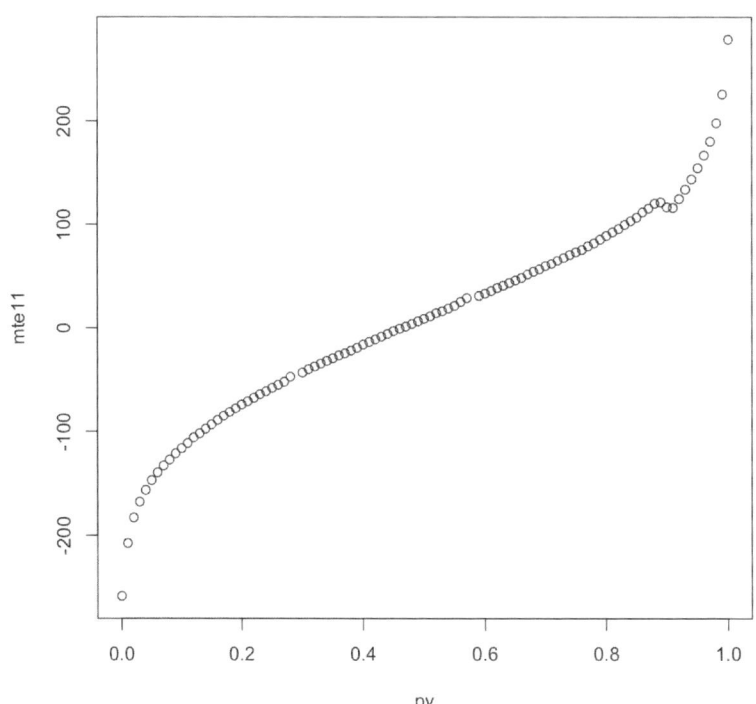

Figure 12.6: $MTE(u_D)$ versus $u_D = p_\nu$ for policy-relevant treatment effect

12.9 Probability as logic and the selection problem

There is a good deal of hand-wringing over the error distribution assignment for the selection equation. We've recounted some of it in our earlier discussions of identification and alternative estimation strategies. Jaynes [2003] suggests this is fuzzy thinking (Jaynes refers to this as an example of the mind projection fallacy) — probabilities represent a state of knowledge or logic, and are not limited to a description of long-run behavior of a physical phenomenon. What we're interested in is a consistent assessment of propositions based on our background knowledge plus new evidence. What we know can be expressed through probability assignment based on the maximum entropy principle (*MEP*). *MEP* incorporates what we know but only what we know — hence, maximum entropy. Then, new evidence is combined with background knowledge via Bayesian updating — the basis for consistent (scientific) reasoning or evaluation of propositions.

For instance, if we know something about variation in the data, then the maximum entropy likelihood function or sampling distribution is Gaussian. This is nearly always the case in a regression (conditional mean) setting. In a discrete choice setting, if we only know choice is discrete then the maximum entropy likelihood is the extreme value or logistic distribution. However, in a selection setting with a discrete selection mechanism, we almost always know something about the variation in the response functions, the variance-covariance matrix for the selection and response errors is positive definite, and, as the variance of the unobservable expected utility associated with selection is not estimable, its variance is normalized (to one). Collectively, then we have bounds on the variance-covariance matrix for the unobservables associated with the choice equation and the response equations. Therefore (and somewhat ironically given how frequently it's maligned), in the selection setting, the maximum entropy likelihood function is typically multivariate Gaussian.[13] Bayesian data augmentation (for missing data) strengthens the argument as we no longer rely on the hazard rate for estimation (or its nonlinearity for identification).

Another pass at what we know may support utilization of a hierarchical model with multivariate Student t conditional posteriors (Li, Poirier, and Tobias [2003]). The posterior probability of treatment conditional on Gaussian (outcome) evidence has a logistic distribution (Kiefer [1980]). Gaussians with uncertain variance go to a noncentral, scaled Student t distribution (on integrating out the variance nuisance parameter to derive the posterior distribution for the mean — the primary concern for average treatment effects). This suggests we assign selection a logistic distribution and outcome a multivariate Student t distribution. Since, the Student t distribution is an excellent approximation for the logistic distribution (Albert and Chib [1993]), we can regard the joint distribution as (approximately) multivariate Student t (O'Brien and Dunson [2003, 2004]). Now, we can

---

[13] If we had knowledge of multimodality, particularly, multimodal unobservable elements in the selection equation, we would be led to a Gaussian mixture likelihood. This is not the case in the regulated report precision setting.

employ a multivariate Student t Gibbs sampler (Li, Poirier, and Tobias [2003]) with degrees of freedom $\nu = 7.3$. This choice of ν minimizes the integrated squared distance between the logistic and Student t densities. If we normalize the scale parameter, σ, for the scaled Student t distributed selection unobservables to $\sigma^2 = \frac{\pi^2}{3} \frac{\nu-2}{\nu}$, then the variances of the scaled Student t and logistic distributions are equal (O'Brien and Dunson [2003]).[14,15]

Perhaps, it's more fruitful to focus our attention on well-posed (framed by theory) propositions, embrace endogeneity and inherent heterogeneity, identification of strong instruments, and, in general, collection of better data, than posturing over what we don't know about sampling distributions. Further, as suggested by Heckman [2001], we might focus on estimation of treatment effects which address questions of import rather than limiting attention to the average treatment effect. Frequently, as our examples suggest, the average treatment effect (*ATE*) is not substantively different from the exogenous effect estimated via *OLS*. If we look only to *ATE* for evidence of endogeneity there's a strong possibility that we'll fail to discover endogeneity even though it's presence is strong. Treatment effects on treated and untreated (*ATT* and *ATUT*) as well as nonconstant marginal treatment effects (*MTE*) are more powerful diagnostics for endogeneity and help characterize outcome heterogeneity.

In this penultimate chapter we discussed Bayesian analysis of selection and identification of treatment effects. In the final chapter, we explore Bayesian identification and inference a bit more broadly reflecting on the potential importance of informed priors.

12.10 Additional reading

Vijverberg [1993] and Koop and Poirier [1997] offer earlier discussions of the use of positive definiteness to bound the unidentified correlation parameter. Chib and Hamilton [2000] also discuss using *McMC* (in particular, the Metropolis-Hastings algorithm) to identify the posterior distribution of counterfactuals where the unidentified correlation parameter is set to zero. Chib and Hamilton [2002] apply general semiparametric *McMC* methods based on a Dirichlet process prior to longitudinal data.

---

[14]The univariate noncentral, scaled Student t density is

$$p(t \mid \mu, \sigma) = \frac{\Gamma\left(\frac{\nu+1}{2}\right)}{\Gamma\left(\frac{\nu}{2}\right)\sqrt{\nu\pi}\sigma} \left[1 + \frac{(t-\mu)^2}{\nu\sigma^2}\right]^{-\frac{\nu+1}{2}}$$

with scale parameter σ and ν degrees of freedom.

[15]Alternatively, instead of fixing the degrees of freedom, ν, we could use a Metropolis-Hastings algorithm to sample the conditional posterior for ν (Poirier, Li, and Tobias [2003]).

13
Informed priors

When building an empirical model we typically attempt to include our understanding of the phenomenon as part of the model. This commonly describes both classical and Bayesian analyses (usually with locally uninformed priors). However, what analysis can we undertake if we have no data (new evidence) on which to apply our model. The above modeling strategy leaves us in a quandary. With no new data, we are not (necessarily) in a state of complete ignorance and this setting suggests the folly of ignoring our background knowledge in standard data analysis. If our model building strategy adequately reflects our state of knowledge plus the new data, we expect inferences from the standard approach described above to match Bayesian inference based on our informed priors plus the new data. If not, we have been logically inconsistent in at least one of the analyses. Hence, at a minimum, Bayesian analysis with informed priors serves as a consistency check on our analysis.

In this section, we briefly discuss maximum entropy priors conditional on our state of knowledge (see Jaynes [2003]). Our state of knowledge is represented by various averages of background knowledge (this includes means, variances, covariances, etc.). This is what we refer to as informed priors. The priors reflect our state of knowledge but no more; hence, maximum entropy conditional on what we know about the problem. Apparently, the standard in physical statistical mechanics for over a century.

13.1 Maximum entropy

What does it mean to be completely ignorant? If we know nothing, then we are unable to differentiate one event or state from another. If we are unable to differentiate events then our probability assignment consistent with this is surely that each event is equally likely. To suggest otherwise, presumes some deeper understanding. In order to deal with informed priors it is helpful to contrast with complete ignorance and its probability assignment. Maximum entropy priors are objective in the sense that two (or more) individuals with the same background knowledge assign the same plausibilities regarding a given set of propositions prior to considering new evidence.

Shannon's [1948] classical information theory provides a measure of our ignorance in the form of entropy. Entropy is defined as

$$H = -\sum_{i=1}^{n} p_i \log p_i$$

where $p_i \geq 0$ and $\sum_{i=1}^{n} p_i = 1$. This can be developed axiomatically from the following conditions.

Condition 13.1 *Some numerical measure $H_n(p_1, \ldots, p_n)$ of "state of knowledge" exists.*

Condition 13.2 *Continuity: $H_n(p_1, \ldots, p_n)$ is a continuous function of p_i.*[1]

Condition 13.3 *Monotonicity: $H_n(p_1, \ldots, p_n)$ is a monotone increasing function of n.*[2]

Condition 13.4 *Consistency: if there is more than one way to derive the value for $H_n(p_1, \ldots, p_n)$, they each produce the same answer.*

Condition 13.5 *Additivity:*[3]

$$H_n(p_1, \ldots p_n) = H_r(p_1, \ldots p_r) + w_1 H_k\left(\frac{p_1}{w_1}, \ldots, \frac{p_k}{w_1}\right)$$
$$+ w_2 H_m\left(\frac{p_{k+1}}{w_2}, \ldots, \frac{p_{k+m}}{w_2}\right) + \cdots$$

Now, we sketch the arguments. Let

$$h(n) \equiv H\left(\frac{1}{n}, \ldots, \frac{1}{n}\right)$$

---

[1] Otherwise, an arbitrarily small change in the probability distribution could produce a large change in $H_n(p_1, \ldots, p_n)$.
[2] Monotonicity provides a sense of direction.
[3] For instance, $H_3(p_1, p_2, p_3) = H_2(p_1, q) + qH_2\left(\frac{p_2}{q}, \frac{p_3}{q}\right)$.

and
$$p_i = \frac{n_i}{\sum_{i=1}^{n} n_i}$$

for integers n_i. Then, combining the above with condition 13.5 implies

$$h\left(\sum_{i=1}^{n} n_i\right) = H(p_1, \ldots, p_n) + \sum_{i=1}^{n} p_i h(n_i)$$

Consider an example where $n = 3$, $n_1 = 3$, $n_2 = 4$, $n_3 = 2$,

$$\begin{aligned}
h(9) &= H\left(\frac{3}{9}, \frac{4}{9}, \frac{2}{9}\right) + \frac{3}{9}h(3) + \frac{4}{9}h(4) + \frac{2}{9}h(2) \\
&= H\left(\frac{3}{9}, \frac{4}{9}, \frac{2}{9}\right) + \frac{3}{9}H\left(\frac{1}{3}, \frac{1}{3}, \frac{1}{3}\right) + \frac{4}{9}H\left(\frac{1}{4}, \frac{1}{4}, \frac{1}{4}, \frac{1}{4}\right) + \frac{2}{9}H\left(\frac{1}{2}, \frac{1}{2}\right) \\
&= H\left(\frac{1}{9}, \ldots, \frac{1}{9}\right)
\end{aligned}$$

If we choose $n_i = m$ then the above collapses to yield

$$h(mn) = h(m) + h(n)$$

and apparently $h(n) = K \log n$, but since we're maximizing a monotone increasing function in p_i we can work with

$$h(n) = \log n$$

then

$$\begin{aligned}
h\left(\sum_{i=1}^{n} n_i\right) &= H(p_1, \ldots, p_n) + \sum_{i=1}^{n} p_i h(n_i) \\
&= H(p_1, \ldots, p_n) + \sum_{i=1}^{n} p_i \log n_i
\end{aligned}$$

Rewriting yields

$$H(p_1, \ldots, p_n) = h\left(\sum_{i=1}^{n} n_i\right) - \sum_{i=1}^{n} p_i \log n_i$$

Substituting $p_i \sum_i n_i$ for n_i yields

$$\begin{aligned} H(p_1, \ldots, p_n) &= h\left(\sum_{i=1}^n n_i\right) - \sum_{i=1}^n p_i \log\left(p_i \sum_i n_i\right) \\ &= h\left(\sum_{i=1}^n n_i\right) - \sum_{i=1}^n p_i \log p_i - \sum_{i=1}^n p_i \log\left(\sum_i n_i\right) \\ &= h\left(\sum_{i=1}^n n_i\right) - \sum_{i=1}^n p_i \log p_i - \log\left(\sum_i n_i\right) \end{aligned}$$

Since $h(n) = \log n$, $h\left(\sum_{i=1}^n n_i\right) = \log\left(\sum_i n_i\right)$, and we're left with Shannon's entropy measure

$$H(p_1, \ldots, p_n) = -\sum_{i=1}^n p_i \log p_i$$

13.2 Complete ignorance

Suppose we know nothing, maximization of H subject to the constraints involves solving the following Lagrangian for p_i, $i = 1, \ldots, n$, and λ_0.[4]

$$-\sum_{i=1}^n p_i \log p_i - (\lambda_0 - 1)\left(\sum_{i=1}^n p_i - 1\right)$$

The first order conditions are

$$-\lambda_0 - \log(p_i) = 0 \quad \text{for all } i$$
$$\sum_{i=1}^n p_i - 1 = 0$$

Then, the solution is

$$p_i = \exp[-\lambda_0] \quad \text{for all } i$$
$$\lambda_0 = \log n$$

In other words, as expected, $p_i = \frac{1}{n}$ for all i. This is the maximum entropy probability assignment.

---

[4] It's often convenient to write the Lagrange multiplier as $(\lambda_0 - 1)$.

13.3 A little background knowledge

Suppose we know a bit more. In particular, suppose we know the mean is F. Now, the Lagrangian is

$$-\sum_{i=1}^{n} p_i \log p_i - (\lambda_0 - 1)\left(\sum_{i=1}^{n} p_i - 1\right) - \lambda_1 \left(\sum_{i=1}^{n} p_i f_i - F\right)$$

where f_i is the realized value for event i. The solution is

$$p_i = \exp\left[-\lambda_0 - f_i \lambda_1\right] \quad \text{for all } i$$

For example, $n = 3$, $f_1 = 1$, $f_2 = 2$, $f_3 = 3$, and $F = 2.5$, the maximum entropy probability assignment and multipliers are[5]

p_1	0.116
p_2	0.268
p_3	0.616
λ_0	2.987
λ_1	-0.834

13.4 Generalization of maximum entropy principle

Suppose variable x can take on n different discrete values $(x_1, \ldots x_n)$ and our background knowledge implies there are m different functions of x

$$f_k(x), \quad 1 \leq k \leq m < n$$

and these have expectations given to us in our statement of the background knowledge

$$E[f_k(x)] = F_k = \sum_{i=1}^{n} p_i f_k(x_i), \quad 1 \leq k \leq m$$

The set of probabilities with maximum entropy that satisfy these m constraints can be identified by Lagrangian methods. As above, the solution is

$$p_i = \exp\left[-\lambda_0 - \sum_{j=1}^{m} \lambda_j f_j(x_i)\right] \quad \text{for all } i$$

and the sum of the probabilities is unity,

$$1 = \sum_{i=1}^{n} p_i = \exp[-\lambda_0] \sum_{i=1}^{n} \exp\left[-\sum_{j=1}^{m} \lambda_j f_j(x_i)\right]$$

---

[5] Of course, if $F = 2$ then $p_i = \frac{1}{3}$ and $\lambda_1 = 0$.

Now define a partition function

$$Z(\lambda_1, \ldots, \lambda_m) \equiv \sum_{i=1}^{n} \exp\left[-\sum_{j=1}^{m} \lambda_j f_j(x_i)\right]$$

and we have

$$1 = \exp[-\lambda_0] Z(\lambda_1, \ldots, \lambda_m)$$

which reduces to

$$\exp[\lambda_0] = Z(\lambda_1, \ldots, \lambda_m)$$

or

$$\lambda_0 = \log[Z(\lambda_1, \ldots, \lambda_m)]$$

Since the average value F_k equals the expected value of $f_k(x)$

$$F_k = \exp[-\lambda_0] \sum_{i=1}^{n} f_k(x_i) \exp\left[-\sum_{j=1}^{m} \lambda_j f_j(x_i)\right]$$

and

$$-\frac{\partial \log[Z(\lambda_1, \ldots, \lambda_m)]}{\partial \lambda_k} = \frac{\sum_{i=1}^{n} f_k(x_i) \exp\left[-\sum_{j=1}^{m} \lambda_j f_j(x_i)\right]}{Z(\lambda_1, \ldots, \lambda_m)}$$

$$= \exp[-\lambda_0] \sum_{i=1}^{n} f_k(x_i) \exp\left[-\sum_{j=1}^{m} \lambda_j f_j(x_i)\right]$$

Therefore,[6]

$$F_k = -\frac{\partial \log Z(\lambda_1, \ldots, \lambda_m)}{\partial \lambda_k}$$

---

[6] Return to the example with $n = 3$, $f_1(x_1) = 1$, $f_1(x_2) = 2$, $f_1(x_3) = 3$, and $F = 2.5$. The partition function is

$$Z(\lambda_1) = \exp[-f_1 \lambda_1] + \exp[-f_2 \lambda_1] + \exp[-f_3 \lambda_1].$$

It is readily verified that $-\frac{\partial \log Z(\lambda_1)}{\partial \lambda_1} = F = 2.5$ on substituting the values of the multipliers.

The maximum value of entropy is

$$\begin{aligned}
H_{\max} &= \max\left[-\sum_{i=1}^{n} p_i \log p_i\right] \\
&= \exp[-\lambda_0] \sum_{i=1}^{n} \exp\left[-\sum_{j=1}^{m} \lambda_j f_j(x_i)\right]\left(\lambda_0 + \sum_{j=1}^{m} \lambda_j f_j(x_i)\right) \\
&= \lambda_0 + \exp[-\lambda_0] \sum_{j=1}^{m} \sum_{i=1}^{n} \lambda_j f_j(x_i) \exp\left[-\sum_{j=1}^{m} \lambda_j f_j(x_i)\right] \\
&= \lambda_0 + \sum_{j=1}^{m} \lambda_j F_j
\end{aligned}$$

To establish support for a global maximum, consider two possible probability distributions

$$\sum_{i=1}^{n} p_i = 1 \quad p_i \geq 0$$

and

$$\sum_{i=1}^{n} u_i = 1 \quad u_i \geq 0$$

Note

$$\log x \leq x - 1 \quad 0 \leq x < \infty$$

with equality if and only if $x = 1$. Accordingly,

$$\sum_{i=1}^{n} p_i \log \frac{u_i}{p_i} \leq \sum_{i=1}^{n} p_i \left(\frac{u_i}{p_i} - 1\right) = \sum_{i=1}^{n} (u_i - p_i) = 0$$

with equality if and only if $p_i = u_i$, $i = 1, \ldots, n$. Rewrite the left hand side in terms of entropy for p_i

$$\begin{aligned}
\sum_{i=1}^{n} p_i \log \frac{u_i}{p_i} &= \sum_{i=1}^{n} p_i \log u_i - \sum_{i=1}^{n} p_i \log p_i \\
&= \sum_{i=1}^{n} p_i \log u_i + H(p_1, \ldots p_n)
\end{aligned}$$

Substitution into the inequality and rearrangement yields

$$H(p_1, \ldots p_n) \leq 0 - \sum_{i=1}^{n} p_i \log u_i$$

or

$$H(p_1, \ldots p_n) \leq \sum_{i=1}^{n} p_i \log \frac{1}{u_i}$$

Let

$$u_i \equiv \frac{1}{Z(\lambda_1, \ldots, \lambda_m)} \exp\left[-\sum_{j=1}^{m} \lambda_j f_j(x_i)\right]$$

where the partition function $Z(\lambda_1, \ldots, \lambda_m)$ effectively serves as a normalizing factor. Now we can write the inequality

$$H(p_1, \ldots p_n) \leq \sum_{i=1}^{n} p_i \log \frac{1}{u_i}$$

as

$$H(p_1, \ldots p_n) \leq \sum_{i=1}^{n} p_i \left[\log Z(\lambda_1, \ldots, \lambda_m) + \sum_{j=1}^{m} \lambda_j f_j(x_i)\right]$$

or

$$H(p_1, \ldots p_n) \leq \log Z(\lambda_1, \ldots, \lambda_m) + \sum_{j=1}^{m} \lambda_j E\left[f_j(x_i)\right]$$

Since p_i can vary over all possible probability distributions and it attains its maximum only when

$$p_i = u_i \equiv \frac{1}{Z(\lambda_1, \ldots, \lambda_m)} \exp\left[-\sum_{j=1}^{m} \lambda_j f_j(x_i)\right]$$

we have a general derivation for the maximum entropy probability assignment subject to background knowledge $F_j, j = 1, \ldots, m$.

13.5 Discrete choice model as maximum entropy prior

From here we can provide a more rigorous argument for the frequent utilization of logistic regression when faced with discrete choice analysis. The logit model for discrete choice D conditional on (regime differences in) covariates X is

$$\begin{aligned}\Pr(D \mid X) &= \frac{1}{1 + \exp[-Y]} \\ &= \frac{1}{1 + \exp[-X\gamma]}\end{aligned}$$

but the basis for this specification is frequently left unanswered. Following Blower [2004], we develop this model specification from the maximum entropy principle.

Bayesian revision yields

$$\Pr(D \mid X) = \frac{\Pr(D, X)}{\Pr(X)}$$

13.5 Discrete choice model as maximum entropy prior

and for treatment selection

$$\Pr(D = 1 \mid X) = \frac{\Pr(D = 1, X)}{\Pr(D = 1, X) + \Pr(D = 0, X)}$$

Rewrite this expression as

$$\Pr(D = 1 \mid X) = \frac{1}{1 + \frac{\Pr(D=0,X)}{\Pr(D=1,X)}}$$

The maximum entropy probability assignments, denoted $\hbar$, for the joint likelihoods, $\Pr(D = 1, X)$ and $\Pr(D = 0, X)$, are

$$\Pr(D = 1, X, \hbar) = \frac{\exp\left[-\sum_{j=1}^{m} \lambda_j f_j(X_1)\right]}{Z(\lambda_1, \ldots, \lambda_m)}$$

and

$$\Pr(D = 0, X, \hbar) = \frac{\exp\left[-\sum_{j=1}^{m} \lambda_j f_j(X_0)\right]}{Z(\lambda_1, \ldots, \lambda_m)}$$

The likelihood ratio is

$$\frac{\Pr(D = 0, X, \hbar)}{\Pr(D = 1, X, \hbar)} = \frac{\exp\left[-\sum_{j=1}^{m} \lambda_j f_j(X_0)\right]}{\exp\left[-\sum_{j=1}^{m} \lambda_j f_j(X_1)\right]}$$

$$= \exp[-Y]$$

where

$$Y = \exp\left[\sum_{j=1}^{m} \lambda_j \{f_j(X_1) - f_j(X_0)\}\right]$$

Hence, we have the logistic regression specification as a maximum entropy probability assignment

$$\Pr(D = 1 \mid X, \hbar) = \frac{1}{1 + \frac{\Pr(D=0,X,\hbar)}{\Pr(D=1,X,\hbar)}}$$

$$= \frac{1}{1 + \exp[-Y]}$$

13.6 Continuous priors

Applying the principle of maximum entropy to continuous prior distributions is more subtle. We sketch Jaynes' [2003, ch. 12] limit arguments by taking the discrete expression of entropy

$$H^d = -\sum_{i=1}^{n} p_i \log p_i$$

to a continuous expression for entropy

$$H_\ell^c = -\int_a^b p(x \mid \Im) \log \frac{p(x \mid \Im)}{m(x)} dx$$

whose terms are defined below.

Let the number of discrete points x_i, $i = 1, \ldots, n$, become very numerous such that

$$\lim_{n \to \infty} \frac{1}{n} (\text{number of points in } a < x < b) = \int_a^b m(x) \, dx$$

and assume this is sufficiently well-behaved that adjacent differences tend to zero such that

$$\lim_{n \to \infty} n(x_{i+1} - x_i) = \frac{1}{m(x_i)}$$

The discrete probability distribution p_i goes into a continuous density, $p(x \mid \Im)$, with background knowledge, $\Im$, via the limiting form of

$$p_i = p(x_i \mid \Im)(x_{i+1} - x_i)$$

or utilizing the limit above

$$p_i \to p(x_i \mid \Im) \frac{1}{nm(x_i)}$$

Since

$$\lim_{n \to \infty} \sum_{i=1}^{n} \frac{1}{nm(x_i)} = \int_a^b dx$$

the limit of discrete entropy is

$$\begin{aligned}
H_\ell^d &\equiv \lim_{n \to \infty} H^d \\
&= -\lim_{n \to \infty} \sum_{i=1}^{n} p_i \log p_i \\
&= -\lim_{n \to \infty} \sum_{i=1}^{n} \frac{p(x_i \mid \Im)}{nm(x_i)} \log \frac{p(x_i \mid \Im)}{nm(x_i)} \\
&= -\int_a^b p(x \mid \Im) \log \frac{p(x \mid \Im)}{nm(x)} dx
\end{aligned}$$

The limit contains an infinite term, $\log n$. Normalize H_ℓ^d by subtracting this term and we have Jaynes' continuous measure of entropy

$$\begin{aligned}H_\ell^c &\equiv \lim_{n\to\infty}\left[H_\ell^d - \log n\right]\\ &= -\int_a^b p(x\mid\Im)\log\frac{p(x\mid\Im)}{m(x)}dx + \int_a^b p(x\mid\Im)\log(n)\,dx - \log n\\ &= -\int_a^b p(x\mid\Im)\log\frac{p(x\mid\Im)}{m(x)}dx\end{aligned}$$

Next, we revisit maximum entropy for continuous prior distributions.

13.6.1 Maximum entropy

The maximum entropy continuous prior is normalized

$$\int_a^b p(x\mid\Im)\,dx = 1$$

and is constrained by m mean values F_k for the various different functions $f_k(x)$ from our background knowledge

$$F_k = \int_a^b f_k(x)\,p(x\mid\Im)\,dx \quad k = 1, 2, \ldots, m$$

Treating $m(x)$ as known, the solution to the Lagrangian identifies the maximum entropy continuous prior

$$p(x\mid\Im) = \frac{m(x)\exp[\lambda_1 f_1(x) + \cdots + \lambda_m f_m(x)]}{Z(\lambda_1, \ldots, \lambda_m)}$$

where the partition function is

$$Z(\lambda_1, \ldots, \lambda_m) = \int_a^b m(x)\exp[\lambda_1 f_1(x) + \cdots + \lambda_m f_m(x)]\,dx$$

and the Lagrange multipliers are determined from

$$F_k = -\frac{\partial \log Z(\lambda_1, \ldots, \lambda_m)}{\partial \lambda_k} \quad k = 1, 2, \ldots, m$$

Then, with the maximum entropy prior in hand, our best estimate (by quadratic loss) of any other function of the parameters, say $q(x)$, is

$$E[q(x)] = \int_a^b q(x)\,p(x\mid\Im)\,dx$$

What is the role of the invariance measure, $m(x)$? First note what $m(x)$ buys us. Inclusion of $m(x)$ in the entropy measure of our state of knowledge means the entropy measure H_ℓ^c, partition function, Lagrange multipliers, and $E[q(x)]$

are invariant under a transformation of parameters, say $x \to y(x)$. What does this imply for ignorance priors? Suppose we only know $a < x < b$, then there are no multipliers and

$$p(x \mid \Im) = \frac{m(x)\exp[0]}{\int_a^b m(x)\exp[0]\,dx}$$

$$= \frac{m(x)}{\int_a^b m(x)\,dx}$$

so that, except for normalizing constant $\frac{1}{\int_a^b m(x)dx}$, $m(x)$ is the prior distribution $p(x \mid \Im)$. Next, we briefly discuss use of transformation groups for resolving the invariance measure, $m(x)$, and fully specifying ignorance priors.

13.6.2 Transformation groups

We focus on ignorance priors since the maximum entropy principle dictates only our background knowledge is included in the prior; this means we must recognize our state of ignorance. Consider one of the most common problems in practice, a two parameter sampling distribution. We observe a sample $x_1, \ldots, x_n$ from a continuous sampling distribution $p(x \mid \nu, \sigma)\,dx = \phi(\nu, \sigma)\,dx$ where ν is a location parameter and σ is a scale parameter and we wish to estimate ν and σ. Suppose we have no knowledge of the location and scale parameters. What is the prior distribution $p(\nu, \sigma \mid \Im)\,d\nu d\sigma = f(\nu, \sigma)\,d\nu d\sigma$? What does it mean to have no knowledge of the location and scale parameters? Jaynes [2003, ch. 12] suggests the following characterization. If a change of location or scale alters our perception of the distribution of the parameters, we must not have been completely ignorant with regard to location and scale. Therefore, the distributions should be invariant to a transformation group.

Suppose we transform the variables as follows

$$\nu' = \nu + b$$
$$\sigma' = a\sigma$$
$$x' - \nu' = a(x - \nu)$$

$-\infty < b < \infty$ and $0 < a < \infty$. Invariance implies the sampling distribution for the transformed variables is the same as the sampling distribution for the original variables

$$p(x' \mid \nu', \sigma')\,dx' = \psi(x', \nu', \sigma')\,dx' = \phi(x, \nu, \sigma)\,dx$$

Similarly, the prior distribution for the transformed parameters, based on the Jacobian, is

$$g(\nu', \sigma') = a^{-1} f(\nu, \sigma)$$

These relations hold irrespective of the distributions $\phi(x, \nu, \sigma)$ and $f(\nu, \sigma)$.

13.6 Continuous priors

If the sampling distribution is invariant under the above transformation group, then the two functions are the same

$$\psi(x,\nu,\sigma) = \phi(x,\nu,\sigma)$$

for all values a and b. Invariance to location and scale implies

$$\phi(x,\nu,\sigma) = \frac{1}{\sigma} h\left(\frac{x-\nu}{\sigma}\right)$$

for arbitrary function $h(\cdot)$.[7] Now, we return to priors.

Consider another problem with sample $x'_1, \ldots, x'_n$ and we wish to estimate ν' and σ' but again have no initial knowledge of the location and scale. Let the prior distribution be $g(\nu',\sigma')$. Since we have two problems with the same background knowledge consistency requires we assign the same prior. Invariance to parameter transformation implies the functions are the same

$$f(\nu,\sigma) = g(\nu,\sigma)$$

Combining

$$g(\nu',\sigma') = a^{-1} f(\nu,\sigma)$$

with the transformation group gives

$$g(\nu + b, a\sigma) = a^{-1} f(\nu,\sigma)$$
$$f(\nu,\sigma) = a g(\nu + b, a\sigma)$$

Now,

$$f(\nu,\sigma) = g(\nu,\sigma)$$
$$f(\nu + b, a\sigma) = g(\nu + b, a\sigma)$$

combining this with the above yields

$$f(\nu,\sigma) = a f(\nu + b, a\sigma)$$

Satisfying this condition implies the prior distribution is

$$f(\nu,\sigma) = \frac{\text{constant}}{\sigma}$$

— this is Jeffrey's prior.

To illustrate, suppose we only know $0 < \nu < 2$ and $1 < \sigma < 2$, then we can assign $m(\nu,\sigma) = \frac{1}{\sigma}$ and $f(\nu,\sigma) = \frac{1}{2\log 2}\frac{1}{\sigma}$. Now, consider the transformation $b = 0.1$, and $a = \frac{1}{2}$, then $af(\nu+b, a\sigma) = \frac{1}{2}f(\nu + 0.1, \frac{1}{2}\sigma) = \frac{1}{2\log 2}\frac{1}{2}\frac{1}{\frac{1}{2}\sigma} = \frac{1}{2\log 2}\frac{1}{\sigma} = f(\nu,\sigma)$ and $m(\nu',\sigma') = \frac{1}{2}\frac{1}{\sigma'} = \frac{1}{2}\frac{1}{\frac{1}{2}\sigma} = \frac{1}{\sigma}$. If we assign $m(\nu',\sigma') = \frac{1}{\sigma'}$, then $m(\nu,\sigma) = 2\frac{1}{\sigma} = 2\frac{1}{2\sigma'} = \frac{1}{\sigma'}$. The key is existence of $m(x)$.

---

[7] This discussion attempts to convey the intuitive implications of transformation groups for maximum entropy. See Jaynes [2003, p. 379] for a more complete discussion.

13.6.3 Uniform prior

Next, we temporarily suppress the invariance measure, $m(x)$, and derive a maximum entropy ignorance prior utilizing differential entropy

$$H = -\int_a^b f(x) \log f(x)\, dx$$

as a measure of continuous entropy. Suppose we're completely ignorant except that x has continuous support over the interval $\{a, b\}$. The maximum entropy prior distribution is surely uniform. Its derivation involves maximization of the limiting form of entropy such that $f(x) \geq 0$ and $\int_a^b f(x)\, dx = 1$. Following Cover and Thomas [1991, ch. 11], formulate the Lagrangian[8]

$$\mathcal{L} = -\int_a^b f(x) \log f(x)\, dx + \lambda_0 \left[\int_a^b f(x)\, dx - 1 \right]$$

Since the partial derivative of the functional $-\int_a^b f(x) \log f(x)\, dx$ with respect to $f(x)$ for each value x is

$$\frac{\partial}{\partial f(x_i)} \left[-\int_a^b f(x) \log f(x)\, dx \right] = -\frac{\partial}{\partial f(x_i)} f(x_i) \log f(x_i)$$
$$= -\log f(x_i) - 1$$

the gradient of the Lagrangian is

$$-\log f(x) - 1 + \lambda_0$$

Solving the first order conditions yields[9]

$$f(x) = \exp\left[-1 + \lambda_0\right]$$

Utilizing the constraint to solve for λ_0 we have

$$\int_a^b f(x)\, dx = 1$$
$$\int_a^b \exp\left[-1 + \lambda_0\right] dx = 1$$
$$\exp\left[-1 + \lambda_0\right](b - a) = 1$$
$$\lambda_0 = 1 - \log(b - a)$$

Now,

$$f(x) = \exp\left[-1 + \lambda_0\right]$$

---

[8] Alternatively, we could begin from the partition function.
[9] Since the second partial derivatives with respect to $f(x)$ are negative for all x, $-\frac{1}{f(x)}$, a maximum is assured.

becomes

$$f(x) = \exp[-1+1-\log(b-a)]$$
$$f(x) = \frac{1}{b-a}$$

The maximum entropy prior with no background knowledge (other than continuity and support) is the uniform distribution. If we return to Jaynes' definition of continuous entropy then we can assign $m(x) = 1$ (an invariance measure exists) and normalization produces $f(x) = \frac{m(x)}{\int_a^b m(x)dx} = \frac{1}{b-a}$, as discussed earlier. Hereafter, we work with differential entropy (for simplicity) and keep in mind the existence of $m(x)$.

13.6.4 Gaussian prior

Suppose our background knowledge is limited to a continuous variable with finite mean μ and finite variance σ^2. Following the development above, the Lagrangian is

$$\mathcal{L} = -\int_{-\infty}^{\infty} f(x) \log f(x)\, dx + \lambda_0 \left(\int_{-\infty}^{\infty} f(x)\, dx - 1 \right)$$
$$+ \lambda_1 \left(\int_{-\infty}^{\infty} x f(x)\, dx - \mu \right) + \lambda_2 \left(\int_{-\infty}^{\infty} (x-\mu)^2 f(x)\, dx - \sigma^2 \right)$$

The first order conditions are

$$-1 - \log f(x) + \lambda_0 + \lambda_1 x + \lambda_2 (x-\mu)^2 = 0$$

or

$$f(x) = \exp\left[-1 + \lambda_0 + \lambda_1 x + \lambda_2 (x-\mu)^2\right]$$

Utilizing the constraints to solve for the multipliers involves

$$\int_{-\infty}^{\infty} \exp\left[-1 + \lambda_0 + \lambda_1 x + \lambda_2 (x-\mu)^2\right] dx = 1$$
$$\int_{-\infty}^{\infty} x \exp\left[-1 + \lambda_0 + \lambda_1 x + \lambda_2 (x-\mu)^2\right] dx = \mu$$
$$\int_{-\infty}^{\infty} (x-\mu)^2 \exp\left[-1 + \lambda_0 + \lambda_1 x + \lambda_2 (x-\mu)^2\right] dx = \sigma^2$$

A solution is[10]

$$\lambda_0 = 1 - \frac{1}{4}\log\left[4\pi^2\sigma^4\right]$$
$$\lambda_1 = 0$$
$$\lambda_2 = -\frac{1}{2\sigma^2}$$

---

[10] The result, $\lambda_1 = 0$, suggests how pivotal variance knowledge is to a Gaussian maximum entropy prior. In fact, for a given variance, the Gaussian distribution has maximum entropy.

Substitution of these values for the multipliers reveals

$$f(x) = \exp\left[-1 + \lambda_0 + \lambda_1 x + \lambda_2 (x-\mu)^2\right]$$

$$f(x) = \frac{1}{\sqrt{2\pi}\sigma} \exp\left[-\frac{1}{2}\frac{(x-\mu)^2}{\sigma^2}\right]$$

Hence, the maximum entropy prior given knowledge of the mean and variance is the Gaussian or normal distribution.

13.6.5 Multivariate Gaussian prior

If multiple variables or parameters are of interest and we have background knowledge of only their means μ and variances σ^2, then we know the maximum entropy prior for each is Gaussian (from above). Further, since we have no knowledge of their interactions, their joint prior is the product of the marginals.

Now, suppose we have background knowledge of the covariances as well. A straightforward line of attack is to utilize the Cholesky decomposition to write the variance-covariance matrix Σ as $\Gamma\Gamma^T$. We may now work with the transformed data $z = \Gamma^{-1}x$, derive the prior for z, and then by transformation of variables identify priors for x. Of course, since the prior for z is the product of marginal Gaussian priors, as before,

$$f(z_1, \ldots, z_k) = f(z_1) \cdots f(z_k)$$

$$= (2\pi)^{-\frac{k}{2}} \prod_{i=1}^{k} \exp\left[-\frac{1}{2}(z_i - \Gamma^{-1}\mu_i)^2\right]$$

where $f(z_i) = \frac{1}{\sqrt{2\pi}}\exp\left[-\frac{1}{2}(z_i - \Gamma^{-1}\mu_i)^2\right]$, the transformation back to the vector $x = \Gamma z$ produces the multivariate Gaussian distribution

$$f(x) = (2\pi)^{-\frac{k}{2}} J \exp\left[-\frac{1}{2}(\Gamma^{-1}x - \Gamma^{-1}\mu)^T (\Gamma^{-1}x - \Gamma^{-1}\mu)\right]$$

$$= (2\pi)^{-\frac{k}{2}} J \exp\left[-\frac{1}{2}(x-\mu)^T \Sigma^{-1}(x-\mu)\right]$$

where J is the Jacobian of the transformation. Since $J = |\Gamma^{-1}| = |\Gamma|^{-1}$ and $\Sigma = \left(LD^{\frac{1}{2}}\right)\left(D^{\frac{1}{2}}L^T\right) = \Gamma\Gamma^T$ is positive definite, $|\Gamma|^{-1} = |\Sigma|^{-\frac{1}{2}}$ where L is a lower triangular matrix and D is a diagonal matrix. Now, the density can be written in standard form

$$f(x) = (2\pi)^{-\frac{k}{2}} |\Sigma|^{-\frac{1}{2}} \exp\left[-\frac{1}{2}(x-\mu)^T \Sigma^{-1}(x-\mu)\right]$$

Hence, the maximum entropy prior when background knowledge is comprised only of means, variances, and covariances for multiple variables or parameters is the multivariate Gaussian distribution.

13.6.6 Exponential prior

Suppose we know the variable of interest has continuous but non-negative support and finite mean β. The Lagrangian is

$$\mathcal{L} = -\int_0^\infty f(x) \log f(x)\, dx + \lambda_0 \left(\int_0^\infty f(x)\, dx - 1 \right)$$
$$+ \lambda_1 \left(\int_0^\infty x f(x)\, dx - \beta \right)$$

The first order conditions are

$$-1 - \log f(x) + \lambda_0 + \lambda_1 x = 0$$

Solving for $f(x)$ produces

$$f(x) = \exp\left[-1 + \lambda_0 + \lambda_1 x\right]$$

Using the constraints to solve for the multipliers involves

$$\int_0^\infty \exp\left[-1 + \lambda_0 + \lambda_1 x\right] dx = 1$$
$$\int_0^\infty x \exp\left[-1 + \lambda_0 + \lambda_1 x\right] dx = \beta$$

and produces

$$\lambda_0 = 1 - \log \beta$$
$$\lambda_1 = -\frac{1}{\beta}$$

Substitution of these multipliers identifies the prior

$$f(x) = \exp\left[-1 + \lambda_0 + \lambda_1 x\right]$$
$$f(x) = \frac{1}{\beta} \exp\left[-\frac{x}{\beta}\right]$$

Hence, the maximum entropy prior is an exponential distribution with mean β.

13.6.7 Truncated exponential prior

If support is shifted to, say, (a, ∞) for $a > 0$ and the mean equals β, the maximum entropy prior is a "truncated" exponential distribution. The first order conditions continue to be

$$-1 - \log f(x) + \lambda_0 + \lambda_1 x = 0$$

Solving for $f(x)$ again produces

$$f(x) = \exp\left[-1 + \lambda_0 + \lambda_1 x\right]$$

But using the constraints to solve for the multipliers involves

$$\int_a^\infty \exp\left[-1 + \lambda_0 + \lambda_1 x\right] dx = 1$$

$$\int_a^\infty x \exp\left[-1 + \lambda_0 + \lambda_1 x\right] dx = \beta$$

and produces

$$\lambda_0 = 1 - \frac{a}{a+\beta} - \log\left[\beta - a\right]$$

$$\lambda_1 = \frac{1}{a-\beta}$$

Substitution of these multipliers identifies the prior

$$f(x) = \exp\left[-1 + \lambda_0 + \lambda_1 x\right]$$

$$f(x) = \frac{1}{\beta - a} \exp\left(-\frac{x-a}{\beta-a}\right)$$

Hence, the maximum entropy prior is a "truncated" exponential distribution with mean β.

13.6.8 Truncated Gaussian prior

Suppose our background knowledge consists of the mean and variance over the limited support region, say (a, ∞), the maximum entropy prior is the truncated Gaussian distribution. This is consistent with the property the Gaussian distribution has maximum entropy of any distribution holding the variance constant.

As an example suppose we compare a mean zero Gaussian with the exponential distribution with variance one (hence, $a = 0$ and the mean of the exponential distribution is also one). If the variance of the truncated Gaussian equals one, then the underlying untruncated Gaussian has variance $\sigma^2 = 2.752$.[11] Entropy for the

---

[11] A general expression for the moments of a truncated Gaussian is

$$E\left[x \mid a \leq x < b\right] = \mu + \frac{\phi\left(\frac{a-\mu}{\sigma}\right) - \phi\left(\frac{b-\mu}{\sigma}\right)}{\Phi\left(\frac{b-\mu}{\sigma}\right) - \Phi\left(\frac{a-\mu}{\sigma}\right)} \sigma$$

$$Var\left[x \mid a \leq x < b\right] = \sigma^2 \left[1 + \frac{\frac{a-\mu}{\sigma}\phi\left(\frac{a-\mu}{\sigma}\right) - \frac{b-\mu}{\sigma}\phi\left(\frac{b-\mu}{\sigma}\right)}{\Phi\left(\frac{b-\mu}{\sigma}\right) - \Phi\left(\frac{a-\mu}{\sigma}\right)} - \left(\frac{\phi\left(\frac{a-\mu}{\sigma}\right) - \phi\left(\frac{b-\mu}{\sigma}\right)}{\Phi\left(\frac{b-\mu}{\sigma}\right) - \Phi\left(\frac{a-\mu}{\sigma}\right)}\right)^2\right]$$

where $\phi(\cdot)$ is the standard normal density function and $\Phi(\cdot)$ is the standard normal cumulative distribution function. For the setting under consideration, we set the variance of the truncated distribution

exponential distribution is

$$H = -\int_0^\infty \exp[-x]\log(\exp[-x])\,dx$$

$$= \int_0^\infty x\exp[-x]\,dx = 1$$

Entropy for the truncated Gaussian distribution is

$$H = -\int_0^\infty \frac{2}{\sqrt{2\pi}\sigma}\exp\left[-\frac{1}{2}\frac{x^2}{\sigma^2}\right]\log\left(\frac{2}{\sqrt{2\pi}\sigma}\exp\left[-\frac{1}{2}\frac{x^2}{\sigma^2}\right]\right)dx$$

$$= -\int_0^\infty \frac{2}{\sqrt{2\pi}\sigma}\exp\left[-\frac{1}{2}\frac{x^2}{\sigma^2}\right]\left[\log\left(\frac{2}{\sqrt{2\pi}\sigma}\right) - \frac{1}{2}\frac{x^2}{\sigma^2}\right]dx$$

$$= 1.232$$

As claimed, a truncated Gaussian distribution with the same variance has greater entropy.

13.7 Variance bound and maximum entropy

A deep connection between maximum entropy distributions and the lower bound of the sampling variance (often called the Cramer-Rao lower bound) can now be demonstrated. Consider a sample of n observations

$$x \equiv \{x_1, x_2, \ldots, x_n\}$$

with sampling distribution dependent on θ, $p(x \mid \theta)$. Let

$$u(x, \theta) \equiv \frac{\partial \log p(x \mid \theta)}{\partial \theta}$$

and

$$(f, g) = \int f(x)\,g(x)\,dx$$

equal to one (equal to the variance of the exponential)

$$1 = \sigma^2 \left[1 - \left(\frac{\phi(0)}{1 - \Phi(0)}\right)^2\right]$$

and solve for σ^2. The mean of the truncated normal distribution is

$$E[x \mid 0 < x < \infty] = 0 + \sigma\frac{\phi(0)}{1 - \Phi(0)} = 1.324$$

13. Informed priors

By the Schwartz inequality we have

$$(f,g)^2 \leq (f,f)(g,g)$$

or, writing it out,

$$\left[\int f(x)g(x)\,dx\right]^2 = \sqrt{\int f(x)f(x)\,dx}\sqrt{\int g(x)g(x)\,dx}$$

where equality holds if and only if $f(x) = qg(x)$, $q = \frac{(f,g)}{(g,g)}$ not a function of x but possibly a function of θ.[12]

Now, choose

$$f(x) = u(x,\theta)\sqrt{p(x\mid\theta)}$$

and

$$g(x) = (\beta(x) - E[\beta])\sqrt{p(x\mid\theta)}$$

then

$$(f,g) = \int u(x,\theta)(\beta(x) - E[\beta])p(x\mid\theta)\,dx$$
$$= E[\beta u] - E[\beta]E[u]$$

---

[12]Clearly, $\int [f(x) - qg(x)]^2\,dx \geq 0$. Now, find q to minimize the integral. The first order condition is

$$0 = \int [f(x) - qg(x)]g(x)\,dx$$
$$0 = \int f(x)f(x)\,dx - q\int g(x)g(x)\,dx$$

solving for q gives

$$q = \frac{(f,g)}{(g,g)}$$

and the inequality becomes an equality

$$\left[\int \frac{(f,g)}{(g,g)}g(x)g(x)\,dx\right]^2 \leq \int \left(\frac{(f,g)}{(g,g)}\right)^2 g(x)g(x)\,dx \int g(x)g(x)\,dx$$
$$\left(\frac{(f,g)}{(g,g)}\right)^2 \left[\int g(x)g(x)\,dx\right]^2 = \left(\frac{(f,g)}{(g,g)}\right)^2 \int g(x)g(x)\,dx \int g(x)g(x)\,dx$$

13.7 Variance bound and maximum entropy

since

$$E[u] = \int u(x,\theta) p(x \mid \theta) \, dx$$
$$= \int \frac{\partial \log p(x \mid \theta)}{\partial \theta} p(x \mid \theta) \, dx$$
$$= \frac{\partial}{\partial \theta} \int p(x \mid \theta) \, dx$$
$$= \frac{\partial}{\partial \theta} [1]$$
$$E[u] = 0$$

we have

$$(f, g) = E[\beta u]$$

We also have

$$(f, f) = \int [u(x,\theta)]^2 p(x \mid \theta) \, dx$$
$$= E[u^2]$$
$$= Var[u]$$

the latter from $E[u] = 0$, and

$$(g, g) = \int (\beta(x) - E[\beta])^2 p(x \mid \theta) \, dx$$
$$= Var[\beta]$$

So the Schwartz inequality simplifies to

$$E[\beta u]^2 \leq Var[\beta] Var[u]$$

or

$$E[\beta u] \leq \sqrt{Var[\beta] Var[u]}$$

But

$$E[\beta u] = \int \beta(x) \frac{\partial \log p(x \mid \theta)}{\partial \theta} p(x \mid \theta) \, dx$$
$$= \int \beta(x) \frac{\partial p(x \mid \theta)}{\partial \theta} \, dx$$
$$= \frac{dE[\beta]}{d\theta}$$
$$= 1 + b'(\theta)$$

13. Informed priors

where $b(\theta) = (E[\beta] - \theta)$, bias in the parameter estimate, and $b'(\theta) = \frac{\partial b(\theta)}{\partial \theta} = \frac{\partial E[\beta]}{\partial \theta} - 1$. This means the inequality can be rewritten as

$$Var[\beta] \geq \frac{E[\beta u]^2}{Var[u]}$$

$$\geq \frac{[1 + b'(\theta)]^2}{\int \left[\frac{\partial \log p(x|\theta)}{\partial \theta}\right]^2 p(x \mid \theta) \, dx}$$

A change of parameters $(\theta \to \tau)$ where $q(\theta) = -\frac{\partial \tau}{\partial \theta}$ and substitution into $f = qg$ yields

$$\frac{\partial \log p(x \mid \theta)}{\partial \theta} \sqrt{p(x \mid \theta)} = -\frac{\partial \tau}{\partial \theta} (\beta(x) - E[\beta]) \sqrt{p(x \mid \theta)}$$

$$\frac{\partial \log p(x \mid \theta)}{\partial \theta} = -\frac{\partial \tau}{\partial \theta} (\beta(x) - E[\beta])$$

Now, integrate over θ

$$\int \frac{\partial \log p(x \mid \theta)}{\partial \theta} d\theta = \int -\tau'(\theta) (\beta(x) - E[\beta]) \, d\theta$$

$$\log p(x \mid \theta) = -\tau(\theta) \beta(x) + \int \frac{\partial \tau}{\partial \theta} E[\beta] \, d\theta$$

$$= -\tau(\theta) \beta(x) + \int E[\beta] \, d\tau + \text{constant}$$

Notice $\int E[\beta] \, d\tau$ is a function of θ, call it $-\log Z(\tau)$. Also, the constant is independent of θ but may depend on x, call it $\log m(x)$. Substitution gives

$$\log p(x \mid \theta) = -\tau(\theta) \beta(x) - \log Z(\tau) + \log m(x)$$

$$p(x \mid \theta) = \frac{m(x)}{Z(\tau)} e^{-\tau(\theta)\beta(x)}$$

This is the maximum entropy distribution with a constraint[13] fixing $E[\beta(x)]$ and $Z(\tau)$ is a normalizing constant such that

$$Z(\tau) = \int m(x) e^{-\tau(\theta)\beta(x)} dx$$

The significance of this connection merits deeper consideration. If the sampling distribution is a maximum entropy distribution then maximal efficiency is achievable in the squared error loss sense, that is, the Cramer-Rao lower bound for the sampling variance is achievable.[14] Bayesian inference consistently processes all information by combining the maximum entropy prior distribution and maximum entropy likelihood function or sampling distribution. This affirms the power of probability as logic (Jaynes [2003]).

13.8 An illustration: Jaynes' widget problem

Jaynes' widget problem is a clever illustration of informed priors (Jaynes [1963], [2003], ch. 14). A manager of a process that produces red (R), yellow (Y), and green (G) widgets must choose between producing R, Y, or G widgets as only 200 of one type of widgets per day can be produced. If this is all that is known (nearly complete ignorance), the manager is indifferent between R, Y, or G. Suppose the manager acquires some background knowledge. For illustrative purposes, we explore stages of background knowledge.

Stage 1: The manager learns the current stock of widgets: 100 red, 150 yellow, and 50 green. With only this background knowledge including no knowledge of the consequences, the manager intuitively chooses to produce green widgets.

Stage 2: The manager learns the average daily orders have been 50 red, 100 yellow, and 10 green widgets. With this background knowledge, the manager may intuitively decide to produce yellow widgets.

---

[13] The constraint is $E[\beta(x)] = -\frac{\partial \log Z(\tau)}{\partial \tau}$ as

$$E[\beta(x)] = \int \beta(x) \frac{m(x)}{Z(\tau)} e^{-\tau(\theta)\beta(x)} dx$$

and

$$-\frac{\partial \log Z(\tau)}{\partial \tau} = -\frac{\partial \int m(x) e^{-\tau(\theta)\beta(x)} dx}{\partial \tau}$$

$$= -\frac{1}{Z(\tau)} \int m(x) e^{-\tau(\theta)\beta(x)} (-\beta(x)) dx$$

$$= \int \beta(x) \frac{m(x)}{Z(\tau)} e^{-\tau(\theta)\beta(x)} dx$$

[14] See Jaynes [2003], p. 520 for exceptions. Briefly, if the sampling distribution does not have the form of a maximum entropy distribution either the lower bound is not achievable or the sampling distribution has discontinuities.

Table 13.1: Jaynes' widget problem: summary of background knowledge by stage

Stage	R	Y	G	Decision
1. in stock	100	150	50	G
2. aver. daily orders	50	100	10	Y
3. aver. individual order size	75	10	20	R
4. specific order	0	0	40	?

Stage 3: The manager learns the average order size has been 75 red, 10 yellow, and 20 green widgets. With this background knowledge, the manager may intuitively switch to producing red widgets.

Stage 4: The manager learns an emergency order for 40 green widgets is imminent. Now, what does the manager decide to produce? It seems common sense is not enough to guide the decision. We'll pursue a formal analysis but first we summarize the problem in table 13.1.

Of course, this is a decision theoretic problem where formally the manager (a) enumerates the states of nature, (b) assigns prior probabilities associated with states conditional on background knowledge, (c) updates beliefs via Bayesian revision (as this framing of the problem involves no new evidence, this step is suppressed), (d) enumerates the possible decisions (produce R, Y, or G), and (e) selects the expected loss minimizing alternative based on a loss function which incorporates background knowledge of consequences.

13.8.1 Stage 1 solution

The states of nature are the number of red, yellow, and green widgets ordered today. Let $n_1 = 0, 1, 2, \ldots$ be the number of red widgets ordered. Similarly, let n_2 and n_3 be the number of yellow and green widgets ordered. If this triple (n_1, n_2, n_3) is known the problem is likely trivial. The maximum entropy prior given only stage 1 background knowledge is

$$\max_{p(n_1,n_2,n_3)} -\sum_{n_1=0}^{\infty}\sum_{n_2=0}^{\infty}\sum_{n_3=0}^{\infty} p(n_1, n_2, n_3) \log p(n_1, n_2, n_3)$$

$$\text{s.t.} \quad \sum_{n_1=0}^{\infty}\sum_{n_2=0}^{\infty}\sum_{n_3=0}^{\infty} p(n_1, n_2, n_3) = 1$$

or solve the Lagrangian

$$\mathcal{L} = -\sum_{n_1=0}^{\infty}\sum_{n_2=0}^{\infty}\sum_{n_3=0}^{\infty} p(n_1, n_2, n_3) \log p(n_1, n_2, n_3)$$

$$- (\lambda_0 - 1) \left[\sum_{n_1=0}^{\infty}\sum_{n_2=0}^{\infty}\sum_{n_3=0}^{\infty} p(n_1, n_2, n_3) - 1 \right]$$

The solution is the improper (uniform) prior

$$p(n_1, n_2, n_3) = \exp[-\lambda_0] \quad \text{for all } (n_1, n_2, n_3)$$

13.8 An illustration: Jaynes' widget problem

where $\lambda_0 = \lim\limits_{n \to \infty} \log n$.

As we have no background knowledge of consequences, the loss function is simply

$$R(x) = \begin{array}{ll} x & x > 0 \\ 0 & x \leq 0 \end{array}$$

and the loss associated with producing red widgets (decision D_1) is

$$L(D_1; n_1, n_2, n_3) = R(n_1 - S_1 - 200) + R(n_2 - S_2) + R(n_3 - S_3)$$

where S_i is the current stock of widget $i = R, Y$, or G. Similarly, the loss associated with producing yellow widgets (decision D_2) is

$$L(D_2; n_1, n_2, n_3) = R(n_1 - S_1) + R(n_2 - S_2 - 200) + R(n_3 - S_3)$$

or green widgets (decision D_3) is

$$L(D_3; n_1, n_2, n_3) = R(n_1 - S_1 - 200) + R(n_2 - S_2) + R(n_3 - S_3 - 200)$$

Then, the expected loss for decision D_1 is

$$\begin{aligned} E[L(D_1)] &= \sum_{n_i} p(n_1, n_2, n_3) L(D_1; n_1, n_2, n_3) \\ &= \sum_{n_1=0}^{\infty} p(n_1) R(n_1 - S_1 - 200) \\ &\quad + \sum_{n_2=0}^{\infty} p(n_2) R(n_2 - S_2) \\ &\quad + \sum_{n_3=0}^{\infty} p(n_3) R(n_3 - S_3) \end{aligned}$$

Expected loss associated with decision D_2 is

$$\begin{aligned} E[L(D_2)] &= \sum_{n_1=0}^{\infty} p(n_1) R(n_1 - S_1) \\ &\quad + \sum_{n_2=0}^{\infty} p(n_2) R(n_2 - S_2 - 200) \\ &\quad + \sum_{n_3=0}^{\infty} p(n_3) R(n_3 - S_3) \end{aligned}$$

and for decision D_3 is

$$E[L(D_3)] = \sum_{n_1=0}^{\infty} p(n_1) R(n_1 - S_1)$$
$$+ \sum_{n_2=0}^{\infty} p(n_2) R(n_2 - S_2)$$
$$+ \sum_{n_3=0}^{\infty} p(n_3) R(n_3 - S_3 - 200)$$

Recognize $p(n_i) = p$ for all n_i, let b any arbitrarily large upper limit such that $p = \frac{1}{b}$, and substitute in the current stock values

$$\begin{aligned}E[L(D_1)] &= \sum_{n_1=0}^{b} pR(n_1 - 300) + \sum_{n_2=0}^{b} pR(n_2 - 150) \\ &\quad + \sum_{n_3=0}^{b} pR(n_3 - 50) \\ &= \frac{(b-300)(b-299)}{2b} + \frac{(b-150)(b-149)}{2b} \\ &\quad + \frac{(b-50)(b-49)}{2b} \\ &= \frac{114500 - 997b + 3b^2}{2b}\end{aligned}$$

$$\begin{aligned}E[L(D_2)] &= \sum_{n_1=0}^{b} pR(n_1 - 100) + \sum_{n_2=0}^{b} pR(n_2 - 350) \\ &\quad + \sum_{n_3=0}^{b} pR(n_3 - 50) \\ &= \frac{(b-100)(b-99)}{2b} + \frac{(b-350)(b-349)}{2b} \\ &\quad + \frac{(b-50)(b-49)}{2b} \\ &= \frac{134500 - 997b + 3b^2}{2b}\end{aligned}$$

$$E[L(D_3)] = \sum_{n_1=0}^{b} pR(n_1 - 100) + \sum_{n_2=0}^{b} pR(n_2 - 150)$$

$$+ \sum_{n_3=0}^{b} pR(n_3 - 250)$$

$$= \frac{(b-100)(b-99)}{2b} + \frac{(b-150)(b-149)}{2b}$$

$$+ \frac{(b-250)(b-249)}{2b}$$

$$= \frac{94500 - 997b + 3b^2}{2b}$$

Since the terms involving b are identical for all decisions, expected loss minimization involves comparison of the constants. Consistent with intuition, the expected loss minimizing decision is D_3.

13.8.2 Stage 2 solution

For stage 2 we know the average demand for widgets. Conditioning on these three averages adds three Lagrange multipliers to our probability assignment. Following the discussion above on maximum entropy probability assignment we have

$$p(n_1, n_2, n_3) = \frac{\exp[-\lambda_1 n_1 - \lambda_2 n_2 - \lambda_3 n_3]}{Z(\lambda_1, \lambda_2, \lambda_3)}$$

where the partition function is

$$Z(\lambda_1, \lambda_2, \lambda_3) = \sum_{n_1=0}^{\infty} \sum_{n_2=0}^{\infty} \sum_{n_3=0}^{\infty} \exp[-\lambda_1 n_1 - \lambda_2 n_2 - \lambda_3 n_3]$$

factoring and recognizing this as a product of three geometric series yields

$$Z(\lambda_1, \lambda_2, \lambda_3) = \prod_{i=1}^{3} (1 - \exp[-\lambda_i])^{-1}$$

Since the joint probability factors into

$$p(n_1, n_2, n_3) = p(n_1) p(n_2) p(n_3)$$

we have

$$p(n_i) = (1 - \exp[-\lambda_i]) \exp[-\lambda_i n_i] \quad \begin{array}{l} i = 1, 2, 3 \\ n_i = 0, 1, 2, \ldots \end{array}$$

$E[n_i]$ is our background knowledge and from the above analysis we know

$$E[n_i] = -\frac{\partial \log Z(\lambda_1, \lambda_2, \lambda_3)}{\partial \lambda_i}$$

$$= \frac{\exp[-\lambda_i]}{1 - \exp[-\lambda_i]}$$

13. Informed priors

Manipulation produces

$$\exp[-\lambda_i] = \frac{E[n_i]}{E[n_i] + 1}$$

substitution finds

$$\begin{aligned} p(n_i) &= (1 - \exp[-\lambda_i]) \exp[-\lambda_i n_i] \\ &= \frac{1}{E[n_i]+1} \left(\frac{E[n_i]}{E[n_i]+1}\right)^{n_i} \quad n_i = 0, 1, 2, \ldots \end{aligned}$$

Hence, we have three exponential distributions for the maximum entropy probability assignment

$$p_1(n_1) = \frac{1}{51}\left(\frac{50}{51}\right)^{n_1}$$

$$p_2(n_2) = \frac{1}{101}\left(\frac{100}{101}\right)^{n_2}$$

$$p_3(n_3) = \frac{1}{11}\left(\frac{10}{11}\right)^{n_3}$$

Now, combine these priors with the uninformed loss function, say for the first component of decision D_1

$$\sum_{n_1=0}^{\infty} p(n_1) R(n_1 - 300) = \sum_{n_1=300}^{\infty} p(n_1)(n_1 - 300)$$

$$= \sum_{n_1=300}^{\infty} p(n_1) n_1 - \sum_{n_1=300}^{\infty} p(n_1) 300$$

By manipulation of the geometric series

$$\begin{aligned}\sum_{n_1=300}^{\infty} p(n_1) n_1 &= (1 - \exp[-\lambda_1]) \\ &\quad \times \frac{\exp[-300\lambda_1](300\exp[\lambda_1] - 299)\exp[-\lambda_1]}{(1 - \exp[-\lambda_1])^2} \\ &= \frac{\exp[-300\lambda_1](300\exp[\lambda_1] - 299)}{\exp[\lambda_1] - 1}\end{aligned}$$

and

$$\begin{aligned}\sum_{n_1=300}^{\infty} p(n_1) 300 &= 300(1 - \exp[-\lambda_1])\frac{\exp[-300\lambda_1]}{1 - \exp[-\lambda_1]} \\ &= 300\exp[-300\lambda_1]\end{aligned}$$

Combining and simplifying produces

$$\sum_{n_1=300}^{\infty} p(n_1)(n_1 - 300) = \frac{\exp[-300\lambda_1](300\exp[\lambda_1] - 299)}{\exp[\lambda_1] - 1}$$

$$- \frac{\exp[-300\lambda_1](300\exp[\lambda_1] - 300)}{\exp[\lambda_1] - 1}$$

$$= \frac{\exp[-300\lambda_1]}{\exp[\lambda_1] - 1}$$

substituting $\exp[-\lambda_1] = \frac{E[n_1]}{E[n_1]+1} = \frac{50}{51}$ yields

$$\sum_{n_1=300}^{\infty} p(n_1)(n_1 - 300) = \frac{\left(\frac{50}{51}\right)^{300}}{\frac{51}{50} - 1} = 0.131$$

Similar analysis of other components and decisions produces the following summary results for the stage 2 decision problem.

$$E[L(D_1)] = \sum_{n_1=0}^{\infty} p(n_1) R(n_1 - 300) + \sum_{n_2=0}^{\infty} p(n_2) R(n_2 - 150)$$

$$+ \sum_{n_3=0}^{\infty} p(n_3) R(n_3 - 50)$$

$$= 0.131 + 22.480 + 0.085 = 22.70$$

$$E[L(D_2)] = \sum_{n_1=0}^{\infty} p(n_1) R(n_1 - 100) + \sum_{n_2=0}^{\infty} p(n_2) R(n_2 - 350)$$

$$+ \sum_{n_3=0}^{\infty} p(n_3) R(n_3 - 50)$$

$$= 6.902 + 3.073 + 10.060 = 10.06$$

$$E[L(D_3)] = \sum_{n_1=0}^{\infty} p(n_1) R(n_1 - 100) + \sum_{n_2=0}^{\infty} p(n_2) R(n_2 - 150)$$

$$+ \sum_{n_3=0}^{\infty} p(n_3) R(n_3 - 250)$$

$$= 6.902 + 22.480 + 4 \times 10^{-10} = 29.38$$

Consistent with our intuition, the stage 2 expected loss minimizing decision is produce yellow widgets.

13.8.3 Stage 3 solution

With average order size knowledge, we are able to frame the problem by enumerating more detailed states of nature. That is, we can account for not only total orders but also individual orders. A state of nature can be described as we receive u_1 orders for one red widget, u_2 orders for two red widgets, etc., we also receive v_y orders for y yellow widgets and w_g orders for g green widgets. Hence, a state of nature is specified by

$$\theta = \{u_1, \ldots, v_1, \ldots, w_1, \ldots\}$$

to which we assign probability

$$p(u_1, \ldots, v_1, \ldots, w_1, \ldots)$$

Today's total demands for red, yellow and green widgets are

$$n_1 = \sum_{r=1}^{\infty} r u_r, \quad n_2 = \sum_{y=1}^{\infty} y u_y, \quad n_3 = \sum_{g=1}^{\infty} g u_g$$

whose expectations from stage 2 are $E[n_1] = 50$, $E[n_2] = 100$, and $E[n_3] = 10$. The total number of individual orders for red, yellow, and green widgets are

$$m_1 = \sum_{r=1}^{\infty} u_r, \quad m_2 = \sum_{y=1}^{\infty} u_y, \quad m_3 = \sum_{g=1}^{\infty} u_g$$

Since we know the average order size for red widgets is 75, for yellow widgets is 10, and for green widgets is 20, we also know the average daily total number of orders for red widgets is $E[m_1] = \frac{E[n_1]}{75} = \frac{50}{75}$, for yellow widgets is $E[m_2] = \frac{E[n_2]}{10} = \frac{100}{10}$, and for green widgets is $E[m_3] = \frac{E[n_3]}{20} = \frac{10}{20}$.

Six averages implies we have six Lagrange multipliers and the maximum entropy probability assignment is

$$p(\theta) = \frac{\exp[-\lambda_1 n_1 - \mu_1 m_1 - \lambda_2 n_2 - \mu_2 m_2 - \lambda_3 n_3 - \mu_3 m_3]}{Z(\lambda_1, \mu_1, \lambda_2, \mu_2, \lambda_3, \mu_3)}$$

Since both the numerator and denominator factor, we proceed as follows

$$\begin{aligned} p(\theta) &= p(u_1, \ldots, v_1, \ldots, w_1, \ldots) \\ &= p_1(u_1, \ldots) p_2(v_1, \ldots) p_3(w_1, \ldots) \end{aligned}$$

where, for instance,

$$\begin{aligned} Z_1(\lambda_1, \mu_1) &= \sum_{u_1=0}^{\infty} \sum_{u_2=0}^{\infty} \cdots \exp[-\lambda_1(u_1 + 2u_2 + 3u_3 + \cdots)] \\ &\quad \times \exp[-\mu_1(u_1 + u_2 + u_3 + \cdots)] \\ &= \prod_{r=1}^{\infty} \frac{1}{1 - \exp[-r\lambda_1 - \mu_1]} \end{aligned}$$

Since
$$E[n_i] = -\frac{\partial \log Z_i(\lambda_i, \mu_i)}{\partial \lambda_i}$$

and
$$E[m_i] = -\frac{\partial \log Z_i(\lambda_i, \mu_i)}{\partial \mu_i}$$

we can solve for, say, λ_1 and μ_1 via

$$\begin{aligned} E[n_i] &= \frac{\partial}{\partial \lambda_1} \sum_{r=1}^{\infty} \log\left(1 - \exp[-r\lambda_1 - \mu_1]\right) \\ &= \sum_{r=1}^{\infty} \frac{r}{\exp[r\lambda_1 + \mu_1] - 1} \end{aligned}$$

and

$$\begin{aligned} E[m_i] &= \frac{\partial}{\partial \mu_1} \sum_{r=1}^{\infty} \log\left(1 - \exp[-r\lambda_1 - \mu_1]\right) \\ &= \sum_{r=1}^{\infty} \frac{1}{\exp[r\lambda_1 + \mu_1] - 1} \end{aligned}$$

The expressions for $E[n_i]$ and $E[m_i]$ can be utilized to numerically solve for λ_i and μ_i to complete the maximum entropy probability assignment (see Tribus and Fitts [1968]), however, as noted by Jaynes [1963, 2003], these expressions converge very slowly. We follow Jaynes by rewriting the expressions in terms of quickly converging sums and then follow Tribus and Fitts by numerically solving for λ_i and μ_i.[15]

For example, use the geometric series

$$\begin{aligned} E[m_1] &= \sum_{r=1}^{\infty} \frac{1}{\exp[r\lambda_1 + \mu_1] - 1} \\ &= \sum_{r=1}^{\infty} \sum_{j=1}^{\infty} \exp[-j(r\lambda_1 + \mu_1)] \end{aligned}$$

Now, evaluate the geometric series over r

$$\sum_{r=1}^{\infty} \sum_{j=1}^{\infty} \exp[-j(r\lambda_1 + \mu_1)] = \sum_{j=1}^{\infty} \frac{\exp[-j(\lambda_1 + \mu_1)]}{1 - \exp[-j\lambda_1]}$$

---

[15] Jaynes [1963] employs approximations rather than computer-based numerical solutions.

Table 13.2: Jaynes' widget problem: stage 3 state of knowledge

Widget	S	$E[n_i]$	$E[m_i]$	λ_i	μ_i
Red	100	50	$\frac{50}{75}$	0.0134	4.716
Yellow	150	100	$\frac{100}{10}$	0.0851	0.514
Green	50	10	$\frac{10}{20}$	0.051	3.657

This expression is rapidly converging (the first term alone is a reasonable approximation). Analogous geometric series ideas apply to $E[n_i]$

$$E[n_1] = \sum_{r=1}^{\infty} \frac{r}{\exp[r\lambda_1 + \mu_1] - 1}$$

$$= \sum_{r=1}^{\infty} \sum_{j=1}^{\infty} r \exp[-j(r\lambda_1 + \mu_1)]$$

$$= \sum_{j=1}^{\infty} \frac{\exp[-j(\lambda_1 + \mu_1)]}{(1 - \exp[-j\lambda_1])^2}$$

Again, this series is rapidly converging. Now, numerically solve for λ_i and μ_i utilizing knowledge of $E[n_i]$ and $E[m_i]$. For instance, solving

$$E[m_1] = \frac{50}{75} = \sum_{j=1}^{\infty} \frac{\exp[-j(\lambda_1 + \mu_1)]}{1 - \exp[-j\lambda_1]}$$

$$E[n_1] = 50 = \sum_{j=1}^{\infty} \frac{\exp[-j(\lambda_1 + \mu_1)]}{(1 - \exp[-j\lambda_1])^2}$$

yields $\lambda_1 = 0.0134$ and $\mu_1 = 4.716$. Other values are determined in analogous fashion and all results are described in table 13.2.[16]

Gaussian approximation

The expected loss depends on the distribution of daily demand, n_i. We compare a Gaussian approximation based on the central limit theorem with the exact distribution for n_i. First, we consider the Gaussian approximation. We can write the

---

[16] Results are qualitatively similar to those reported by Tribus and Fitts [1968].

expected value for the number of orders of, say, size r as

$$E[u_r] = \sum_{u_r=0}^{\infty} p_1(u_r) u_r$$

$$= \sum_{u_r=0}^{\infty} \frac{\exp[-(r\lambda_1 + \mu_1)u_r]}{Z(\lambda_1, \mu_1)} u_r$$

$$= \sum_{u_r=0}^{\infty} \frac{\exp[-(r\lambda_1 + \mu_1)u_r]}{\frac{1}{1-\exp[-r\lambda_1-\mu_1]}} u_r$$

$$= (1 - \exp[-r\lambda_1 - \mu_1]) \frac{\exp[-r\lambda_1 - \mu_1]}{(1 - \exp[-r\lambda_1 - \mu_1])^2}$$

$$= \frac{1}{\exp[r\lambda_1 + \mu_1] - 1}$$

and the variance of u_r as

$$Var[u_r] = E[u_r^2] - E[u_r]^2$$

$$E[u_r^2] = \sum_{u_r=0}^{\infty} \frac{\exp[-(r\lambda_1 + \mu_1)u_r]}{\frac{1}{1-\exp[-r\lambda_1-\mu_1]}} u_r^2$$

$$= \sum_{u_r=0}^{\infty} (1 - \exp[-r\lambda_1 - \mu_1])$$

$$\times \frac{\exp[-(r\lambda_1 + \mu_1)] + \exp[-2(r\lambda_1 + \mu_1)]}{(1 - \exp[-r\lambda_1 - \mu_1])^3}$$

$$= \frac{\exp[r\lambda_1 + \mu_1] + 1}{(\exp[r\lambda_1 + \mu_1] - 1)^2}$$

Therefore,

$$Var[u_r] = \frac{\exp[r\lambda_1 + \mu_1]}{(\exp[r\lambda_1 + \mu_1] - 1)^2}$$

Since n_1 is the sum of independent random variables

$$n_1 = \sum_{r=1}^{\infty} r u_r$$

the probability distribution for n_1 has mean $E[n_1] = 50$ and variance

$$Var[n_1] = \sum_{r=1}^{\infty} r^2 Var[u_r]$$

$$= \sum_{r=1}^{\infty} \frac{r^2 \exp[r\lambda_1 + \mu_1]}{(\exp[r\lambda_1 + \mu_1] - 1)^2}$$

Table 13.3: Jaynes' widget problem: stage 3 state of knowledge along with standard deviation

Widget	S	$E[n_i]$	$E[m_i]$	λ_i	μ_i	σ_i
Red	100	50	$\frac{50}{75}$	0.0134	4.716	86.41
Yellow	150	100	$\frac{100}{10}$	0.0851	0.514	48.51
Green	50	10	$\frac{10}{20}$	0.051	3.657	19.811

We convert this into the rapidly converging sum[17]

$$\sum_{r=1}^{\infty} \frac{r^2 \exp[r\lambda_1 + \mu_1]}{(\exp[r\lambda_1 + \mu_1] - 1)^2} = \sum_{r=1}^{\infty} \sum_{j=1}^{\infty} jr^2 \exp[-j(r\lambda_1 + \mu_1)]$$

$$= \sum_{j=1}^{\infty} j \frac{\exp[-j(\lambda_1 + \mu_1)] + \exp[-j(2\lambda_1 + \mu_1)]}{(1 - \exp[-j\lambda])^3}$$

Next, we repeat stage 3 knowledge updated with the numerically-determined standard deviation of daily demand, σ_i, for the three widgets in table 13.3.[18,19]

The central limit theorem applies as there are many ways for large values of n_i to arise.[20] Then the expected loss of failing to meet today's demand given current stock, S_i, and today's production, $P_i = 0$ or 200, is

$$\sum_{n_i=1}^{\infty} p(n_i) R(n_i - S_i - P_i)$$

$$\approx \frac{1}{\sqrt{2\pi}\sigma_i} \int_{S_i + P_i}^{\infty} (n_i - S_i - P_i) \exp\left[-\frac{1}{2}\frac{(n_i - E[n_i])^2}{\sigma_i^2}\right] dn_i$$

Numerical evaluation yields the following expected unfilled orders conditional on decision D_i.

$$E[L(D_1)] = 0.05 + 3.81 + 0.16 = 4.02$$
$$E[L(D_2)] = 15.09 + 0.0 + 0.16 = 15.25$$
$$E[L(D_3)] = 15.09 + 3.81 + 0.0 = 18.9$$

Clearly, producing red widgets is preferred given state 3 knowledge based on our central limit theorem (Gaussian) approximation. Next, we follow Tribus and Fitts [1968] and revisit the expected loss employing exact distributions for n_i.

---

[17] For both variance expressions, $Var[u_r]$ and $Var[n_1]$, we exploit the idea that the converging sum $\sum_{j=1}^{\infty} j^2 \exp[-jx] = \frac{\exp[-x] + \exp[-2x]}{(1-\exp[-x])^3}$.

[18] Jaynes [1963] employs the quite good approximation $Var[n_i] \approx \frac{2}{\lambda_i} E[n_i]$.

[19] Results are qualitatively similar to those reported by Tribus and Fitts [1968].

[20] On the other hand, when demand is small, say, $n_i = 2$, there are only two ways for this to occur, $u_1 = 2$ or $u_2 = 1$.

13.8 An illustration: Jaynes' widget problem

Exact distributions

We derive the distribution for daily demand given stage 3 knowledge, $p(n_r \mid \Im_3)$, from the known distribution of daily orders $p(u_1, \ldots \mid \Im_3)$ by appealing to Bayes' rule

$$\begin{aligned}
p(n_r \mid \Im_3) &= \sum_{u_1=0}^{\infty} \sum_{u_2=0}^{\infty} \cdots p(n_r u_1 u_2 \ldots \mid \Im_3) \\
&= \sum_{u_1=0}^{\infty} \sum_{u_2=0}^{\infty} \cdots p(n_r \mid u_1 u_2 \ldots \Im_3) p(u_1 u_2 \ldots \mid \Im_3)
\end{aligned}$$

We can write

$$p(n_r \mid u_1 u_2 \ldots \Im_3) = \delta\left(n_r - \sum_{j=1}^{\infty} j u_j\right)$$

where $\delta(x) = 1$ if $x = 0$ and $\delta(x) = 0$ otherwise. Using independence of u_i, we have

$$p(n_r \mid \Im_3) = \sum_{u_1=0}^{\infty} \sum_{u_2=0}^{\infty} \cdots \delta\left(n_r - \sum_{j=1}^{\infty} j u_j\right) \prod_{i=1}^{\infty} p(u_i \mid \Im_3)$$

Definition 13.1 *Define the z transform as follows. For $f(n)$ a function of the discrete variable n, the z transform $F(z)$ is*

$$F(z) \equiv \sum_{n=0}^{\infty} f(n) z^n \quad 0 \leq z \leq 1$$

Let $P(z)$ be the z transform of $p(n_r \mid \Im_3)$

$$\begin{aligned}
P(z) &= \sum_{n_r=0}^{\infty} \sum_{u_1=0}^{\infty} \sum_{u_2=0}^{\infty} \cdots z^{n_r} \delta\left(n_r - \sum_{j=1}^{\infty} j u_j\right) \prod_{i=1}^{\infty} p(u_i \mid \Im_3) \\
&= \sum_{u_1=0}^{\infty} \sum_{u_2=0}^{\infty} \cdots z^{\sum_{j=1}^{\infty} j u_j} \prod_{i=1}^{\infty} p(u_i \mid \Im_3) \\
&= \sum_{u_1=0}^{\infty} \sum_{u_2=0}^{\infty} \cdots \prod_{i=1}^{\infty} p(u_i \mid \Im_3) z^{i u_i} \\
&= \prod_{i=1}^{\infty} \sum_{u_i=0}^{\infty} z^{i u_i} p(u_i \mid \Im_3)
\end{aligned}$$

Substituting $p(u_i \mid \Im_3) = (1 - \exp[-i\lambda_1 - \mu_1]) \exp[-u_i(i\lambda_1 + \mu_1)]$ yields

$$P(z) = \prod_{i=1}^{\infty} (1 - \exp[-i\lambda_1 - \mu_1]) \prod_{i=1}^{\infty} \sum_{u_i=0}^{\infty} (z^i \exp[-i\lambda_1 - \mu_1])^{u_i}$$

Since $P(0) = \prod_{i=1}^{\infty} (1 - \exp[-i\lambda_1 - \mu_1])$, we can write

$$P(z) = P(0) \prod_{i=1}^{\infty} \sum_{u_i=0}^{\infty} \left(z^i \exp[-i\lambda_1 - \mu_1]\right)^{u_i}$$

The first few terms in the product of sums is

$$\begin{aligned}\frac{P(z)}{P(0)} &= \prod_{i=1}^{\infty} \sum_{u_i=0}^{\infty} \left(z^i \exp[-i\lambda_1 - \mu_1]\right)^{u_i} \\ &= 1 + \left(ze^{-\lambda_1}\right)e^{-\mu_1} + \left(ze^{-\lambda_1}\right)^2 \left[e^{-\mu_1} + e^{-2\mu_1}\right] \\ &\quad + \left(ze^{-\lambda_1}\right)^3 \left[e^{-\mu_1} + e^{-2\mu_1} + e^{-3\mu_1}\right] + \cdots\end{aligned}$$

Or, write

$$\frac{P(z)}{P(0)} = \sum_{n=0}^{\infty} C_n \left(ze^{-\lambda_1}\right)^n$$

where the coefficients C_n are defined by $C_0 = 1$ and

$$C_n = \sum_{j=1}^{n} C_{j,n} e^{-j\mu_1}, \quad \sum_{i=1}^{\infty} u_i = j, \quad \sum_{i=1}^{\infty} i u_i = n$$

and

$$C_{j,n} = C_{j-1,n-1} + C_{j,n-j}$$

with starting values $C_{1,1} = C_{1,2} = C_{1,3} = C_{1,4} = C_{2,2} = C_{2,3} = C_{3,3} = C_{3,4} = C_{4,4} = 1$ and $C_{2,4} = 2$.[21]

Let $p_0 \equiv p(n = 0 \mid \Im_3)$. Then, the inverse transform of $P(z)$ yields the distribution for daily demand

$$p(n \mid \Im_3) = p_0 C_n e^{-n\lambda_1}$$

We utilize this expression for $p(n \mid \Im_3)$, the coefficients $C_n = \sum_{j=1}^{n} C_{j,n} e^{-j\mu_1}$, the recursion formula $C_{j,n} = C_{j-1,n-1} + C_{j,n-j}$, and the earlier-derived Lagrange multipliers to numerically derive the distributions for daily demand for red, yellow, and green widgets. The distributions are plotted in figure 13.1.

As pointed out by Tribus and Fitts, daily demand for yellow widgets is nearly symmetric about the mean while daily demand for red and green widgets is "hit

---

[21] $C_{j,j} = 1$ for all j and $C_{j,n} = 0$ for all $n < j$. See the appendix of Tribus and Fitts [1968] for a proof of the recursion expression.

13.8 An illustration: Jaynes' widget problem

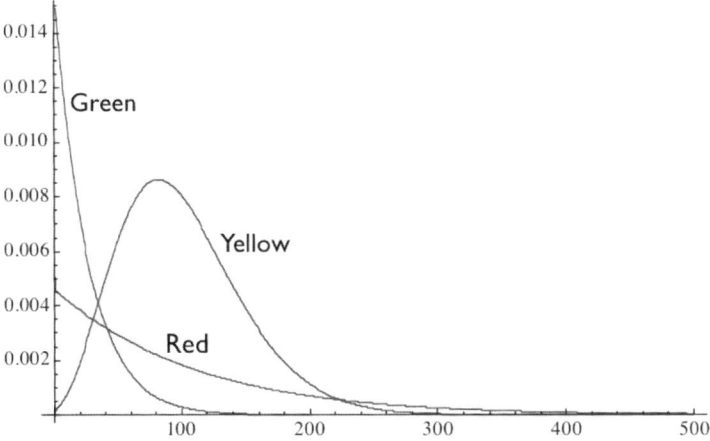

Figure 13.1: "Exact" distributions for daily widget demand

or miss." Probabilities of zero orders for the widgets are

$$p(n_1 = 0) = 0.51$$
$$p(n_2 = 0) = 0.0003$$
$$p(n_3 = 0) = 0.61$$

Next, we recalculate the minimum expected loss decision based on the "exact" distributions. The expected loss of failing to meet today's demand given current stock, S_i, and today's production, $P_i = 0$ or 200, is

$$\sum_{n_i=1}^{\infty} p(n_i \mid \Im_3) R(n_i - S_i - P_i) = \sum_{S_i+P_i}^{\infty} (n_i - S_i - P_i) p(n_i \mid \Im_3)$$

Numerical evaluation yields the following expected unfilled orders conditional on decision D_i.

$$E[L(D_1)] = 2.35 + 5.07 + 1.31 = 8.73$$

$$E[L(D_2)] = 18.5 + 0.0 + 1.31 = 19.81$$

$$E[L(D_3)] = 18.5 + 5.07 + 0.0 = 23.58$$

While the Gaussian approximation for the distribution of daily widget demand and numerical evaluation of the "exact" distributions produce somewhat different expected losses, the both demonstrably support production of red widgets today.

13.8.4 Stage 4 solution

Stage 4 involves knowledge of an imminent order of 40 green widgets. This effectively changes the stage 3 analysis so that the current stock of green widgets is 10 rather than 50. Expected losses based on the Gaussian approximation are

$$E[L(D_1)] = 0.05 + 3.81 + 7.9 = 11.76$$

$$E[L(D_2)] = 15.09 + 0.0 + 7.9 = 22.99$$

$$E[L(D_3)] = 15.09 + 3.81 + 0.0 = 18.9$$

On the other hand, expected losses based on the "exact" distributions are

$$E[L(D_1)] = 2.35 + 5.07 + 6.70 = 14.12$$

$$E[L(D_2)] = 18.5 + 0.0 + 6.70 = 25.20$$

$$E[L(D_3)] = 18.5 + 5.07 + 0.0 = 23.58$$

While stage 4 knowledge shifts production in favor of green relative to yellow widgets, both distributions for daily widget demand continue to support producing red widgets today. Next, we explore another probability assignment puzzle.

13.9 Football game puzzle

Jaynes [2003] stresses consistent reasoning as the hallmark of the maximum entropy principle. Sometimes, surprisingly simple settings can pose a challenge. Consider the following puzzle posed by Walley [1991, pp. 270-271]. A football match-up between two football rivals produces wins (W), losses (L), or draws (D) for the home team. If this is all we know then the maximum entropy prior for the home team's outcome is uniform $\Pr(W, L, D) = \left(\frac{1}{3}, \frac{1}{3}, \frac{1}{3}\right)$. Suppose we know the home team wins half the time. Then, the maximum entropy prior is $\Pr(W, L, D) = \left(\frac{1}{2}, \frac{1}{4}, \frac{1}{4}\right)$. Suppose we learn the game doesn't end in a draw. The posterior distribution is $\Pr(W, L, D) = \left(\frac{2}{3}, \frac{1}{3}, 0\right)$.[22]

Now, we ask what is the maximum entropy prior if the home team wins half the time and the game is not a draw. The maximum entropy prior is $\Pr(W, L, D) = \left(\frac{1}{2}, \frac{1}{2}, 0\right)$. What is happening? This appears to be inconsistent reasoning. Is there something amiss with the maximum entropy principle?

We suggest two different propositions are being evaluated. The former involves a game structure that permits draws but we gain new evidence that a particular game did not end in a draw. On the other hand, the latter game structure precludes draws. Consequently, the information regarding home team performance has a very different implication (three states of nature, W vs. L or D, compared with

---

[22] We return to this puzzle later when we discuss Jaynes' A_p distribution.

two states of nature, W vs. L). This is an example of what Jaynes [2003, pp. 470-3] calls "the marginalization paradox," where nuisance parameters are integrated out of the likelihood in deriving the posterior. If we take care to recognize these scenarios involve different priors and likelihoods, there is no contradiction. In Jaynes' notation where we let $\varsigma = W$, $y = $ not D, and $z = $ null, the former involves posterior $p(\varsigma \mid y, z, \Im_1)$ with prior $\Im_1$ permitting W, L, or D, while the latter involves posterior $p(\varsigma \mid z, \Im_2)$ with prior $\Im_2$ permitting only W or L. Evaluation of propositions involves joint consideration of priors and likelihoods, if either changes there is no surprise when our conclusions are altered.

The example reminds us of the care required in formulating the proposition being evaluated. The next example revisits an accounting issue where informed priors are instrumental to identification and inference.

13.10 Financial statement example

13.10.1 Under-identification and Bayes

If we have more parameters to be estimated than data, we often say the problem is under-identified. However, this is a common problem in accounting. To wit, we often ask what activities did the organization engage in based on our reading of their financial statements. We know there is a simple linear relation between the recognized accounts and transactions

$$Ay = x$$

where A is an $m \times n$ matrix of ± 1 and 0 representing simple journal entries in its columns and adjustments to individual accounts in its rows, y is the transaction amount vector, and x is the change in the account balance vector over the period of interest (Arya, et al [2000]). Since there are only $m - 1$ linearly independent rows (due to the balancing property of accounting) and m (the number of accounts) is almost surely less than n (the number of transactions we seek to estimate) we're unable to invert from x to recover y. Do we give up? If so, we might be forced to conclude financial statements fail even this simplest of tests.

Rather, we might take a page from physicists (Jaynes [2003]) and allow our prior knowledge to assist estimation of y. Of course, this is what decision theory also recommends. If our prior or background knowledge provides a sense of the first two moments for y, then the Gaussian or normal distribution is our maximum entropy prior. Maximum entropy implies that we fully utilize our background knowledge but don't use background knowledge we don't have (Jaynes [2003], ch. 11). That is, maximum entropy priors combined with Bayesian revision make efficient usage of both background knowledge and information from the data (in this case, the financial statements). As in previously discussed accounting examples, background knowledge reflects potential equilibria based on strategic interaction of various, relevant economic agents and accounting recognition choices for summarizing these interactions.

Suppose our background knowledge $\Im$ is completely summarized by

$$E[y \mid \Im] = \mu$$

and

$$Var[y \mid \Im] = \Sigma$$

then our maximum entropy prior distribution is

$$p(y \mid \Im) \sim N(\mu, \Sigma)$$

and the posterior distribution for transactions, y, conditional on the financial statements, x, is

$$p(y \mid x, \Im)$$
$$\sim N\left(\mu + \Sigma A_0^T \left(A_0 \Sigma A_0^T\right)^{-1} A_0 (y^p - \mu), \Sigma - \Sigma A_0^T \left(A_0 \Sigma A_0^T\right)^{-1} A_0 \Sigma\right)$$

where $N(\cdot)$ refers to the Gaussian or normal distribution with mean vector denoted by the first term, and variance-covariance matrix denoted by the second term, A_0 is A after dropping one row and y^p is any consistent solution to $Ay = x$ (for example, form any spanning tree from a directed graph of $Ay = x$ and solve for y^p). For the special case where $\Sigma = \sigma^2 I$ (perhaps unlikely but nonetheless illuminating), this simplifies to

$$p(y \mid x, \Im) \sim N\left(P_{R(A)} y^p + \left(I - P_{R(A)}\right) \mu, \sigma^2 \left(I - P_{R(A)}\right)\right)$$

where $P_{R(A)} = A_0^T \left(A_0 A_0^T\right)^{-1} A_0$ (projection into the rowspace of A), and then $I - P_{R(A)}$ is the projection into the nullspace of A.[23]

---

[23]In the general case, we could work with the subspaces (and projections) of $A_0 \Gamma$ where $\Sigma = \Gamma \Gamma^T$ (the Cholesky decomposition of Σ) and the transformed data $z \equiv \Gamma^{-1} y \sim N\left(\Gamma^{-1} \mu, I\right)$ (Arya, Fellingham, and Schroeder [2000]). Then, the posterior distribution of z conditional on the financial statements x is

$$p(z \mid x, \Im) \sim N\left(P_{R(A_0 \Gamma)} z^p + \left(I - P_{R(A_0 \Gamma)}\right) \mu_z, I - P_{R(A_0 \Gamma)}\right)$$

where $z^p = \Gamma^{-1} y^p$ and $\mu_z = \Gamma^{-1} \mu$. From this we can recover the above posterior distribution of y conditional on x via the inverse transformation $y = \Gamma z$.

13.10.2 Numerical example

Suppose we observe the following financial statements.

Balance sheets	Ending balance	Beginning balance
Cash	110	80
Receivables	80	70
Inventory	30	40
Property & equipment	110	100
Total assets	330	290
Payables	100	70
Owner's equity	230	220
Total equities	330	290

Income statement	for period
Sales	70
Cost of sales	30
SG&A	30
Net income	10

Let x be the change in account balance vector where credit changes are negative. The sum of x is zero; a basis for the left nullspace of A is a vector of ones.

change in account	amount
Δ cash	30
Δ receivables	10
Δ inventory	(10)
Δ property & equipment	10
Δ payables	(30)
sales	(70)
cost of sales	30
sg&a expenses	30

We envision the following transactions associated with the financial statements and are interested in recovering their magnitudes y.

transaction	amount
collection of receivables	y_1
investment in property & equipment	y_2
payment of payables	y_3
bad debts expense	y_4
sales	y_5
depreciation - period expense	y_6
cost of sales	y_7
accrued expenses	y_8
inventory purchases	y_9
depreciation - product cost	y_{10}

374   13. Informed priors

A crisp summary of these details is provided by a directed graph as depicted in figure 13.2.

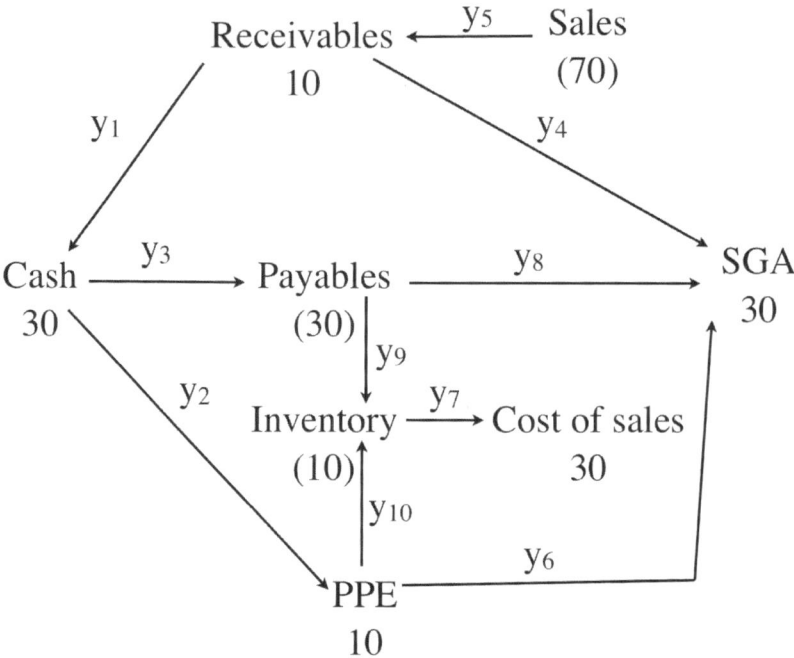

Figure 13.2: Directed graph of financial statements

The A matrix associated with the financial statements and directed graph where credits are denoted by -1 is

$$A = \begin{bmatrix} 1 & -1 & -1 & 0 & 0 & 0 & 0 & 0 & 0 & 0 \\ -1 & 0 & 0 & -1 & 1 & 0 & 0 & 0 & 0 & 0 \\ 0 & 0 & 0 & 0 & 0 & 0 & -1 & 0 & 1 & 1 \\ 0 & 1 & 0 & 0 & 0 & -1 & 0 & 0 & 0 & -1 \\ 0 & 0 & 1 & 0 & 0 & 0 & 0 & -1 & -1 & 0 \\ 0 & 0 & 0 & 0 & -1 & 0 & 0 & 0 & 0 & 0 \\ 0 & 0 & 0 & 0 & 0 & 0 & 1 & 0 & 0 & 0 \\ 0 & 0 & 0 & 1 & 0 & 1 & 0 & 1 & 0 & 0 \end{bmatrix}$$

and a basis for the nullspace is immediately identified by any set of linearly independent loops in the graph, for example,

$$N = \begin{bmatrix} 1 & 0 & 1 & -1 & 0 & 0 & 0 & 1 & 0 & 0 \\ 0 & 1 & -1 & 0 & 0 & 0 & 0 & 0 & -1 & 1 \\ 0 & 0 & 0 & 0 & 0 & 1 & 0 & -1 & 1 & -1 \end{bmatrix}$$

13.10 Financial statement example

A consistent solution y^p is readily identified by forming a spanning tree and solving for the remaining transaction amounts. For instance, let $y_3 = y_6 = y_9 = 0$, the spanning tree is depicted in figure 13.3.

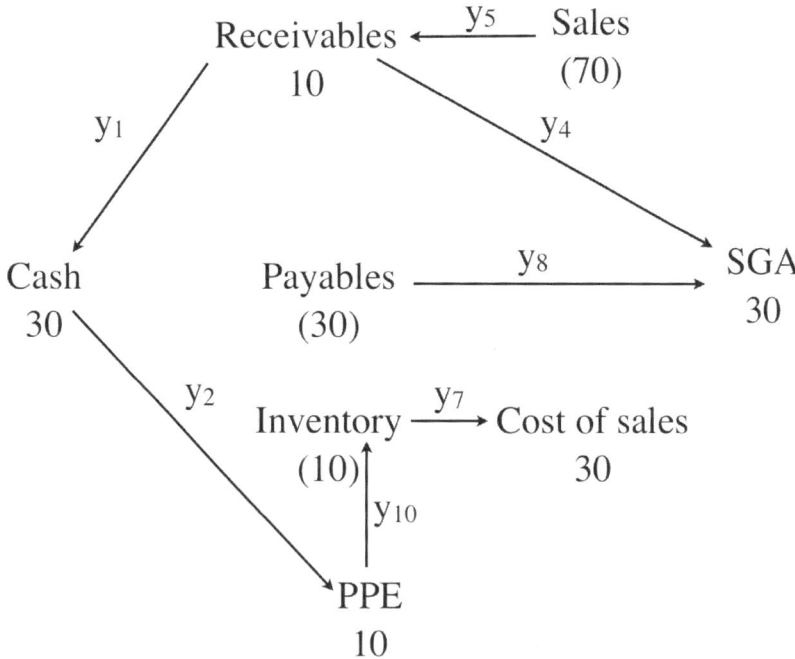

Figure 13.3: Spanning tree

Then, $(y^p)^T = \begin{bmatrix} 60 & 30 & 0 & 0 & 70 & 0 & 30 & 30 & 0 & 20 \end{bmatrix}$.

Now, suppose background knowledge $\Im$ regarding transactions is described by the first two moments

$$E\left[y^T \mid \Im\right] = \mu^T = \begin{bmatrix} 60 & 20 & 25 & 2 & 80 & 5 & 40 & 10 & 20 & 15 \end{bmatrix}$$

and

$$Var\left[y \mid \Im\right] = \Sigma = \begin{bmatrix} 10 & 0 & 0 & 0 & 5 & 0 & 0 & 0 & 0 & 0 \\ 0 & 1 & 0 & 0 & 0 & 0.2 & 0 & 0 & 0 & 0.2 \\ 0 & 0 & 1 & 0 & 0 & 0 & 0 & 0.2 & 0 & 0 \\ 0 & 0 & 0 & 0.5 & 0.1 & 0 & 0 & 0 & 0 & 0 \\ 5 & 0 & 0 & 0.1 & 10 & 0 & 3.5 & 0 & 0 & 0 \\ 0 & 0.2 & 0 & 0 & 0 & 1 & 0 & 0 & 0 & 0 \\ 0 & 0 & 0 & 0 & 3.5 & 0 & 5 & 0 & 0.2 & 0 \\ 0 & 0 & 0.2 & 0 & 0 & 0 & 0 & 1 & 0 & 0 \\ 0 & 0 & 0 & 0 & 0 & 0 & 0.2 & 0 & 1 & 0 \\ 0 & 0.2 & 0 & 0 & 0 & 0 & 0 & 0 & 0 & 1 \end{bmatrix}$$

maximum entropy priors for transactions are normally distributed with parameters described by the above moments.

Given financial statements x and background knowledge $\Im$, posterior beliefs regarding transactions are normally distributed with $E\left[y^T \mid x, \Im\right] =$

$$[\ 58.183 \quad 15.985 \quad 12.198 \quad 1.817 \quad 70 \quad 5.748 \quad 30 \quad 22.435 \quad 19.764 \quad 0.236\ ]$$

and $Var\left[y \mid x, \Im\right] =$

$$\begin{bmatrix}
0.338 & 0.172 & 0.167 & -0.338 & 0 & 0.164 & 0 & 0.174 & -0.007 & 0.007 \\
0.172 & 0.482 & -0.310 & -0.172 & 0 & 0.300 & 0 & -0.128 & -0.182 & 0.182 \\
0.167 & -0.310 & 0.477 & -0.167 & 0 & -0.135 & 0 & 0.302 & 0.175 & -0.175 \\
-0.338 & -0.172 & -0.167 & 0.338 & 0 & -0.164 & 0 & -0.174 & 0.007 & -0.007 \\
0 & 0 & 0 & 0 & 0 & 0 & 0 & 0 & 0 & 0 \\
0.164 & 0.300 & -0.135 & -0.164 & 0 & 0.445 & 0 & -0.281 & 0.145 & -0.145 \\
0 & 0 & 0 & 0 & 0 & 0 & 0 & 0 & 0 & 0 \\
0.174 & -0.128 & 0.302 & -0.174 & 0 & -0.281 & 0 & 0.455 & -0.153 & 0.153 \\
-0.007 & -0.182 & 0.175 & 0.007 & 0 & 0.145 & 0 & -0.153 & 0.328 & -0.328 \\
0.007 & 0.182 & -0.175 & -0.007 & 0 & -0.145 & 0 & 0.153 & -0.328 & 0.328
\end{bmatrix}$$

As our intuition suggests, the posterior mean of transactions is consistent with the financial statements, $A\left(E\left[y \mid x, \Im\right]\right) = x$, and there is no residual uncertainty regarding transactions that are not in loops, sales and cost of sales are $y_5 = 70$ and $y_7 = 30$, respectively. Next, we explore accounting accruals as a source of both valuation and evaluation information.

13.11 Smooth accruals

Now, we explore valuation and evaluation roles of smooth accruals in a simple, yet dynamic setting with informed priors regarding the initial mean of cash flows.[24] Accruals smooth cash flows to summarize the information content regarding expected cash flows from the past cash flow history. This is similar in spirit to Arya et al [2002]. In addition, we show in a moral hazard setting that the foregoing accrual statistic can be combined with current cash flows and non-accounting contractible information to efficiently (subject to *LEN* model restrictions[25]) supply incentives to replacement agents via sequential spot contracts. Informed priors regarding the permanent component of cash flows facilitates performance evaluation. The *LEN* (linear exponential normal) model application is similar to Arya et al [2004]. It is not surprising that accruals can serve as statistics for valuation or evaluation, rather the striking contribution here is that the same accrual statistic can serve both purposes without loss of efficiency.

---

[24] These examples were developed from conversations with Joel Demski, John Fellingham, and Haijin Lin.

[25] See Holmstrom and Milgrom [1987], for details on the strengths and limitations of the *LEN* (linear exponential normal) model.

13.11.1 DGP

The data generating process (DGP) is as follows. Period t cash flows (excluding the agent's compensation s) includes a permanent component m_t that derives from productive capital, the agent's contribution a_t, and a stochastic error e_t.

$$cf_t = m_t + a_t + e_t$$

The permanent component (mean) is subject to stochastic shocks.

$$m_t = g\, m_{t-1} + \epsilon_t$$

where m_0 is common knowledge (strongly informed priors), g is a deterministic growth factor, and stochastic shock ϵ_t. In addition, there exists contractible, non-accounting information that is informative of the agent's action a_t with noise μ_t.

$$y_t = a_t + \mu_t$$

Variance knowledge for the errors, e, ϵ, and μ, leads to a joint normal probability assignment with mean zero and variance-covariance matrix Σ. The DGP is common knowledge to management and the auditor. Hence, the auditor's role is simply to assess manager's reporting compliance with the predetermined accounting system.[26]

The agent has reservation wage RW and is evaluated subject to moral hazard. The agent's action is binary $a \in \{a_H, a_L\}$, $a_H > a_L$, with personal cost $c(a), c(a_H) > c(a_L)$, and the agent's preferences for payments s and actions are $CARA$ $U(s,a) = -exp\{-r[s - c(a)]\}$. Payments are linear in performance measures w_t (with weights γ_t) plus flat wage δ_t, $s_t = \delta_t + \gamma_t^T w_t$.

The valuation role of accruals is to summarize next period's unknown expected cash flow m_{t+1} based on the history of cash flows through time t (restricted recognition). The incentive-induced equilibrium agent action a_t^* is effectively known for valuation purposes. Hence, the observable cash flow history at time t is $\{cf_1 - a_1^*, cf_2 - a_2^*, \ldots, cf_t - a_t^*\}$.

13.11.2 Valuation results

For the case $\Sigma = D$ where D is a diagonal matrix comprised of $\sigma_e^2, \sigma_\epsilon^2$, and σ_μ^2 (appropriately aligned), the following OLS regression identifies the most efficient valuation usage of the past cash flow history.

$$\widehat{m}_t = (H^T H)^{-1} H^T z,$$

---

[26] Importantly, this eliminates strategic reporting considerations typically associated with equilibrium earnings management.

$$H = \begin{bmatrix} -\nu & 0 & 0 & 0 & 0 \\ 1 & 0 & 0 & 0 & 0 \\ \nu g & -\nu & 0 & 0 & 0 \\ 0 & 1 & 0 & 0 & 0 \\ \vdots & \vdots & \ddots & \vdots & \vdots \\ 0 & 0 & \cdots & \nu g & -\nu \\ 0 & 0 & \cdots & 0 & 1 \end{bmatrix}, z = \begin{bmatrix} -\nu g\, m_0 \\ cf_1 - a_1^* \\ 0 \\ cf_2 - a_2^* \\ \vdots \\ 0 \\ cf_t - a_t^* \end{bmatrix}, \text{ and } \nu = \frac{\sigma_e}{\sigma_\varepsilon}.[27]$$

Can accruals supply a sufficient summary of the cash flow history for the cash flow mean?[28]

We utilize difference equations to establish accruals as a valuation statistic. Let $m_t = g\, m_{t-1} + \epsilon_t$, $\nu = \frac{\sigma_e}{\sigma_\varepsilon}$, and $\phi = \frac{\sigma_e}{\sigma_\mu}$. Also, $B = \begin{bmatrix} 1+\nu^2 & \nu^2 \\ g^2 & g^2\nu^2 \end{bmatrix} = S\Lambda S^{-1}$ where

$$\Lambda = \begin{bmatrix} \frac{1+\nu^2+g^2\nu^2-\sqrt{(1+\nu^2+g^2\nu^2)^2-4g^2\nu^4}}{2} & 0 \\ 0 & \frac{1+\nu^2+g^2\nu^2+\sqrt{(1+\nu^2+g^2\nu^2)^2-4g^2\nu^4}}{2} \end{bmatrix}$$

and

$$S = \begin{bmatrix} \frac{1+\nu^2-g^2\nu^2-\sqrt{(1+\nu^2+g^2\nu^2)^2-4g^2\nu^4}}{2g^2} & \frac{1+\nu^2-g^2\nu^2+\sqrt{(1+\nu^2+g^2\nu^2)^2-4g^2\nu^4}}{2g^2} \\ 1 & 1 \end{bmatrix}.$$

Now, define the difference equations by

$$\begin{bmatrix} den_t \\ num_t \end{bmatrix} = B^t \begin{bmatrix} den_0 \\ num_0 \end{bmatrix} = S\Lambda^t S^{-1} \begin{bmatrix} 1 \\ 0 \end{bmatrix}.$$

The primary result for accruals as a valuation statistic is presented in proposition 13.1.[29]

Proposition 13.1 *Let $m_t = g\, m_{t-1} + e_t$, $\Sigma = D$, and $\nu = \frac{\sigma_e}{\sigma_\varepsilon}$. Then, $accruals_{t-1}$ and cf_t are, collectively, sufficient statistics for the mean of cash flows m_t based on the history of cash flows and $g^{t-1} accruals_t$ is an efficient statistic for m_t*

$$[\widehat{m}_t | cf_1, ..., cf_t] = g^{t-1} accruals_t$$
$$= \frac{1}{den_t} \left[\frac{num_t}{g^2}(cf_t - a_t^*) + g^{t-1}\nu^2 den_{t-1} accruals_{t-1} \right]$$

where $accruals_0 = m_0$, and $\begin{bmatrix} den_t \\ num_t \end{bmatrix} = B^t \begin{bmatrix} den_0 \\ num_0 \end{bmatrix} = S\Lambda^t S \begin{bmatrix} 1 \\ 0 \end{bmatrix}$. The variance of accruals is equal to the variance of the estimate of the mean of cash

---

[27] Other information, y_t, is suppressed as it isn't informative for the cash flow mean.

[28] As the agent's equilibrium contribution a^* is known, expected cash flow for the current period is estimated by $\widehat{m}_t + a_t^*$ and next period's expected cash flow is predicted by $g\,\widehat{m}_t + a_{t+1}^*$.

[29] All proofs are included in the end of chapter appendix.

flows multiplied by $g^{2(t-1)}$; the variance of the estimate of the mean of cash flows equals the coefficient on current cash flow multiplied by σ_e^2, $Var[\widehat{m}_t] = \frac{num_t}{den_t g^2}\sigma_e^2$.

Hence, the current accrual equals the estimate of the current mean of cash flows scaled by g^{t-1}, $accruals_t = \frac{1}{g^{t-1}}\widehat{m}_t$.

Tidy accruals

To explore the tidiness property of accruals in this setting it is instructive to consider the weight placed on the most recent cash flow as the number of periods becomes large. This limiting result is expressed in corollary 13.2.

Corollary 13.2 *As t becomes large, the weight on current cash flows for the efficient estimator of the mean of cash flows approaches*

$$\frac{2}{1+(1-g^2)\nu^2+\sqrt{(1+(1+g^2)\nu^2)^2-4g^2\nu^4}}$$

and the variance of the estimate approaches

$$\frac{2}{1+(1-g^2)\nu^2+\sqrt{(1+(1+g^2)\nu^2)^2-4g^2\nu^4}}\sigma_e^2.$$

Accruals, as identified above, are tidy in the sense that each period's cash flow is ultimately recognized in accounting income or remains as a "permanent" amount on the balance sheet.[30] This permanent balance is approximately

$$\sum_{t=1}^{k-1} cf_t \left[1 - \frac{num_t}{num_k} - num_t \sum_{n=t}^{k-1} \frac{g^{n-t-2}\nu^{2(n-1)}}{g^{n-1}den_n}\right]$$

where k is the first period where $\frac{num_t}{g^2 den_t}$ is well approximated by the asymptotic rate identified in corollary 1 and the estimate of expected cash flow $\widehat{m}_t$ is identified from tidy accruals as $g^{t-1}accruals_t$.[31]

In the benchmark case ($\Sigma = \sigma_e^2 I$, $\nu = \phi = 1$, and $g = 1$), this balance reduces to

$$\sum_{t=1}^{k-1} cf_t \left[1 - \frac{F_{2t}}{F_{2k}} - F_{2t}\sum_{n=t}^{k-1}\frac{1}{F_{2n+1}}\right]$$

where the estimate of expected cash flow $\widehat{m}_t$ is equal to tidy $accruals_t$.

---

[30] The permanent balance is of course settled up on disposal or dissolution.
[31] Cash flows beginning with period k and after are fully accrued as the asymptotic rate effectively applies each period. Hence, a convergent geometric series is formed that sums to one. On the other hand, the permanent balance arises as a result of the influence of the common knowledge initial expected cash flow m_0.

13.11.3 Performance evaluation

On the other hand, the evaluation role of accruals must regard a_t as unobservable while previous actions of this or other agents are at the incentive-induced equilibrium action a^*, and all observables are potentially (conditionally) informative: $\{cf_1 - a_1^*, cf_2 - a_2^*, \ldots, cf_t\}$, and $\{y_1 - a_1^*, y_2 - a_2^*, \ldots, y_t\}$.[32]

For the case $\Sigma = D$, the most efficient linear contract can be found by determining the incentive portion of compensation via *OLS* and then plugging a constant δ to satisfy individual rationality.[33] The (linear) incentive payments are equal to the *OLS* estimator, the final element of $\widehat{a}_t$, multiplied by $\Delta = \frac{c(a_H) - c(a_L)}{a_H - a_L}$, $\gamma_t = \Delta\, \widehat{a}_t$ where[34]

$$\widehat{a}_t = (H_a^T H_a)^{-1} H_a^T w_t,$$

$$H_a = \begin{bmatrix} -\nu & 0 & 0 & 0 & 0 & 0 \\ 1 & 0 & 0 & 0 & 0 & 0 \\ \nu g & -\nu & 0 & 0 & 0 & 0 \\ 0 & 1 & 0 & 0 & 0 & 0 \\ \vdots & \vdots & \ddots & \vdots & \vdots & \vdots \\ 0 & 0 & \cdots & \nu g & -\nu & 0 \\ 0 & 0 & \cdots & 0 & 1 & 1 \\ 0 & 0 & \cdots & 0 & 0 & \phi \end{bmatrix}, \quad w_t = \begin{bmatrix} -\nu g\, m_0 \\ cf_1 - a_1^* \\ 0 \\ cf_2 - a_2^* \\ \vdots \\ 0 \\ cf_t \\ \phi y_t \end{bmatrix}, \quad \text{and } \phi = \frac{\sigma_e}{\sigma_\varepsilon}.$$

Further, the variance of the incentive payments equals the last row, column element of $\Delta^2 (H_a^T H_a)^{-1} \sigma_e^2$.

In a moral hazard setting, the incentive portion of the *LEN* contract based on cash flow and other monitoring information history is identified in proposition 13.3. Incentive payments depend only on two realizations: unexpected cash flow and other monitoring information for period t. Unexpected cash flow at time t is

$$\begin{aligned} cf_t - E[cf_t | cf_1, \ldots, cf_{t-1}] &= cf_t - g^{t-1} accruals_{t-1} \\ &= cf_t - \widehat{m}_{t-1} \\ &= cf_t - [\widehat{m}_t | cf_1, \ldots, cf_{t-1}]. \end{aligned}$$

As a result, sequential spot contracting with replacement agents has a particularly streamlined form. Accounting accruals supply a convenient and sufficient summary of the cash flow history for the cash flow mean. Hence, the combination of last period's accruals with current cash flow yields the pivotal unexpected cash flow variable.

---

[32] For the case $\Sigma = D$, past y's are uninformative of the current period's act.

[33] Individual rationality is satisfied if

$\delta = RW - \{E[incentive payments | a] - \frac{1}{2} rVar[s] - c(a)\}$.

[34] The nuisance parameters (the initial $2t$ elements of $\widehat{a}_t$) could be avoided if one employs *GLS* in place of *OLS*.

Proposition 13.3 *Let $m_t = g\, m_{t-1} + e_t$, $\Sigma = D$, $\nu = \frac{\sigma_e}{\sigma_\epsilon}$, and $\phi = \frac{\sigma_e}{\sigma_\mu}$. Then, $accruals_{t-1}$, cf_t, and y_t, collectively, are sufficient statistics for evaluating the agent with incentive payments given by*

$$\gamma_t^T w_t = \Delta \left[\frac{1}{\nu^2 den_{t-1} + \phi^2 den_t} \cdot \frac{\phi^2 den_t y_t}{} + \nu^2 den_{t-1} \left(cf_t - g^{t-1} accruals_{t-1}\right) \right]$$

and variance of payments equal to

$$Var[\gamma_t^T w_t] = \Delta^2 \frac{den_t}{\nu^2 den_{t-1} + \phi^2 den_t} \sigma_e^2$$

where $\Delta = \frac{c(a_H) - c(a_L)}{a_H - a_L}$, and $accruals_{t-1}$ and den_t are as defined in proposition 13.1.

Benchmark case

Suppose $\Sigma = \sigma_e^2 I$ ($\nu = \phi = 1$) and $g = 1$. This benchmark case highlights the key informational structure in the data. Corollary 13.4 identifies the linear combination of current cash flows and last period's accruals employed to estimate the current cash flow mean conditional on cash flow history for this benchmark case.

Corollary 13.4 *For the benchmark case $\Sigma = \sigma_e^2 I$ ($\nu = \phi = 1$) and $g = 1$, accruals at time t are an efficient summary of past cash flow history for the cash flow mean if*

$$[\widehat{m}_t | cf_1, ..., cf_t] = accruals_t$$
$$= \frac{F_{2t}}{F_{2t+1}}(cf_t - a_t^*) + \frac{F_{2t-1}}{F_{2t+1}} accruals_{t-1}$$

where $F_n = F_{n-1} + F_{n-2}$, $F_0 = 0$, $F_1 = 1$ (the Fibonacci series), and the sequence is initialized with $accruals_0 = m_0$ (common knowledge mean beliefs). Then, variance of accruals equals $Var[\widehat{m}_t] = \frac{F_{2t}}{F_{2t+1}} \sigma_e^2$.

For the benchmark case, the evaluation role of accruals is synthesized in corollary 13.5.

Corollary 13.5 *For the benchmark case $\Sigma = \sigma_e^2 I$ ($\nu = \phi = 1$) and $g = 1$, $accruals_{t-1}$, cf_t, and y_t are, collectively, sufficient statistics for evaluating the agent with incentive payments given by*

$$\gamma_t^T w_t = \Delta \left\{ \frac{F_{2t+1}}{L_{2t}} y_t + \frac{F_{2t-1}}{L_{2t}} (cf_t - accruals_{t-1}) \right\}$$

and variance of payments equals $\Delta^2 \frac{F_{2t+1}}{L_{2t}} \sigma_e^2$ where $accruals_{t-1}$ is as defined in corollary 2, $L_n = L_{n-1} + L_{n-2}$, $L_0 = 2$, $L_1 = 1$ (the Lucas series), and $\Delta = \frac{c(a_H) - c(a_L)}{a_H - a_L}$.[35]

---

[35] The Lucas and Fibonacci series are related by $L_n = F_{n-1} + F_{n+1}$, for $n = 1, 2, ...$.

13.11.4 Summary

A positive view of accruals is outlined above. Accruals combined with current cash flow can serve as sufficient statistics of the cash flow history for the mean of cash flows. Further, in a moral hazard setting accruals can be combined with current cash flow and other monitoring information to efficiently evaluate replacement agents via sequential spot contracts. Informed priors regarding the contaminating permanent component facilitates this performance evaluation exercise. Notably, the same accrual statistic serves both valuation and evaluation purposes.

Next, we relax common knowledge of the *DGP* by both management and the auditor to explore strategic reporting equilibria albeit with a simpler *DGP*. That is, we revisit earnings management with informed priors and focus on Bayesian separation of signal (regarding expected cash flows) from noise.

13.12 Earnings management

We return to the earnings management setting introduced in chapter 2 and continued in chapter 3.[36] Now, we focus on belief revision with informed priors. First, we explore stochastic manipulation, as before, and, later on, selective manipulation.

13.12.1 Stochastic manipulation

The analyst is interested in uncovering the mean of accruals $E[x_t] = \mu$ (for all t) from a sequence of reports $\{y_t\}$ subject to stochastic manipulation by management. Earnings management is curbed by the auditor such that manipulation is limited to δ. That is, reported accruals y_t equal privately observed accruals x_t when there is no manipulation $I_t = 0$ and add δ when there is manipulation $I_t = 1$

$$y_t = x_t \quad \Pr(I_t = 0) = 1 - \alpha$$
$$y_t = x_t + \delta \quad \Pr(I_t = 1) = \alpha$$

The (prior) probability of manipulation α is known as well as the variance of x_t, σ_d^2. Since the variance is known, the maximum entropy likelihood function for the data is Gaussian with unknown, but finite and constant, mean. Background knowledge regarding the mean of x_t is that the mean is μ_0 with variance σ_0^2. Hence, the maximum entropy prior distribution for the mean is also Gaussian. And, the analysts' interests focus on the mean of the posterior distribution for x, $E\left[\mu \mid \mu_0, \sigma_0^2, \sigma_d^2, \{y_t\}\right]$.

Consider the updating of beliefs when the first report is observed, y_1. The analyst knows

$$y_1 = x_1 \quad I_1 = 0$$
$$y_1 = x_1 + \delta \quad I_1 = 1$$

---

[36] These examples were developed from conversations with Joel Demski and John Fellingham.

13.12 Earnings management

plus the prior probability of manipulation is α. The report contains evidence regarding the likelihood of manipulation. Thus, the posterior probability of manipulation[37] is

$$p_1 \equiv \Pr\left(I_1 = 1 \mid \mu_0, \sigma_0^2, \sigma_d^2, y_1\right)$$

$$= \frac{\alpha \phi\left(\frac{y_1 - \delta - \mu_0}{\sqrt{\sigma_d^2 + \sigma_0^2}}\right)}{\alpha \phi\left(\frac{y_1 - \delta - \mu_0}{\sqrt{\sigma_d^2 + \sigma_0^2}}\right) + (1-\alpha) \phi\left(\frac{y_1 - \mu_0}{\sqrt{\sigma_d^2 + \sigma_0^2}}\right)}$$

where $\phi(\cdot)$ is the standard Normal (Gaussian) density function. The density functions are, of course, conditional on manipulation or not and the random variable of interest is $x_1 - \mu_0$ which is Normally distributed with mean zero and variance $\sigma_d^2 + \sigma_0^2 = \sigma_d^2\left(1 + \frac{1}{\nu^2}\right)$ where $\nu = \frac{\sigma_d}{\sigma_0}$.

Bayesian updating of the mean following the first report is

$$\mu_1 = \mu_0 + \sigma_1^2 \frac{1}{\sigma_d^2} \left(p_1 (y_1 - \delta) + (1 - p_1) y_1 - \mu_0\right)$$

$$= \frac{1}{\nu^2 + 1} \left[\nu^2 \mu_0 + p_1 (y_1 - \delta) + (1 - p_1) y_1\right]$$

where the variance of the estimated mean is

$$\sigma_1^2 = \frac{1}{\frac{1}{\sigma_0^2} + \frac{1}{\sigma_d^2}}$$

$$= \frac{\sigma_d^2}{\nu^2 + 1}$$

Since

$$Var\left[p_t (y_t - \delta \mid I_t = 1) + (1 - p_t)(y_t \mid I_t = 0)\right] = Var[x_t] \equiv \sigma_d^2 \quad \text{for all } t$$

$\sigma_1^2, \ldots, \sigma_t^2$ are known in advance of observing the reported data. That is, the information matrix is updated each period in a known way.

---

[37] The posterior probability is logistic distributed (see Kiefer [1980]).

$$p_t = \frac{1}{1 + Exp\left[a_t + b_t y_t\right]}$$

where

$$a_t = \ln\left(\frac{1-\alpha}{\alpha}\right) + \frac{1}{2\left(\sigma_d^2 + \sigma_{t-1}^2\right)}\left[(\delta + \mu_{t-1})^2 - \mu_{t-1}^2\right]$$

and

$$b_t = \frac{1}{\left(\sigma_d^2 + \sigma_{t-1}^2\right)}\left[\mu_{t-1}^2 - (\delta + \mu_{t-1})^2\right]$$

This updating is repeated each period.[38] The posterior probability of manipulation given the series of observed reports through period t is

$$p_t \equiv \Pr\left(I_t = 1 \mid \mu_0, \sigma_0^2, \sigma_d^2, \{y_t\}\right)$$

$$= \frac{\alpha\phi\left(\frac{y_t - \delta - \mu_{t-1}}{\sqrt{\sigma_d^2 + \sigma_{t-1}^2}}\right)}{\alpha\phi\left(\frac{y_t - \delta - \mu_{t-1}}{\sqrt{\sigma_d^2 + \sigma_{t-1}^2}}\right) + (1-\alpha)\phi\left(\frac{y_1 - \mu_{t-1}}{\sqrt{\sigma_d^2 + \sigma_{t-1}^2}}\right)}$$

where the random variable of interest is $x_t - \mu_{t-1}$ which is Normally distributed with mean zero and variance $\sigma_d^2 + \sigma_{t-1}^2$. The updated mean is

$$\mu_t = \mu_{t-1} + \sigma_t^2 \frac{1}{\sigma_d^2}\left(p_t\left(y_t - \delta\right) + (1 - p_t)y_t - \mu_{t-1}\right)$$

$$= \frac{1}{\nu^2 + t}\left(\nu^2 \mu_0 + \sum_{k=1}^{t} p_k(y_k - \delta) + (1 - p_k)y_k\right)$$

and the updated variance of the mean is[39]

$$\sigma_t^2 = \frac{1}{\frac{1}{\sigma_0^2} + t\frac{1}{\sigma_d^2}}$$

$$= \frac{\sigma_d^2}{\nu^2 + t}$$

---

[38] To see this as a standard conditional Gaussian distribution result, suppose there is no manipulation so that $x_1, \ldots, x_t$ are observed and we're interested in $E[\mu \mid x_1, \ldots, x_t]$ and $Var[\mu \mid x_1, \ldots, x_t]$. The conditional distribution follows immediately from the joint distribution of

$$\mu = \mu_0 + \eta_0$$
$$x_1 = \mu + \varepsilon_1 = \mu_0 + \eta_0 + \varepsilon_1$$

and so on

$$x_t = \mu + \varepsilon_t = \mu_0 + \eta_0 + \varepsilon_t$$

The joint distribution is multivariate Gaussian

$$N\left(\begin{bmatrix} \mu_0 \\ \mu_0 \\ \vdots \\ \mu_0 \end{bmatrix}, \begin{bmatrix} \sigma_0^2 & \sigma_0^2 & \sigma_0^2 & \cdots & \sigma_0^2 \\ \sigma_0^2 & \sigma_0^2 + \sigma_d^2 & \sigma_0^2 & \cdots & \sigma_0^2 \\ \vdots & & & \ddots & \vdots \\ \sigma_0^2 & \sigma_0^2 & \sigma_0^2 & \cdots & \sigma_0^2 + \sigma_d^2 \end{bmatrix}\right)$$

With manipulation, the only change is x_t is replaced by $(y_t - \delta \mid I_t = 1)$ with probability p_t and $(y_t \mid I_t = 0)$ with probability $1 - p_t$.

[39] Bayesian updating of the mean can be thought of as a stacked weighted projection exercise where the prior "sample" is followed by the new evidence. For period t, the updated mean is

$$\mu_t \equiv E[\mu \mid \mu_0, \{y_t\}] = \left(X_t^T X_t\right)^{-1} X_t^T Y_t$$

and the updated variance of the mean is

$$\sigma_t^2 \equiv Var[\mu \mid \mu_0, \{y_t\}] = \left(X_t^T X_t\right)^{-1}$$

13.12 Earnings management

Now, it's time to look at some data.

Experiment

Suppose the prior distribution for x has mean $\mu_0 = 100$ and standard deviation $\sigma_0 = 25$, then it follows (from maximum entropy) the prior distribution is Gaussian. Similarly, x_t is randomly sampled from a Gaussian distribution with mean μ, where the value of μ is determined by a random draw from the prior distribution $N(100, 25)$, and standard deviation $\sigma_d = 20$. Reports y_t are stochastically manipulated as $x_t + \delta$ with likelihood $\alpha = 0.2$, where $\delta = 20$, and $y_t = x_t$ otherwise.

Results

Two plots summarize the data. The first data plot, figure 13.4, depicts the mean of 100 simulated samples of $t = 100$ observations and the mean of the 95% interval estimates of the mean along with the baseline (dashed line) for the randomly drawn mean μ of the data. As expected, the mean estimates converge toward the baseline as t increases and the interval estimates narrow around the baseline.

The second data plot, figure 13.5, shows the incidence of manipulation along with the assessed posterior probability of manipulation (multiplied by δ) based on the report for a representative draw. The graph depicts a reasonably tight correspondence between incidence of manipulation and posterior beliefs regarding manipulation.

Scale uncertainty

Now, we consider a setting where the variance (scale parameter) associated with privately observed accruals, σ_d^2, and the prior, σ_0^2, are uncertain. Suppose we only

---

where

$$Y_t = \begin{bmatrix} \frac{1}{\sigma_0} \mu_0 \\ \frac{\sqrt{p_1}}{\sigma_d}(y_1 - \delta) \\ \frac{\sqrt{1-p_1}}{\sigma_d} y_1 \\ \vdots \\ \frac{\sqrt{p_t}}{\sigma_d}(y_t - \delta) \\ \frac{\sqrt{1-p_t}}{\sigma_d} y_t \end{bmatrix}$$

and

$$X_t = \begin{bmatrix} \frac{1}{\sigma_0} \\ \frac{\sqrt{p_1}}{\sigma_d} \\ \frac{\sqrt{1-p_1}}{\sigma_d} \\ \vdots \\ \frac{\sqrt{p_t}}{\sigma_d} \\ \frac{\sqrt{1-p_t}}{\sigma_d} \end{bmatrix}$$

386   13. Informed priors

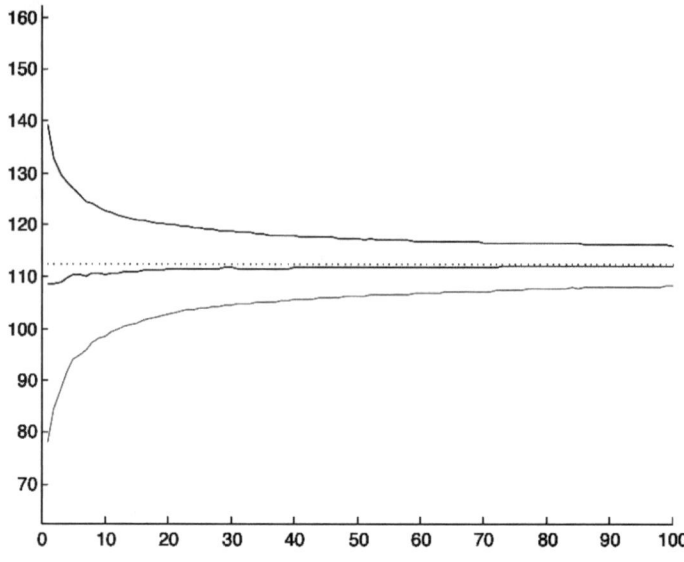

Figure 13.4: Stochastic manipulation σ_d known

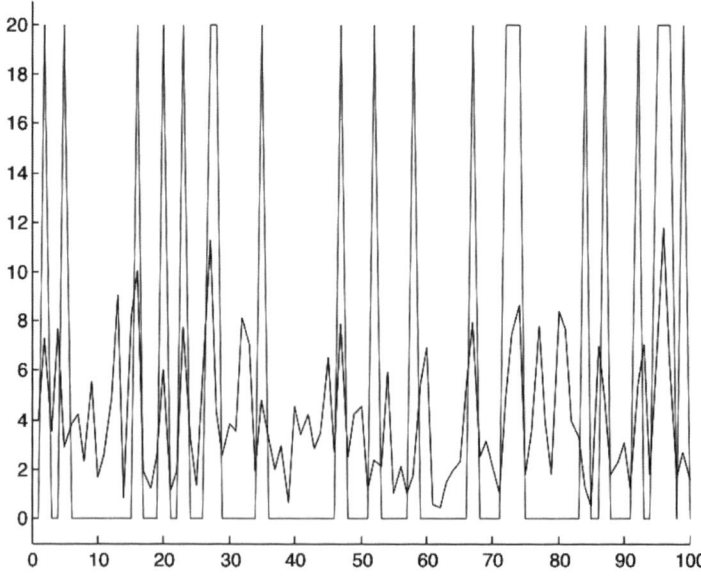

Figure 13.5: Incidence of stochastic manipulation and posterior probability

know $\nu = \frac{\sigma_d}{\sigma_0}$ and σ_d^2 and σ_0^2 are positive. Then, Jeffreys' prior distribution for scale is proportional to the square root of the determinant of the information matrix for t reports $\{y_t\}$ (see Jaynes [2003]),

$$f(\sigma_d) \propto \sqrt{\frac{\nu^2 + t}{\sigma_d^2}}$$

Hence, the prior for scale is proportional to $\frac{1}{\sigma_d}$

$$f(\sigma_d) \propto \frac{1}{\sigma_d}$$

With the mean μ and scale σ_d^2 unknown, following Box and Tiao [1973, p. 51], we can write the likelihood function (with priors on the mean incorporated as above) as

$$\ell\left(\mu, \sigma_d^2 \mid \{y_t\}\right) = \left(\sigma_d^2\right)^{-\frac{t+1}{2}} \exp\left[-\frac{1}{2\sigma_d^2}(Y - X\mu)^T(Y - X\mu)\right]$$

Now, rewrite[40]

$$\begin{aligned}(Y - X\mu)^T(Y - X\mu) &= \left(Y - \widehat{Y}\right)^T\left(Y - \widehat{Y}\right) \\ &\quad + \left(\widehat{Y} - X\mu\right)^T\left(\widehat{Y} - X\mu\right) \\ &= ts_t^2 + (\mu - \mu_t)^T X^T X (\mu - \mu_t)\end{aligned}$$

---

[40] The decomposition is similar to decomposition of mean square error into variance and squared bias but without expectations. Expand both sides of

$$(Y - X\mu)^T(Y - X\mu) = \left(Y - \widehat{Y}\right)^T\left(Y - \widehat{Y}\right) + \left(\widehat{Y} - X\mu\right)^T\left(\widehat{Y} - X\mu\right)$$

The left hand side is

$$Y^T Y - 2Y^T X\mu + \mu^T X^T X\mu$$

The right hand side is

$$Y^T Y - 2Y^T X\widehat{\mu} + \widehat{\mu}^T X^T X\widehat{\mu} + \mu^T X^T X\mu - 2\mu^T X^T X\widehat{\mu} + \widehat{\mu}^T X^T X\widehat{\mu}$$

Now show

$$-2Y^T X\mu = -2Y^T X\widehat{\mu} + 2\widehat{\mu}^T X^T X\widehat{\mu} - 2\widehat{\mu}^T X^T X\mu$$

Rewriting yields

$$Y^T X(\widehat{\mu} - \mu) = \widehat{\mu}^T X^T X(\widehat{\mu} - \mu)$$

or combining

$$\begin{aligned}(Y - X\widehat{\mu})^T X(\widehat{\mu} - \mu) &= 0 \\ \widehat{\varepsilon}^T X(\widehat{\mu} - \mu) &= 0\end{aligned}$$

The last expression is confirmed as $X^T\widehat{\varepsilon} = 0$ by least squares estimator construction (the residuals $\widehat{\varepsilon}$ are chosen to be orthogonal to the columns of X).

13. Informed priors

where

$$s_t^2 = \frac{1}{t}\left(Y - \widehat{Y}\right)^T \left(Y - \widehat{Y}\right)$$

$$Y^T = \begin{bmatrix} \nu\mu_0 & \sqrt{p_1}y_1 & \sqrt{1-p_1}y_1 & \cdots & \sqrt{p_t}y_t & \sqrt{1-p_t}y_t \end{bmatrix}$$

$$\widehat{Y} = X\mu_t$$

$$X^T = \begin{bmatrix} \nu & \sqrt{p_1} & \sqrt{1-p_1} & \cdots & \sqrt{p_t} & \sqrt{1-p_t} \end{bmatrix}$$

$$\mu_t = \left(X^T X\right)^{-1} X^T Y$$

Hence,

$$\ell\left(\mu, \sigma_d^2 \mid \{y_t\}\right) = \left(\sigma_d^2\right)^{-\frac{t+1}{2}} \exp\left[-\frac{ts_t^2}{2\sigma_d^2} - \frac{(\mu - \mu_t)^T X^T X (\mu - \mu_t)}{2\sigma_d^2}\right]$$

$$= \left(\sigma_d^2\right)^{-\frac{t+1}{2}} \exp\left[-\frac{ts_t^2}{2\sigma_d^2}\right] \exp\left[-\frac{(\mu - \mu_t)^T X^T X (\mu - \mu_t)}{2\sigma_d^2}\right]$$

The posterior distribution for the unknown parameters is then

$$f\left(\mu, \sigma_d^2 \mid \{y_t\}\right) \propto \ell\left(\mu, \sigma_d^2 \mid \{y_t\}\right) f\left(\sigma_d\right)$$

substitution from above gives

$$f\left(\mu, \sigma_d^2 \mid \{y_t\}\right) \propto \left(\sigma_d^2\right)^{-\left(\frac{t}{2}+1\right)} \exp\left[-\frac{ts_t^2}{2\sigma_d^2}\right]$$

$$\times \exp\left[-\frac{(\mu - \mu_t)^T X^T X (\mu - \mu_t)}{2\sigma_d^2}\right]$$

The posterior decomposes into

$$f\left(\mu, \sigma_d^2 \mid \{y_t\}\right) = f\left(\sigma_d^2 \mid s_t^2\right) f\left(\mu \mid \mu_t, \sigma_d^2\right)$$

where

$$f\left(\mu \mid \mu_t, \sigma_d^2\right) \propto \exp\left[-\frac{(\mu - \mu_t)^T X^T X (\mu - \mu_t)}{2\sigma_d^2}\right]$$

is the multivariate Gaussian kernel, and

$$f\left(\sigma_d^2 \mid s_t^2\right) \propto \left(\sigma_d^2\right)^{-\left(\frac{t}{2}+1\right)} \exp\left[-\frac{ts_t^2}{2\sigma_d^2}\right], \quad t \geq 1$$

is the inverted chi-square kernel, which is conjugate prior to the variance of a Gaussian distribution. Integrating out σ_d^2 yields the marginal posterior for μ,

$$f\left(\mu \mid \{y_t\}\right) = \int_0^\infty f\left(\mu, \sigma_d^2 \mid \{y_t\}\right) d\sigma_d^2$$

which has a noncentral, scaled-Student $t\left(\mu_t, s_t^2 \left(X^T X\right)^{-1}, t\right)$ distribution. In other words,

$$T = \frac{\mu - \mu_t}{\frac{s_t}{\sqrt{\nu^2 + t}}}$$

has a Student $t(t)$ distribution, for $t \geq 1$ (see Box and Tiao [1973, p. 117-118]).[41]

Now, the estimate for μ conditional on reports to date is the posterior mean[42]

$$\begin{aligned}
\mu_t &\equiv E\left[\mu \mid \{y_t\}\right] \\
&= \frac{\int_{-\infty}^{\infty} \mu f\left(\mu \mid \{y_t\}\right) d\mu}{\int_{-\infty}^{\infty} f\left(\mu \mid \{y_t\}\right) d\mu} \\
&= \frac{\nu^2 \mu_0 + p_1 y_1 + (1 - p_1) y_1 + \cdots + p_t y_t + (1 - p_t) y_t}{\nu^2 + t}
\end{aligned}$$

from the above posterior distribution and p_t is defined below. The variance of the estimate for μ is

$$\begin{aligned}
\sigma_t^2 &\equiv Var\left[\mu_t\right] \\
&= \tilde{s}_t^2 \left(X^T X\right)^{-1} = \frac{\tilde{s}_t^2}{t + \nu^2}, \quad t \geq 1
\end{aligned}$$

where $\tilde{s}_t^2$ is the estimated variance of the posterior distribution for x_t (see discussion below under a closer look at the variance). Hence, the highest posterior density (most compact) interval for μ with probability p is

$$\mu_t \pm t\left(t; 1 - \frac{p}{2}\right) \sigma_t$$

$$\frac{\nu^2 \mu_0 + p_1 y_1 + (1-p_1) y_1 + \cdots + p_t y_t + (1-p_t) y_t}{\nu^2 + t}$$
$$\pm t\left(t; 1 - \frac{p}{2}\right) \frac{\tilde{s}_t}{\sqrt{t + \nu^2}} \qquad t \geq 1$$

---

[41] This follows from a transformation of variables,

$$z = \frac{A}{2\sigma_d^2}$$

where

$$A = ts^2 + (\mu - \mu_t)^T X^T X (\mu - \mu_t)$$

that produces the kernel of a scaled Student t times the integral of a gamma distribution (see Gelman et al [2004], p.76). Or, for $a > 0, p > 0$,

$$\int_0^\infty x^{-(p+1)} e^{-\frac{a}{x}} dx = a^{-p} \Gamma(p)$$

where

$$\Gamma(z) = \int_0^\infty t^{z-1} e^{-t} dt$$

and for n a positive integer

$$\Gamma(n) = (n-1)!$$

a constant which can be ignored when identifying the marginal posterior (see Box and Tiao [1973, p. 144]).

[42] For emphasis, we write the normalization factor in the denominator of the expectations expression.

390   13. Informed priors

The probability the current report y_t, for $t \geq 2$,[43] is manipulated conditional on the history of reports (and manipulation probabilities) is

$$p_t \equiv \Pr\left(I_t = 1 \mid \nu, \{y_t\}, \{p_{t-1}\}, \mu_{t-1}, \sigma^2_{t-1}\right), \, t \geq 2$$

$$= \frac{\alpha \int \int f\left(y_t \mid D_t = 1, \mu, \sigma^2_d\right) f\left(\mu \mid \mu_{t-1}, \sigma^2_d\right) f\left(\sigma^2_d \mid s^2_{t-1}\right) d\mu d\sigma^2_d}{den(p_t)}$$

$$= \frac{\alpha \int \int (\sigma^2_d)^{-\frac{1}{2}} \exp\left[-\frac{(y_t - \delta - \mu)^2}{2\sigma^2_d}\right] (\sigma^2_d)^{-\frac{1}{2}} \exp\left[-\frac{(\mu - \mu_{t-1})^X T X (\mu - \mu_{t-1})}{2\sigma^2_d}\right] d\mu}{den(p_t)}$$

$$\times (\sigma^2_d)^{-\left(\frac{t}{2}+1\right)} \exp\left[-\frac{(t-1)s^2_{t-1}}{2\sigma^2_d}\right] d\sigma^2_d$$

$$+ (1-\alpha) \int \frac{1}{\sqrt{\sigma^2_d + \sigma^2_{t-1}}} \exp\left[-\frac{1}{2}\frac{(y_t - \mu_{t-1})^2}{\sigma^2_d + \sigma^2_{t-1}}\right]$$

$$\times (\sigma^2_d)^{-\left(\frac{t}{2}+1\right)} \exp\left[-\frac{(t-1)s^2_{t-1}}{2\sigma^2_d}\right] d\sigma^2_d$$

where

$$den(p_t) = \alpha \int \frac{1}{\sqrt{\sigma^2_d + \sigma^2_{t-1}}} \exp\left[-\frac{1}{2}\frac{(y_t - \delta - \mu_{t-1})^2}{\sigma^2_d + \sigma^2_{t-1}}\right]$$

$$\times (\sigma^2_d)^{-\left(\frac{t}{2}+1\right)} \exp\left[-\frac{(t-1)s^2_{t-1}}{2\sigma^2_d}\right] d\sigma^2_d$$

$$+ (1-\alpha) \int \frac{1}{\sqrt{\sigma^2_d + \sigma^2_{t-1}}} \exp\left[-\frac{1}{2}\frac{(y_t - \mu_{t-1})^2}{\sigma^2_d + \sigma^2_{t-1}}\right]$$

$$\times (\sigma^2_d)^{-\left(\frac{t}{2}+1\right)} \exp\left[-\frac{(t-1)s^2_{t-1}}{2\sigma^2_d}\right] d\sigma^2_d$$

Now, we have

$$f(y_t - \delta \mid D_t = 1) = \int_0^\infty \frac{1}{\sqrt{\sigma^2_d + \sigma^2_{t-1}}} \exp\left[-\frac{1}{2}\frac{(y_t - \delta - \mu_{t-1})^2}{\sigma^2_d + \sigma^2_{t-1}}\right]$$

$$\times (\sigma^2_d)^{-\left(\frac{t}{2}+1\right)} \exp\left[-\frac{(t-1)s^2_{t-1}}{2\sigma^2_d}\right] d\sigma^2_d$$

and

$$f(y_t \mid D_t = 0) = \int_0^\infty \frac{1}{\sqrt{\sigma^2_d + \sigma^2_{t-1}}} \exp\left[-\frac{1}{2}\frac{(y_t - \mu_{t-1})^2}{\sigma^2_d + \sigma^2_{t-1}}\right]$$

$$\times (\sigma^2_d)^{-\left(\frac{t}{2}+1\right)} \exp\left[-\frac{(t-1)s^2_{t-1}}{2\sigma^2_d}\right] d\sigma^2_d$$

---

[43]For $t = 1$, $p_t \equiv \Pr(D_t = 1 \mid y_t) = \alpha$ as the distribution for $(y_t \mid D_t)$ is so diffuse (s_0^2 has zero degrees of freedom) the report y_t is uninformative.

13.12 Earnings management

are noncentral, scaled-Student $t\left(\mu_{t-1}, s_{t-1}^2 + s_{t-1}^2 \left(X^T X\right)^{-1}, t-1\right)$ distributed. In other words,

$$T = \frac{y_t - \mu_{t-1}}{\sqrt{s_{t-1}^2 + \frac{s_{t-1}^2}{\nu^2 + t - 1}}}$$

has a Student $t(t-1)$ distribution for $t \geq 2$.

A closer look at the variance.

Now, we more carefully explore what s^2 estimates. We're interested in estimates of μ and $\frac{\sigma_d^2}{\nu^2 + t}$ and we have the following relations:

$$\begin{aligned} x_t &= \mu + \varepsilon_t \\ &= y_t - \delta D_t \end{aligned}$$

If D_t is observed then x_t is effectively observed and estimates of $\mu = E[x]$ and $\sigma_d^2 = Var[x]$ are, by standard methods, $\bar{x}$ and s^2. However, when manipulation D_t is not observed, estimation is more subtle. $x_t = y_t - \delta D_t$ is estimated via $y_t - \delta p_t$ which deviates from x_t by $\eta_t = -\delta(D_t - p_t)$. That is, $x_t = y_t - \delta p_t + \eta_t$. where

$$E[\eta_t \mid y_t] = -\delta[p_t(1 - p_t) + (1 - p_t)(0 - p_t)] = 0$$

and

$$\begin{aligned} Var[\eta_t \mid y_t] &= \delta^2 \left[p_t(1 - p_t)^2 + (1 - p_t)(0 - p_t)^2\right] \\ &= \delta^2 p_t(1 - p_t) = \delta^2 Var[D_t \mid y_t] \end{aligned}$$

s^2 estimates $E\left[\widehat{\varepsilon}_t^T \widehat{\varepsilon}_t \mid y_t\right]$ where $\widehat{\varepsilon}_t = y_t - \delta p_t - \mu_t$. However, $\sigma_d^2 = E\left[\varepsilon_t^T \varepsilon_t\right]$ is the object of interest. We can write

$$\begin{aligned} \widehat{\varepsilon}_t &= y_t - \delta p_t - \mu_t \\ &= (\delta D_t + \mu + \varepsilon_t) - \delta p_t - \mu_t \\ &= \varepsilon_t + \delta(D_t - p_t) + (\mu - \mu_t) \end{aligned}$$

In other words,

$$\varepsilon_t + (\mu - \mu_t) = \widehat{\varepsilon}_t - \delta(D_t - p_t)$$

Since $E\left[X^T \varepsilon_t\right] = 0$ (the regression condition) and μ_t is a linear combination of X, $Cov[\varepsilon_t, (\mu - \mu_t)] = 0$. Then, the variance of the left-hand side is a function of σ_d^2, the parameter of interest.

$$\begin{aligned} Var\left[\varepsilon_t + (\mu - \mu_t) \mid y_t\right] &= Var[\varepsilon_t] + Var[\mu - \mu_t \mid y_t] \\ &= \sigma_d^2 + \sigma_d^2 \frac{1}{\nu^2 + t} \\ &= \frac{\nu^2 + t + 1}{\nu^2 + t} \sigma_d^2 \end{aligned}$$

13. Informed priors

As D_t is stochastic
$$E\left[\widehat{\varepsilon}_t\left(D_t - p_t\right) \mid y_t\right] = 0$$

the variance of the right-hand side is

$$\begin{aligned}
Var\left[\widehat{\varepsilon}_t - \delta\left(D_t - p_t\right) \mid y_t\right] &= Var\left[\widehat{\varepsilon}_t \mid y_t\right] + \delta^2 Var\left[(D_t - p_t) \mid y_t\right] \\
&\quad - 2\delta Cov\left[\widehat{\varepsilon}_t, (D_t - p_t) \mid y_t\right] \\
&= Var\left[\widehat{\varepsilon}_t \mid y_t\right] \\
&\quad + \delta^2 \left[p_t(1-p_t)^2 + (1-p_t)(0-p_t)^2\right] \\
&= Var\left[\widehat{\varepsilon}_t \mid y_t\right] + \delta^2 p_t(1-p_t)
\end{aligned}$$

As $Var\left[\widehat{\varepsilon}_t\right]$ is consistently estimated via s^2, we can estimate σ_d^2 by

$$\begin{aligned}
\widehat{\sigma}_d^2 &= \frac{\nu^2 + t}{\nu^2 + t + 1}\left(s^2 + \delta^2 p_t(1-p_t)\right) \\
\widehat{\sigma}_d^2 &= \frac{\nu^2 + t}{\nu^2 + t + 1}\widetilde{s}^2
\end{aligned}$$

where $p_t(1-p_t)$ is the variance of D_t and $\widetilde{s}^2 = s^2 + \delta^2 p_t(1-p_t)$ estimates the variance of $\widehat{\varepsilon}_t + \eta_t$ given the data $\{y_t\}$.

Experiment

Repeat the experiment above except now we account for variance uncertainty as described above.[44]

Results

For 100 simulated samples of $t = 100$, we generate a plot, figure 13.6, of the mean and average 95% interval estimates. As expected, the mean estimates converge toward the baseline (dashed line) as t increases and the interval estimates narrow around the baseline but not as rapidly as the known variance setting.

---

[44] Another (complementary) inference approach involves creating the posterior distribution via conditional posterior simulation. Continue working with prior $p\left(\sigma_d^2 \mid X\right) \propto \frac{1}{\sigma_d^2}$ to generate a posterior distribution for the variance

$$p\left(\sigma_d^2 \mid X, \{y_t\}\right) \sim Inv-\chi^2\left(t, \widehat{\sigma}_d^2\right)$$

and conditional posterior distribution for the mean

$$p\left(\mu \mid \sigma_d^2, X, \{y_t\}\right) \sim N\left(\left(X^TX\right)^{-1}X^TY, \sigma_d^2\left(X^TX\right)^{-1}\right)$$

That is, draw σ_d^2 from the inverted, scaled chi-square distribution with t degrees of freedom and scale parameter $\widehat{\sigma}_d^2$. Then draw μ from a Gaussian distribution with mean $\left(X^TX\right)^{-1}X^TY$ and variance equal to the draw for $\sigma_d^2\left(X^TX\right)^{-1}$ from the step above.

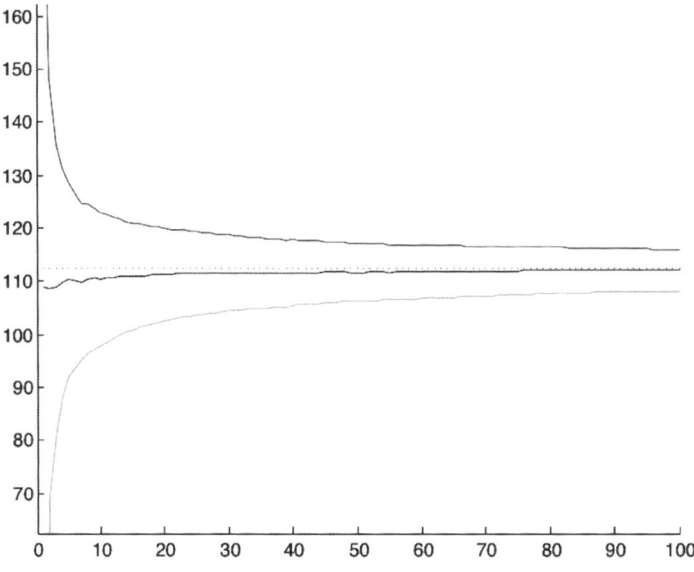

Figure 13.6: Stochastic manipulation σ_d unknown

13.12.2 Selective earnings management

Suppose earnings are manipulated whenever privately observed accruals x_t lie below prior periods' average reported accruals $\bar{y}_{t-1}$. That is,

$$\begin{array}{ll} x_t < \bar{y}_{t-1} & I_t = 1 \\ \text{otherwise} & I_t = 0 \end{array}$$

where $\bar{y}_0 = \mu_0$; for simplicity, μ_0 and $\bar{y}_t$ are commonly observed.[45] The setting differs from stochastic earnings management only in the prior and posterior probabilities of manipulation. The prior probability of manipulation is

$$\begin{aligned} \alpha_t &\equiv \Pr\left(x_t < \bar{y}_{t-1} \mid \mu_0, \sigma_0^2, \sigma_d^2, \{y_{t-1}\}\right) \\ &= \Phi\left(\frac{\bar{y}_{t-1} - \mu_{t-1}}{\sqrt{\sigma_d^2 + \sigma_{t-1}^2}}\right) \end{aligned}$$

where $\Phi(\cdot)$ represents the cumulative distribution function for the standard normal. Updated beliefs are informed by reported results even though they may be manipulated. If reported results exceed average reported results plus δ, then we

---

[45] This assumption could be relaxed or, for example, interpreted as an unbiased forecast conveyed via the firm's prospectus.

394   13. Informed priors

know there is no manipulation. Or, if reported results are less than average reported results less δ, then we know there is certain manipulation. Otherwise, there exists a possibility the reported results are manipulated or not. Therefore, the posterior probability of manipulation is

$$p_t \equiv \begin{array}{c} \Pr\left(I_t = 1 \mid \mu_0, \sigma_0^2, \sigma_d^2, \{y_t\}, y_t > \overline{y}_{t-1} + \delta\right) = 0 \\ \Pr\left(I_t = 1 \mid \mu_0, \sigma_0^2, \sigma_d^2, \{y_t\}, y_t < \overline{y}_{t-1} - \delta\right) = 1 \\ \Pr\left(I_t = 1 \mid \mu_0, \sigma_0^2, \sigma_d^2, \{y_t\}, \overline{y}_{t-1} - \delta \leq y_t \leq \overline{y}_{t-1} + \delta\right) \end{array}$$

$$= \frac{\alpha_t \phi\left(\frac{y_t - \delta - \mu_{t-1}}{\sqrt{\sigma_d^2 + \sigma_{t-1}^2}}\right)}{\alpha_t \phi\left(\frac{y_t - \delta - \mu_{t-1}}{\sqrt{\sigma_d^2 + \sigma_{t-1}^2}}\right) + (1 - \alpha_t) \phi\left(\frac{y_1 - \mu_{t-1}}{\sqrt{\sigma_d^2 + \sigma_{t-1}^2}}\right)}$$

As before, the updated mean is

$$\mu_t = \mu_{t-1} + \rho_t^2 \frac{1}{\sigma_d^2} \left(p_t (y_t - \delta) + (1 - p_t) y_t - \mu_{t-1} \right)$$

$$= \frac{1}{\nu^2 + t} \left(\nu^2 \mu_0 + \sum_{k=1}^{t} p_k (y_k - \delta) + (1 - p_k) y_k \right)$$

and the updated variance of the mean is

$$\sigma_t^2 = \frac{1}{\frac{1}{\sigma_0^2} + t \frac{1}{\sigma_d^2}}$$

$$= \frac{\sigma_d^2}{\nu^2 + t}$$

Time for another experiment.

Experiment

Suppose the prior distribution for x has mean $\mu_0 = 100$ and standard deviation $\sigma_0 = 25$, then it follows (from maximum entropy) the prior distribution is Gaussian. Similarly, x_t is randomly sampled from a Gaussian distribution with mean μ, a random draw from the prior distribution $N(100, 25)$, and standard deviation $\sigma_d = 20$. Reports y_t are selectively manipulated as $x_t + \delta$ when $x_t < \overline{y}_{t-1}$, where $\delta = 20$, and $y_t = x_t$ otherwise.

Results

Again, two plots summarize the data. The first data plot, figure 13.7, depicts the mean and average 95% interval estimates based on 100 simulated samples of $t = 100$ observations along with the baseline (dashed line) for the randomly drawn mean μ of the data. As expected, the mean estimates converge toward the baseline as t increases and the interval estimates narrow around the baseline. The second data plot, figure 13.8, shows the incidence of manipulation along with the assessed posterior probability of manipulation (multiplied by δ) based on the report for a representative draw. The graph depicts a reasonably tight correspondence between incidence of manipulation and posterior beliefs regarding manipulation.

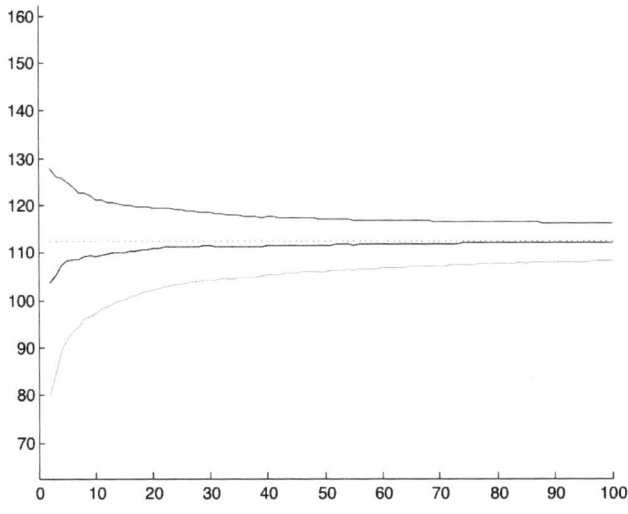

Figure 13.7: Selective manipulation σ_d known

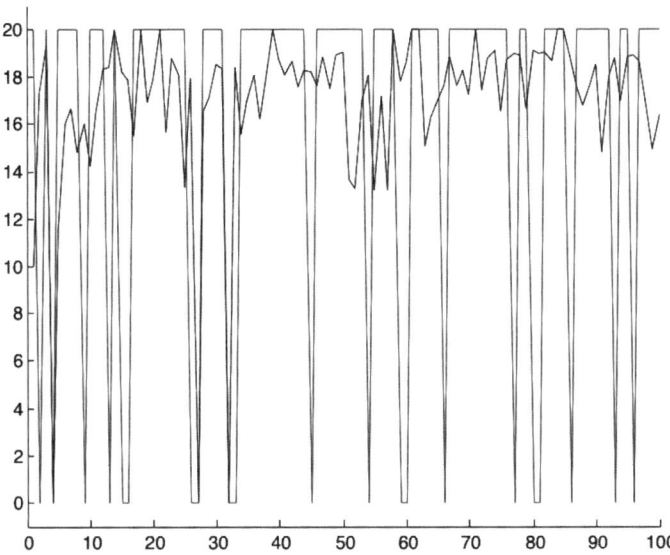

Figure 13.8: Incidence of selective manipulation and posterior probability

Scale uncertainty

Again, we consider a setting where the variance (scale parameter) associated with privately observed accruals σ_d^2 is uncertain but manipulation is selective. The only changes from the stochastic manipulation setting with uncertain scale involve the probabilities of manipulation.

The prior probability of manipulation is

$$\alpha_t \equiv \int \Pr\left(x_t < \bar{y}_{t-1} \mid \mu_0, \nu, \sigma_d^2, \{y_{t-1}\}\right) f\left(\sigma_d^2 \mid s_{t-1}^2\right) d\sigma_d^2$$

$$= \int_0^\infty \int_{-\infty}^{\bar{y}_{t-1}} f\left(x_t \mid \mu_0, \nu, \sigma_d^2, \{y_{t-1}\}\right) dx_t f\left(\sigma_d^2 \mid s_{t-1}^2\right) d\sigma_d^2, t \geq 2$$

On integrating σ_d^2 out, the prior probability of manipulation then simplifies as

$$\alpha_t = \int_{-\infty}^{\bar{y}_{t-1}} f\left(x_t \mid \mu_0, \nu, \{y_{t-1}\}\right) dx_t, t \geq 2$$

a cumulative noncentral, scaled-Student $t\left(\mu_{t-1}, s_{t-1}^2 + s_{t-1}^2 \left(X^T X\right)^{-1}, t - 1\right)$ distribution; in other words,

$$T = \frac{x_t - \mu_{t-1}}{\sqrt{s_{t-1}^2 + \frac{s_{t-1}^2}{\nu^2 + t - 1}}}$$

has a Student $t(t-1)$ distribution, $t \geq 2$.[46]

Following the report y_t, the posterior probability of manipulation is

$$p_t \equiv \begin{array}{l} \Pr\left(I_t = 1 \mid \mu_0, \nu, \{y_t\}, y_t > \bar{y}_{t-1} + \delta\right) = 0 \\ \Pr\left(I_t = 1 \mid \mu_0, \nu, \{y_t\}, y_t < \bar{y}_{t-1} - \delta\right) = 1 \\ \Pr\left(I_t = 1 \mid \mu_0, \nu, \{y_t\}, \bar{y}_{t-1} - \delta \leq y_t \leq \bar{y}_{t-1} + \delta\right) \\ = \frac{\alpha_t \int f\left(y_t \mid I_t = 1, \Im_{t-1}, \sigma_d^2\right) f\left(\sigma_d^2 \mid s_{t-1}^2\right) d\sigma_d^2}{\int f\left(y_t \mid \Im_{t-1}, \sigma_d^2\right) f\left(\sigma_d^2 \mid s_{t-1}^2\right) d\sigma_d^2}, t \geq 2 \end{array}$$

where $\Im_{t-1} = [\mu_0, \nu, \{y_{t-1}\}]$,

$$f\left(y_t \mid \Im_{t-1}, \sigma_d^2\right) = \alpha_t f\left(y_t - \delta \mid I_t = 1, \Im_{t-1}, \sigma_d^2\right)$$
$$+ (1 - \alpha_t) f\left(y_t \mid I_t = 0, \Im_{t-1}, \sigma_d^2\right)$$

$f\left(y_t - \delta \mid I_t = 1, \Im_{t-1}, \sigma_d^2\right)$ and $f\left(y_t \mid I_t = 0, \Im_{t-1}, \sigma_d^2\right)$ are noncentral, scaled-Student $t\left(\mu_{t-1}, s_{t-1}^2 + s_{t-1}^2 \left(X^T X\right)^{-1}, t - 1\right)$ distributed. In other words,

$$T = \frac{y_t - \mu_{t-1}}{\sqrt{s_{t-1}^2 + \frac{s_{t-1}^2}{\nu^2 + t - 1}}}$$

has a Student $t(t-1)$ distribution for $t \geq 2$.

---

[46] The prior probability of manipulation is uninformed or $p_t = \frac{1}{2}$ for $t < 2$.

13.12 Earnings management

A closer look at the variance.

In the selective manipulation setting,

$$\begin{aligned}
Var\left[\widehat{\varepsilon}_t - \delta\left(D_t - p_t\right) \mid y_t\right] &= Var\left[\widehat{\varepsilon}_t \mid y_t\right] + \delta^2 Var\left[\left(D_t - p_t\right) \mid y_t\right] \\
&\quad - 2\delta E\left[\widehat{\varepsilon}_t\left(D_t - p_t\right) \mid y_t\right] \\
&= Var\left[\widehat{\varepsilon}_t \mid y_t\right] + \delta^2 p_t\left(1 - p_t\right) \\
&\quad - 2\delta E\left[\widehat{\varepsilon}_t\left(D_t - p_t\right) \mid y_t\right]
\end{aligned}$$

The last term differs from the stochastic setting as selective manipulation produces truncated expectations. That is,

$$\begin{aligned}
2\delta E\left[\widehat{\varepsilon}_t\left(D_t - p_t\right) \mid y_t\right] &= 2\delta\{p_t E\left[\widehat{\varepsilon}_t\left(1 - p_t\right) \mid y_t, D_t = 1\right] \\
&\quad + \left(1 - p_t\right) E\left[\widehat{\varepsilon}_t\left(0 - p_t\right) \mid y_t, D_t = 0\right]\} \\
&= 2\delta\{p_t E\left[\widehat{\varepsilon}_t\left(1 - p_t\right) \mid y_t, x_t < \overline{y}_{t-1}\right] \\
&\quad + \left(1 - p_t\right) E\left[\widehat{\varepsilon}_t\left(0 - p_t\right) \mid y_t, x_t > \overline{y}_{t-1}\right]\} \\
&= 2\delta\{p_t E\left[\widehat{\varepsilon}_t\left(1 - p_t\right) \mid y_t, \widehat{\varepsilon}_t + \eta_t < \overline{y}_{t-1} - \mu_t\right] \\
&\quad + \left(1 - p_t\right) E\left[\widehat{\varepsilon}_t\left(0 - p_t\right) \mid y_t, \widehat{\varepsilon}_t + \eta_t > \overline{y}_{t-1} - \mu_t\right]\} \\
&= 2\delta\left\{p_t E\left[\widehat{\varepsilon}_t \mid y_t, \widehat{\varepsilon}_t + \eta_t < \overline{y}_{t-1} - \mu_t\right] - E\left[\widehat{\varepsilon}_t p_t \mid y_t\right]\right\} \\
&= 2\delta\left\{p_t \int\int \sigma\phi\left(\frac{\overline{y}_{t-1} - \mu}{\sigma}\right)\mid \mu, \sigma\right) f\left(\mu, \sigma\right) d\mu d\sigma - 0\right\} \\
&= -2\delta p_t \widetilde{s} f\left(\frac{\overline{y}_{t-1} - \mu_t}{\widetilde{s}}\right)
\end{aligned}$$

where $\widetilde{s}^2 = s^2 + \delta^2 p_t\left(1 - p_t\right)$ estimates the variance of $\widehat{\varepsilon}_t + \eta_t$ with no truncation, σ^2. The extra term, $\widetilde{s} f\left(\frac{\overline{y}_{t-1} - \mu_t}{\widetilde{s}}\right)$, arises from truncated expectations induced by selective manipulation rather than random manipulation. As both μ and σ are unknown, we evaluate this term by integrating out μ and σ where $f\left(\frac{\overline{y}_{t-1} - \mu_t}{\widetilde{s}}\right)$ has a Student $t(t)$ distribution. Hence, we can estimate σ_d^2 by

$$\begin{aligned}
\widehat{\sigma}_d^2 &= \frac{\nu^2 + t}{\nu^2 + t + 1}\left(s^2 + \delta^2 p_t\left(1 - p_t\right) + 2\delta p_t \widetilde{s} f\left(\frac{\overline{y}_{t-1} - \mu_t}{\widetilde{s}}\right)\right) \\
&= \frac{\nu^2 + t}{\nu^2 + t + 1}\left(\widetilde{s}^2 + 2\delta p_t \widetilde{s} f\left(\frac{\overline{y}_{t-1} - \mu_t}{\widetilde{s}}\right)\right)
\end{aligned}$$

conditional on the data $\{y_t\}$.

Experiment

Repeat the experiment above except now we account for variance uncertainty.

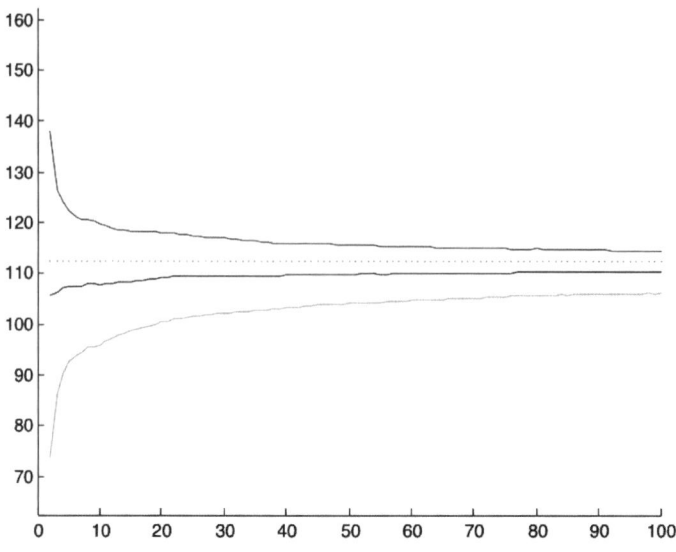

Figure 13.9: Selective manipulation σ_d unknown

Results

For 100 simulated samples of $t = 100$ observations, we generate a plot, figure 13.9, of the mean and average 95% interval estimates to summarize the data. As expected, the mean estimates converge toward the baseline (dashed line) as t increases and the interval estimates narrow around the baseline but not as rapidly as the known variance setting.

13.13 Jaynes' A_p distribution

Our story is nearly complete. However, consistent reasoning regarding propositions involves another, as yet unaddressed, element. For clarity, consider binary propositions. We might believe the propositions are equally likely but we also may be very confident of these probabilities, somewhat confident, or not confident at all. Jaynes [2003, ch. 18] compares propositions regarding heads or tails from a coin flip with life ever existing on Mars. He suggests that the former is very stable in light of additional evidence while the latter is very instable when faced with new evidence. Jaynes proposes a self-confessed odd proposition or distrib-

13.13 Jaynes' A_p distribution

ution (depending on context) denoted A_p to tidily handle consistent reasoning.[47] The result is tidy in that evaluation of new evidence based on background knowledge (including A_p) follows from standard rules of probability theory — Bayes' theorem.

This new proposition is defined by

$$\Pr(A \mid A_p, E, \Im) \equiv p$$

where A is the proposition of interest, E is any additional evidence, $\Im$ mathematically relevant background knowledge, and A_p is something like regardless of anything else the probability of A is p. The propositions are mutually exclusive and exhaustive. As this is surely an odd proposition or distribution over a probability, let the distribution for A_p be denoted (A_p). High instability or complete ignorance leads to

$$(A_p \mid \Im) = 1 \quad 0 \leq p \leq 1$$

Bayes' theorem leads to

$$\begin{aligned}(A_p \mid E, \Im) &= (A_p \mid \Im) \frac{\Pr(E \mid A_p, \Im) \Pr(\Im)}{\Pr(E \mid \Im) \Pr(\Im)} \\ &= (A_p \mid \Im) \frac{\Pr(E \mid A_p, \Im)}{\Pr(E \mid \Im)}\end{aligned}$$

Given complete ignorance, this simplifies as

$$\begin{aligned}(A_p \mid E, \Im) &= (1) \frac{\Pr(E \mid A_p, \Im)}{\Pr(E \mid \Im)} \\ &= \frac{\Pr(E \mid A_p, \Im)}{\Pr(E \mid \Im)}\end{aligned}$$

Also, integrating out A_p we have

$$\Pr(A \mid E, \Im) = \int_0^1 (A, A_p \mid E, \Im)\, dp$$

expanding the integrand gives

$$\Pr(A \mid E, \Im) = \int_0^1 \Pr(A \mid A_p, E, \Im)(A_p \mid E, \Im)\, dp$$

from the definition of A_p, the first factor is simply p, leading to

$$\Pr(A \mid E, \Im) = \int_0^1 p \times (A_p \mid E, \Im)\, dp$$

Hence, the probability assigned to the proposition A is just the first moment or expected value of the distribution for A_p conditional on the new evidence. The key feature involves accounting for our uncertainty via the joint behavior of the prior and the likelihood.

---

[47] Jaynes' A_p distribution is akin to over-dispersed models. That is, hierarchical generalized linear models that allow dispersion beyond the assigned sampling distribution.

13.13.1 Football game puzzle revisited

Reconsider the football game puzzle posed by Walley [1991, pp. 270-271]. Recall the puzzle involves a football match-up between two football rivals which produces either a win (W), a loss (L), or a draw (D) for the home team. Suppose we know the home team wins *more than* half the time and we gain evidence the game doesn't end in a draw. Utilizing Jaynes' A_p distribution, the posterior distribution differs from the earlier case where the prior probability of a win is one-half, $\Pr(W, L, D) = \left(\frac{3}{4}, \frac{1}{4}, 0\right)$. The reasoning for this is as follows. Let A be the proposition the home team wins (the argument applies analogously to a loss) and we know only the probability is at least one-half, then

$$(A_p \mid \Im_1) = 2 \quad \tfrac{1}{2} \leq p \leq 1$$

and

$$(A_p \mid E, \Im_1) = (2) \frac{\Pr(E \mid A_p, \Im)}{\Pr(E \mid \Im)}$$

Since, $\Pr(E = \text{not } D \mid \Im_1) = \Pr(E = \text{not } D \mid A_p, \Im_1) = \tfrac{3}{4}$ if draws are permitted, or $\Pr(E = \text{not } D \mid \Im_2) = \Pr(E = \text{not } D \mid A_p, \Im_2) = 1$ if draws are not permitted by the game structure.

$$(A_p \mid E, \Im_1) = (2) \frac{\tfrac{3}{4}}{\tfrac{3}{4}} = 2$$

$$(A_p \mid E, \Im_2) = (2) \frac{1}{1} = 2$$

Hence,

$$\Pr(A = W \mid E, \Im_j) = \int_{\tfrac{1}{2}}^{1} p \cdot (A_p \mid E, \Im_j) \, dp$$

$$= \int_{\tfrac{1}{2}}^{1} (2p) \, dp = \frac{3}{4}$$

Here the puzzle is resolved by careful interpretation of prior uncertainty combined with consistent reasoning enforced by Jaynes' A_p distribution.[48] Prior instability forces us to reassess the evaluation of new evidence; consistent evaluation of the evidence is the key. Some alternative characterizations of our confidence in the prior probability the home team wins are illustrated next.

How might we reconcile Jaynes' A_p distribution and Walley's $\left\{\tfrac{2}{3}, \tfrac{1}{3}, 0\right\}$ or $\left\{\tfrac{1}{2}, \tfrac{1}{2}, 0\right\}$ probability conclusion. The former follows from background knowledge that the home team wins more than half the time with one-half most likely

---

[48]For a home team loss, we have

$$\Pr(A = L \mid E, \Im) = \int_0^{\tfrac{1}{2}} 2p \, dp = \frac{1}{4}$$

and monotonically declining toward one. A_p in this case is triangular $8 - 8p$ for $\frac{1}{2} \leq p \leq 1$. The latter case is supported by background knowledge that the home team wins about half the time but no other information regarding confidence in this claim. Then, A_p is uniform for $0 \leq p \leq 1$.

13.14 Concluding remarks

Now that we're "fully" armed, it's time to re-explore the accounting settings in this and previous chapters as well as other settings, collect data, and get on with the serious business of evaluating accounting choice. But this monograph must end somewhere, so we hope the reader will find continuation of this project a worthy task. We anxiously await the blossoming of an evidentiary archive and new insights.

13.15 Additional reading

There is a substantial and growing literature on maximum entropy priors. Jaynes [2003] is an excellent starting place. Cover and Thomas [1991, ch. 12] expand the maximum entropy principle via minimization of relative entropy in the form of a conditional limit theorem. Also, Cover and Thomas [1991, ch. 11] discuss maximum entropy distributions for time series data including Burg's theorem (Cover and Thomas [1991], pp. 274-5) stating the Gaussian distribution is the maximum entropy error distribution given autocovariances. Walley [1991] critiques the precise probability requirement of Bayesian analysis, the potential for improper ignorance priors, and the maximum entropy principle while arguing in favor of an upper and lower probability approach to consistent reasoning (see Jaynes' [2003] comment in the bibliography).

Financial statement inferences are extended to bounding transactions amounts and financial ratios in Arya, Fellingham, Mittendorf, and Schroeder [2004]. Earnings management implications for performance evaluation are discussed in path breaking papers by Arya, Glover, and Sunder [1998] and Demski [1998]. Arya et al discuss earnings management as a potential substitute for (lack of) commitment in conveying information about the manager's input. Demski discusses accruals smoothing as a potential means of conveying valuable information about the manager's talent and input. Demski, Fellingham, Lin, and Schroeder [2008] discuss the corrosive effects on organizations of excessive reliance on individual performance measures.

13.16 Appendix

This appendix supplies proofs to the propositions and corollaries for the smooth accruals discussion.

Proposition 13.1. *Let* $m_t = g\, m_{t-1} + e_t$, $\Sigma = D$, *and* $\nu = \frac{\sigma_e}{\sigma_\epsilon}$. *Then, accruals$_{t-1}$ and cf$_t$ are, collectively, sufficient statistics for the mean of cash flows* m_t *based on the history of cash flows and* g^{t-1}accruals$_t$ *is an efficient statistic for* m_t

$$[\widehat{m}_t | cf_1, \dots, cf_t] = g^{t-1} \text{accruals}_t$$
$$= \frac{1}{den_t} \left\{ \frac{num_t}{g^2}(cf_t - a_t^*) + g^{t-1}\nu^2 den_{t-1} \text{accruals}_{t-1} \right\}$$

where accruals$_0 = m_0$, *and* $\frac{den_t}{num_t} = B^t \frac{den_0}{num_0} = S\Lambda^t S \begin{bmatrix} 1 \\ 0 \end{bmatrix}$. *The variance of accruals is equal to the variance of the estimate of the mean of cash flows multiplied by* $g^{2(t-1)}$; *the variance of the estimate of the mean of cash flows equals the coefficient on current cash flow multiplied by* σ_e^2, $Var[\widehat{m}_t] = \frac{num_t}{den_t g^2} \sigma_e^2$.

Proof. Outline of the proof:

1. Since the data are multivariate normally distributed, *BLU* estimation is efficient (achieves the Cramer-Rao lower bound amongst consistent estimators; see Greene [1997], p. 300-302).

2. *BLU* estimation is written as a recursive least squares exercise (see Strang [1986], p. 146-148).

3. The proof is completed by induction. That is, the difference equation solution is shown, by induction, to be equivalent to the recursive least squares estimator. A key step is showing that the information matrix $\Im$ and its inverse can be derived in recursive fashion via LDL^T decomposition (i.e., $D^{-1}L^{-1}\Im = L^T$).

Recursive least squares. Let $H_1 = \begin{bmatrix} -\nu \\ 1 \end{bmatrix}$ (a 2 by 1 matrix), $H_2 = \begin{bmatrix} g\nu & -\nu \\ 0 & 1 \end{bmatrix}$ (a 2 by 2 matrix), $H_t = \begin{bmatrix} 0 & \cdots & 0 & g\nu & -\nu \\ 0 & \cdots & 0 & 0 & 1 \end{bmatrix}$ (a 2 by t matrix with $t-2$ leading columns of zeroes), $z_1 = \begin{bmatrix} -g\nu m_0 \\ cf_1 - a_1^* \end{bmatrix}$, $z_2 = \begin{bmatrix} 0 \\ cf_2 - a_2^* \end{bmatrix}$, and $z_t = \begin{bmatrix} 0 \\ cf_t - a_t^* \end{bmatrix}$. The information matrix for a t-period cash flow history is

$$\Im_t = \Im_{t-1}^a + H_t^T H_t$$

$$= \begin{bmatrix} 1+\nu^2+g^2\nu^2 & -g\nu^2 & 0 & \cdots & 0 \\ -g\nu^2 & 1+\nu^2+g^2\nu^2 & -g\nu^2 & \ddots & \vdots \\ 0 & -g\nu^2 & \ddots & -g\nu^2 & 0 \\ \vdots & \ddots & -g\nu^2 & 1+\nu^2+g^2\nu^2 & -g\nu^2 \\ 0 & \cdots & 0 & -g\nu^2 & 1+\nu^2 \end{bmatrix}$$

a symmetric tri-diagonal matrix, where $\Im_{t-1}^a$ is $\Im_{t-1}$ augmented with a row and column of zeroes to conform with $\Im_t$. For instance, $\Im_1 = [1 + \nu^2]$ and $\Im_1^a = \begin{bmatrix} 1+\nu^2 & 0 \\ 0 & 0 \end{bmatrix}$. The estimate of the mean of cash flows is derived recursively as

$$b_t = b_{t-1}^a + k_t \left(z_t - H_t b_{t-1}^a\right)$$

for $t > 1$ where $k_t = \Im_t^{-1} H_t^T$, the gain matrix, and b_{t-1}^a is b_{t-1} augmented with a zero to conform with b_t. The best linear unbiased estimate of the current mean is the last element in the vector b_t and its variance is the last row-column element of $\Im_t^{-1}$ multiplied by σ_e^2.

Difference equations. The difference equations are

$$\begin{bmatrix} den_t \\ num_t \end{bmatrix} = \begin{bmatrix} 1+\nu^2 & \nu^2 \\ g^2 & g^2\nu^2 \end{bmatrix} \begin{bmatrix} den_{t-1} \\ num_{t-1} \end{bmatrix}$$

with $\begin{bmatrix} den_0 \\ num_0 \end{bmatrix} = \begin{bmatrix} 1 \\ 0 \end{bmatrix}$. The difference equations estimator for the current mean of cash flows and its variance are

$$\widehat{m}_t = \frac{1}{den_t}\left(\frac{num_t}{g^2}(cf_t - a_t^*) + g\nu^2 den_{t-1}\widehat{m}_{t-1}\right)$$
$$= g^{t-1} accruals_t$$
$$= \frac{1}{den_t}\left(\frac{num_t}{g^2}(cf_t - a_t^*) + g^{t-1}\nu^2 den_{t-1} accruals_{t-1}\right)$$

where $accruals_0 = m_0$, and

$$Var\left[\widehat{m}_t\right] = g^{2(t-1)} Var\left[accruals_t\right] = \sigma_e^2 \frac{num_t}{g^2 den_t}.$$

Induction steps. Assume

$$\widehat{m}_t = \frac{1}{den_t}\left(\frac{num_t}{g^2}(cf_t - a_t^*) + g\nu^2 den_{t-1}\widehat{m}_{t-1}\right)$$
$$= g^{t-1} accruals_t$$
$$= \frac{1}{den_t}\left(\frac{num_t}{g^2}(cf_t - a_t^*) + g^{t-1}\nu^2 den_{t-1} accruals_{t-1}\right)$$
$$= \left[b_{t-1}^a + k_t\left(z_t - H_t b_{t-1}^a\right)\right][t]$$

and

$$Var\left[\widehat{m}_t\right] = g^{2(t-1)} Var\left[accruals_t\right] = Var\left[b_t\right][t,t]$$

where $[t]$ ($[t,t]$) refers to element t (t,t) in the vector (matrix). The above is clearly true for the base case, $t = 1$ and $t = 2$. Now, show

$$\widehat{m}_{t+1} = \frac{1}{den_{t+1}}\left(\frac{num_{t+1}}{g^2}(cf_{t+1} - a_{t+1}^*) + g^t\nu^2 den_t accruals_t\right)$$
$$= \left[b_t^a + k_{t+1}\left(z_{t+1} - H_{t+1} b_t^a\right)\right][t+1].$$

Recall $z_{t+1} = \begin{bmatrix} 0 \\ c_{t+1} - a_{t+1}^* \end{bmatrix}$ and $H_{t+1} = \begin{bmatrix} 0 & \cdots & 0 & g\nu & -\nu \\ 0 & \cdots & 0 & 0 & 1 \end{bmatrix}$. From LDL^T decomposition of $\Im_{t+1}$ (recall $L^T = D^{-1}L^{-1}\Im$ where L^{-1} is simply products of matrices reflecting successive row eliminations - no row exchanges are involved due to the tri-diagonal structure and D^{-1} is the reciprocal of the diagonal elements remaining following eliminations) the last row of $\Im_{t+1}^{-1}$ is

$$\begin{bmatrix} \frac{g^{t-1}\nu^{2(t-1)}num_1}{g^2 den_{t+1}} & \cdots & \frac{g^2\nu^4 num_{t-1}}{g^2 den_{t+1}} & \frac{g\nu^2 num_t}{g^2 den_{t+1}} & \frac{num_{t+1}}{g^2 den_{t+1}} \end{bmatrix}.$$

This immediately identifies the variance associated with the estimator as the last term in $\Im_{t+1}^{-1}$ multiplied by the variance of cash flows, $\frac{num_{t+1}}{g^2 den_{t+1}}\sigma_e^2$. Hence, the difference equation and the recursive least squares variance estimators are equivalent.

Since $H_{t+1}^T z_{t+1} = \begin{bmatrix} 0 \\ \vdots \\ 0 \\ cf_{t+1} - a_{t+1}^* \end{bmatrix}$, the lead term on the RHS of the $[t+1]$

mean estimator is $\frac{num_{t+1}}{g^2 den_{t+1}}(cf_{t+1} - a_{t+1}^*)$ which is identical to the lead term on the left hand side (LHS). Similarly, the second term on the RHS (recall the focus is on element t, the last element of b_t^a is 0) is

$$[(I - k_{t+1}H_{t+1})b_t^a][t+1]$$

$$= \left[\left(I - \Im_{t+1}^{-1} \begin{bmatrix} 0 & 0 & \cdots & 0 & 0 \\ 0 & \vdots & \ddots & \vdots & \vdots \\ \vdots & 0 & \ddots & 0 & 0 \\ 0 & \cdots & 0 & g^2\nu^2 & -g\nu^2 \\ 0 & \cdots & 0 & -g\nu^2 & 1+\nu^2 \end{bmatrix} \right) b_t^a \right] [t+1]$$

$$= \left(\frac{-g^3\nu^4 num_t}{g^2 den_{t+1}} + \frac{g\nu^2 num_{t+1}}{g^2 den_{t+1}} \right) \widehat{m}_t$$

$$= \left(\frac{-g^3\nu^4 num_t + g\nu^2 num_{t+1}}{g^2 den_{t+1}} \right) g^{t-1} accruals_t.$$

The last couple of steps involve substitution of $\widehat{m}_t$ for $b_t^a[t+1]$ followed by $g^{t-1}accruals_t$ for $\widehat{m}_t$ on the right hand side (RHS) The difference equation relation, $num_{t+1} = g^2 den_t + g^2\nu^2 num_t$, implies

$$\frac{-g^3\nu^4 num_t + g\nu^2 num_{t+1}}{g^2 den_{t+1}} \widehat{m}_t = \frac{1}{den_{t+1}} g\nu^2 den_t \, \widehat{m}_t$$

$$= \frac{1}{den_{t+1}} g^t \nu^2 den_t \, accruals_t$$

the second term on the LHS. This completes the induction steps. ∎

13.16 Appendix

Corollary 13.2. *As t becomes large, the weight on current cash flows for the efficient estimator of the mean of cash flows approaches*

$$\frac{2}{1+(1-g^2)\nu^2+\sqrt{(1+(1+g^2)\nu^2)^2-4g^2\nu^4}}$$

and the variance of the estimate approaches

$$\frac{2}{1+(1-g^2)\nu^2+\sqrt{(1+(1+g^2)\nu^2)^2-4g^2\nu^4}}\sigma_e^2.$$

Proof. The difference equations

$$\begin{bmatrix} den_t \\ num_t \end{bmatrix} = S\Lambda^t S^{-1} \begin{bmatrix} den_0 \\ num_0 \end{bmatrix}$$

$$= S\Lambda^t S^{-1} \begin{bmatrix} 1 \\ 0 \end{bmatrix} = S\Lambda^t c$$

imply

$$c = S^{-1} \begin{bmatrix} den_0 \\ num_0 \end{bmatrix} = \begin{bmatrix} \frac{-g^2}{1+(1+g^2)\nu^2+\sqrt{(1+(1+g^2)\nu^2)^2-4g^2\nu^4}} \\ \frac{g^2}{1+(1+g^2)\nu^2+\sqrt{(1+(1+g^2)\nu^2)^2-4g^2\nu^4}} \end{bmatrix}$$

Thus,

$$\begin{bmatrix} den_t \\ num_t \end{bmatrix} = S \begin{bmatrix} \lambda_1^t & 0 \\ 0 & \lambda_2^t \end{bmatrix} c$$

$$= \frac{1}{\sqrt{(1+(1+g^2)\nu^2)^2-4g^2\nu^4}}$$

$$\times \begin{bmatrix} \frac{1}{2} \left\{ \begin{array}{l} \lambda_2^t \left(1+(1-g^2)\nu^2+\sqrt{(1+(1+g^2)\nu^2)^2-4g^2\nu^4}\right) \\ -\lambda_1^t \left(1+(1-g^2)\nu^2-\sqrt{(1+(1+g^2)\nu^2)^2-4g^2\nu^4}\right) \end{array} \right\} \\ g^2 \left(\lambda_2^t - \lambda_1^t\right) \end{bmatrix}$$

Since λ_2 is larger than λ_1, λ_1^t contributes negligibly to $\begin{bmatrix} den_t \\ num_t \end{bmatrix}$ for arbitrarily large t. Hence,

$$\lim_{t \to \infty} \frac{num_t}{g^2 den_t} = \frac{2}{1+(1-g^2)\nu^2+\sqrt{(1+(1+g^2)\nu^2)^2-4g^2\nu^4}}.$$

13. Informed priors

From proposition 13.1, the variance of the estimator for expected cash flow is $\frac{num_t}{g^2 den_t}\sigma_e^2$. Since

$$\lim_{t\to\infty} \frac{num_t}{g^2 den_t} = \frac{2}{1+(1-g^2)\nu^2 + \sqrt{(1+(1+g^2)\nu^2)^2 - 4g^2\nu^4}}.$$

the asymptotic variance is

$$\frac{2}{1+(1-g^2)\nu^2 + \sqrt{(1+(1+g^2)\nu^2)^2 - 4g^2\nu^4}}\sigma_e^2.$$

This completes the asymptotic case. ∎

Proposition 13.2. *Let* $m_t = g\,m_{t-1} + e_t$, $\Sigma = D$, $\nu = \frac{\sigma_e}{\sigma_\epsilon}$, *and* $\phi = \frac{\sigma_e}{\sigma_\mu}$. *Then,* $accruals_{t-1}$, cf_t, *and* y_t, *collectively, are sufficient statistics for evaluating the agent with incentive payments given by*

$$\gamma_t^T w_t = \Delta \frac{1}{\nu^2 den_{t-1} + \phi^2 den_t} \times \left[\phi^2 den_t y_t + \nu^2 den_{t-1}\left(cf_t - g^{t-1} accruals_{t-1}\right)\right]$$

and variance of payments equal to

$$Var[\gamma_t^T w_t] = \Delta^2 \frac{den_t}{\nu^2 den_{t-1} + \phi^2 den_t}\sigma_e^2$$

where $\Delta = \frac{c(a_H) - c(a_L)}{a_H - a_L}$, *and* $accruals_{t-1}$ *and* den_t *are as defined in proposition 13.1.*

Proof. Outline of the proof:

1. Show that the "best" linear contract is equivalent to the *BLU* estimator of the agent's current act rescaled by the agent's marginal cost of the act.

2. The *BLU* estimator is written as a recursive least squares exercise (see Strang [1986], p. 146-148).

3. The proof is completed by induction. That is, the difference equation solution is shown, by induction, to be equivalent to the recursive least squares estimator. Again, a key step involves showing that the information matrix $\Im_a$ and its inverse can be derived in recursive fashion via LDL^T decomposition (i.e., $D^{-1}L^{-1}\Im_a = L^T$).

"Best" linear contracts. The program associated with the optimal a_H-inducing *LEN* contract written in certainty equivalent form is

$$\min_{\delta,\gamma} \delta + E\left[\gamma^T w | a_H\right]$$

subject to

$$\delta + E\left[\gamma^T w | a_H\right] - \frac{r}{2} Var\left[\gamma^T w\right] - c(a_H) \geq RW \quad \text{(IR)}$$

$$\delta + E\left[\gamma^T w | a_H\right] - \frac{r}{2} Var\left[\gamma^T w\right] - c(a_H)$$
$$\geq \delta + E\left[\gamma^T w | a_L\right] - \frac{r}{2} Var\left[\gamma^T w\right] - c(a_L) \quad \text{(IC)}$$

As demonstrated in Arya et al [2004], both *IR* and *IC* are binding and γ equals the *BLU* estimator of a based on the history w (the history of cash flows cf and other contractible information y) rescaled by the agent's marginal cost of the act $\Delta = \frac{c(a_H) - c(a_L)}{a_H - a_L}$. Since *IC* is binding,

$$\delta + E\left[\gamma^T w | a_H\right] - \frac{r}{2} Var\left[\gamma^T w\right] - \left(\delta + E\left[\gamma^T w | a_L\right] - \frac{r}{2} Var\left[\gamma^T w\right]\right)$$
$$= c(a_H) - c(a_L)$$

$$E\left[\gamma^T w | a_H\right] - E\left[\gamma^T w | a_L\right] = c(a_H) - c(a_L)$$
$$\gamma^T \left\{E\left[w | a_H\right] - E\left[w | a_L\right]\right\} = c(a_H) - c(a_L)$$
$$(a_H - a_L) \gamma^T \hbar = c(a_H) - c(a_L)$$

where

$$w = \begin{bmatrix} cf_1 - m_0 - a_1^* \\ cf_2 - m_0 - a_2^* \\ \vdots \\ cf_t - m_0 \\ y_t \end{bmatrix}$$

and $\hbar$ is a vector of zeroes except the last two elements are equal to one, and

$$\gamma^T \hbar = \frac{c(a_H) - c(a_L)}{a_H - a_L}.$$

Notice, the sum of the last two elements of γ equals one, $\gamma^T \hbar = 1$, is simply the unbiasedness condition associated with the variance minimizing estimator of a based on design matrix H_a. Hence, $\gamma^T w$ equals the *BLU* estimator of a rescaled by Δ, $\gamma_t^T w_t = \Delta \widehat{a}_t$. As δ is a free variable, *IR* can always be exactly satisfied by setting

$$\delta = RW - \left\{E\left[\gamma^T w | a_H\right] - \frac{r}{2} Var\left[\gamma^T w\right] - c(a_H)\right\}.$$

Recursive least squares. H_t remains as defined in the proof of proposition 13.1. Let $H_{a1} = \begin{bmatrix} -\nu & 0 \\ 1 & 1 \\ 0 & \phi \end{bmatrix}$ (a 3 by 2 matrix), $H_{a2} = \begin{bmatrix} g\nu & -\nu & 0 \\ 0 & 1 & 1 \\ 0 & 0 & \phi \end{bmatrix}$ (a 3 by 3 matrix), $H_{at} = \begin{bmatrix} 0 & \cdots & 0 & g\nu & -\nu & 0 \\ 0 & \cdots & 0 & 0 & 1 & 1 \\ 0 & \cdots & 0 & 0 & 0 & \phi \end{bmatrix}$ (a 3 by $t+1$ matrix with leading zeroes), $\widetilde{w}_1 = \begin{bmatrix} -g\nu m_0 \\ cf_1 \\ y_1 \end{bmatrix}$, $\widetilde{w}_2 = \begin{bmatrix} 0 \\ cf_2 \\ y_2 \end{bmatrix}$, and $\widetilde{w}_t = \begin{bmatrix} 0 \\ cf_t \\ y_t \end{bmatrix}$. The information matrix for a t-period cash flow and other monitoring information history is

$$\Im_{at} = \Im_{t-1}^{aa} + H_{at}^T H_{at} =$$

$$\begin{bmatrix} 1+\nu^2+g^2\nu^2 & -g\nu^2 & 0 & 0 & \cdots & 0 \\ -g\nu^2 & 1+\nu^2+g^2\nu^2 & -g\nu^2 & \ddots & \cdots & 0 \\ 0 & -g\nu^2 & \ddots & \ddots & 0 & \vdots \\ 0 & \ddots & \ddots & 1+\nu^2+g^2\nu^2 & -g\nu^2 & 0 \\ \vdots & \cdots & 0 & -g\nu^2 & 1+\nu^2 & 1 \\ 0 & 0 & \cdots & 0 & 1 & 1+\phi^2 \end{bmatrix}$$

a symmetric tri-diagonal matrix where $\Im_{t-1}^{aa}$ is $\Im_{t-1}^{a}$ (the augmented information matrix employed to estimate the cash flow mean in proposition 13.1) augmented with an additional row and column of zeroes (i.e., the information matrix from proposition 13.1, $\Im_{t-1}$, is augmented with two columns of zeroes on the right and two rows of zeroes on the bottom). The recursive least squares estimator is

$$b_{at} = \left[b_{t-1}^{aa} + k_{at} \left(\widetilde{w}_t - H_{at} b_{t-1}^{aa} \right) \right]$$

for $t > 1$ where b_{t-1}^{aa} is b_{t-1} (the accruals estimator of m_{t-1} from proposition 13.1) augmented with two zeroes and $k_{at} = \Im_{at}^{-1} H_{at}^T$. The best linear unbiased estimate of the current act is the last element in the vector b_{at} and its variance is the last row-column element of $\Im_{at}^{-1}$ multiplied by σ_e^2. Notice, recursive least squares applied to the performance evaluation exercise utilizes the information matrix $\Im_{t-1}^{aa}$ (the information matrix employed in proposition 13.1 augmented with two trailing rows-columns of zeroes) and estimator b_{t-1}^{aa} (the accruals estimator of m_{t-1} from proposition 13.1 augmented with the two trailing zeroes). This accounts for the restriction on the parameters due to past actions already having been motivated in the past (i.e., past acts are at their equilibrium level a^*). Only the current portion of the design matrix H_{at} and the current observations w_t (in place of z_t) differ from the setup for accruals (in proposition 13.1).

Difference equations. The difference equations are

$$\begin{bmatrix} den_t \\ num_t \end{bmatrix} = \begin{bmatrix} 1+\nu^2 & \nu^2 \\ g^2 & g^2\nu^2 \end{bmatrix} \begin{bmatrix} den_{t-1} \\ num_{t-1} \end{bmatrix}$$

with $\begin{bmatrix} den_0 \\ num_0 \end{bmatrix} = \begin{bmatrix} 1 \\ 0 \end{bmatrix}$. The difference equations estimator for the linear incentive payments $\gamma^T w$ is

$$\begin{aligned}
\gamma_t^T w_t &= \Delta \frac{1}{\nu^2 den_{t-1} + \phi^2 den_t} \left[\phi^2 den_t y_t + \nu^2 den_{t-1} \left(cf_t - g\, \widehat{m}_{t-1} \right) \right] \\
&= \Delta \frac{1}{\nu^2 den_{t-1} + \phi^2 den_t} \\
&\quad \times \left[\phi^2 den_t y_t + \nu^2 den_{t-1} \left(cf_t - g^{t-1}\, accruals_{t-1} \right) \right]
\end{aligned}$$

and the variance of payments is

$$Var\left[\gamma^T w\right] = \Delta^2 \frac{den_t}{\nu^2 den_{t-1} + \phi^2 den_t} \sigma_e^2.$$

Induction steps. Assume

$$\begin{aligned}
\gamma_t^T w_t &= \Delta \frac{1}{\nu^2 den_{t-1} + \phi^2 den_t} \left[\phi^2 den_t y_t + \nu^2 den_{t-1} \left(cf_t - g\, \widehat{m}_{t-1} \right) \right] \\
&= \Delta \frac{1}{\nu^2 den_{t-1} + \phi^2 den_t} \\
&\quad \times \left[\phi^2 den_t y_t + \nu^2 den_{t-1} \left(cf_t - g^{t-1}\, accruals_{t-1} \right) \right] \\
&= \Delta \left[b_{t-1}^a + k_{at} \left(w_t - H_{at} b_{t-1}^a \right) \right] [t+1]
\end{aligned}$$

and

$$Var\left[\gamma_t^T w_t\right] = \Delta^2 Var\left[\widehat{a}_t\right][t+1, t+1]$$

where $[t+1]$ ($[t+1, t+1]$) refers to element $t+1$ ($t+1, t+1$) in the vector (matrix). The above is clearly true for the base case, $t = 1$ and $t = 2$. Now, show

$$\begin{aligned}
&\Delta \frac{1}{\nu^2 den_t + \phi^2 den_{t+1}} \left[\phi^2 den_{t+1} y_{t+1} + \nu^2 den_t \left(cf_{t+1} - g\, \widehat{m}_t \right) \right] \\
&= \Delta \frac{1}{\nu^2 den_t + \phi^2 den_{t+1}} \left[\phi^2 den_{t+1} y_{t+1} + \nu^2 den_t \left(cf_{t+1} - g^t\, accruals_t \right) \right] \\
&= \Delta \left[b_t^a + k_{at+1} \left(\widetilde{w}_{t+1} - H_{at+1} b_t^a \right) \right] [t+2].
\end{aligned}$$

Recall $\widetilde{w}_{t+1} = \begin{bmatrix} 0 \\ cf_{t+1} \\ \phi y_{t+1} \end{bmatrix}$ and $H_{at+1} = \begin{bmatrix} 0 & \cdots & 0 & g\nu & \nu & 0 \\ 0 & \cdots & 0 & 0 & 1 & 1 \\ 0 & \cdots & 0 & 0 & 0 & \phi \end{bmatrix}$. From LDL^T decomposition of $\Im_{at+1}$ (recall $L^T = D^{-1}L^{-1}\Im_a$ where L^{-1} is simply products of matrices reflecting successive row eliminations - no row exchanges are involved due to the tri-diagonal structure and D^{-1} is the reciprocal of the remaining elements remaining after eliminations) the last row of $\Im_{at+1}^{-1}$ is

$$\frac{1}{\nu^2 den_t + \phi^2 den_{t+1}} \begin{bmatrix} -g^{t-1}\nu^{2(t-1)}den_1 \\ \vdots \\ -g\nu^2\left(den_{t-1} + \nu^2 num_{t-1}\right) \\ -\left(den_t + \nu^2 num_t\right) \\ den_{t+1} \end{bmatrix}^T .$$ [49]

This immediately identifies the variance associated with the estimator as the last term in $\Im_{at+1}^{-1}$ multiplied by the product of the agent's marginal cost of the act squared and the variance of cash flows, $\Delta^2 \frac{den_{t+1}}{\nu^2 den_t + \phi^2 den_{t+1}} \sigma_e^2$. Hence, the difference equation and the recursive least squares variance of payments estimators are equivalent.

Since $H_{at+1}^T \widetilde{w}_{t+1} = \begin{bmatrix} 0 \\ \vdots \\ 0 \\ cf_{t+1} \\ cf_{t+1} + y_{t+1} \end{bmatrix}$ and the difference equation implies

$den_{t+1} = (1 + \nu^2)\, den_t + \nu^2 num_t$, the lead term on the RHS is

$$\frac{den_{t+1}}{\nu^2 den_t + \phi^2 den_{t+1}} (y_{t+1} + cf_{t+1}) - \frac{den_t + \nu^2 num_t}{\nu^2 den_t + \phi^2 den_{t+1}} cf_{t+1}$$

$$= \frac{den_{t+1}}{\nu^2 den_t + \phi^2 den_{t+1}} y_{t+1} - \frac{\nu^2 den_t}{\nu^2 den_t + \phi^2 den_{t+1}} cf_{t+1}$$

which equals the initial expression on the LHS of the $[t+2]$ incentive payments. Similarly, the $\widehat{m}_t = g^{t-1}\, accruals_t$ term on the RHS (recall the focus is on element $t+2$) is

$$\left[(I - k_{at+1}H_{at+1})\, b_t^a\right][t+2]$$

$$= \left[\left(I - \Im_{at+1}^{-1} \begin{bmatrix} 0 & 0 & \cdots & 0 & 0 & 0 \\ 0 & \vdots & \ddots & \vdots & \vdots & \vdots \\ \vdots & 0 & \cdots & 0 & 0 & 0 \\ 0 & \cdots & 0 & g^2\nu^2 & -g\nu^2 & 0 \\ 0 & \cdots & 0 & -g\nu^2 & 1+\nu^2 & 1 \\ 0 & \cdots & 0 & 0 & 1 & 1+\phi^2 \end{bmatrix}\right) b_t^a \right][t+2]$$

$$= -\frac{g\nu^2 den_t}{\nu^2 den_t + \phi^2 den_{t+1}} \widehat{m}_t$$

$$= -\frac{g^t \nu^2 den_t}{\nu^2 den_t + \phi^2 den_{t+1}} accruals_t.$$

---

[49] Transposed due to space limitations.

13.16 Appendix   411

Combining terms and simplifying produces the result

$$\frac{1}{\nu^2 den_t + \phi^2 den_{t+1}} \left[\phi^2 den_{t+1} y_{t+1} + \nu^2 den_t \left(cf_{t+1} - g\, \widehat{m}_t \right) \right]$$

$$= \frac{1}{\nu^2 den_t + \phi^2 den_{t+1}} \left[\phi^2 den_{t+1} y_{t+1} + \nu^2 den_t \left(cf_{t+1} - g^t\, accruals_t \right) \right].$$

Finally, recall the estimator $\widehat{a}_t$ (the last element of b_{at}) rescaled by the agent's marginal cost of the act identifies the "best" linear incentive payments

$$\gamma_t^T w_t = \Delta \widehat{a}_t$$

$$= \Delta \frac{1}{\nu^2 den_{t-1} + \phi^2 den_t} \left[\phi^2 den_t y_t + \nu^2 den_{t-1} \left(cf_t - g\, \widehat{m}_{t-1} \right) \right]$$

$$= \Delta \frac{1}{\nu^2 den_{t-1} + \phi^2 den_t}$$

$$\times \left[\phi^2 den_t y_t + \nu^2 den_{t-1} \left(cf_t - g^{t-1} accruals_{t-1} \right) \right].$$

This completes the induction steps. ∎

Corollary 13.4. *For the benchmark case* $\Sigma = \sigma_e^2 I$ ($\nu = \phi = 1$) *and* $g = 1$, *accruals at time* t *are an efficient summary of past cash flow history for the cash flow mean if*

$$[\widehat{m}_t | cf_1, ..., cf_t] = accruals_t$$

$$= \frac{F_{2t}}{F_{2t+1}} (cf_t - a_t^*) + \frac{F_{2t-1}}{F_{2t+1}} accruals_{t-1}$$

where $F_n = F_{n-1} + F_{n-2}$, $F_0 = 0$, $F_1 = 1$ *(the Fibonacci series), and the sequence is initialized with* $accruals_0 = m_0$ *(common knowledge mean beliefs). Then, variance of accruals equals* $Var[\widehat{m}_t] = \frac{F_{2t}}{F_{2t+1}} \sigma_e^2$.

Proof. Replace $g = \nu = 1$ in proposition 13.1. Hence,

$$\begin{bmatrix} den_t \\ num_t \end{bmatrix} = B \begin{bmatrix} den_{t-1} \\ num_{t-1} \end{bmatrix}$$

reduces to

$$\begin{bmatrix} den_t \\ num_t \end{bmatrix} = \begin{bmatrix} 2 & 1 \\ 1 & 1 \end{bmatrix} \begin{bmatrix} den_{t-1} \\ num_{t-1} \end{bmatrix}.$$

Since

$$\begin{bmatrix} F_{n+1} \\ F_n \end{bmatrix} = \begin{bmatrix} 1 & 1 \\ 1 & 0 \end{bmatrix} \begin{bmatrix} F_n \\ F_{n-1} \end{bmatrix}$$

and

$$\begin{bmatrix} F_{n+2} \\ F_{n+1} \end{bmatrix} = \begin{bmatrix} 1 & 1 \\ 1 & 0 \end{bmatrix} \begin{bmatrix} 1 & 1 \\ 1 & 0 \end{bmatrix} \begin{bmatrix} F_n \\ F_{n-1} \end{bmatrix} = \begin{bmatrix} 2 & 1 \\ 1 & 1 \end{bmatrix} \begin{bmatrix} F_n \\ F_{n-1} \end{bmatrix},$$

$den_t = F_{2t+1}, num_t = F_{2t}, den_{t-1} = F_{2t-1}$, and $num_{t-1} = F_{2t-2}$.
For $g = \nu = 1$, the above implies

$$\widehat{m}_t = g^{t-1} accruals_t$$
$$= \frac{1}{den_t} \left(\frac{num_t}{g^2} (cf_t - a_t^*) + g^{t-1}\nu^2 den_{t-1} accruals_{t-1} \right)$$

reduces to

$$\frac{F_{2t}}{F_{2t+1}} (cf_t - a_t^*) + \frac{F_{2t-1}}{F_{2t+1}} accruals_{t-1}$$

and variance of accruals equals $\frac{F_{2t}}{F_{2t+1}} \sigma_e^2$. ∎

Corollary 13.5 *For the benchmark case $\Sigma = \sigma_e^2 I$ ($\nu = \phi = 1$) and $g = 1$, $accruals_{t-1}$, cf_t, and y_t are, collectively, sufficient statistics for evaluating the agent with incentive payments given by*

$$\gamma_t^T w_t = \Delta \left\{ \frac{F_{2t+1}}{L_{2t}} y_t + \frac{F_{2t-1}}{L_{2t}} (cf_t - accruals_{t-1}) \right\}$$

and variance of payments equals $\Delta^2 \frac{F_{2t+1}}{L_{2t}} \sigma_e^2$ where $accruals_{t-1}$ is as defined in corollary 13.4 and $L_n = L_{n-1} + L_{n-2}$, $L_0 = 2$, and $L_1 = 1$ (the Lucas series), and $\Delta = \frac{c(a_H) - c(a_L)}{a_H - a_L}$.

Proof. Replace $g = \nu = \phi = 1$ in proposition 13.3. Hence,

$$\begin{bmatrix} den_t \\ num_t \end{bmatrix} = B \begin{bmatrix} den_{t-1} \\ num_{t-1} \end{bmatrix}$$

reduces to

$$\begin{bmatrix} den_t \\ num_t \end{bmatrix} = \begin{bmatrix} 2 & 1 \\ 1 & 1 \end{bmatrix} \begin{bmatrix} den_{t-1} \\ num_{t-1} \end{bmatrix}.$$

Since

$$\begin{bmatrix} F_{n+1} \\ F_n \end{bmatrix} = \begin{bmatrix} 1 & 1 \\ 1 & 0 \end{bmatrix} \begin{bmatrix} F_n \\ F_{n-1} \end{bmatrix}$$

and

$$\begin{bmatrix} F_{n+2} \\ F_{n+1} \end{bmatrix} = \begin{bmatrix} 1 & 1 \\ 1 & 0 \end{bmatrix} \begin{bmatrix} 1 & 1 \\ 1 & 0 \end{bmatrix} \begin{bmatrix} F_n \\ F_{n-1} \end{bmatrix} = \begin{bmatrix} 2 & 1 \\ 1 & 1 \end{bmatrix} \begin{bmatrix} F_n \\ F_{n-1} \end{bmatrix}$$

$den_t = F_{2t+1}, num_t = F_{2t}, den_{t-1} = F_{2t-1}, num_{t-1} = F_{2t-2}$, and $L_t = F_{t+1} + F_{t-1}$. For $g = \nu = \phi = 1$, the above implies

$$\gamma_t^T w_t = \Delta \frac{1}{\nu^2 den_{t-1} \phi^2 den_t} \left[\phi^2 den_t y_t + \nu^2 den_{t-1} \left(cf_t - g^{t-1} accruals_{t-1} \right) \right]$$

reduces to

$$\Delta \left\{ \frac{F_{2t-1}}{L_{2t}} (cf_t - accruals_{t-1}) + \frac{F_{2t+1}}{L_{2t}} y_t \right\}$$

and variance of payments equals $\Delta^2 \frac{F_{2t+1}}{L_{2t}} \sigma_e^2$. ∎

Appendix A
Asymptotic theory

Approximate or asymptotic results are an important foundation of statistical inference. Some of the main ideas are discussed below. The ideas center around the fundamental theorem of statistics, laws of large numbers (*LLN*), and central limit theorems (*CLT*). The discussion includes definitions of convergence in probability, almost sure convergence, convergence in distribution and rates of stochastic convergence.

The *fundamental theorem of statistics* states that if we sample randomly with replacement from a population, the empirical distribution function is consistent for the population distribution function (Davidson and MacKinnon [1993], p. 120-122). The fundamental theorem sets the stage for the remaining asymptotic theory.

A.1 Convergence in probability (laws of large numbers)

Definition A.1 *Convergence in probability.*
x_n *converges in probability to constant* c *if* $\lim_{n \to \infty} \Pr(|x_n - c| > \varepsilon) = 0$ *for all* $\varepsilon > 0$. *This is written* $p \lim (x_n) = c$.

A frequently employed special case is convergence in quadratic mean.

Theorem A.1 *Convergence in quadratic mean (or mean square).*
If x_n *has mean* μ_n *and variance* σ_n^2 *such that ordinary limits of* μ_n *and* σ_n^2 *are* c *and* 0, *respectively, then* x_n *converges in mean square to* c *and* $p \lim (x_n) = c$.

A proof follows from Chebychev's Inequality.

Theorem A.2 *Chebychev's Inequality.*
If x_n is a random variable and c_n and ε are constants then

$$Pr\left(|x_n - c_n| > \varepsilon\right) \leq E\left[(x_n - c_n)^2\right]/\varepsilon^2$$

A proof follows from Markov's Inequality.

Theorem A.3 *Markov's Inequality.*
If y_n is a nonnegative random variable and δ is a positive constant then

$$Pr\left(y_n \geq \delta\right) \leq E\left[y_n\right]/\delta$$

Proof.

$$E[y_n] = Pr(y_n < \delta) E[y_n \mid y_n < \delta] + Pr(y_n \geq \delta) E[y_n \mid y_n \geq \delta]$$

Since $y_n \geq 0$ both terms are nonnegative.
Therefore, $E[y_n] \geq Pr(y_n \geq \delta) E[y_n \mid y_n \geq \delta]$.
Since $E[y_n \mid y_n \geq \delta]$ must be greater than δ, $E[y_n] \geq Pr(y_n \geq \delta)\delta$. ∎

Proof. To prove Theorem A.2, let $y_n = (x_n - c)^2$ and $\delta = \varepsilon^2$ then

$$(x_n - c)^2 > \delta$$

implies $|x - c| > \varepsilon$. ∎

Proof. Now consider a special case of Chebychev's Inequality. Let $c = \mu_n$, $Pr\left(|x_n - \mu_n| > \varepsilon\right) \leq \sigma^2/\varepsilon^2$. Now, if $\lim_{n\to\infty} E[x_n] = c$ and $\lim_{n\to\infty} Var[x_n] = 0$, then $\lim_{n\to\infty} Pr\left(|x_n - \mu_n| > \varepsilon\right) \leq \lim_{n\to\infty} \sigma^2/\varepsilon^2 = 0$. The proof of Theorem A.1 is completed by Definition A.1 $p\lim(x_n) = \mu_n$. ∎

We have shown convergence in mean square implies convergence in probability.

A.1.1 Almost sure convergence

Definition A.2 *Almost sure convergence.*
$\bar{z}_n \xrightarrow{a.s.} z$ if $Pr\left(\lim_{n\to\infty} |\bar{z}_n - z| < \varepsilon\right) = 1$ *for all $\varepsilon > 0$.*
That is, there is large enough n such that the probability of the joint event $Pr\left(|\bar{z}_{n+1} - z| > \varepsilon, |\bar{z}_{n+2} - z| > \varepsilon, ...\right)$ diminishes to zero.

Theorem A.4 *Markov's strong law of large numbers.*
If $\{zj\}$ is sequence of independent random variables with $E[z_j] = \mu_j < \infty$ and

if for some $\delta > 0$, $\frac{E\left[|z_j - \mu_j|^{1+\delta}\right]}{j^{1+\delta}} < \infty$ then $\bar{z}_n - \bar{\mu}_n$ converges almost surely to 0, where $\bar{z}_n = n^{-1} \sum_{j=1}^{n} z_j$ and $\bar{\mu}_n = n^{-1} \sum_{j=1}^{n} \mu_j$.
This is denoted $\bar{z}_n - \bar{\mu}_n \xrightarrow{as} 0$.

Kolmogorov's law is somewhat weaker as it employs $\delta = 1$.

Theorem A.5 *Kolmogorov's strong law of large numbers.*
If $\{z\}$ is sequence of independent random variables with $E[z_j] = \mu_j < \infty$, $Var[z_j] = \sigma_j^2 < \infty$ and $\sum_{j=1}^{n} \frac{\sigma_j^2}{j^2} < \infty$ then $\bar{z}_n - \bar{\mu}_n \xrightarrow{as} 0$.

Both of the above theorems allow variances to increase but slowly enough that sums of variances converge. Almost sure convergence states that the behavior of the mean of sample observations is the same as the behavior of the average of the population means (not that the sample means converge to anything specific).

The following is a less general result but adequate for most econometric applications. Further, Chebychev's law of large numbers differs from Kinchine's in that Chebychev's does not assume *iid* (independent, identical distributions).

Theorem A.6 *Chebychev's weak law of large numbers.*
If $\{z\}$ is sequence of uncorrelated random variables with $E[z_j] = \mu_j < \infty$, $Var[z_j] = \sigma_j^2 < \infty$, and $\lim_{n \to \infty} n^{-2} \sum_{j=1}^{\infty} \sigma_j^2 < \infty$, then $\bar{z}_n - \bar{\mu}_n \xrightarrow{p} 0$.

Almost sure convergence implies convergence in probability (but not necessarily the converse).

A.1.2 Applications of convergence

Definition A.3 *Consistent estimator.*
An estimator $\hat{\theta}$ of parameter θ is a consistent estimator iff $p \lim \left(\hat{\theta}\right) = \theta$.

Theorem A.7 *Consistency of sample mean.*
The mean of a random sample from any population with finite mean μ and finite variance σ^2 is a consistent estimator of μ.

Proof. $E[\bar{x}] = \mu$ and $Var[\bar{x}] = \frac{\sigma^2}{n}$, therefore by Theorem A.1 (convergence in quadratic mean) $p \lim (\bar{x}) = \mu$. ∎
An alternative theorem with weaker conditions is Kinchine's weak law of large numbers.

Theorem A.8 *Kinchine's theorem (weak law of large numbers).*
Let $\{x_j\}$, $j = 1, 2, ..., n$, be a random sample (iid) and assume $E[x_j] = \mu$ (a finite constant) then $\bar{x} \xrightarrow{p} \mu$.

The Slutsky Theorem is an extremely useful result.

Theorem A.9 *Slutsky Theorem.*
For continuous function $g(x)$ that is not a function of n,

$$p\lim(g(x_n)) = g(p\lim(x_n))$$

A proof follows from the implication rule.

Theorem A.10 *The implication rule.*
Consider events E and F_j, $j = 1, ..., k$, such that $E \supset \cap_{j=1,k} F_j$.
Then $\Pr(\bar{E}) \leq \sum_{j=1}^{k} \Pr(\bar{F}_j)$.

Notation: $\bar{E}$ is the complement to E, $A \supset B$ means event B implies event A (inclusion), and $A \cap B \equiv AB$ means the intersection of events A and B.

Proof. A proof of the implication rule is from Lukacs [1975].
1. $\Pr(A \cup B) = \Pr(A) + \Pr(B) - \Pr(AB)$.
2. $\Pr(\bar{A}) = 1 - \Pr(A)$.

from 1

3. $\Pr(A \cup B) \leq \Pr(A) + \Pr(B)$
4. $\Pr(\cup_{j=1,\infty} A_j) \leq \sum_{j=1}^{k} \Pr(A_j)$

1 and 2 imply $\Pr(AB) = \Pr(A) - \Pr(B) + 1 - \Pr(A \cup B)$. Since $1 - \Pr(A \cup B) \geq 0$, we obtain

5. $\Pr(AB) \geq \Pr(A) - \Pr(\bar{B}) = 1 - \Pr(\bar{A}) - \Pr(\bar{B})$ (*Boole's Inequality*).
$\Pr(\cap_j A_j) \geq 1 - \Pr(\bar{A}_1) - \Pr(\overline{\cap_{j=2,\infty} A_j}) = 1 - \Pr(\bar{A}_1) - \Pr(\cup_{j=2,\infty} \bar{A}_j)$.

This inequality and 4 imply

6. $\Pr(\cap_{j=1,k} A_j) \geq 1 - \sum_{j=1}^{k} \Pr(\bar{A}_j)$ (*Boole's Generalized Inequality*).

5 can be rewritten as

7. $\Pr(\bar{A}) + \Pr(\bar{B}) \geq 1 - \Pr(AB) = \Pr(\overline{AB}) = \Pr(\bar{A} \cup \bar{B})$.

Now let C be an event implied by AB, that is $C \supset AB$, then $\bar{C} \subset \bar{A} \cup \bar{B}$ and

8. $\Pr(\bar{C}) \leq \Pr(\bar{A} \cup \bar{B})$.

Combining 7 and 8 obtains
The Implication Rule.

Let A, B, and C be three events such that $C \supset AB$, then
$\Pr(\bar{C}) \leq \Pr(\bar{A}) + \Pr(\bar{B})$. ∎

Proof. Slutsky Theorem (White [1984])
Let $g_j \in g$. For every $\varepsilon > 0$, continuity of g implies there exists $\delta(\varepsilon) > 0$ such that if $|x_{nj}(w) - x_j| < \delta(\varepsilon)$, $j = 1, ..., k$, then $|g_j(x_n(w)) - g_j(x)| < \varepsilon$. Define events

$$Fj \equiv [w : |x_{nj}(w) - x_j| < \delta(\varepsilon)]$$

and

$$E \equiv [w : |g_j(x_{nj}(w)) - g_j(x)| < \varepsilon]$$

Then $E \supset \cap_{j=1,k} F_j$, by the implication rule, leads to $\Pr(\bar{E}) \leq \sum_{j=1}^{k} \Pr(\bar{F}_j)$. Since $x_n \xrightarrow{p} x$ for arbitrary $\eta > 0$ and all n sufficiently large, $\Pr(F_j) \leq \eta$. Thus, $\Pr(\bar{E}) \leq k\eta$ or $\Pr(E) \geq 1 - k\eta$. Since $\Pr[E] \leq 1$ and η is arbitrary, $\Pr(E) \longrightarrow 1$ as $n \longrightarrow \infty$. Hence, $g_j(x_n(w)) \xrightarrow{p} g_j(x)$. Since this holds for all $j = 1, ..., k$, $g(x_n(w)) \xrightarrow{p} g(x)$. ∎

Comparison of Slutsky Theorem with Jensen's Inequality highlights the difference between the expectation of a random variable and probability limit.

Theorem A.11 *Jensen's Inequality.*
If $g(x_n)$ is a concave function of x_n then $g(E[x_n]) \geq E[g(x)]$.

The comparison between the Slutsky theorem and Jensen's inequality helps explain how an estimator may be consistent but not be unbiased.[1]

Theorem A.12 *Rules for probability limits.*
If x_n and y_n are random variables with $p\lim(x_n) = c$ and $p\lim(y_n) = d$ then
a. $p\lim(x_n + y_n) = c + d$ *(sum rule)*
b. $p\lim(x_n y_n) = cd$ *(product rule)*
c. $p\lim\left(\frac{x_n}{y_n}\right) = \frac{c}{d}$ *if $d \neq 0$ (ratio rule)*

If W_n is a matrix of random variables and if $p\lim(W_n) = \Omega$ then

d. $p\lim(W_n^{-1}) = \Omega^{-1}$ *(matrix inverse rule)*

If X_n and Y_n are random matrices with $p\lim(X_n) = A$ and $p\lim(Y_n) = B$ then

e. $p\lim(X_n Y_n) = AB$ *(matrix product rule).*

A.2 Convergence in distribution (central limit theorems)

Definition A.4 *Convergence in distribution.*
x_n *converges in distribution to random variable x with CDF $F(x)$ if* $\lim_{n \to \infty} |F(x_n) - F(x)| = 0$ *at all continuity points of $F(x)$.*

---

[1] Of course, Jensen's inequality is exploited in the construction of concave utility functions to represent risk aversion.

418   Appendix A. Asymptotic theory

Definition A.5 *Limiting distribution.*
If x_n converges in distribution to random variable x with CDF $F(x)$ then $F(x)$ is the limiting distribution of x_n; this is written $x_n \xrightarrow{d} x$.

Example A.1 $t_{n-1} \xrightarrow{d} N(0,1)$.

Definition A.6 *Limiting mean and variance.*
The limiting mean and variance of a random variable are the mean and variance of the limiting distribution assuming the limiting distribution and its moments exist.

Theorem A.13 *Rules for limiting distributions.*
(a) If $x_n \xrightarrow{d} x$ and $p \lim (y_n) = c$, then $x_n y_n \xrightarrow{d} xc$.
Also, $x_n + y_n \xrightarrow{d} x + c$, and
$\frac{x_n}{y_n} \xrightarrow{d} \frac{x}{c}$, $c \neq 0$.
(b) If $x_n \xrightarrow{d} x$ and $g(x)$ is a continuous function then $g(x_n) \xrightarrow{d} g(x)$ (this is the analog to the Slutsky theorem).
(c) If y_n has limiting distribution and $p \lim (x_n - y_n) = 0$, then x_n has the same limiting distribution as y_n.

Example A.2 $F(1,n) \xrightarrow{d} \chi^2(1)$.

Theorem A.14 *Lindberg-Levy Central Limit Theorem (univariate).*
If $x_1, ..., x_n$ are a random sample from probability distribution with finite mean μ and finite variance σ^2 and $\bar{x} = n^{-1} \sum_{t=1}^{n} x_t$, then $\sqrt{n}(\bar{x} - \mu) \xrightarrow{d} N(0, \sigma^2)$.

Proof. (Rao [1973], p. 127)
Let $f(t)$ be the characteristic function of $X_t - \mu$.[2] Since the first two moments exist,
$$f(t) = 1 - \frac{1}{2}\sigma^2 t^2 + o(t^2)$$
The characteristic function of $Y_n = \frac{1}{\sqrt{n}\sigma} \sum_{i=1}^{n} (X_i - \mu)$ is
$$f_n(t) = \left[f\left(\frac{t}{\sigma\sqrt{n}}\right) \right]^n = \left[1 - \frac{1}{2}\sigma^2 t^2 + o(t^2) \right]^n$$

---

[2]The characteristic function $f(t)$ is the complex analog to the moment generating function
$$f(t) = \int e^{itx} dF(x)$$
$$= \int \cos(tx) dF(x) + i \int \sin(tx) dF(x)$$
where $i = \sqrt{-1}$ (Rao [1973], p. 99).

A.2 Convergence in distribution (central limit theorems)

And
$$\log\left[1 - \frac{1}{2}\sigma^2 t^2 + o\left(t^2\right)\right]^n = n\log\left[1 - \frac{1}{2}\sigma^2 t^2 + o\left(t^2\right)\right]^n \to -\frac{t^2}{2}$$

That is, as $n \to \infty$
$$f_n(t) \to e^{-\frac{t^2}{2}}$$

Since the limiting distribution is continuous, the convergence of the distribution function of Y_n is uniform, and we have the more general result

$$\lim_{n \to \infty}\left[F_{Y_n}(x_n) - \Phi(x_n)\right] \to 0$$

where x_n may depend on n in any manner. This result implies that the distribution function of $\bar{X}_n$ can be approximated by that of a normal random variable with mean μ and variance $\frac{\sigma^2}{n}$ for sufficiently large n. ∎

Theorem A.15 *Lindberg-Feller Central Limit Theorem (unequal variances).* Suppose $\{x_1, ..., x_n\}$ is a set of random variables with finite means μ_j and finite variance σ_j^2. Let $\bar{\mu} = n^{-1}\sum_{t=1}^{n}\mu_t$ and $\bar{\sigma}_n^2 = n^{-1}\left(\sigma_1^2 + \sigma_2^2 + ...\right)$. If no single term dominates the average variance ($\lim_{n \to \infty}\frac{\max(\sigma_j)}{n\bar{\sigma}_n} = 0$), if the average variance converges to a finite constant ($\lim_{n \to \infty}\bar{\sigma}_n^2 = \bar{\sigma}^2$), and $\bar{x} = n^{-1}\sum_{t=1}^{n}x_t$, then $\sqrt{n}(\bar{x} - \bar{\mu}) \xrightarrow{d} N(0, \bar{\sigma}^2)$.

Multivariate versions apply to both; the multivariate version of the Lindberg-Levy CLT follows.

Theorem A.16 *Lindberg-Levy Central Limit Theorem (multivariate).*
If $X_1, ..., X_n$ are a random sample from multivariate probability distribution with finite mean vector μ and finite covariance matrix Q, and $\bar{x} = n^{-1}\sum_{t=1}^{n}x_t$, then $\sqrt{n}(\bar{X} - \mu) \xrightarrow{d} N(0, Q)$.

Delta method.
The "Delta method" is used to justify usage of linear Taylor series approximation to analyze distributions and moments of a function of random variables. It combines Theorem A.9 Slutsky's probability limit, Theorem A.13 limiting distribution, and the Central Limit Theorems A.14-A.16.

Theorem A.17 *Limiting normal distribution of a function.*
If $\sqrt{n}(z_n - \mu) \xrightarrow{d} N(0, \sigma^2)$ and if $g(z_n)$ is a continuous function not involving n, then
$$\sqrt{n}(g(z_n) - g(\mu)) \xrightarrow{d} N\left(0, \{g'(\mu)\}^2 \sigma^2\right)$$

A key insight for the Delta method is the mean and variance of the limiting distribution are the mean and variance of a *linear* approximation evaluated at μ, $g(z_n) \approx g(\mu) + g\prime(\mu)(z_n - \mu)$.

Theorem A.18 *Limiting normal distribution of a set of functions (multivariate).*
If z_n is a $K \times 1$ sequence of vector-valued random variables such that

$$\sqrt{n}(z_n - \mu_n) \xrightarrow{d} N(0, \Sigma)$$

and if $c(z_n)$ is a set of J continuous functions of z_n not involving n, then

$$\sqrt{n}(c(z_n) - c(\mu_n)) \xrightarrow{d} N(0, C\Sigma C^T)$$

where C is a $J \times K$ matrix with jth row a vector of partial derivatives of jth function with respect to z_n, $\frac{\partial c(z_n)}{\partial z_n^T}$.

Definition A.7 *Asymptotic distribution.*
An asymptotic distribution is a distribution used to approximate the true finite sample distribution of a random variable.

Example A.3 *If $\sqrt{n}\left[\frac{x_n - \mu}{\sigma}\right] \xrightarrow{d} N(0,1)$, then approximately or asymptotically $\bar{x}_n \sim N\left(\mu, \frac{\sigma^2}{n}\right)$. This is written $\bar{x}_n \xrightarrow{d} N\left(\mu, \frac{\sigma^2}{n}\right)$.*

Definition A.8 *Asymptotic normality and asymptotic efficiency.*
An estimator $\widehat{\theta}$ is asymptotically normal if $\sqrt{n}\left(\widehat{\theta} - \theta\right) \xrightarrow{d} N(0, V)$. An estimator is asymptotically efficient if the covariance matrix of any other consistent, asymptotically normally distributed estimator exceeds $n^{-1}V$ by a nonnegative definite matrix.

Example A.4 *Asymptotic inefficiency of median in normal sampling.*
In sampling from a normal distribution with mean μ and variance σ^2, both the sample mean $\bar{x}$ and median M are consistent estimators of μ. Their asymptotic properties are $\bar{x}_n \xrightarrow{a} N\left(\mu, \frac{\sigma^2}{n}\right)$ and $M \xrightarrow{a} N\left(\mu, \frac{\pi}{2}\frac{\sigma^2}{n}\right)$. Hence, the sample mean is a more efficient estimator for the mean than the median by a factor of $\pi/2 \approx 1.57$.

This result for the median follows from the next theorem (see Mood, Graybill, and Boes [1974], p. 257).

Theorem A.19 *Asymptotic distribution of order statistics.*
Let $x_1, ..., x_n$ be iid random variables with density f and cumulative distribution function F. F is strictly monotone. Let ξ_p be a unique solution in x of $F(x) = p$ for some $0 < p < 1$ (ξ_p is the pth quantile). Let p_n be such that np_n is an integer and $n|p_n - p|$ is bounded. Let $y_{np_n}^{(n)}$ denote (np)th order statistic for a random sample of size n. Then $y_{np_n}^{(n)}$ is asymptotically distributed as a normal distribution with mean ξ_p and variance $\frac{p(1-p)}{n[f(\xi_p)]^2}$.

A.2 Convergence in distribution (central limit theorems)

Example A.5 *Sample median.*
Let $p = \frac{1}{2}$ (implies ξ_p = sample median). The sample median

$$M \xrightarrow{a} N\left(\xi_p, \frac{1}{4n\left[f\left(1/2\right)\right]^2}\right)$$

Since $\xi_{\frac{1}{2}} = \mu$, $f\left(\xi_{\frac{1}{2}}\right)^2 = \left(2\pi\sigma^2\right)^{-1}$ and the variance is $\frac{\frac{1}{2}\left(\frac{1}{2}\right)}{nf\left(\xi_{\frac{1}{2}}\right)^2} = \frac{\pi}{2}\frac{\sigma^2}{n}$, the result asserted above.

Theorem A.20 *Asymptotic distribution of nonlinear function.*
If $\widehat{\theta}$ is a vector of estimates such that $\widehat{\theta} \xrightarrow{a} N\left(\theta, n^{-1}V\right)$ and if $c(\theta)$ is a set of J continuous functions not involving n, then

$$c\left(\widehat{\theta}\right) \xrightarrow{a} N\left(c\left(\theta\right), n^{-1}C\left(\theta\right)VC\left(\theta\right)^T\right)$$

where $C(\theta) = \frac{\partial c(\theta)}{\partial \theta^T}$.

Example A.6 *Asymptotic distribution of a function of two estimates.*
Suppose b and t are estimates of β and θ such that

$$\begin{bmatrix} b \\ t \end{bmatrix} \xrightarrow{a} N\left(\begin{bmatrix} \beta \\ \theta \end{bmatrix}, \begin{bmatrix} \sigma_{\beta\beta} & \sigma_{\beta\theta} \\ \sigma_{\theta\beta} & \sigma_{\theta\theta} \end{bmatrix}\right)$$

We wish to find the asymptotic distribution for $c = \frac{b}{1-t}$. Let $\gamma = \frac{\beta}{1-\theta}$ – the true parameter of interest. By the Slutsky Theorem and consistency of the sample mean, c is consistent for γ. Let $\gamma_\beta = \frac{\partial \gamma}{\partial \beta} = \frac{1}{1-\theta}$ and $\gamma_\theta = \frac{\partial \gamma}{\partial \theta} = \frac{\beta}{(1-\theta)^2}$. The asymptotic variance is

$$Asy.Var\left[c\right] = \begin{bmatrix} \gamma_\beta & \gamma_\theta \end{bmatrix} \Sigma \begin{bmatrix} \gamma_\beta \\ \gamma_\theta \end{bmatrix} = \gamma_\beta \sigma_{\beta\beta} + \gamma_\theta \sigma_{\theta\theta} + 2\gamma_\beta \gamma_\theta \sigma_{\beta\theta}$$

Notice this is simply the variance of a linear approximation $\widehat{\gamma} \approx \gamma + \gamma_\beta \left(b - \beta\right) + \gamma_\theta \left(t - \theta\right)$.

Theorem A.21 *Asymptotic normality of MLE Theorem*
MLE, $\widehat{\theta}$, for strongly asymptotically identified model represented by log-likelihood function $\ell(\theta)$, when it exists and is consistent for θ, is asymptotically normal if
(i) contributions to log-likelihood $\ell_t(y, \theta)$ are at least twice continuously differentiable in θ for almost all y and all θ,
(ii) component sequences $\{D^2_{\theta\theta}\ell_t(y, \theta)\}_{t=1,\infty}$ satisfy WULLN (weak uniform law) on θ,
(iii) component sequences $\{D_\theta \ell_t(y, \theta)\}_{t=1,\infty}$ satisfy CLT.

A.3 Rates of convergence

Definition A.9 *Order $1/n$ (big-O notation).*
If f and g are two real-valued functions of positive integer variable n, then the notation $f(n) = O(g(n))$ *(optionally as $n \to \infty$)* means there exists a constant $k > 0$ (independent of n) and a positive integer N such that $\left|\frac{f(n)}{g(n)}\right| < k$ for all $n < N$. ($f(n)$ is of same order as $g(n)$ asymptotically).

Definition A.10 *Order less than $1/n$ (little-o notation).*
If f and g are two real-valued functions of positive integer variable n, then the notation $f(n) = o(g(n))$ means the $\lim_{n \to \infty} \frac{f(n)}{g(n)} = 0$ ($f(n)$ is of smaller order than $g(n)$ asymptotically).

Definition A.11 *Asymptotic equality.*
If f and g are two real-valued functions of positive integer variable n such that $\lim_{n \to \infty} \frac{f(n)}{g(n)} = 1$, then $f(n)$ and $g(n)$ are asymptotically equal. This is written $f(n) \stackrel{a}{=} g(n)$.

Definition A.12 *Stochastic order relations.*
If $\{a_n\}$ is a sequence of random variables and $g(n)$ is a real-valued function of positive integer argument n, then
(1) $a_n = o_p(g(n))$ means $\lim_{n \to \infty} \frac{a_n}{g(n)} = 0$,
(2) similarly, $a_n = O_p(g(n))$ means there is a constant k such that (for all $\varepsilon > 0$) there is a positive integer N such that $\Pr\left(\left|\frac{a_n}{g(n)}\right| > k\right) < \varepsilon$ for all $n > N$, and
(3) If $\{b_n\}$ is a sequence of random variables, then the notation $a_n \stackrel{a}{=} b_n$ means $\lim_{n \to \infty} \frac{a_n}{b_n} = 1$.

Comparable definitions apply to almost sure convergence and convergence in distribution (though these are infrequently used).

Theorem A.22 *Order rules:*

$$O(n^p) \pm O(n^q) = O\left(n^{max(p,q)}\right)$$

$$o(n^p) \pm o(n^q) = o\left(n^{max(p,q)}\right)$$

$$\begin{aligned} O(n^p) \pm o(n^q) &= O\left(n^{max(p,q)}\right) && \text{if } p \geq q \\ &= o\left(n^{max(p,q)}\right) && \text{if } p < q \end{aligned}$$

$$O(n^p) O(n^q) = O(n^{p+q})$$

$$o(n^p) o(n^q) = o(n^{p+q})$$

$$O(n^p) o(n^q) = o(n^{p+q})$$

Example A.7 *Square-root n convergence.*

(1) If each $x = O(1)$ has mean μ and the central limit theorem applies $\sum_{t=1}^{n} x_t = O(n)$ and $\sum_{t=1}^{n}(x_t - \mu) = O(\sqrt{n})$.

(2) Let $\Pr(y_t = 1) = 1/2$, $\Pr(y_t = 0) = 1/2$, $z_t = y_t - 1/2$, and $b_n = \sqrt{n}\sum_{t=1}^{n} z_t$. $Var[b_n] = n^{-1}Var[z_t] = n^{-1}(1/4)$. $\sqrt{n}b_n = n^{-\frac{1}{2}}\sum_{t=1}^{n} z_t$.

$$E\left[\sqrt{n}b_n\right] = 0$$

and

$$Var\left[\sqrt{n}b_n\right] = 1/4$$

Thus, $\sqrt{n}b_n = O(1)$ which implies $b_n = O\left(n^{-\frac{1}{2}}\right)$.

These examples represent common econometric results. That is, the average of n centered quantities is $O\left(n^{-\frac{1}{2}}\right)$, and is referred to as square-root n convergence.

A.4 Additional reading

Numerous books and papers including Davidson and MacKinnon [1993, 2003], Greene [1997], and White [1984] provide in depth review of asymptotic theory. Hall and Heyde [1980] reviews limit theory (including laws of large numbers and central limit theorems) for martingales.

Bibliography

[1] Abadie, J. Angrist, and G. Imbens. 1998. "Instrumental variable estimation of quantile treatment effects," *National Bureau of Economic Research* no. 229.

[2] Abadie, J. 2000. "Semiparametric estimation of instrumental variable models for causal effects," *National Bureau of Economic Research* no. 260.

[3] Abadie, J., and G. Imbens. 2006. "Large sample properties of matching estimators for average treatment effects," *Econometrica* 74(1). 235-267.

[4] Abbring, J. and J. Heckman. 2007. "Econometric evaluation of social programs, part III: Distributional treatment effects, dynamic treatment effects, dynamic discrete choice, and general equilibrium policy evaluation," *Handbook of Econometrics* Volume 6B. J. Heckman and E. Leamer, eds. 5145-5306.

[5] Admati, A. 1985. "A rational expectations equilibrium for multi-asset securities markets," *Econometrica* 53 (3). 629-658.

[6] Ahn, H. and J. Powell. 1993. "Semiparametric estimation of censored selection models with a nonparametric selection mechanism," *Journal of Econometrics* 58. 3-29.

[7] Aitken, A. 1935. "On least squares and linear combinations of observations," *Proceedings of the Royal Statistical Society* 55. 42-48.

[8] Albert, J. and S. Chib. 1993. "Bayesian analysis of binary and polychotomous response data," *Journal of the American Statistical Association* 88 (422). 669-679.

[9] Amemiya, T. 1978. "The estimation of a simultaneous equation generalized probit model," *Econometrica* 46 (5). 1193-1205.

[10] Amemiya, T.1985. *Advanced Econometrics*. Cambridge, MA: Harvard University Press.

[11] Andrews, D. and M. Schafgans. 1998. "Semiparametric estimation of the intercept of a sample selection model," *The Review of Economic Studies* 65 (3). 497-517.

[12] Angrist, J., G. Imbens, and D. Rubin. 1996. "Identification of causal effects using instrumental variables," *Journal of the American Statistical Association* 91 (434). 444-485.

[13] Angrist, J. and A. Krueger. 1998. "Empirical strategies in labor economics," Princeton University working paper (prepared for the *Handbook of Labor Economics*, 1999).

[14] Angrist, J. and V. Lavy. 1999. "Using Maimonides' rule to estimate the effect of class size on scholastic achievement," *The Quarterly Journal of Economics* 114 (2). 533-575.

[15] Angrist, J. 2001. "Estimation of limited dependent variable models with dummy endogenous regressors: Simple strategies for empirical practice," *Journal of Business & Economic Statistics* 190(1). 2-16.

[16] Angrist, J. and J. Pischke. 2009. *Mostly Harmless Econometrics*. Princeton, N. J.: Princeton University Press.

[17] Antle, R., J. Demski, and S. Ryan. 1994. "Multiple sources of information, valuation, and accounting earnings," *Journal of Accounting, Auditing & Finance* 9 (4). 675-696.

[18] Arabmazar, A. and P. Schmidt. 1982. "An investigation of the robustness of the Tobit estimator to nonnormality," *Econometrica* 50 (4). 1055-1063.

[19] Arya, A., J. Glover, and S. Sunder. 1998. "Earnings management and the revelation principle," *Review of Accounting Studies* 3.7-34.

[20] Arya, A., J. Fellingham, and D. Schroeder. 2000. "Accounting information, aggregation, and discriminant analysis," *Management Science* 46 (6). 790-806.

[21] Arya, A., J. Fellingham, J. Glover, D. Schroeder, and G. Strang. 2000. "Inferring transactions from financial statements," *Contemporary Accounting Research* 17 (3). 365-385.

[22] Arya, A., J. Fellingham, J. Glover, and D. Schroeder. 2002. "Depreciation in a model of probabilistic investment," *The European Accounting Review* 11 (4). 681-698.

[23] Arya, A., J. Fellingham, and D. Schroeder. 2004. "Aggregation and measurement errors in performance evaluation," *Journal of Management Accounting Research* 16. 93-105.

[24] Arya, A., J. Fellingham, B. Mittendorf, and D. Schroeder. 2004. "Reconciling financial information at varied levels of aggregation," *Contemporary Accounting Research* 21 (2). 303-324.

[25] Bagnoli, M., H. Liu, and S. Watts. 2006. "Family firms, debtholder-shareholder agency costs and the use of covenants in private debt," Purdue University working paper, forthcoming in *Annals of Finance*.

[26] Ben-Akiva, M. and B. Francois 1983. "Mu-homogenous generalized extreme value model," working paper, Department of Civil Engineering, MIT.

[27] Bernardo, J. and A. Smith. 1994. *Bayesian Theory*. New York, NY: John Wiley & Sons.

[28] Berndt, E., B. Hall, R. Hall, and J. Hausman. 1974. "Estimation and inference in nonlinear structural models," *Annals of Economic and Social Measurement* 3/4. 653-665.

[29] Berry, S. 1992. "Estimation of a model of entry in the airline industry," *Econometrica* 60 (4). 889-917.

[30] Besag, J. 1974. "Spatial interaction and the statistical analysis of lattice systems," *Journal of the Royal Statistical Society, Series B* 36 (2). 192-236.

[31] Bhat, C. 2001. "Quasi-random maximum simulated likelihood estimation of the mixed multinomial logit model', *Transportation Research B: Methodological* 35 (7). 677–693.

[32] Bhat, C. 2003. "Simulation estimation of mixed discrete choice models using randomized and scrambled Halton sequences," *Transportation Research B: Methodological* 37 (9). 837-855.

[33] Bjorklund, A. and R. Moffitt. 1987. "The estimation of wage gains and welfare gains in self-selection models," *The Review of Economics and Statistics* 69 (1). 42-49.

[34] Blackwell, D. 1953. "Equivalent comparisons of experiments," *The Annals of Mathematical Statistics* 24 (2). 265-272.

[35] Blackwell, D. and M. Girshick. 1954. *Theory of Games and Statistical Decision*. New York, NY: Dover Publications, Inc.

[36] Blower, D. 2004. "An easy derivation of logistic regression from the Bayesian and maximum entropy perspective," *BAYESIAN INFERENCE AND MAXIMUM ENTROPY METHODS IN SCIENCE AND ENGINEERING. 23rd International Workshop on Bayesian inference and maximum entropy methods in science and engineering. AIP Conference Proceedings volume 707.* 30-43.

[37] Bound, J., C. Brown, and N. Mathiowetz. 2001. "Measurement error in survey data," *Handbook of Econometrics Volume 5.* J. Heckman and E. Leamer, eds. 3705-3843.

[38] Box. G. and G. Jenkins. 1976. *Time Series Analysis: Forecasting and Control.* San Francisco, CA: Holden-Day, Inc.

[39] Box. G. and G. Tiao. 1973. *Bayesian Inference in Statistical Analysis.* Reading, MA: Addison-Wesley Publishing Co.

[40] Bresnahan, R. and P. Reiss. 1990. "Entry into monopoly markets," *Review of Economic Studies* 57 (4). 531-553.

[41] Bresnahan, R. and P. Reiss. 1991. "Econometric models of discrete games," *Journal of Econometrics* 48 (1/2). 57-81.

[42] Burgstahler, D. and I. Dichev. 1997. "Earnings management to avoid earnings decreases and losses," *Journal of Accounting and Economics* 24. 99-126.

[43] Campbell, D and J. Stanley. 1963. *Experimental And Quasi-experimental Designs For Research.* Boston, MA: Houghton Mifflin Company.

[44] Cameron, A. C. and P. Trivedi. 2005. *Microeconometrics: Methods and Applications.* New York, NY: Cambridge University Press.

[45] Campolieti, M. 2001. "Bayesian semiparametric estimation of discrete duration models: An application of the Dirichlet process prior," *Journal of Applied Econometrics* 16 (1). 1-22.

[46] Card, D. 2001. "Estimating the returns to schooling: Progress on some persistent econometric problems," *Econometrica* 69 (5), 1127-1160.

[47] Carneiro, P., K. Hansen, and J. Heckman. 2003. "Estimating distributions of treatment effects with an application to the returns to schooling and measurement of the effects of uncertainty of college choice," *International Economic Review* 44 (2). 361-422.

[48] Casella, G. and E. George. 1992. "Explaining the Gibbs sampler," *The American Statistician* 46 (3). 167-174.

[49] Chamberlain, G. 1980. "Analysis of covariance with qualitative data," *Review of Economic Studies* 47 (1). 225-238.

[50] Chamberlain, G. 1982. "Multivariate regression models for panel data," *Journal of Econometrics* 18. 5-46.

[51] Chamberlain, G. 1984. "Panel Data," *Handbook of Econometrics* Volume 2. Z. Griliches and M. Intriligator, eds. 1247-1320.

[52] Chenhall, R. and R. Moers. 2007a. "The issue of endogeneity within theory-based, quantitative management accounting research," *The European Accounting Review* 16 (1). 173–195.

[53] Chenhall, R. and F. Moers. 2007b. "Endogeneity: A reply to two different perspectives," *The European Accounting Review* 16 (1). 217-221.

[54] Chib, S. and E. Greenberg. 1995 "Understanding the Metropolis-Hastings algorithm," *The American Statistician* 49 (4). 327-335.

[55] Chib, S. and B. Hamilton. 2000. "Bayesian analysis of cross-section and clustered data treatment models," *Journal of Econometrics* 97. 25-50.

[56] Chib, S. and B. Hamilton. 2002. "Semiparametric Bayes analysis of longitudinal data treatment models," *Journal of Econometrics* 110. 67 – 89.

[57] Christensen, J. and J. Demski. 2003. *Accounting Theory: An Information Content Perspective*. Boston, MA: McGraw-Hill Irwin.

[58] Christensen, J. and J. Demski. 2007. "Anticipatory reporting standards," *Accounting Horizons* 21 (4). 351-370.

[59] Cochran, W. 1965. "The planning of observational studies of human populations," *Journal of the Royal Statistical Society, Series A (General)*. 128 (2). 234-266.

[60] Coslett, S. 1981. "Efficient estimation of discrete choice models," *Structural Analysis of Discrete Choice Data with Econometric Applications*. C. Manski and D. McFadden, eds. Cambridge, MA: The MIT Press.

[61] Coslett, S. 1983. "Distribution-free maximum likelihood estimator of the binary choice model," *Econometrica* 51 (3). 765-782.

[62] Cover, T. and J. Thomas. 1991. *Elements of Information Theory*. New York, NY: John Wiley & Sons, Inc.

[63] Cox, D. 1972. "Regression models and life-tables," *Journal of the Royal Statistical Society, Series B (Methodological)* 34 (2). 187-200.

[64] Craven, P. and G. Wahba. 1979. "Smoothing noisy data with spline functions," *Numerische Mathematik 31* (4). 377-403.

[65] Cox, D. 1958. *Planning of Experiments*. New York, NY: Wiley.

[66] Davidson, R. and J. MacKinnon. 1993. *Estimation and Inference in Econometrics*. New York, NY: Oxford University Press.

[67] Davidson, R. and J. MacKinnon. 2003. *Econometric Theory and Methods*. New York, NY: Oxford University Press.

[68] Dawid, A. P. 2000. "Causal inference without counterfactuals," *Journal of the American Statistical Association* 95 (450). 407-424.

[69] Demski, J. 1973. "The general impossibility of normative accounting standards," *The Accounting Review* 48 (4). 718-723.

[70] Demski, J. 1994. *Managerial Uses of Accounting Information*. Boston, MA: Kluwer Academic Publishers.

[71] Demski, J. 1998. "Performance measure manipulation," *Contemporary Accounting Research* 15 (3). 261-285.

[72] Demski, J. and D. Sappington. 1999. "Summarization with errors: A perspective on empirical investigations of agency relationships," *Management Accounting Research* 10. 21-37.

[73] Demski, J. 2004. "Endogenous expectations," *The Accounting Review* 79 (2). 519-539.

[74] Demski, J. 2008. *Managerial Uses of Accounting Information*. revised edition. New York, NY: Springer.

[75] Demski, J., D. Sappington, and H. Lin. 2008. "Asset revaluation regulations with multiple information sources," *The Accounting Review* 83 (4). 869-891.

[76] Demski, J., J. Fellingham, H. Lin, and D. Schroeder. 2008. "Interaction between productivity and measurement," *Journal of Management Accounting Research* 20. 169-190.

[77] Draper, D., J. Hodges, C. Mallows, and D. Pregibon. 1993. "Exchangeability and data analysis," *Journal of the Royal Statistical Society* series A 156 (part 1). 9-37.

[78] Dubin, J. and D. Rivers. 1989. "Selection bias in linear regression, logit and probit models," *Sociological Methods and Research* 18 (2,3). 361-390.

[79] Duncan, G. 1983. "Sample selectivity as a proxy variable problem: On the use and misuse of Gaussian selectivity corrections," *Research in Labor Economics*, Supplement 2. 333-345.

[80] Dye, R. 1985. *"Disclosure of nonproprietary information," Journal of Accounting Research* 23 (1). 123-145.

[81] Dye, R. and S. Sridar. 2004. "Reliability-relevance trade-offs and the efficiency of aggregation," *Journal of Accounting Research* 42 (1). 51-88.

[82] Dye, R. and S. Sridar. 2007. "The allocational effects of reporting the precision of accounting estimates," *Journal of Accounting Research* 45 (4). 731-769.

[83] Ebbes, P. 2004. "Latent instrumental variables – A new approach to solve for endogeneity," Ph.D. dissertation. University of Michigan.

[84] Efron, B. 1979. "Bootstrapping methods: Another look at the jackknife," *The Annals of Statistics* 7 (1). 1-26.

[85] Efron, B. 2000. "The Bootstrap and modern statistics," *Journal of the American Statistical Association* 95 (452). 1293-1296.

[86] Ekeland, I., J. Heckman, and L. Nesheim. 2002. "Identifying hedonic models," *American Economic Review* 92 (4). 304-309.

[87] Ekeland, I., J. Heckman, and L. Nesheim. 2003. "Identification and estimation of hedonic models," *IZA* discussion paper 853.

[88] Evans, W. and R. Schwab. 1995. "Finishing high school and starting college: Do Catholic schools make a difference," *The Quarterly Journal of Economics* 110 (4). 941-974.

[89] Ferguson, T. 1973. "A Bayesian analysis of some nonparametric problems," *The Annals of Statistics* 1 (2). 209-230.

[90] Fisher, R. 1966. *The Design of Experiments*. New York, NY: Hafner Publishing.

[91] Florens, J., J. Heckman, C. Meghir, and E. Vytlacil. 2003. "Instrumental variables, local instrumental variables and control functions," working paper Institut d'Économie Industrielle (IDEI), Toulouse.

[92] Florens, J., J. Heckman, C. Meghir, and E. Vytlacil. 2008. "Identification of treatment effects using control functions in models with continuous endogenous treatment and heterogeneous effects," *National Bureau of Economic Research* no. 14002.

[93] Freedman. D. 1981. "Bootstrapping regression models," *The Annals of Statistics* 9 (6). 1218-1228.

[94] Freedman. D. and S. Peters. 1984. "Bootstrapping a regression equation: Some empirical results," *Journal of the American Statistical Association* 79 (385). 97-106.

[95] Frisch, R. and F. Waugh. 1933. "Partial time regressions as compared with individual trends," *Econometrica* 1 (4). 387-401.

[96] Galton, F. 1886. "Regression towards mediocrity in hereditary stature," *Journal of the Anthropological Institute* 15. 246-263.

[97] Gauss, K. 1809. *Theoria Motus Corporum Celestium*. Hamburg: Perthes; English translation, *Theory of the Motion of the Heavenly Bodies About the Sun in Conic Sections*. New York, NY: Dover Publications, Inc. 1963.

[98] Gelfand, A. and A. Smith. 1990. "Sampling-based approaches to calculating marginal densities," *Journal of the American Statistical Association* 85 (410). 398-409.

[99] Gelman, A., J. Carlin, H. Stern, and D. Rubin. 2003. *Bayesian Data Analysis*. Boca Raton, FL: Chapman and Hall/CRC.

[100] Godfrey, and Wickens 1982. "A simple derivation of the limited information maximum likelihood estimator," *Economics Letters* 10. 277-283.

[101] Goldberger, A. 1972. "Structural equation methods in the social sciences," *Econometrica* 40 (6). 979-1001.

[102] Goldberger, A. 1983. "Abnormal selection bias," in Karlin, S., T. Amemiya, and L. Goodman (editors). *Studies in Econometrics, Time Series and Multivariate Statistics*. New York: Academic Press, Inc. 67-84.

[103] Graybill, F. 1976. *Theory and Application of the Linear Model*. North Scituate, MA: Duxbury Press.

[104] Greene, W. 1997. *Econometric Analysis*. Upper Saddle River, NJ: Prentice-Hall.

[105] Griliches, Z. 1986. "Economic data issues," *Handbook of Econometrics* Volume 3. Z. Griliches and M. Intriligator, eds. 1465-1514.

[106] Gronau, R. 1974. "Wage comparisons - a selectivity bias," *Journal of Political Economy* 82 (6). 1119-1143.

[107] Hall, P. and C. Heyde. 1980. *Martingale Limit Theory and Its Application*. New York, NY: Academic Press.

[108] Hammersley, J. and P. Clifford. 1971. "Markov fields on finite graphs and lattices," unpublished working paper, Oxford University.

[109] Hardle, W. 1990. *Applied Nonparametric Regression*. Cambridge, U.K.: Cambridge University Press.

[110] Hausman, J. 1978. "Specification tests in econometrics," *Econometrica* 46 (6). 1251-1271.

[111] Hausman, J. 2001. "Mismeasured variables in econometric analysis: Problems from the right and problems from the left," *Journal of Economic Perspectives* 15 (4). 57-67.

[112] Heckman, J. 1974. "Shadow prices, market wages and labor supply," *Econometrica* 42 (4). 679-694.

[113] Heckman, J. 1976. "The common structure of statistical models of truncation, sample selection and limited dependent variables and a simple estimator for such models," *The Annals of Economic and Social Measurement* 5 (4). 475-492.

[114] Heckman, J. 1978. "Dummy endogenous variables in a simultaneous equation system," *Econometrica* 46 (4). 931-959.

[115] Heckman, J. 1979. "Sample selection bias as a specification error," *Econometrica* 47 (1). 153-162.

[116] Heckman, J. and B. Singer. 1985. "Social science duration analysis," *Longitudinal Analysis of Labor Market Data*. J. Heckman and B. Singer, eds. 39-110; also Heckman, J. and B. Singer. 1986. "Econometric analysis of longitudinal data," *Handbook of Econometrics* Volume 3. Z. Griliches and M. Intriligator, eds. 1689-1763.

[117] Heckman, J. and R. Robb. 1986. "Alternative methods for solving the problem of selection bias in evaluating the impact of treatments on outcomes," *Drawing Inferences from Self-Selected Samples*. H. Wainer, ed. New York, NY: Springer-Verlag.

[118] Heckman, J. and B. Honore. 1990. "The empirical content of the Roy model," *Econometrica* 58 (5). 1121-1149.

[119] Heckman, J. and J. Smith. 1995. "Assessing the case for social experiments," *Journal of Economic Perspectives* 9 (2). 85-110.

[120] Heckman, J. 1996. "Randomization as an instrumental variable," *The Review of Economics and Statistics* 78 (1). 336-341.

[121] Heckman, J. H. Ichimura, and P. Todd. 1997. "Matching as an econometric evaluation estimator: Evidence from evaluating a job training program," *Review of Economic Studies* 64 (4). 605-654.

[122] Heckman, J. 1997. "Instrumental variables: A study of implicit behavioral assumptions used in making program evaluations," *The Journal of Human Resources* 32 (3). 441-462.

[123] Heckman, J. H. Ichimura, and P. Todd. 1998. "Matching as an econometric evaluation estimator," *Review of Economic Studies* 65 (2). 261-294.

[124] Heckman, J. H. Ichimura, J. Smith, and P. Todd. 1998. "Characterizing selection bias using experimental data," *Econometrica* 66 (5). 1017-1098.

[125] Heckman, J. and E. Vytlacil. 1998. "Instrumental variables methods for the correlated random coefficient model," *The Journal of Human Resources* 33 (4). 974-987.

[126] Heckman, J., L. Lochner, and C. Taber. 1998a. "Explaining rising wage inequality: Explorations with a dynamic general equilibrium model of labor earnings with heterogeneous agents," *Review of Economic Dynamics* 1 (1). 1-58.

[127] Heckman, J., L. Lochner, and C. Taber. 1998b. "Tax policy and human-capital formation," *The American Economic Review* 88 (2). 293-297.

[128] Heckman, J., L. Lochner, and C. Taber. 1998c. "General equilibrium treatment effects: A study of tuition policy," *The American Economic Review* 88 (2). 381-386.

[129] Heckman, J., R. LaLonde, and J. Smith. 1999. "The economics and econometrics of active labor market programs," *Handbook of Labor Economics* Volume 3. A. Ashenfelter and D. Card., eds. 1865-2097.

[130] Heckman, J. 2000. "Causal parameters and policy analysis in economics: A twentieth century retrospective," *The Quarterly Journal of Economics* 115 (1). 45-97.

[131] Heckman J. and E. Vytlacil. 2000. "The relationship between treatment parameters within a latent variable framework," *Economics Letters* 66. 33-39.

[132] Heckman, J. 2001. "Micro data, heterogeneity, and the evaluation of public policy: Nobel lecture," *Journal of Political Economy* 109 (4). 673-747.

[133] Heckman J. and E. Vytlacil. 2001. "Policy-relevant treatment effects," *The American Economic Review* 91 (2). 107-111.

[134] Heckman, J., R. Matzkin, and L. Nesheim. 2003a. "Simulation and estimation of hedonic models," *IZA* discussion paper 843.

[135] Heckman, J., R. Matzkin, and L. Nesheim. 2003b. "Simulation and estimation of nonadditive hedonic models," *National Bureau of Economic Research* working paper 9895.

[136] Heckman, J. and S. Navarro-Lozano. 2004. "Using matching, instrumental variables, and control functions to estimate economic choice models," *Review of Economics and Statistics* 86 (1). 30-57.

[137] Heckman, J., R. Matzkin, and L. Nesheim. 2005. "Nonparametric estimation of nonadditive hedonic models," University College London working paper.

[138] Heckman, J. and E. Vytlacil. 2006. "Structural equations, treatment effects and econometric policy evaluation," *Econometrica* 73 (3). 669-738.

[139] Heckman J., S. Urzua, and E. Vytlacil. 2006. "Understanding instrumental variables in models with essential heterogeneity," *The Review of Economics and Statistics* 88 (3). 389-432.

[140] Heckman, J. and E. Vytlacil. 2007a. "Econometric evaluation of social programs, part I: Causal models, structural models and econometric policy evaluation," *Handbook of Econometrics* Volume 6B. J. Heckman and E. Leamer, eds. 4779-4874.

[141] Heckman, J. and E. Vytlacil. 2007b. "Econometric evaluation of social programs, part II: Using the marginal treatment effect to organize alternative econometric estimators to evaluate social programs, and to forecast their effects in new environments," *Handbook of Econometrics* Volume 6B. J. Heckman and E. Leamer, eds. 4875-5144.

[142] Hildreth, C. and J. Houck. 1968. "Estimators for a linear model with random coefficients," *Journal of the American Statistical Association* 63 (322). 584-595.

[143] Hoeffding, W. 1948. "A class of statistics with asymptotically Normal distribution," *Annals of Mathematical Statistics* 19 (3). 293-325.

[144] Holmstorm, B. and P. Milgrom. 1987. "Aggregation and linearity in the provision of intertemporal incentives," *Econometrica* 55, 303-328.

[145] Horowitz, J. 1991. "Reconsidering the multinomial probit model," *Transportation Research B* 25. 433–438.

[146] Horowitz, J. and W. Hardle. 1996. "Direct semiparametric estimation of single-index models with discrete covariates," *Journal of the American Statistical Association* 91 (436). 1632-1640.

[147] Horowitz, J. 1999. "Semiparametric estimation of a proportional hazard model with unobserved heterogeneity," *Econometrica* 67 (5). 1001-1028.

[148] Horowitz, J. 2001. "The bootstrap," *Handbook of Econometrics* Volume 5. J. Heckman and E. Leamer, eds. 3159-3228.

[149] Hurwicz, L. 1962. "On the structural form of interdependent systems," *Logic, Methodology and Philosophy of Science*, Proceedings of the 1960 International Congress. E. Nagel, P. Suppes, and A. Tarski, eds. Stanford, CA: Stanford University Press.

[150] Ichimura, H. and P. Todd. 2007. "Implementing nonparametric and semiparametric estimators," *Handbook of Econometrics* Volume 6B. J. Heckman and E. Leamer, eds. 5369-5468.

[151] Imbens, G. and J. Angrist. 1994. "Identification and estimation of local average treatment effects," *Econometrica* 62 (2). 467-475.

[152] Imbens, G. and D. Rubin. 1997. "Estimating outcome distributions for compliers in instrumental variables models," *Review of Economic Studies* 64 (4). 555-574.

[153] Imbens, G. 2003. "Sensitivity to exogeneity assumptions in program evaluation," *American Economic Review* 93 (2). 126-132.

[154] Jaynes, E. T. 2003. *Probability Theory: The Logic of Science*. New York, NY: Cambridge University Press.

[155] Kiefer, N. 1980. "A note on switching regressions and logistic discrimination," *Econometrica* 48 (4). 1065-1069.

[156] Kingman, J. 1978. "Uses of exchangeability," *The Annals of Probability* 6 (2). 183-197.

[157] Koenker, R. and G. Bassett. 1978. "Regression quantiles," *Econometrica* 46 (1). 33-50.

[158] Koenker, R. 2005. *Quantile Regression*. New York, NY: Cambridge University Press.

[159] Koenker, R. 2009. "Quantile regression in R: A vignette," http://cran.r-project.org/web/packages/quantreg/index.html.

[160] Koop, G. and D. Poirier. 1997. "Learning about the across-regime correlation in switching regression models," *Journal of Econometrics* 78. 217-227.

[161] Koop, G., D. Poirier, and J. Tobias. 2007. *Bayesian Econometric Methods*. New York, NY: Cambridge University Press.

[162] Kreps, D. 1988. *Notes on the Theory of Choice*. Boulder, CO: Westview Press.

[163] LaLonde, R. 1986. "Evaluating the econometric evaluations of training programs with experimental data," *The American Economic Review* 76 (4). 604-620.

[164] Lambert, R., C. Leuz, and R. Verrecchia. 2007. "Accounting information, disclosure, and the cost of capital," *Journal of Accounting Research* 45 (2). 385-420.

[165] Larcker, D. and , T. Rusticus. 2004. "On the use of instrumental variables in accounting research," working paper, University of Pennsylvania, forthcoming in *Journal of Accounting and Economics*.

[166] Larcker, D. and , T. Rusticus. 2007. "Endogeneity and empirical accounting research," *The European Accounting Review* 16 (1). 207-215.

[167] Lee, C. 1981. "1981, Simultaneous equation models with discrete and censored dependent variables," *Structural Analysis of Discrete Choice Data with Econometric Applications*. C. Manski and D. McFadden, eds. Cambridge, MA: The MIT Press.

[168] Lewbel, A. 1997. "Constructing instruments for regressions with measurement error when no additional data are available, with an application to patents and R & D," *Econometrica* 65 (5). 1201-1213.

[169] Li, M., D. Poirier, and J. Tobias. 2004. "Do dropouts suffer from dropping out? Estimation and prediction of outcome gains in generalized selection models," *Journal of Applied Econometrics* 19. 203-225.

[170] Lintner, J. 1965. "The valuation of risk assets and the selection of risky investments in stock portfolios and capital budgets," *Review of Economic and Statistics* 47 (1). 13-37.

[171] Lovell, M. 1963. "Seasonal adjustment of economic time series," *Journal of the American Statistical Association* 58 (304). 993-1010.

[172] Lucas, R. 1976. "Econometric policy evaluation: A critique," *The Phillips Curve and Labor Markets*. vol. 1 Carnegie-Rochester Conference Series on Public Policy. K. Brunner and A. Meltzer, eds. Amsterdam, The Netherlands: North-Holland Publishing Company. 19-46.

[173] Luce, R. 1959. *Individual Choice Behavior*. New York, NY: John Wiley & Sons.

[174] MacKinnon, J. 2002. "Bootstrap inference in econometrics," *The Canadian Journal of Economics* 35 (4). 615-645.

[175] Madansky, A. 1959. "The fitting of straight lines when both variables are subject to error," *Journal of the American Statistical Association* 54 (285). 173-205.

[176] Manski, C. 1993. "Identification of endogenous social effects: The reflection problem," *Review of Economic Studies* 60 (3). 531-542.

[177] Manski, C. 2007. *Identification for Prediction and Decision*. Cambridge, MA: Harvard University Press.

[178] Marschak, J. 1953. "Economic measurements for policy and prediction," *Studies in Econometric Method* by Cowles Commission research staff members, W. Hood and T. Koopmans, eds.

[179] Marschak, J. 1960. "Binary choice constraints on random utility indicators," *Proceedings of the First Stanford Symposium on Mathematical Methods in the Social Sciences, 1959*. K. Arrow, S. Karlin, and P. Suppes, ed. Stanford, CA: Stanford University Press. 312-329.

[180] Marschak, J. and K. Miyasawa. 1968. "Economic comparability of information systems," *International Economic Review* 9 (2). 137-174.

[181] Marshall, A. 1961. *Principles of Economics*. London, U.K.: Macmillan.

[182] R. Matzkin. 2007. "Nonparametric identification," *Handbook of Econometrics* Volume 6B. J. Heckman and E. Leamer, eds. 5307-5368.

[183] McCall, J. 1991. "Exchangeability and its economic applications," *Journal of Economic Dynamics and Control* 15 (3). 549-568.

[184] McFadden, D. 1978. "Modeling the choice of residential location," in A. Karlqvist, L. Lundqvist, F. Snickars, and J. Weibull, eds., *Spatial Interaction Theory and Planning Models*, Amsterdam, The Netherlands: North-Holland. pp. 75–96.

[185] McFadden, D. 1981. "Econometric models of probabilistic choice," *Structural Analysis of Discrete Choice Data with Econometric Applications*. C. Manski and D. McFadden, eds. Cambridge, MA: The MIT Press.

[186] McFadden, D. and K. Train. 2000. "Mixed MNL models for discrete response," *Journal of Applied Econometrics* 15 (5). 447-470.

[187] McFadden, D. 2001. "Economic choices," *The American Economic Review* 91 (3). 351-378.

[188] McKelvey, R. and T. Palfrey. 1995 "Quantal response equilibria for normal form games," *Games and Economic Behavior* 10. 6-38.

[189] McKelvey, R. and T. Palfrey. 1998. "Quantal response equilibria for extensive form games," *Experimental Economics* 1. 9-41.

[190] Mossin, J. 1966. "Equilibrium in a capital asset market," *Econometrica* 24 (4). 768-783.

[191] Morgan, S. and C. Winship. 2007. *Counterfactuals and Causal Inference*. New York, NY: Cambridge University Press.

[192] Mullahy, J. 1997. "Instrumental-variable estimation of count data models: Applications to models of cigarette smoking behavior," *The Review of Economics and Statistics* 79 (4). 586-593.

[193] Mundlak, Y. 1978. "On the pooling of time series and cross section data," *Econometrica* 46 (1). 69-85.

[194] Newey, W. 1985. "Maximum likelihood specification testing and conditional moment tests," *Econometrica* 53 (5). 1047-1070.

[195] Newey, W. and J. Powell. 2003. "Instrumental variable estimation of nonparametric models," *Econometrica* 71 (5). 1565-1578.

[196] Newey, W. 2007. "Locally linear regression," course materials for 14.386 New Econometric Methods, Spring 2007. MIT OpenCourseWare (http://ocw.mit.edu), Massachusetts Institute of Technology.

[197] Neyman, J. 1923. "Statistical problems in agricultural experiments," *Journal of the Royal Statistical Society* II (supplement, 2). 107-180.

[198] Nikolaev, V. and L. Van Lent. 2005. "The endogeneity bias in the relation between cost-of-debt capital and corporate disclosure policy," *The European Accounting Review* 14 (4). 677 724.

[199] Nobile, A. 2000. "Comment: Bayesian multinomial probit models with a normalization constraint," *Journal of Econometrics* 99. 335-345.

[200] O'Brien, S. and D. Dunson. 2003. "Bayesian multivariate logistic regression," MD A3-03, National Institute of Environmental Health Sciences.

[201] O'Brien, S. and D. Dunson. 2004. "Bayesian multivariate logistic regression," *Biometrics* 60 (3). 739-746.

[202] Pagan, A. and F. Vella. 1989. Diagnostic tests for models based on individual data: A survey," *Journal of Applied Econometrics* 4 (Supplement). S29-S59.

[203] Poirier, D. 1995. *Intermediate Statistics and Econometrics*. Cambridge, MA: The MIT Press.

[204] Poirier, D. and J. Tobias. 2003. "On the predictive distributions of outcome gains in the presence of an unidentified parameter," *Journal of Business and Economic Statistics* 21 (2). 258-268.

[205] Powell, J., J. Stock, and T. Stober. 1989. "Semiparametric estimation of index coefficients," *Econometrica* 57 (6). 1403-1430.

[206] Quandt, R. 1972. "A new approach to estimating switching regressions," *Journal of the American Statistical Association* 67 (338). 306-310.

[207] **R** Development Core Team. 2009. *R: A language and environment for statistical computing*. **R** Foundation for Statistical Computing, Vienna, Austria. ISBN 3-900051-07-0, URL http://www.R-project.org.

[208] Rao, C. R. 1965. "The theory of least squares when the parameters are stochastic and its application to the analysis of growth curves," *Biometrika* 52 (3/4). 447-458.

[209] Rao, C. R. 1986, "Weighted Distributions", *A Celebration of Statistics*. Feinberg, S. ed. Berlin, Germany: Springer-Verlag. 543-569

[210] Rao, C. R. 1973. *Linear Statistical Inference and Its Applications*. New York, NY: John Wiley & Sons.

[211] Rivers, D. and Q. Vuong. 1988 "Limited information estimators and exogeneity tests for simultaneous probit models," *Journal of Econometrics* 39. 347-366.

[212] Robinson, C. 1989. "The joint determination of union status and union wage effects: Some tests of alternative models," *Journal of Political Economy* 97 (3). 639-667.

[213] Robinson, P. 1988. "Root-N-consistent semiparametric regression," *Econometrica* 56 (4). 931-954.

[214] Roll, R. 1977. "A critique of the asset pricing theory's tests: Part I: On past and potential testability of the theory," *Journal of Financial Economics* 4. 129-176.

[215] Rosenbaum, P. and D. Rubin. 1983a. "The central role of the propensity score in observational studies for causal effects," *Biometrika* 70(1). 41-55.

[216] Rosenbaum, P. and D. Rubin. 1983b. "Assessing sensitivity to an unobserved binary covariate in an observational study with binary outcome," *Journal of the Royal Statistical Society*, Series B 45(2). 212-218.

[217] Ross, S. 1976. "The arbitrage theory of capital asset pricing," *Journal of Economic Theory* 13. 341-360.

[218] Rossi, P., G. Allenby, and R. McCulloch. 2005. *Bayesian Statistics and Marketing*. New York, NY: John Wiley & Sons.

[219] Roy, A. 1951. "Some thoughts on distribution of earnings," *Oxford Economic Paper* 3 (2). 135-46.

[220] Rubin, D. 1974. "Estimating causal effects of treatments in randomized and nonrandomized studies," *Journal of Educational Psychology* 66(5). 688-701.

[221] Ruud, P. 1984. "Tests of specification in econometrics," *Econometric Reviews* 3 (2). 211-242.

[222] Savage, L. 1972. *The Foundations of Statistics*. New York, NY: Dover Publications, Inc.

[223] Sekhon, J. 2008. "Multivariate and propensity score matching software with automated balance optimization: The matching package for R," *Journal of Statistical Software*. forthcoming.

[224] Sharpe, W. 1964. "Capital asset prices: A theory of market equilibrium under conditions of risk," *Journal of Finance* 19 (3). 425-442.

[225] Shugan, S. and D. Mitra. 2009. "Metrics — when and why nonaveraging statistics work," *Management Science* 55 (1), 4-15.

[226] Signorino, C. 2002. "Strategy and selection in international relations," *International Interactions* 28. 93-115.

[227] Signorino, C. 2003. "Structure and uncertainty in discrete choice models," *Political Analysis* 11 (4). 316-344.

[228] Sims, C. 1996. "Macroeconomics and Methodology," *Journal of Economic Perspectives* 10 (1). 105-120.

[229] Spiegelhalter D., A. Thomas, N. Best, and D. Lunn. 2003. "WinBUGS Version 1.4 Users Manual," MRC Biostatistics Unit, Cambridge University. URL http://www.mrc- bsu.cam.ac.uk/bugs/.

[230] Stiger, S. 2007. "The epic story of maximum likelihood," *Statistical Science* 22 (4). 598-620.

[231] Stoker, T. 1991. *Lectures on Developments in Semiparametric Econometrics*. CORE Lecture Series. Universite Catholique de Louvain.

[232] Strang, G. 1986. *Introduction to Applied Mathematics*. Wellesley, MA: Wellesley-Cambridge Press.

[233] Strang, G. 1988. *Linear Algebra and its Applications*. Wellesley, MA: Wellesley-Cambridge Press.

[234] Swamy, P. 1970. "Efficient inference in a random coefficient regression model," *Econometrica* 38 (2). 311-323.

[235] Tamer, E. 2003. "Incomplete simultaneous discrete response model with multiple equilibria," *The Review of Economic Studies* 70 (1). 147-165.

[236] Tanner, M. and W. Wong. 1987. "The calculation of posterior distributions by data augmentation," *Journal of the American Statistical Association* 82 (398). 528-540.

[237] Theil, H. 1971. *Principles of Econometrics*. New York, NY: John Wiley & Sons.

[238] Tobin, J. 1958. "Estimation of relationships for limited dependent variables," *Econometrica* 26 (1). 635-641.

[239] Train, K. 2003. *Discrete Choice Models with Simulation*. Cambridge, U. K.: Cambridge University Press.

[240] Tribus, M. and G. Fitts. 1968. "The widget problem revisited," *IEEE Transactions on Systems Science and Cybernetics* SSC-4 (2). 241-248.

[241] Trochim, W. 1984. *Research Design for Program Evaluation: The Regression-Discontinuity Approach*. Beverly Hills, CA: Sage Publications.

[242] van der Klaauw, W. 2002. "Estimating the effect of financial aid offers on college enrollment: A regression-discontinuity approach," *International Economic Review* 43 (4). 1249-1287.

[243] Van Lent, L. 2007. "Endogeneity in management accounting research: A comment," *The European Accounting Review* 16 (1). 197–205.

[244] Vella, F. and M. Verbeek. 1999. "Estimating and interpreting models with endogenous treatment effects," *Journal of Business & Economic Statistics* 17 (4). 473-478.

[245] Vijverberg, W. 1993. "Measuring the unidentified parameter of the extended Roy model of selectivity," *Journal of Econometrics* 57. 69-89.

[246] Vuong, Q.1984. "Two-stage conditional maximum likelihood estimation of econometric models," working paper California Institute of Technology.

[247] Vytlacil, E. 2002. "Independence, Monotonicity, and Latent Index Models: An Equivalence Result," *Econometrica* 70 (1). 331-341.

[248] Vytlacil, E. 2006. "A note on additive separability and latent index models of binary choice: Representation results," *Oxford Bulletin of Economics and Statistics* vol. 68 (4). 515-518.

[249] Wald, A. 1947. "A note on regressions analysis," *The Annals of Mathematical Statistics* 18 (4). 586-589.

[250] Walley, P. 1991. *Statistical Reasoning with Imprecise Probabilities,* London: Chapman and Hall.

[251] White, H. 1984. *Asymptotic Theory for Econometricians.* Orlando, FL: Academic Press.

[252] Willis, R. and S. Rosen. 1979. "Education and self-selection," *Journal of Political Economy* 87 (5, part 2: Education and income distribution). S7-S36.

[253] Wooldridge, J. 1997. "On two stage least squares estimation of the average treatment effect in a random coefficient model," *Economics Letters* 56. 129-133.

[254] Wooldridge, J. 2003. "Further results on instrumental variables estimation of average treatment effects in the correlated random coefficient model," *Economics Letters* 79. 185-191.

[255] Wooldridge, J. 2002. *Econometric Analysis of Cross Section and Panel Data.* Cambridge, MA: The MIT Press.

[256] Yitzhaki, S. 1996. "On using linear regressions in welfare economics," *Journal of Business and Economic Statistics* 14 (4). 478–486..

[257] Zheng, J. 1996. "A consistent test of functional form via nonparametric estimation techniques," *Journal of Econometrics* 75. 263-289.

Index

accounting
    disclosure quality, 129
    information, 123
    recognition, 17
    regulation, 12, 16
    reserves, 11
accounting & other information sources, 45–48
    ANOVA, 46, 48
    complementarity, 45
    DGP, 47, 48
    information content, 45
    misspecification, 46, 48
    proxy variable, 45
    restricted recognition, 45
    saturated design, 46, 48
    valuation, 45
accounting choice
    information system design, 9
additive separability, 173
aggregation, 4, 6
Aitken estimator, 22
all or nothing loss, 59, 61

analyst, 2, 123
ANOVA theorem, 56
artificial regression, 62, 64, 89
    specification test, 88
asset revaluation regulation, 12–14, 175–202
    average treatment effect (ATE), 178
    average treatment effect on treated (ATT), 178
    average treatment effect on untreated (ATUT), 178
    average treatment effect sample statistics, 179
    certification cost, 175
    certification cost type, 179
    equilibrium, 175
    full certification, 177
    fuzzy regression discontinuity design, 198
    2SLS-IV identification, 198, 200
    binary instrument, 199
    DGP, 198
    missing "factual" data, 200

propensity score, 198
homogeneity, 183
identification, 179
OLS estimates, 178, 180
outcome, 177–180, 202
propensity score matching treatment effect, 182
propensity score treatment effect, 181, 189
regression discontinuity design, 181
selective certification, 177–188
    ATE, 186, 187
    ATT, 184, 187
    ATUT, 185, 187
    conditional average treatment effect, 188
    data augmentation, 193
    DGP, 190
    heterogeneity, 186
    identification, 194
    missing "factual" data, 193
    OLS estimand, 185
    OLS estimates, 191
    outcome, 187, 190
    propensity score, 191
    propensity score matching, 190, 192
    selection bias, 186
sharp regression discontinuity design, 196
    full certification, 196
    missing "factual" data, 197
    selective certification, 197
treatment
    investment, 175, 177
treatment effect
    common support, 201
    uniform distribution, 201
asymptotic results, 21
asymptotic theory, 413–424
    Boole's generalized inequality, 416
    Boole's inequality, 416
    convergence in distribution (central limit theorems), 417–421
        asymptotic distribution, 420
        asymptotic distribution of nonlinear function, 421
        asymptotic distribution of order statistics, 420
        asymptotic inefficiency of median in normal sample, 420
        asymptotic normality and efficiency, 420
        asymptotic normality of MLE theorem, 421
        limiting distribution, 418
        limiting mean and variance, 418
        limiting normal distribution of a function, 419
        limiting normal distribution of a set of functions (multivariate), 420
        Lindberg-Feller CLT (unequal variances), 419
        Lindberg-Levy CLT (multivariate), 419
        Lindberg-Levy CLT (univariate), 418
        rules for limiting distributions, 418
    convergence in probability (laws of large numbers), 413–417
        almost sure convergence, 414
        Chebychev's inequality, 414
        Chebychev's weak law of large numbers, 415
        consistent estimator, 415
        convergence in quadratic mean, 413
        Kinchine's theorem (weak law of large numbers), 415
        Kolmogorov's strong law of large numbers, 415
        Markov's inequality, 414
        Markov's strong law of large numbers, 414
        rules for probability limits, 417
        Slutsky theorem, 416
delta method, 419

fundamental theorem of statistics, 413
implication rule, 416
Jensen's inequality, 417
Jensen's inequality and risk aversion, 417
rates of convergence, 422–423
    asymptotic equality, 422
    order 1/n (big-O notation)(, 422
    order less than 1/n (little-o notation), 422
    order rules, 422
    square-root n convergence, 423
    stochastic order relations, 422
asymptotic variance, 69
auditor, 11, 12, 17
authors
    Aakvik, 288
    Abadie, 234
    Abbring, 236, 276, 291–293, 300, 301
    Admati, 2
    Ahn, 143
    Albert, 94, 118, 301, 330
    Allenby, 119
    Amemiya, 54, 68, 76, 133, 135, 156, 204, 273
    Andrews, 105
    Angrist, 54, 55, 129, 130, 156, 181, 196, 198, 218, 234
    Antle, 45
    Arya, 17, 371, 376, 401, 407
    Bagnoli, 131
    Bassett, 76
    Ben-Akiva, 81
    Bernardo, 113
    Berndt, 68
    Berry, 135
    Besag, 117
    Best, 120
    Bhat, 120–122
    Bjorklund, 275
    Blackwell, 3
    Bock, 71
    Boes, 420
    Bound, 54
    Box, 155, 387, 389
    Bresnahan, 135, 141
    Brown, 54
    Cameron, 54, 122, 156
    Campbell, 276
    Campolieti, 144, 146
    Carlin, 113, 120
    Carneiro, 289
    Casella, 122
    Chamberlain, 129
    Chenhall, 155
    Chib, 94, 118, 122, 301, 330, 331
    Christensen, 3, 9, 14, 18, 54
    Clifford, 117
    Cochran, 130
    Coslett, 95
    Cover, 401
    Cox, 145, 146, 208, 277
    Craven, 104
    Davidson, 38, 54, 62, 63, 76, 86, 87, 89, 95, 108, 122, 413, 423
    de Finetti, 107
    Demski, 3, 9, 10, 12, 14, 18, 45, 48, 54, 123, 156, 175, 376, 382, 401
    Dubin, 130
    Dunson, 330
    Dye, 10, 12, 175
    Ebbes, 147
    Efron, 108
    Eicker, 22
    Evans, 131
    Fellingham, 17, 371, 376, 382, 401, 407
    Ferguson, 115
    Fisher, 76, 130, 208, 277
    Fitts, 363, 368
    Florens, 289
    Francois, 81
    Freedman, 108, 109
    Frisch, 22
    Galton, 56
    Gauss, 33, 76
    Gelfand, 118

Gelman, 113, 120
George, 122
Girshick, 3
Glover, 17, 371, 376, 401
Godfrey, 43, 132
Goldberger, 124
Graybill, 54, 420
Greene, 54, 68, 76, 78, 86, 130, 131, 402, 423
Griliches, 155
Hall, 68, 423
Hamilton, 122, 331
Hammersley, 117
Hansen, 289
Hardle, 97, 99, 104, 105
Hausman, 26, 54, 68
Heckman, 123, 125, 133, 135, 142, 148, 155, 156, 172, 173, 182, 205, 208, 210, 220, 233, 236–238, 245, 249, 275–278, 280, 282, 283, 286–289, 291–293, 300, 301, 331
Heyde, 423
Hildreth, 31
Hoeffding, 100
Holmstrom, 376
Honore, 208, 210, 277
Horowitz, 86, 105, 122, 144, 146
Houck, 31
Huber, 22
Hutton, 43
Ichimura, 172, 173, 182, 205
Imbens, 218
James, 70
Jaynes, 3, 4, 8, 33, 35, 36, 55, 330, 333, 342, 344, 345, 355, 363, 369–371, 387, 398, 401
Judge, 71
Kiefer, 54, 330, 383
Koenker, 76
Koop, 331
Kreps, 122
Krueger, 129, 130, 156
Lalonde, 129, 130
Lambert, 2

Larcker, 126, 155
Lavy, 181, 198
Lee, 132
Leuz, 2
Lewbel, 156
Li, 301, 306, 330
Lin, 12, 175, 376, 401
Lintner, 2
Liu, 131
Lochner, 293
Lovell, 22
Luce, 78
Lukacs, 416
Lunn, 120
MacKinnon, 38, 54, 62, 63, 76, 87, 89, 95, 107–109, 122, 413, 423
Madansky, 156
Marschak, 3, 78
Marshall, 125
Mathiowetz, 54
McCall, 122
McCulloch, 119
McFadden, 79, 81, 95
McKelvey, 135
Meghir, 289
Milgrom, 376
Mitra, 76
Mittendorf, 401
Miyasawa, 3
Moers, 155
Moffitt, 275
Mood, 420
Mossin, 2
Mullahy, 95
Navarro-Lozano, 205, 286
Newey, 101, 103, 105
Neyman, 208, 277
Nikolaev, 129
Nobile, 304
O'Brien, 330
Pagan, 101
Palfrey, 135
Peters, 108
Pischke, 54, 55, 181, 196, 198
Poirier, 76, 95, 301, 306, 330, 331

448 INDEX

Powell, 100, 101, 105, 143
Quandt, 208, 277
R Development Core Team, 118
Rao, 54, 76, 276, 280, 418
Reiss, 135, 141
Rivers, 130, 131, 133
Robb, 210, 278, 291
Robinson, 99, 128, 129
Roll, 2
Rosenbaum, 172
Ross, 2
Rossi, 119
Roy, 208, 210, 277
Rubin, 113, 120, 172, 208, 218, 277
Rusticus, 126, 155
Ruud, 101
Ryan, 45
Sappington, 12, 175
Schafgans, 105
Schroeder, 17, 371, 376, 401, 407
Schwab, 131
Sekhon, 182
Shannon, 334
Sharpe, 2
Shugan, 76
Signorino, 135
Singer, 156
Smith, 113, 118, 205
Spiegelhalter, 120
Sridar, 10
Stanley, 276
Stein, 70, 72
Stern, 113, 120
Stigler, 76
Stock, 100, 101, 105, 143
Stoker, 92, 100, 101, 105, 128, 143
Strang, 17, 35, 371, 402, 406
Sunder, 401
Swamy, 31
Tabor, 293
Tamer, 141
Tanner, 122
Theil, 54, 68, 76
Thomas, 120, 401
Tiao, 387, 389

Tobias, 301, 306, 330
Tobin, 94
Todd, 172, 173, 182, 205
Train, 80, 81, 83, 95, 119, 120, 122
Tribus, 362, 368
Trivedi, 54, 122, 156
Trochim, 196
Urzua, 287, 300
van der Klaauw, 198
Van Lent, 129, 155
Vella, 101
Verrecchia, 2
Vijverberg, 331
Vuong, 131, 133
Vytlacil, 123, 148, 208, 220, 233, 236–238, 249, 275–277, 280, 282, 283, 287–289, 300
Wahba, 104
Wald, 31
Walley, 369, 400, 401
Watts, 131
Waugh, 22
White, 22, 417, 423
Wickens, 132
Wong, 122
Wooldridge, 54, 129, 156, 204, 212, 216, 237, 238, 246, 269, 270, 272, 273
Yitzhaki, 276, 280
Zheng, 101

bandwidth, 93, 98, 101, 104
Bayesian analysis, 4
Bayesian data augmentation, 118, 146
　discrete choice model, 94
Bayesian regression, 117
Bayesian statistics, 17
Bayesian treatment effect, 301–331
　binary regulated report precision, 313–316
　　McMC estimated average treatment effects, 314
　　McMC estimated average treatment effects from MTE, 314
　bounds, 301, 302

conditional posterior distribution, 303
  augmented outcome, 303
  latent utility, 303
  Nobile's algorithm, 304
  parameters, 304
  SUR (seemingly-unrelated regression), 304
  truncated normal distribution, 303
  Wishart distribution, 304
counterfactual, 302
data augmentation, 301
DGP, 302
Dirichlet distribution, 331
distribution, 301
Gibbs sampler, 303
identification, 301
latent utility, 302
Metropolis-Hastings algorithm, 331
mixture of normals distribution, 306
  Dirichlet distribution, 307
  likelihood function, 306
  multinomial distribution, 307
  prior, 307
MTE, 309
  weights, 309
outcome, 302
predictive distribution, 305
  Rao-Blackwellization, 306
prior distribution
  normal, 305
  Wishart, 305
probability as logic, 330
  evidence of endogeneity, 331
  maximum entropy principle (MEP), 330
  maximum entropy principle (MEP) & Gaussian distribution, 330
  maximum entropy principle (MEP) & Student t distribution, 330
prototypical example, 307
  average treatment effect sample statistics, 310
  DGP, 307
  latent utility, 307
  McMC estimated average treatment effects, 310
  McMC estimated marginal treatment effects, 311
  McMC MTE-weighted average treatment effects, 311
  outcome, 307
  simulation, 308
regulated continuous but observed binary report precision, 316–319
  instrument, 317
  latent utility, 317
  McMC estimated average treatment effects, 318
  McMC estimated average treatment effects from MTE, 319
regulated continuous, nonnormal but observed binary report precision, 319–323
  latent utility, 321
  McMC estimated average treatment effects, 321
  McMC estimated average treatment effects from MTE, 321
  nonnormality, 319
  policy-relevant treatment effect, 326
  stronger instrument, 323
  stronger instrument McMC average treatment effect estimates, 323
  stronger instrument McMC average treatment effect estimates from MTE, 323
regulated report precision, 311
  latent utility, 311
  outcome, 312
selection, 302
Bernoulli distribution, 240
best linear predictor theorem (regression justification II), 57
beta distribution, 113
BHHH estimator, 66, 70
binomial distribution, 65, 94, 107, 113

block matrix, 44
block one factorization, 24
BLU estimator, 21
bootstrap, 108
    paired, 109
    panel data regression, 109
    regression, 108
    wild, 109
BRMR, 89
BRMR (binary response model regression), 88
burn-in, 118, 120, 121

CAN, 21, 102, 212
CARA, 15
causal effect, 1, 275, 277
    definition, 125
    general equilibrium, 293
        policy invariance, 293
        uniformity, 293
causal effects, 14, 22, 105, 123, 128
CEF decomposition theorem, 55
CEF prediction theorem, 55
central limit theorem, 35
certainty equivalent, 15
certified value, 12
chi-square distribution, 74
chi-squared statistic, 38
Cholesky decomposition, 21, 119
classical statistics, 4, 17
CLT, 41, 69, 70
completeness, 3
conditional expectation function (CEF), 55
conditional likelihood ratio statistic, 135
conditional mean independence, 158, 164, 167, 204
conditional mean independence (redundancy), 213
conditional mean redundancy, 213, 214, 238
conditional moment tests, 101
conditional posterior distribution, 95, 117–119
conditional score statistic, 135

conditional stochastic independence, 158, 165, 172
conjugacy, 111
conjugate prior, 111
consistency, 42, 45, 70, 102, 155, 213
consistent, 108
contribution to gradient (CG) matrix, 67
control function, 203
    inverse Mills, 130
convergence, 98
convolution, 35
cost of capital, 129
counterfactual, 142, 173, 207, 215, 218, 275
covariance for MLE, 66
Cramer-Rao lower bound, 67
critical smoothing, 93

data, 1–3, 11, 123, 155
delta method, 41
density-weighted average derivative estimator
    instrumental variable, 143
DGP, 19, 21, 22, 38, 41–44, 68, 89, 91, 123, 147, 148, 155, 159, 167, 181, 207, 210
diagnostic checks, 3
differences-in-differences, 129
Dirichlet distribution, 115, 146
duration model, 143–145
    Bayesian semiparametric, 144
    proportional hazard, 144, 145
    semiparametric proportional hazard, 144

earnings inequality, 210
earnings management, 10–12, 16, 48–54, 382–397, 401–412
    equilibrium reporting behavior, 382
    logistic distribution, 383
    performance evaluation, 401
        accruals smoothing, 401
        limited commitment, 401
    selective manipulation, 393–397
        closer look at the variance, 397

noncentral scaled t distribution, 396
scale uncertainty, 396
scale uncertainty simulation, 397
simulation, 394
truncated expectations, 397
stacked weighted projection, 384
stochastic manipulation, 382–392
closer look at variance, 391
inverted chi-square distribution, 388
noncentral scaled t distribution, 388, 391
scale uncertainty, 385
scale uncertainty simulation, 392
simulation, 385
Eicker-Huber-White estimator, 22
empirical model building, 2
empiricist, 1
endogeneity, 1, 3, 9, 14, 123, 130, 148, 155
endogenous causal effects, 9
endogenous regressor, 43, 126
entropy, 334
equilibrium, 9, 11, 12, 15–17
equilibrium earnings management, 10–12, 16, 48–54, 382–397, 401–412
analyst, 52
Bernoulli distribution, 50
data, 54
endogeneity, 48
equilibrium, 49
fair value accruals, 48
information advantage, 52
instrument, 48
model specification, 54
nonlinear price-accruals relation, 50
omitted variable, 51
private information, 48
propensity score, 48, 51, 53, 54
logit, 54
saturated design, 51
social scientist, 52
theory, 54

unobservable, 53, 54
error cancellation, 35
error components model, 26
estimand, 123, 218
evidentiary archive, 401
exact tests, 108
exchangeable, 107, 110, 146
exclusion restriction, 207, 218, 221
expected squared error loss, 71, 72
expected utility, 13, 16, 77
exponential distribution, 113, 145
external validity, 276
extreme value (logistic) distribution, 79

F statistic, 21, 37, 38, 40
fair value accruals, 10
financial statement example
directed graph, 374
Gaussian distribution, 371
left nullspace, 372
linear independence, 371
nullspace, 374
posterior distribution, 375
spanning tree, 374
under-identification and Bayes, 370
financial statement inference, 17, 370–375, 401
bounding, 401
fineness, 3
first order considerations, 3
fixed effects, 127
lagged dependent variable, 129
fixed effects model, 26–30
between-groups (BG) estimator, 27
FWL, 26
projection matrix, 27
individual effects, 26
OLS, 26
time effects, 26
within-groups (WG) estimator, 27
fixed vs. random effects, 26–30
consistency & efficiency considerations, 26
equivalence of GLS and fixed effects, 30

equivalence of GLS and OLS, 30
Hausman test, 26
flexible fit, 97
football game puzzle
    marginalization paradox, 370
    probability as logic, 369
    proposition, 370
Frisch-Waugh-Lovell (FWL), 22, 36, 38, 39, 99, 126, 128
fundamental principle of probabilistic inference, 4
fundamental theorem of statistics, 108

gains to trade, 14
gamma distribution, 113
Gauss-Markov theorem, 19
Gauss-Newton regression (GNR), 62–65
Gaussian distribution, 33, 55, *see* normal distribution
Gaussian function, 35
    convergence, 35
    convolution, 35
    Fourier transform, 35
    maximum entropy given variance, 35
    preservation, 35
    product, 35
GCV (generalized cross-validation), 104
general equilibrium, 276
generalized least squares (GLS), 21
Gibbs sampler, 95, 118
global concavity, 66
GLS, 110
GNR, 88
gradient, 40, 63, 67
Gumbel distribution, 80, 83

Halton draw
    random, 121
Halton sequences, 120
Hammersley-Clifford theorem, 117
Hausman test, 43
hazard function
    conditional, 145
    integrated, 144
    unconditional, 144
Heckman's two-stage procedure, 214
    standard errors, 214
Hessian, 62, 63, 66, 67, 70
    positive definite, 63
heterogeneity, 129, 144, 146, 162, 210
heterogeneous outcome, 14
heteroskedastic-consistent covariance estimator, 65
homogeneity, 146, 159, 210, 211
homogeneous outcome, 14

identification, 78, 79, 83, 87, 107, 142, 158, 207, 210, 211, 238
    Bayes, 165
    Bayes' sum and product rules, 219
    control functions, 213
    LATE
        binary instrument, 218
    nonparametric, 164
    propensity score, 169
    propensity score and linearity, 172
ignorable treatment (selection on observables), 148, 149, 157–204, 207
independence of irrelevant alternatives (IIA), 78, 80–83, 92
index sufficiency, 172
inferring transactions from financial statements, 17, 370–375
information matrix, 67, 68
    asymptotic or limiting, 67
    average, 67
informational complementarities, 3
informed priors, 333–401
instrument, 41, 42, 142, 143, 164, 211
    binary, 218
instrumental variable (IV), 41, 43, 95, 100, 105, 126
    2SLS-IV, 126, 211
    exclusion restriction, 277
    linear
        exclusion restrictions, 211
    local (LIV), 276

over-identifying tests of restrictions, 211
internal validity, 276
interval estimation, 36
    normal (Gaussian) distribution, 36
intervention, 130
inverse Mills, 142
inverse-gamma distribution, 116
inverse-Wishart distribution, 116
iterated expectations, 172, 216

James-Stein shrinkage estimator, 70–75
Jaynes' Ap distribution, 398
    Bayes' theorem, 399
    football game puzzle revisited, 400
Jaynes' widget problem, 355–369
    stage 1 knowledge, 356
        expected loss, 357
    stage 2 knowledge, 358
    stage 3 knowledge, 361
        exact distributions, 366
        Gaussian approximation, 363
        rapidly converging sum, 365
        z transform, 366
    stage 4 knowledge, 369
joint density, 6

kernel, 111
kernel density, 92, 98, 102, 146
    Gaussian, 100
    leave one out, 98

Lagrange Multiplier (LM) statistic, 40
Lagrange multiplier (LM) statistic, 38
Lagrange multiplier (LM) test, 43
latent IV, 147
latent utility, 77, 79, 83, 86–88, 95
latent variable, 119
law of iterated expectations, 55
LEN model, 15
likelihood, 65
likelihood function, 111
likelihood ratio (LR) statistic, 38, 40
likelihood ratio (LR) test, 87
limited information estimation, 131
Lindberg-Feller CLT, 40

linear CEF theorem (regression justification I), 56
linear conditional expectation function Mathiowetz, 54
linear loss, 61
linear probability model, 78
link function, 77
LIV
    estimation, 286
        common support, 288
        nonparametric FWL (double residual regression), 287
        nonparametric kernel regression, 287
        propensity score, 287
LLN, 63, 69, 70
log-likelihood, 34, 40, 62, 65, 67, 70, 83, 86, 87, 94, 131–134, 146
logistic distribution, 66, 79
logit, 66–84
    binary, 79
    conditional, 80, 82, 84
    generalized extreme value (GEV), 81
    generalized nested (GNL), 82, 84
    multinomial, 82
    multinomial , 80
    nested, 82, 84
    nested (NL), 82
    nests, 81
lognormal distribution, 145

marginal density, 6
marginal posterior distribution, 117
marginal probability effect, 78, 86, 87, 138, 140
marginal treatment effect (MTE), 275–300
    discrete outcome
        identification, 288
    FORTRAN program, 300
    heterogeneity, 288
    identification, 278
        Bayes' theorem, 279

policy-relevant treatment effect, 279
local instrumental variable (LIV), 279
support, 280
uniform distribution, 279
market portfolio, 2
maximum entropy, 334–354
   background knowledge, 337
   Cholesky decomposition, 348
   continuous distributions
      exponential, 349
      Gaussian, 347
      multivariate Gaussian, 348
      truncated exponential, 349
      truncated Gaussian, 350
      uniform, 346
   continuous support, 342
   discrete choice logistic regression (logit), 340
   generalization, 337
   ignorance, 334, 336
   Lagrangian, 337
   partition function, 338
   probability as logic, 355
   probability assignment, 334
   transformation groups, 344
      invariance, 344
      Jeffrey's prior, 345
      location and scale knowledge, 344
   variance bound, 351
      Cramer-Rao, 351
      Schwartz inequality, 353
maximum entropy principle (MEP), 4, 8, 17
maximum entropy priors, 333
maximum likelihood estimation (MLE), 33, 65, 68, 69, 89, 214
McMC (Markov chain Monte Carlo), 95, 117-120
   tuning, 120
mean conditional independence, 43
measurement error, 41
Metropolis-Hastings (MH) algorithm, 118–120
   random walk, 119
   tuning, 120
minimum expected loss, 59
minimum mean square error (MMSE), 55
missing data, 97
mixed logit
   robust choice model, 92
model specification, 1, 54, 86, 89, 123, 155
MSE (mean squared error), 103, 104
multinomial distribution, 115

Nash equilibrium, see equilibrium
natural experiment, 129
negative-binomial distribution, 113
Newey-West estimator, 22
Newton's method, 62
non-averaging statistics, 76
noncentrality parameter, 71
nonlinear least squares, 66
nonlinear regression, 62, 88
nonlinear restrictions, 41
nonnormal distribution, 147
nonparametric, 146
nonparametric discrete choice regression
   robust choice model, 93
nonparametric kernel matching, 182
nonparametric model, 143
nonparametric regression, 97–105
   fuzzy discontinuity design, 198
   leave one out, 104
   locally linear, 103
   objectives, 97
   specification test, 101
normal distribution, 86, 92, 94, 116, 132, 138, 139
   bivariate, 130
nR-squared test, 40, 43

observationally equivalent, 208
OLS, 20–23, 41–43, 45, 64, 101, 110, 124, 133, 150
   exogenous dummy variable regression, 158, 166

omitted variable, 3, 41, 43, 123, 128, 149, 203, 207, 211
omitted variables, 99
OPG estimator, 70
ordinal, 78
ordinary least squares (OLS), 19
orthogonal matrix, 75
orthogonality, 19, 41
outcome, 14, 16, 147, 157, 207, 208, 211, 219
    objective, 275
    subjective, 275
outerproduct of gradient (OPG), 66
over-identifying restriction, 43

panel data, 26, 129
    regression, 26
partition, 173
partitioned matrix, 126
perfect classifier, 79
perfect regressor, 159
performance evaluation
    excessive individual measures, 401
pivotal statistic, 108
    asymptotically, 111
pivots, 21
Poisson distribution, 73
poisson distribution, 113
policy evaluation, 155, 205, 275
policy invariance, 276
    utility, 276
positive definite matrix, 72
posterior distribution, 111
posterior mean, 59
posterior mode, 59, 62
posterior quantile, 59, 61
prior distribution, 111
private information, 2, 3, 15, 175
probability as logic, 3, 4, 8, 333-401
probability assignment, 5, 6
probability limit (plim), 126
probit, 66, 86, 118, 142
    Bayesian, 95
    bivariate, 130
    conditionally-heteroskedastic, 86, 90
    simultaneous, 131
        2SCML (two-stage conditional maximum likelihood), 133
        G2SP (generalized two-stage simultaneous probit), 133
        IVP (instrumental variable probit), 132
        LIML (limited information maximum likelihood), 132
projection, 36, 38, 64, 98
proxy variables, 43, 44
public information, 2, 15

quadratic loss, 60

R program
    Matching library, 182
R project
    bayesm package, 119
random coefficients, 31–33
    correlated
        average treatment effect, 33
        panel data regression, 33
        stochastic regressors identification, 32
        nonstochastic regressors identification, 31
random effects model, 26–30
    consistency, 30
    GLS, 28
random utility model (RUM), 78, 92, 150
Rao-Blackwell theorem, 21
recursive, 135
recursive least squares, 117
reduced form, 124
regression CEF theorem (regression justification III), 57
regression justification, 55
regulated report precision, 14–17, 239–272, 293–299, 311–329
    adjusted outcome, 244
    binary treatment, 239
    causal effect, 239
    continous treatment

balanced panel data, 267
endogeneity, 239
equilibrium selection, 244
expected utility, 239
heterogeneity, 244
  average treatment effects, 249
  common support, 245
  continuous choice but observed binary, 253
  DGP, 246
  estimated average treatment effect, 246
  estimated average treatment effect on treated, 246
  estimated average treatment effect on untreated, 246
  inverse Mills IV control function, 252, 257, 261
  OLS estimate, 246
  OLS estimated average treatment effects, 254
  ordinate IV control function, 250, 257, 260
  poor instruments, 246
  propensity score, 249, 254, 259
  propensity score matching, 250, 256, 260
  stronger instrument, 249, 259
  treatment effect on treated, 245
  treatment effect on untreated, 246
  unobservable, 245
  weak instruments, 248
homogeneity, 244
MTE, 293
nonnormality, 293
  inverse Mills IV control function, 296
  MTE via LIV, 296
  nonparametric selection, 298
  OLS estimated average treatment effects, 294
  ordinate IV control function, 294
  stronger instrument and inverse Mills IV, 298
  stronger instrument and LIV, 298
  stronger instrument and ordinate IV control, 298
  unobservable, 297
  weak instruments, 297
observed continous treatment equilibrium, 267
  OLS estimated average treatment effect on treated, 268
  outcome, 267
observed continuous treatment, 266
  2SLS-IV estimated average treatment effect on treated, 269, 270, 272
OLS estimand, 241
OLS selection bias, 241
outcome, 239
perfect predictor of treatment, 243
saturated regression, 244
Simpson's paradox, 262
  inverse Mills IV control function, 265
  OLS estimated treatment effects, 263
  ordinate IV control function, 263
transaction design, 16
treatment effect
  sample statistics, 242
treatment effect on treated, 240
treatment effect on untreated, 241
unobservable cost, 239
representative utility, 77
restricted least squares, 181
ridge regression, 104
risk preference, 14
Roy model
  DGP, 278
  extended, 277, 278
  generalized, 210, 277
  gross benefit, 278
  observable cost, 278
  observable net benefit, 278
  original (basic), 210, 277, 278
  treatment cost, 278
  unobservable net benefit, 278

sample selection, 142, 148
scale, 78, 87, 91, 98
scientific reasoning, 4
score vector, 67
selection, 14, 41, 105, 147, 149, 157, 207, 208, 210
    logit, 212, 213, 217
    probit, 212–214, 217, 249
selection bias, 205
selective experimentation, 149
semiparametric regression
    density-weighted average derivative, 100
    discrete regressor, 105
    partial index, 101, 128, 143
    partial linear regression, 99, 128
    single index regression, 99
semiparametric single index discrete choice models
    robust choice model, 92
Simpson's paradox, 148, 149
simulation
    Bayesian, 111
    McMC (Markov chain Monte Carlo), 117
    Monte Carlo, 108
simultaneity, 41, 124
simultaneity bias, 125
singular value decomposition (SVD), 29
    eigenvalues, 29
    eigenvectrors, 29
        orthogonal matrix, 29
Slutsky's theorem, 42, 416
smooth accruals, 376-381, 401-412
    accruals as a statistic, 376
        tidiness, 379
    appendix, 401-412
        BLU estimation, 402
        Cramer-Rao lower bound, 402
        difference equations, 403
        Fibonacci series, 411
        induction, 403
        information matrix, 402
        LDL' factorization, 402
        LEN contract, 406

Lucas series, 412
proofs, 401
DGP, 376
LEN model, 376
performance evaluation role, 379
valuation role, 377
social scientist, 2
squared bias, 105
statistical sufficiency, 3
stochastic regressors, 20
strategic choice model, 135–141
    analyst error, 135
    expected utility, 135
    logistic distribution, 135
    normal distribution, 135
    private information, 135
    quantal response equilibrium, 135
    sequential game, 135
    unique equilibrium, 135
strategic interaction, 17
strategic probit, 139
strong ignorability, 172
structural, 124, 275
structural model, 237
Student t distribution, 114
    multivariate, 116
sum of uniform random variables, 4
SUR (seemingly unrelated regression), 26
survivor function, 145

t statistic, 21
Taylor series, 62, 69
tests of restrictions, 22, 38
theory, 1, 11, 123, 155, 275
three-legged strategy, 1, 3
Tobit (censored regression), 94
trace, 72
trace plots, 118
treatment, 14, 16, 276
    continuous, 236
    cost, 210
    gross benefit, 210
    latent utility, 210
    observable cost, 210

observable net benefit, 210
uniformity, 233
unobservable net benefit, 210
treatment effect, 130, 147, 157, 207
    ANOVA (nonparametric regression), 168
    average, 147, 280
    average (ATE), 147, 150, 158, 159, 165, 166, 169, 204, 208, 212, 216
    average on treated, 280
    average on treated (ATE), 158
    average on treated (ATT), 147, 150, 169, 173, 204, 209, 212, 216
        heterogeneity, 217
    average on untreated, 280
    average on untreated (ATUT), 148, 150, 158, 169, 173, 204, 209, 216
        heterogeneity, 217
    average on utreated (ATUT), 212
    common support, 205, 214
    compared with structural modeling, 155
    comparison of identification strategies
        control function, 284
        linear IV, 284
        LIV (local IV), 284
        matching, 284
    conditional average (ATE(X)), 166, 167, 216
    conditional average on treated (ATT(X)), 215
    conditional average on untreated (ATUT(X)), 216
    continuous, 289, 291
        2SLS-IV, 238
    DGP, 301
    distributions, 291
        factor model, 291
    dynamic timing, 292
        duration model, 292
        outcome, 292
        policy invariance, 292
    endogenous dummy variable IV model, 211
    heterogeneity, 212
    heterogeneous outcome, 277
    identification
        control functions, 286
        linear IV, 286
        local IV, 286
        matching, 286
        perfect classifier, 286
    identification via IV, 148
    inverse Mills IV control function
        heterogeneity, 214
    LATE
        always treated, 222
        complier, 221, 222
        defier, 221
        never treated, 222
    LATE 2SLS-IV estimation, 221
    LATE = ATT, 221
    LATE = ATUT, 221
    linear IV weight, 283
    LIV identification
        common propensity score support, 286
    local average, 281
    local average (LATE), 198, 209, 218
        censored regression, 234
        discrete marginal treatment effect, 218
    local IV (LIV), 233
    marginal (MTE), 209, 212, 233, 276
    multilevel discrete, 289
    ordered choice, 289
        unordered choice, 290
    OLS bias for ATE, 151
    OLS bias for ATT, 150
    OLS bias for ATUT, 151
    OLS inconsistent, 211
    OLS weight, 284
    ordinate IV control function
        heterogeneity, 213
    outcome, 277
    policy invariance, 282

policy-relevant treatment effect versus policy effect, 282
probit, 281
propensity score, 169
    estimands, 169
propensity score IV, 212
propensity score IV identification heterogeneity, 213
propensity score matching, 172
propensity score matching versus general matching, 173
quantile, 234
selection, 277
Tuebingen example
    case 8-1, 151
    case 8-2, 151
    case 8-3, 153
    case 8-4 Simpson's paradox, 153
Tuebingen example with regressors
    case 9-1, 159
    case 9-2, 160
    case 9-3, 161
    case 9-4 Simpson's paradox, 162
Tuebingen IV example
    case 10-1 ignorable treatment, 223
    case 10-1b uniformity fails, 223
    case 10-2 heterogeneous response, 225
    case 10-2b LATE = ATT, 227
    case 10-3 more heterogeneity, 228
    case 10-3b LATE = ATUT, 228
    case 10-4 Simpsons' paradox, 229
    case 10-4b exclusion restriction violated, 230
    case 10-5 lack of common support, 231
    case 10-5b minimal common support, 232
Tuebingen-style, 149
uniformity, 221, 277, 282
truncated normal distribution, 95, 119, 215

U statistic, 100
unbiasedness, 20

uniform distribution, 83, 120, 138, 179
union status, 129
unobservable heterogeneity, 208, 218, 233
unobservables, 14, 17, 78, 84, 86, 87, 89, 123, 129, 146

variance, 105
variance for MLE, *see* covariance for MLE

Wald statistic, 38, 40, 41, 43
    modified, 134
Weibull distribution, 145
weighted distribution, 280
weighted least squares, 236
winBUGs, 120